VOLUME SIXTY NINE

CURRENT TOPICS IN MEMBRANES

Metal Transporters

Current Topics in Membranes, Volume 69

Series Editors

ROBERT BALABAN
National Heart, Lung and Blood Institute,
National Institute of Health,
Bethesda, Maryland

SIDNEY A. SIMON
Department of Physiology and Biophysics,
University of Alabama,
Birmingham, Alabama

VOLUME SIXTY NINE

CURRENT TOPICS IN MEMBRANES

Metal Transporters

Edited by

SVETLANA LUTSENKO
Department of Physiology
Johns Hopkins University
Baltimore, MD, USA

JOSÉ M. ARGÜELLO
Department of Chemistry and Biochemistry
Life Sciences and Bioengineering Center
Worcester Polytechnic Institute
Worcester, MA, USA

ELSEVIER

AMSTERDAM • BOSTON • HEIDELBERG • LONDON
NEW YORK • OXFORD • PARIS • SAN DIEGO
SAN FRANCISCO • SYDNEY • TOKYO
Academic Press is an imprint of Elsevier

Academic Press is an imprint of Elsevier
525 B Street, Suite 1900, San Diego, CA 92101-4495, USA
225 Wyman Street, Waltham, MA 02451, USA
32, Jamestown Road, London NW1 7BY, UK
The Boulevard, Langford Lane, Kidlington, Oxford, OX51GB, UK
Radarweg 29, PO Box 211, 1000 AE Amsterdam, The Netherlands

First edition 2012

ISBN: 978-0-12-394390-3
ISSN: 1063-5823

For information on all Academic Press publications
visit our website at store.elsevier.com

Printed and bound by CPI Group (UK) Ltd, Croydon, CR0 4YY

Transferred to digital print 2012

CONTENTS

CONTRIBUTORS

José M. Argüello
Department of Chemistry and Biochemistry, Worcester Polytechnic Institute, Worcester, MA, USA

Ninian J. Blackburn
Department of Science and Engineering, Oregon Health and Sciences University, Beaverton, OR, USA

Karl Brillet
UMR 7242, Université de Strasbourg-CNRS, ESBS, Boulevard Sébastien Brant, Strasbourg, France

Mathieu F. M. Cellier
Institut National de la Research Scientifique–Institut Armand-Frappier, Laval, QC, Canada

Sarah S. Conte
Department of Biology, University of Massachusetts, Amherst, MA, USA

Robert E. Dempski
Department of Chemistry and Biochemistry, Worcester Polytechnic Institute, Worcester, MA, USA

Caroline A. Enns
Department of Cell and Developmental Biology, Oregon Health & Sciences University, Oregon, USA

Said Eshaghi
School of Biological Sciences, Nanyang Technological University, Singapore, Republic of Singapore

Adrian G. Flores
Department of Molecular Biosciences and Chemistry of Life Processes Institute, Northwestern University, Evanston, IL, USA

Hsueh-Liang Fu
Department of Pathology, Wayne State University School of Medicine, Detroit, MI, USA

Albert Guskov
School of Biological Sciences, Nanyang Technological University, Singapore, Republic of Singapore

Nesrin M. Hasan
Department of Physiology, Johns Hopkins University, Baltimore, MD, USA

Taiho Kambe
Division of Integrated Life Science, Graduate School of Biostudies, Kyoto University, Kyoto, Japan

Jack H. Kaplan
Department of Biochemistry and Molecular Genetics, University of Illinois College of Medicine, Chicago, IL, USA

Yung-Feng Lin
Department of Medical Laboratory Science and Biotechnology, Taipei Medical University, Taipei, Taiwan

Svetlana Lutsenko
Department of Physiology, Johns Hopkins University, Baltimore, MD, USA

Anne B. Mason
Department of Biochemistry, University of Vermont, College of Medicine, Burlington, VT, USA

Megan M. McEvoy
Department of Chemistry and Biochemistry, University of Arizona, Tucson, AZ, USA

Tiffany D. Mealman
Department of Chemistry and Biochemistry, University of Arizona, Tucson, AZ, USA

Gaëtan L. A. Mislin
UMR 7242, Université de Strasbourg-CNRS, ESBS, Boulevard Sébastien Brant, Strasbourg, France

Christopher R. Pope
Department of Molecular Biosciences and Chemistry of Life Processes Institute, Northwestern University, Evanston, IL, USA

Barry P. Rosen
Department of Cellular Biology and Pharmacology, Florida International University College of Medicine, Miami, FL, USA

Amy C. Rosenzweig
Department of Molecular Biosciences and Department of Chemistry, Northwestern University, Evanston, IL, USA

Isabelle J. Schalk
UMR 7242, Université de Strasbourg-CNRS, ESBS, Boulevard Sébastien Brant, Strasbourg, France

Aaron D. Smith
Department of Pharmaceutical Sciences, School of Pharmacy, University of Maryland, Baltimore, USA

Ashley N. Luck*
Department of Biochemistry, University of Vermont, College of Medicine, Burlington, VT, USA

* Previously Ashley N. Steere. Current address: New England Biolabs, Inc., 240 County Road, Ipswich, MA, USA

Vinzenz M. Unger
Department of Molecular Biosciences and Chemistry of Life Processes Institute, Northwestern University, Evanston, IL, USA

Elsbeth L. Walker
Department of Biology, University of Massachusetts, Amherst, MA, USA

Angela Wilks
Department of Pharmaceutical Sciences, School of Pharmacy, University of Maryland, Baltimore, USA

Hung-Chi Yang
Department of Medical Biotechnology and Laboratory Sciences, Chang-Gung University, Tao-Yuan, Taiwan

Ningning Zhao
Department of Cell and Developmental Biology, Oregon Health & Sciences University, Oregon, USA

PREFACE

The Blossoming Field of Metal Transport Enters the Era of Mechanistic Research

Metals, such as iron, copper, zinc, cobalt, and manganese, play an important role in the biology of living organisms. Although required only in small quantities, these metals are indispensible for numerous processes as fundamental as respiration and radical defense and as intricate as cell-to-cell signaling, wound healing, and antimicrobial defense. In recent years, the additional roles for metals in cell physiology have emerged, rapidly expanding the diversity of metal functions in cells. These functional discoveries have coincided with the identification of the major cellular systems that regulate the entry and distribution of metals within cells as well as prompt removal of excess metals to avoid toxicity. The combination of rich biology and the established molecular identity of metal transporters makes the metal transport field an exciting area of modern research.

The complexity of cellular machinery involved in the transport of metals is remarkable. In the case of zinc alone, its homeostasis in cells requires numerous influx and efflux transporters from a least three distinct protein families in addition to regulatory molecules, presumably to fine-tune the distribution of zinc between various cell compartments. In the case of copper and iron, the need to transport metals to their proper destinations in cells or tissues is coupled with the need to protect cells against the electron transfer reactions in which these metals can get easily engaged. These requirements necessitated evolution of specific classes of the metal storage and metal carrier proteins. The function of such metal carriers is tightly coupled to the function of transport proteins, which in turn are controlled through the transcriptional, translational, and posttranslational mechanisms. This volume illustrates the complexity of metal transport systems in the chapters that discuss regulation of transport and interactions between different components of the metal transport machinery.

The vast field of metal transport biology is ripe for mechanistic research. Recent years have yielded exciting information about the biochemistry and structure of several metal transporters. This progress has highlighted the diverse array of structural/mechanistic solutions that cells have developed to transport transition metals. The areas of the transmembrane transport of copper, arsenic, and zinc have been particularly fruitful, as the high-resolution structures have become available alongside the detailed mechanistic descriptions. The number of systems, which show significant sequence variability and for which molecular information is very limited, underscores not only the transporters' diversity but also wonderful opportunities to discover novel structures and associated transport mechanisms. The most recent advances in these areas are summarized in this volume. These and other pioneering studies revealed unique molecular features of the metal ion transporters and raised challenging questions about the mechanisms of metal translocation across biologic membranes.

It is abundantly clear that the recent findings represent the very tip of the iceberg. Currently, for many metal transport system, the exact mechanisms on metal ion selectivity, the driving forces behind metal ion transfer and release, and the cross talk between various components of the metal transport machinery remain to be characterized. Successes in dissecting the mechanisms of arsenic, zinc, iron, heme, cobalt, magnesium, and copper translocation that are discussed in this volume illustrate that the exploration of metal ion transport in cells is an exciting and rewarding journey. We hope the readers of this volume would enjoy it.

Svetlana Lutsenko and José M. Argüello

PREVIOUS VOLUMES IN SERIES

Current Topics in Membranes and Transport

Volume 23 Genes and Membranes: Transport Proteins and Receptors* (1985)
Edited by Edward A. Adelberg and Carolyn W. Slayman

Volume 24 Membrane Protein Biosynthesis and Turnover (1985)
Edited by Philip A. Knauf and John S. Cook

Volume 25 Regulation of Calcium Transport across Muscle Membranes (1985)
Edited by Adil E. Shamoo

Volume 26 Na^{+}–H^{+} Exchange, Intracellular pH, and Cell Function* (1986)
Edited by Peter S. Aronson and Walter F. Boron

Volume 27 The Role of Membranes in Cell Growth and Differentiation (1986)
Edited by Lazaro J. Mandel and Dale J. Benos

Volume 28 Potassium Transport: Physiology and Pathophysiology* (1987)
Edited by Gerhard Giebisch

Volume 29 Membrane Structure and Function (1987)
Edited by Richard D. Klausner, Christoph Kempf, and Jos van Renswoude

Volume 30 Cell Volume Control: Fundamental and Comparative Aspects in Animal Cells (1987)
Edited by R. Gilles, Arnost Kleinzeller, and L. Bolis

Volume 31 Molecular Neurobiology: Endocrine Approaches (1987)
Edited by Jerome F. Strauss, III, and Donald W. Pfaff

Volume 32 Membrane Fusion in Fertilization, Cellular Transport, and Viral Infection (1988)
Edited by Nejat Düzgünes and Felix Bronner

Volume 33 Molecular Biology of Ionic Channels* (1988)
Edited by William S. Agnew, Toni Claudio, and Frederick J. Sigworth

*Part of the series from the Yale Department of Cellular and Molecular Physiology

Volume 34 Cellular and Molecular Biology of Sodium Transport* (1989)
Edited by Stanley G. Schultz

Volume 35 Mechanisms of Leukocyte Activation (1990)
Edited by Sergio Grinstein and Ori D. Rotstein

Volume 36 Protein–Membrane Interactions* (1990)
Edited by Toni Claudio

Volume 37 Channels and Noise in Epithelial Tissues (1990)
Edited by Sandy I. Helman and Willy Van Driessche

Current Topics in Membranes

Volume 38 Ordering the Membrane Cytoskeleton Trilayer* (1991)
Edited by Mark S. Mooseker and Jon S. Morrow

Volume 39 Developmental Biology of Membrane Transport Systems (1991)
Edited by Dale J. Benos

Volume 40 Cell Lipids (1994)
Edited by Dick Hoekstra

Volume 41 Cell Biology and Membrane Transport Processes* (1994)
Edited by Michael Caplan

Volume 42 Chloride Channels (1994)
Edited by William B. Guggino

Volume 43 Membrane Protein–Cytoskeleton Interactions (1996)
Edited by W. James Nelson

Volume 44 Lipid Polymorphism and Membrane Properties (1997)
Edited by Richard Epand

Volume 45 The Eye's Aqueous Humor: From Secretion to Glaucoma (1998)
Edited by Mortimer M. Civan

Volume 46 Potassium Ion Channels: Molecular Structure, Function, and Diseases (1999)
Edited by Yoshihisa Kurachi, Lily Yeh Jan, and Michel Lazdunski

Volume 47 Amiloride-Sensitive Sodium Channels: Physiology and Functional Diversity (1999)
Edited by Dale J. Benos

Volume 48 Membrane Permeability: 100 Years since Ernest Overton (1999)
Edited by David W. Deamer, Arnost Kleinzeller, and Douglas M. Fambrough

Volume 49 Gap Junctions: Molecular Basis of Cell Communication in Health and Disease
Edited by Camillo Peracchia

Volume 50 Gastrointestinal Transport: Molecular Physiology
Edited by Kim E. Barrett and Mark Donowitz

Volume 51 Aquaporins
Edited by Stefan Hohmann, Søren Nielsen and Peter Agre

Volume 52 Peptide–Lipid Interactions
Edited by Sidney A. Simon and Thomas J. McIntosh

Volume 53 Calcium-Activated Chloride Channels
Edited by Catherine Mary Fuller

Volume 54 Extracellular Nucleotides and Nucleosides: Release, Receptors, and Physiological and Pathophysiological Effects
Edited by Erik M. Schwiebert

Volume 55 Chemokines, Chemokine Receptors, and Disease
Edited by Lisa M. Schwiebert

Volume 56 Basement Membranes: Cell and Molecular Biology
Edited by Nicholas A. Kefalides and Jacques P. Borel

Volume 57 The Nociceptive Membrane
Edited by Uhtaek Oh

Volume 58 Mechanosensitive Ion Channels, Part A
Edited by Owen P. Hamill

Volume 59 Mechanosensitive Ion Channels, Part B
Edited by Owen P. Hamill

Volume 60 Computational Modelling of Membrane Bilayers
Edited by Scott E. Feller

Volume 61 Free Radical Effects on Membranes
Edited by Sadis Matalon

Volume 62 The Eye's Aqueous Humor
Edited by Mortimer M. Civan

Volume 63 Membrane Protein Crystallization
Edited by Larry DeLucas

Volume 64 Leukocyte Adhesion
Edited by Klaus Ley

Volume 65 Claudins
Edited by Alan S. L. Yu

Volume 66 Structure and Function of Calcium Release Channels
Edited by Irina I. Serysheva

Volume 67 Advances in Adrenergic Receptor Biology
Edited by Qin Wang

Volume 68 Membrane Fusion
Edited by Leonid V. Chernomordik and Michael M. Kozlov

SECTION ONE

Iron Delivery to Cells

CHAPTER ONE

Transferrin-Mediated Cellular Iron Delivery

Ashley N. Luck* and Anne B. Mason[1]
Department of Biochemistry, University of Vermont, College of Medicine, Burlington, VT, USA
*Previously Ashley N. Steere. Current address: New England Biolabs, Inc., 240 County Road, Ipswich, MA, USA
[1]Corresponding author: E-mail: anne.mason@uvm.edu

Contents

Abstract

Essential to iron homeostasis is the transport of iron by the *bilobal* protein human serum transferrin (hTF). Each lobe (N- and C-lobe) of hTF forms a deep cleft which binds a single Fe^{3+}. Iron-bearing hTF in the blood binds tightly to the specific transferrin

Current Topics in Membranes, Volume 69
ISSN 1063-5823,
http://dx.doi.org/10.1016/B978-0-12-394390-3.00001-X

receptor (TFR), a homodimeric transmembrane protein. After undergoing endocytosis, acidification of the endosome initiates the release of Fe^{3+} from hTF in a TFR-mediated process. Iron-free hTF remains tightly bound to the TFR at acidic pH; following recycling back to the cell surface, it is released to sequester more iron. Efficient delivery of iron is critically dependent on hTF/TFR interactions. Therefore, identification of the pH-specific contacts between hTF and the TFR is crucial. Recombinant protein production has enabled deconvolution of this complex system. The studies reviewed herein support a model in which pH-induced interrelated events control receptor-stimulated iron release from each lobe of hTF.

1. IRON

As a transition element, iron can assume a number of different oxidation states, from −2 to +6. However, the ferrous (Fe^{2+}) and ferric (Fe^{3+}) states are most common and are easily exchangeable through the transfer of one electron (Aisen, Enns, & Wessling-Resnick, 2001). Due to its inherent redox properties, iron is critical to a number of biological processes, including oxygen and electron transport, making it essential for most life on Earth (Aisen et al., 2001). Only two organisms have been identified that do not require iron: *Borrelia burgdorferi*, the bacteria responsible for Lyme disease, and a bacterium found in soil, *Lactobacillus plantarum* (Posey & Gherardini, 2000). These organisms substitute other metals, such as manganese, for iron to maintain activity of biologically important molecules. The same redox properties that provide biological versatility also make iron potentially dangerous. In oxygen-rich environments, Fe^{3+} forms extremely insoluble iron oxides (i.e. rust), while Fe^{2+} catalyzed Fenton reactions that can be toxic to living cells (Fenton, 1876, 1893). Specifically, reduction of O_2 by Fe^{2+} generates superoxide, which can ultimately lead to the formation of the hydroxyl radical, a powerful oxidant known to damage DNA, proteins and lipids (Park, Bacon, Brittenham, & Tavill, 1987). In an evolutionary conundrum, aerobic life on Earth requires iron while the oxygen-rich environment of the planet maintains most of it in a chemically inert state (Fe^{3+}). However, it must be remembered that oxygen conditions on Earth have changed. When the first single-cell organisms began to appear on Earth, the atmospheric composition was drastically different, with very little dissolved O_2. The ready availability of Fe^{2+} at such low oxygen concentrations allowed iron to procure a strong foothold in essential biological processes. Hence, organisms that had evolved biological pathways utilizing Fe^{2+} were presented with a dilemma when high oxygen conditions limited iron bioavailability. Life has adapted to the ever-changing conditions of the atmosphere by manipulating the metallome in a manner that allows for the endless redox cycling that takes place in modern day organisms.

2. IRON HOMEOSTASIS

2.1. Iron Uptake

It is estimated that the adult human body contains ~3.5 g of iron (Theil & Goss, 2009). To prevent detrimental effects due to the redox potential of the Fe^{2+}/Fe^{3+} pair, iron must be tightly regulated within the body. Since no physiological excretion mechanism exists (McCance & Widdowson, 1938), uptake of iron into the body is strictly controlled. Erythroid precursor cells are the largest consumer of bodily iron, utilizing nearly 1000 million iron atoms a day (ca. two third of the total body iron content) to maintain hemoglobin levels (Andrews & Schmidt, 2007; Theil & Goss, 2009). A large portion of the iron contained in hemoglobin is recycled from aged and damaged erythrocytes that are phagocytosed by tissue macrophages in the spleen (Andrews & Schmidt, 2007). However, 1–2 mg of iron is lost daily due to normal sloughing of epithelial cells (from skin, gastrointestinal and urinary tracts). Therefore, this same amount (~1 to 2 mg) must be offset by iron acquired from the diet every day.

The diet of omnivorous humans generally contains both heme and non-heme iron. Of the non-heme sources, only ferrous iron can be directly absorbed at the apical surface of duodenal enterocytes. Therefore, any Fe^{3+} in the diet must first be reduced by the ferric reductase, duodenal cytochrome b, before being transported into the enterocyte by the divalent metal transporter 1 (DMT1) (Fleming et al., 1997; Gunshin et al., 1997, 2005). An energy-dependent transporter, DMT1, transports a number of other divalent metals and protons as well as Fe^{2+} (Gunshin et al., 1997). Although the majority of the iron in most diets is non-heme iron (~90 to 95%), it is absorbed rather inefficiently (only ~2 to 20%) (Sharp & Srai, 2007). On the other hand, heme iron accounts for only 5–10% of the iron in the diet, but is absorbed more efficiently (~20 to 30%) (Hallberg, Bjorn-Rasmussen, Howard, & Rossander, 1979). While the mechanism of non-heme iron absorption is fairly well characterized, far less is understood about the mechanism of heme transport into duodenal enterocytes. It is thought that the intact porphyrin of heme is absorbed. Although a number of different potential heme transporters have been suggested (Sharp, 2010), only one, heme carrier protein-1 (HCP1), has been identified in the duodenum (Shayeghi et al., 2005). However, more recent studies have revealed that HCP1 displays ~100-fold lower Km for folate than heme (Qiu et al., 2006), suggesting HCP1 is actually a specific cell membrane transporter for folate, not heme. Thus, the mechanism of heme absorption in the duodenal enterocyte remains unknown. Iron absorbed as heme is freed within the enterocytic endosomes by the enzyme heme oxygenase 1 (Raffin, Woo, Roost, Price, & Schmid, 1974). Iron acquired from either the heme or non-heme pathway becomes part of the same intracellular iron pool, where it can either be stored in the enterocyte by ferritin (see below) or trafficked to the basolateral surface of the enterocyte for transport into the blood.

The transfer of Fe^{2+} from the basolateral membrane of the enterocyte into the circulation is mediated by ferroportin (FPN) (Donovan et al., 2000), the only transporter involved in iron efflux from the enterocyte identified to date (Sharp, 2010). The iron released from the enterocytes of the duodenum via FPN is in the more dangerous ferrous form. To avoid Fenton chemistry and the production of damaging free radicals, iron must be oxidized to the ferric form. This oxidation is accomplished by the membrane-bound multi-copper-containing ferroxidase hephaestin (Vulpe et al., 1999), located on the basolateral surface of the enterocyte. A homologue of hephaestin is the soluble ferroxidase, ceruloplasmin (Gubler, Lahey, Chase, Cartwright, & Wintrobe, 1952; Lahey, Gubler, Chase, Cartwright, & Wintrobe, 1952), which circulates in the serum. Critically, following the degradation of senescent red blood cells, Fe^{2+} released from macrophages (via FPN) is oxidized by ceruloplasmin to Fe^{3+}. Following oxidation, Fe^{3+} is carefully chaperoned through the body by the bilobal iron binding and transport protein human serum transferrin (hTF) (Kaplan & O'Halloran, 1996).

3. TRANSFERRIN

Synthesized in the liver and secreted into the blood plasma, hTF is an ~80 kDa bilobal (N- and C-lobes) glycoprotein that binds Fe^{3+} very tightly (K_d ~10^{22}/M) (Aisen, Leibman, & Zweier, 1978), yet reversibly. Since hTF can bind iron in both lobes, four different hTF species may circulate in the blood, differing only in iron content. The ~25 to 50 μM (~2.5 mg/ml) hTF in the serum (Sun, Li, & Sadler, 1999) appears to be nonrandomly distributed among diferric (Fe_2hTF, ~11 to 27%), monoferric N-lobe hTF (Fe_NhTF, ~21 to 23%), monoferric C-lobe hTF (Fe_ChTF, ~11 to 17%)

and iron-free hTF (apohTF, ~40 to 51%) (Makey & Seal, 1976; Williams & Moreton, 1980) (see below). At any given time only ~30% of the hTF in circulation is saturated with iron although even within an individual the range is considerable with variables including diet and time of day. Importantly, the iron-binding sites in the unsaturated 70% of hTF remain available and provide significant buffering capacity in times of iron influx or overload. Interestingly, although the C-lobe has a greater affinity for Fe^{3+} at neutral pH than the N-lobe (Cannon & Chasteen, 1975; Lestas, 1976; Princiotto & Zapolski, 1975), there appears to be a somewhat greater percentage of Fe_NhTF in the serum than Fe_ChTF (Williams & Moreton, 1980). The source of the unequal distribution between lobes (if it exists, see for example Huebers, Josephson, Huebers, Csiba, & Finch, 1984) could include either incomplete loading of hTF at the level of uptake from the enterocytes and/or incomplete iron removal within cells. Although the distribution appears to be nonrandom and to favor the kinetically weaker N-lobe, several observations should be considered: 1) Using electrophoretic measurements, it is challenging to determine the distribution in fresh serum without disturbing the equilibrium. Most studies employ urea gel analysis alone or in combination with other techniques (Leibman & Aisen, 1979); 2) The small number of individual serum samples in combination with the large differences in hTF saturation may belie the generality of the conclusions that are derived from such data; 3) The substantial concentration of hTF in the serum, in addition to the difference in the binding affinity of Fe_2hTF in comparison to either monoferric species, is probably sufficient to guarantee that binding of Fe_2hTF will be favored. Thus, the concentration of the monoferric species may be somewhat irrelevant.

3.1. The Transferrin Cycle

At the pH of the serum (~7.4) iron-bearing hTF binds with nanomolar affinity to the specific transferrin receptor (TFR, also referred to as TFR1), located on the cell surface of all iron-requiring cells. It is important to note that at pH 7.4, although Fe_2hTF binds to the TFR with the highest affinity (K_d ~4 nM), the two monoferric hTFs (Fe_NhTF and Fe_ChTF) also form a high-affinity stable complex with the TFR (K_d ~36 nM and K_d ~32 nM, respectively) (Mason et al., 2009) while apohTF binds very weakly at this pH. Following clathrin-dependent endocytosis of the hTF/TFR complex (Morgan & Appleton, 1969), the pH within the endosome is lowered through the action of ATP-dependent H^+ pumps (Morgan, 1981). Salt and a currently unidentified chelator, along with the significantly lower pH (~ 5.6) within the endosome, initiate receptor-stimulated iron release from hTF. Although the TFR purportedly influences the redox potential of Fe^{3+} bound to hTF (Dhungana et al., 2004), the recent discovery of an endosomal ferrireductase (Steap3) in erythroid cells suggests that the reduction of Fe^{3+} to Fe^{2+} is accomplished only after Fe^{3+} is released from hTF within the endosome (Ohgami et al., 2005). Critical to the hTF endocytic cycle, apohTF remains bound to the TFR with high affinity at endosomal pH. This allows the apohTF/TFR complex to return to the cell surface instead of being targeted to late endosomes and avoid degradation within the lysosome. Upon returning to the cell surface, apohTF is released into the serum, either through dissociation from the TFR or displacement by an iron-containing hTF (Leverence, Mason, & Kaltashov, 2010), where it is free to bind more Fe^{3+}. The process of internalization and recycling of the hTF/TFR complex has been so extensively studied that it has become the classic example of clathrin-dependent receptor-mediated endocytosis and often serves as a positive control for other systems (Grant & Donaldson, 2009).

3.2. Cellular Iron Utilization

Once inside the cell, iron is compartmentalized in a number of different ways. These include incorporation into heme, iron–sulfur clusters, and other iron-requiring enzymes/proteins, formation of an intracellular labile iron pool or storage in the iron storage protein ferritin. The majority of heme biosynthesis occurs within the mitochondria, with a few intermediary steps taking place in the cytosol (Andrews & Schmidt, 2007). The molecular pathway utilized to traffic iron from the endosome to the mitochondria has been a subject of debate for many years. One model proposes that after leaving the endosome via DMT1, Fe^{2+} enters a cytosolic pool of iron composed of low molecular weight iron–chelate complexes (Greenberg & Wintrobe, 1946; Jacobs, 1977). A combination of pulse-chase experiments (Ponka, Borova, Neuwirt, & Fuchs, 1979) and cytosolic fractionation studies (Weaver & Pollack, 1989) were first utilized to suggest that such a chelatable intracellular pool of iron might exist. However, a major drawback is that these studies required cell lysis that can often lead to decompartmentalization and iron release from organelles such as the mitochondria. More recently, cell permeable fluorescent iron chelators have detected low (~2.5 to 5 μM), but significant, levels of free cytosolic iron (Kakhlon & Cabantchik, 2002). The concept of a labile iron pool is very controversial given the redox-active chemical properties of the Fe^{2+}/Fe^{3+} pair. Another theory, supported by experiments in the Ponka laboratory and termed the "kiss and run" hypothesis (Sheftel, Mason, & Ponka, 2012) proposes that, at least in reticulocytes, iron bypasses the cytosol and is instead directly transferred from hTF-laden endosomes to the mitochondria through direct interaction of the two organelles. Given the considerable iron requirements of reticulocytes, alternative iron trafficking pathways may be utilized in other cell types (Sheftel et al., 2012). Further study is required to delineate exactly how iron is trafficked once inside the cell.

Iron not utilized in the formation of heme or other iron-containing proteins is stored in the cytosolic iron storage protein ferritin. First identified and crystallized from horse spleen (Laufberger, 1937), 24 subunits (H- and L-chains) form the large cage-like structure (~80 Å diameter) of ferritin, within which up to 4500 Fe^{3+} atoms may be stored (Harrison & Arosio, 1996). Ferritin also possesses ferroxidase activity that is critical to consuming Fe^{2+} released by endosomes via DMT1 and storing the less toxic Fe^{3+} (Arosio & Levi, 2010). However, the bioavailability of the iron stored in ferritin remains questionable (Sheftel et al., 2012). Some suggest that the ferric iron is readily bioavailable via eight hydrophilic channels in the protein shell of ferritin (Bou-Abdallah, McNally, Liu, & Melman, 2011), while other evidence suggests that at least partial degradation of the protein shell must occur before iron can be released from ferritin (Sheftel et al., 2012). Although the mechanism of iron release from ferritin remains unclear, its biological importance is emphasized by its conservation among various life forms, e.g. some form of ferritin has been identified in nearly every biological classification, except for yeast which stores excess iron in vacuoles (Arosio & Levi, 2010).

3.3. Transferrin Structural Overview

As mentioned previously, hTF is a bilobal protein comprised of 679 amino acids. Connected by a seven amino acid bridge (residues 332–338), the N- and C-lobes are ~40% identical and probably arose from gene duplication followed by fusion (Park et al., 1985). Each lobe folds into two subdomains (termed N1 and N2 and C1 and C2, Fig. 1). The N1 and C1 subdomains are discontinuous in primary sequence (N1: residues 1–93 and 247–331, C1: 339–425 and 573–679), while the N2 and C2 subdomains are continuous (N2: residues 94–246, C2: 426–572). The Fe^{3+} binds in a deep cleft formed between the two subdomains of a lobe. The subdomains are comprised of similar α/β folds centered around a β-sheet (Hall et al., 2002; Noinaj et al., 2012). A characteristic structural feature of most transferrins (except insect

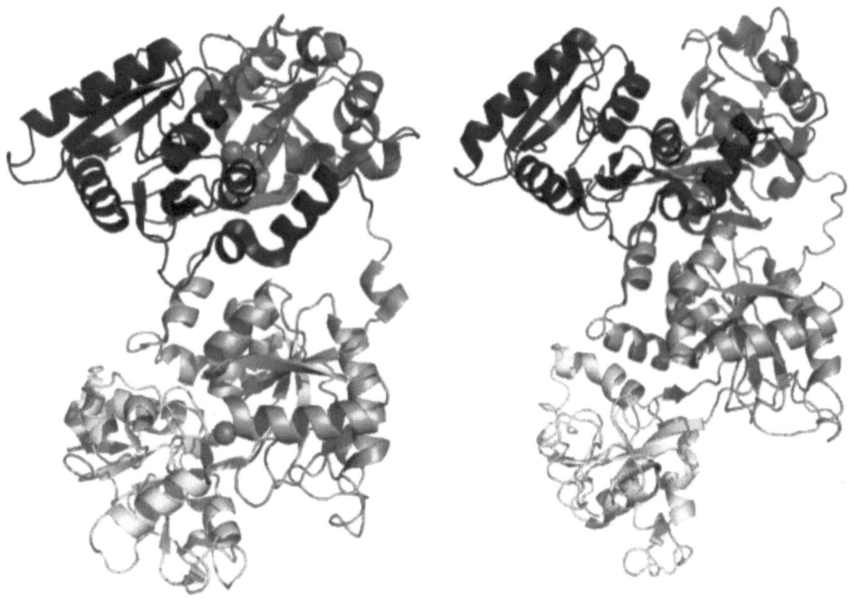

Figure 1 hTF. Structures of diferric hTF (Left, PDB ID: 3V83) and apohTF (Right, PDB ID: 2HAU). Individual subdomains are colored accordingly: N1-blue, N2-red, C1-green, C2-yellow. The two Fe^{3+} atoms in the diferric transferrin are shown as orange spheres. Figure generated using PyMOL (Delano, 2002). See the color plate.

transferrins) is a classical γ turn in the N1 and C1 subdomains (Lambert, Perri, Halbrooks, & Mason, 2005). The strain on the Leu–Leu–Phe sequence (which forms the γ turn) places the second leucine residue in the disallowed region of the Ramachandran plot and may help to correctly position pH-sensitive elements critical to iron release from each lobe (Eckenroth, Mason, McDevitt, Lambert, & Everse, 2010). Iron release from each lobe is accompanied by large rigid body movements of the subdomains (opening of ~50°) (Hall et al., 2002; Wally et al., 2006). A "hinge" region, consisting of two β-strands connecting the subdomains of each lobe, mediates opening of the cleft (Jeffrey et al., 1998; MacGillivray et al., 1998; Wally et al., 2006). Additional stability is provided by the participation of all 38 cysteine residues in hTF in the formation of 19 disulfide bonds (eight in the N-lobe and 11 in the C-lobe, Fig. 2). Native hTF is a glycoprotein with two N-linked biantennary glycans in the C-lobe (at residues Asn413 and Asn611). While glycosylation appears to have no effect on TFR binding or iron uptake or release by hTF (Mason et al., 1993), it has been suggested that loss of the terminal sialic acids over time may play a role in clearance of hTF from the blood by asialo receptors in the liver (Regoeczi, Bolyos, & Chindemi, 1989).

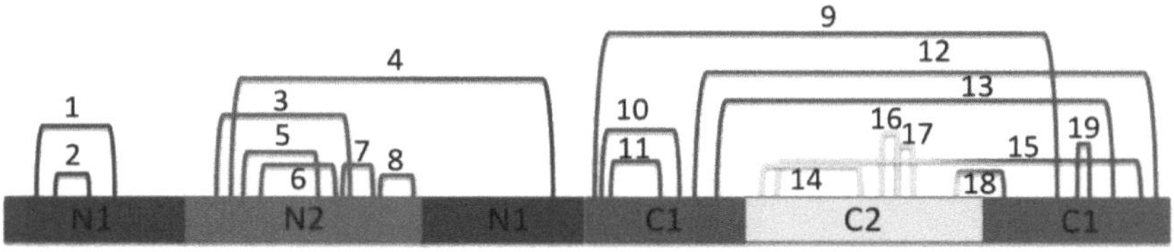

	N-lobe			C-lobe	
Disulfide Bond #	Cys Residue 1	Cys Residue 2	Disulfide Bond #	Cys Residue 1	Cys Residue 2
			9	339	596
1	9	48	10	345	377
2	19	39	11	355	368
			12	402	674
			13	418	637
3	118	194	14	450	523
4	137	331	15	474	665
5	158	174	16	484	498
6	161	177			
7	171	179	17	495	506
8	227	241	18	563	577
			19	615	620

Figure 2 Disulfide bonds of hTF. The 19 disulfide bonds formed within hTF are shown numbered according to primary sequence. Equivalent disulfide bonds from each lobe are listed on the same line. Individual subdomains are colored accordingly: N1, blue; N2, red; C1, green; C2, yellow. See the color plate.

3.4. Transferrin Iron-Binding Ligands

For many years, the identity of the iron-binding residues of hTF was unknown and the subject of much conjecture. Based on various biochemical studies, histidine, tyrosine and tryptophan residues were suggested as candidates. It was not until the first crystal structure of the transferrin family member lactoferrin (LTF) was determined by the Baker laboratory (Anderson et al., 1987) that the iron-binding ligands were finally identified. The Fe^{3+} is coordinated by identical ligands in both lobes of hTF: two tyrosine residues, one aspartic acid and one histidine residue (Tyr95, Tyr188, Asp63 and His249 in the N-lobe; Tyr426, Tyr517, Asp392 and His585 in the C-lobe, hTF numbering) (Fig. 3). The remaining two ligands which complete the distorted octahedral coordination of the iron are provided by carbonate, which is anchored in place by a conserved arginine residue (Arg124 in the N-lobe and Arg456 in the C-lobe). Interestingly, synergistic anion binding is an absolute requirement for high-affinity Fe^{3+} binding by hTF (Price & Gibson, 1972b). The initial binding of carbonate may be necessary to bring the remaining iron-binding ligands of hTF, which are very

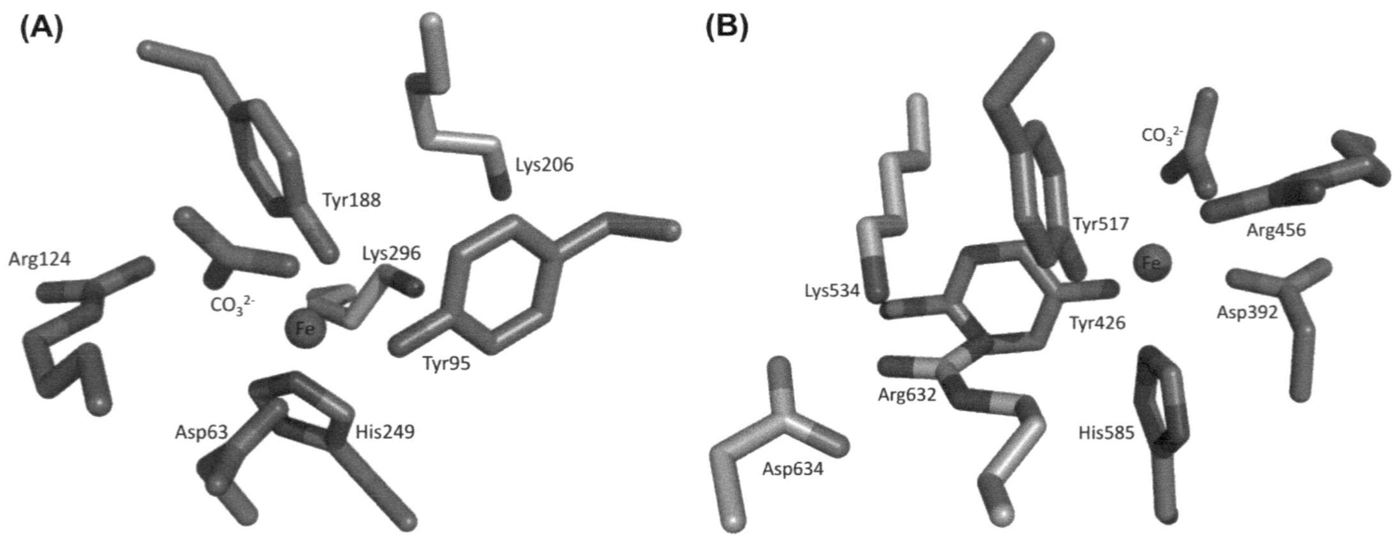

Figure 3 Close-up of ligands and second-shell residues in the hTF N- and C-lobes of Fe_2hTF (PDB 3V83). The ligands are colored orange. The second-shell residues are depicted in pink. The conserved arginine is in green. The figure was produced by Dr Stephen J. Everse using PyMOL. See the color plate.

distant in the primary sequence of the protein, into proximity (Gaber, Miskowski, & Spiro, 1974).

Given its relatively high concentration in the serum, carbonate is the physiologically relevant synergistic anion (Schade, Reinhart, & Levy, 1949); however, other molecules (oxalate, glycolate, malonate, etc.) can substitute for carbonate to promote high-affinity Fe^{3+} binding to hTF (Schlabach & Bates, 1975). All functional synergistic anions appear to follow an interlocking sites model (Schlabach & Bates, 1975) in which the carboxylate group of the anion is available to bind the anchoring arginine residue in each lobe of hTF (Arg124 and Arg456), as well as a proximal electron donor group 1–2 carbon atoms away to complete the distorted octahedral coordination of Fe^{3+}(Baker, 1994).

The coordination of Fe^{3+} by the two tyrosine residues in each lobe of hTF not only produces an intense ligand-to-metal charge transfer band but also disrupts the π–$\pi^\star$ transition of the two tyrosines. This causes an increase in the UV absorbance at 280 nm that overlaps with and strongly quenches the tryptophan fluorescence through what is thought to be radiationless transfer of excited-state energy (Lehrer, 1969; Patch & Carrano, 1981). Importantly, the overall effect is that Fe^{3+} binding by hTF significantly quenches the intrinsic tryptophan fluorescence of the protein (Lehrer, 1969) such that upon transitioning from Fe_2hTF to apohTF a 368% increase in the intrinsic tryptophan fluorescence is observed (Fig. 4) (Byrne & Mason, 2009). This spectral property has been invaluable in the study of the kinetics of iron release from hTF.

3.5. Second-Shell Residues

Although the iron-binding ligands are completely conserved in each lobe of hTF, the mechanism of iron release from each lobe is distinctly different, largely because of differences in "second-shell" residues (these residues are not directly involved in the coordination of the Fe^{3+} but form an intricate hydrogen bonding network with the primary Fe^{3+}-binding ligands of hTF). Notably, the two lobes differ in their response to pH (Baldwin & de Sousa, 1981; Lestas, 1976; Princiotto & Zapolski, 1975), anions (Baldwin, 1980; Baldwin & de Sousa, 1981), the TFR (Bali & Aisen, 1991; Byrne, Chasteen, Steere, & Mason, 2010; Zak & Aisen, 2003), and the conformation of the other lobe in part due to differences in the composition of the second-shell residues in each lobe.

The first suggestion of a potential iron release mechanism from the N-lobe of hTF was provided by the crystal structure of the N-lobe of hen ovotransferrin (Dewan, Mikami, Hirose, & Sacchettini, 1993). Two

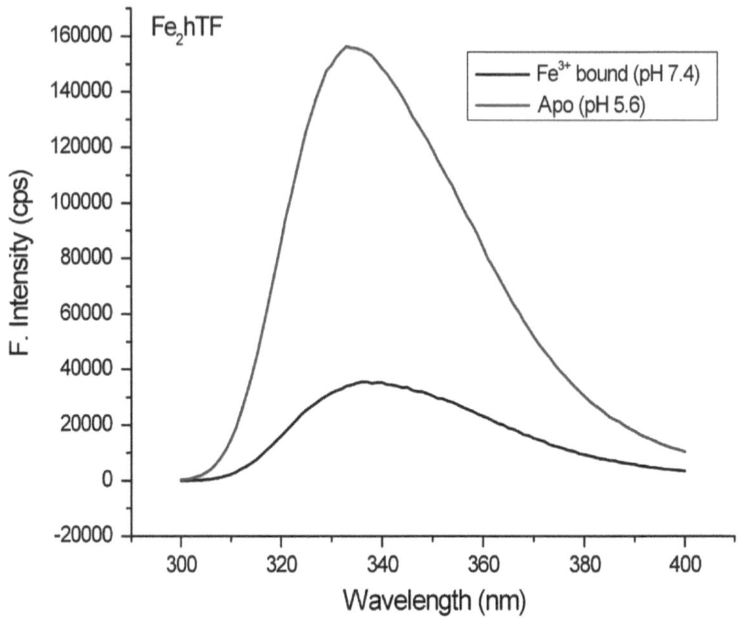

Figure 4 Steady-state tryptophan fluorescence of Fe_2hTF. Steady-state emission spectra Fe_2hTF before (iron containing 100 mM HEPES buffer (4-(2-hydroxyethyl)-1-piperazineethansulfonic acid), pH 7.4) and after treatment with iron removal buffer (apohTF 100 mM MES (2-(N-morpholino)ethanesulfonic acid), pH 5.6, containing 300 mM KCl and 4 mM EDTA for 15 min) following constant excitation at 280 nm. For color version of this figure, the reader is referred to the online version of this book.

second-shell lysine residues (one from each subdomain) in the N-lobe, Lys206 and Lys296, are close enough to share a hydrogen bond (3.0 Å) and constitute what is referred to as the dilysine trigger (Dewan et al., 1993) (Fig. 3). The formation of this hydrogen bond is attributed to an unusually low pK_a value of either one or both of the lysine residues due to their location within a hydrophobic environment. Protonation of one of the lysine residues due to a reduction in pH causes the positively charged lysines to repel each other (moving at least 9 Å apart in the apo N-lobe structure; Jeffrey et al., 1998), and literally triggers cleft opening to promote iron release. Mutation of either Lys206 or Lys296 to a glutamate or alanine drastically slows the rate of iron removal, validating the critical nature of the dilysine trigger in the mechanism of iron release from the N-lobe of hTF (He & Mason, 2002).

Until recently, the mechanism of iron release from the C-lobe has been unclear. As described above, crystal structures of iron-bound hen oTF and apohTF helped clarify the action of the dilysine trigger in the N-lobe. However, no crystal structure of an iron-containing C-lobe of hTF was available. Using sequence alignments and an unpublished crystal structure of Fe_ChTF (Zuccola, 1993), Dewan et al. (1993) proposed a slightly different mechanism of iron release from the C-lobe in which the dilysine trigger is replaced by a triad of residues (Lys534, Arg632 and Asp634) (Fig. 3). Similar to the dilysine trigger, Lys534 and Arg632 may share a hydrogen bond that is stabilized by Asp634, which upon protonation would trigger iron release from the C-lobe; however, in the structure of porcine transferrin, the NZ and NE group of the homologous lysine (Lys543) and arginine (Arg641) are ~4.1 Å apart, seemingly too far to share a hydrogen bond (Hall et al., 2002). Lacking a crystal structure of an iron-containing C-lobe of hTF, the precise mechanism by which the triad might trigger iron release from the C-lobe remains unclear. However, it has been shown that mutation of Lys534 or Arg632 to an alanine severely retards iron release from that lobe, essentially locking iron in the C-lobe (Halbrooks et al., 2003). It has recently become clear that the triad is less relevant to the mechanism of iron release from the C-lobe in the presence of the TFR (Eckenroth, Steere, Chasteen, Everse, & Mason, 2011; Steere, Byrne, et al., 2010) (see discussion of hTF residue His349 below).

3.6. Kinetically Significant Anion-Binding Sites

Along with synergistic anion-binding, non-synergistic anions are known to bind to hTF as well. By definition non-synergistic anions bind to sites on hTF but do not facilitate high-affinity Fe^{3+} binding. As originally detailed by Folajtar and Chasteen (1982) the allosteric binding of non-synergistic anions to hTF follows the lyotropic series ($SCN^- > ClO_4^- >$ pyrophosphate (PP_i) $>$ ATP $> Cl^- \gg BF_4^-$) and is thought to induce structural changes in hTF that perturb the Fe^{3+}-binding center and thereby influence iron release. Furthermore, as clearly demonstrated by Kretchmar and Raymond (1988), binding of non-synergistic anions to hTF is absolutely required for iron release: iron release from hTF ceases as the ionic strength is extrapolated to zero at pH 7.4. In order to emphasize their important allosteric effect on iron release from hTF, these non-synergistic anion-binding sites were designated "kinetically significant anion-binding" or KISAB sites by Egan et al. (Egan, Ross, Purves, & Adams, 1992; Egan, Zak, & Aisen, 1993; Marques, Egan, & Pattrick, 1990; Marques, Watson, & Egan, 1991).

In the past, many different spectroscopic techniques, including electron paramagnetic resonance spectroscopy (Chasteen, 1983; Folajtar & Chasteen,

1982; Grady, Mason, Woodworth, & Chasteen, 1995; Price & Gibson, 1972a; Thompson, McCarty, & Chasteen, 1986), nuclear magnetic resonance spectroscopy (Kubal, Mason, Patel, Sadler, & Woodworth, 1993) and UV-difference spectra (Harris, 1985; Harris & Bali, 1988; Harris, Cafferty, Abdollahi, & Trankler, 1998; Harris, Nesset-Tollefson, Stenback, & Mohamed-Hani, 1990; Pecoraro, Harris, Carrano, & Raymond, 1981) have been employed to detect anion binding to hTF. It was clearly demonstrated (Williams, Chasteen, & Moreton, 1982) that at pH 7.4, the presence of salt retards the rate of iron release from the N-lobe while facilitating iron release from the C-lobe. The UV-difference technique applied by the Harris laboratory provided an estimate of the binding strength of various anions to apohTF (Bali & Harris, 1990; Harris, 1985; Harris et al., 1990). In addition, the Sadler group published a review summarizing the equilibrium binding constants for 13 non-synergistic anions to hTF (Sun et al., 1999).

Importantly, many of these earlier studies monitoring anion binding to hTF were performed at neutral pH (~7.4), and in the absence of the TFR. Our laboratory has suggested that in order to be designated as kinetically significant, non-synergistic anion binding must correlate with low pH; i.e. anion binding to a KISAB site must exert maximal effect at endosomal pH (~5.6) (Byrne, Steere, Chasteen, & Mason, 2010). Although it is well documented that non-synergistic anions are required for iron release from hTF, the specific identification of these binding sites has remained elusive. Obvious choices for potential KISAB sites are positively charged (lysine, arginine and histidine) residues. The most likely KISAB sites were long thought to be iron liganding residues, second-shell residues and residues in the hinging β-strands that facilitate opening and closing of each lobe of hTF (Egan et al., 1993; Grady et al., 1995; Harris & Bali, 1988). Although anions may bind to these residues (normally buried within the cleft of each lobe when Fe^{3+} is bound), it is likely that this only occurs after iron is released. Thus, residues within these buried regions of hTF do not qualify as legitimate KISAB sites, but may function to stabilize the apo conformation of each lobe.

4. THE TFR

In the 1960s, it was proposed that a membrane-bound receptor for transferrin might exist (Jandl & Katz, 1963). However, it was not until 1977 that (Leibman and Aisen 1977) verified the presence of the TFR1 in a detergent-solubilized fraction of lysed rabbit reticulocytes that had been previously incubated with ^{125}I-labeled transferrin. Soon after, (Seligman, Schleicher, and Allen 1979) were the first to isolate and characterize the TFR from human placental tissue. In 1984, two laboratories published the primary amino acid sequence of the TFR1 based on the messenger RNA (mRNA) sequence (Kuhn, McClelland, & Ruddle, 1984; McClelland, Kuhn, & Ruddle, 1984; Schneider, Owen, Banville, & Williams, 1984). Intriguingly, the human TFR gene is located on chromosome 3, as are the genes for hTF and melanotransferrin. The significance of the proximity of these related genes within the genome remains unclear, but raises some interesting evolutionary questions. The ubiquitously expressed TFR1 is the main route of entry for iron into most cells. Importantly, TFR1 expression is controlled at the translational level by five iron responsive element (IREs) in the 3′ UTR of the mRNA, which allows the quantity of TFR1 at the cell surface to be modulated by intracellular iron concentrations (Subramaniam, Summerville, & Wallace, 2002). Due to neurological defects and impaired erythropoiesis, TFR1 knockout mice do not survive past embryonic day 12.5, emphasizing the essential nature of the TFR1 (Levy, Jin, Fujiwara, Kuo, & Andrews, 1999).

TFR1 is a type II transmembrane homodimeric receptor. Two ~90 kDa monomers form the TFR homodimer and are comprised of a short cytoplasmic tail (residues 1–67) with an internalization motif, a membrane-spanning portion (residues 68–88), a stalk region (residues 89–120) which contains intermolecular two disulfide bonds (Cys89 and Cys98), which covalently link the two TFR monomers and a large extracellular ectodomain (residues 121-760) (McClelland et al., 1984). The ectodomain is further subdivided into three domains: the protease-like domain (residues 121–188 and 384–606), the apical domain (residues 189–383) and the helical domain (residues 607–760) responsible for dimerization (Fig. 5) (Lawrence et al., 1999). Each TFR monomer binds one hTF molecule such that a 2:2 complex is formed (Enns & Sussman, 1981).

TFR1 is extensively posttranslationally modified. The ectodomain contains three N-linked glycosylation sites (Asn251, Asn317 and Asn727) and one O-linked glycosylation site (Thr104). The N-linked glycosylation is important for proper folding and translocation of the TFR to the cell surface (Hayes, Williams, Lucas, & Enns, 1997). However, mutagenesis studies performed in our laboratory determined that only N-linked glycosylation at Asn317 was critical to TFR function. The N317D TFR mutation decreased the affinity of the TFR for hTF, affected the rate of iron release from the C-lobe of hTF and decreased the production of the recombinant soluble portion of the TFR (sTFR) (Byrne et al., 2006).

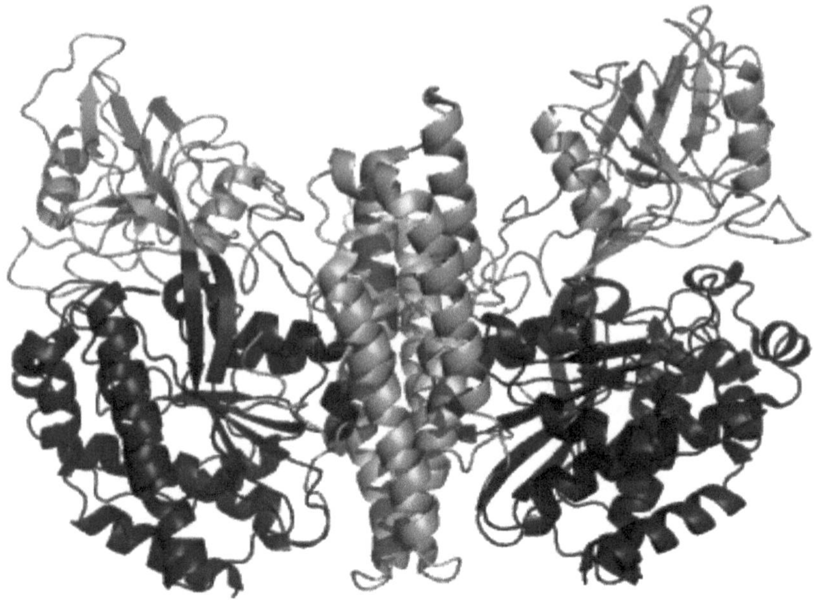

Figure 5 Ectodomain of the TFR. Crystal structure of the homodimeric ectodomain of the TFR (PDB ID: 1CX8) is oriented with the cell membrane at the bottom (Lawrence et al., 1999). The domains of each monomer are colored as follows: protease-like domain, blue; apical domain, orange; helical domain, green. Figure created using PyMOL (Delano, 2002). See the color plate

Intracellularly, phosphorylation occurs at residue Ser24 on the TFR. It is thought that phosphorylation at this location may affect endocytic trafficking of the TFR (Daniels, Delgado, Rodriguez, Helguera, & Penichet, 2006); however, the definitive function of this posttranslational modification remains unclear. Additionally, it has been reported that the tetrapeptide sequence YTRF (residues 20–23) serves as an internalization signal for endocytosis of the TFR (Collawn et al., 1993). Palmitoylation occurs at two positions (Cys62and Cys67) within the intracellular portion and may aid in anchoring the TFR to the cell membrane (Aisen, 2004).

5. RECOMBINANT EXPRESSION

5.1. Recombinant Expression and Purification of hTF

Isolation of hTF is economically feasible and relatively simple given the relative abundance in human serum (~25 to 50 μM) (Sun et al., 1999).

However, the potential for exposure to blood-borne pathogens and the inability to introduce mutations in serum-derived hTF has prompted the recombinant production of hTF in a number of expression systems. Naturally, one of the first expression platforms utilized to produce recombinant hTF is also one of the most common, *Escherichia coli*. The numerous advantages of using a bacterial expression system are well established. However, production of hTF in *E. coli* met with extremely limited success (de Smit et al., 1995; Hershberger et al., 1991; Hoefkens et al., 1996; Ikeda, Bowman, Yang, & Lokey, 1992; Steinlein & Ikeda, 1993). Due to an inability to correctly form 19 disulfide bonds (Fig. 2), attempts to produce hTF in *E. coli* yielded low amounts of improperly folded protein.

Another common expression system for recombinant proteins is the yeast, *Pichia pastoris*. Attempts to produce the isolated N-lobe of hTF in this eukaryotic system produced significant amounts (50–250 mg/l) of functional protein (Mason et al., 1996; Steinlein, Graf, & Ikeda, 1995). Although no effect on function was observed, analysis of the purified protein by electrospray mass spectrometry identified an O-linked glycosylation on residue Ser32, not present in the naturally derived protein (Mason et al., 1996). While the *P. pastoris* expression system offers several advantages (high yields and low cost), until recently no significant amount of full-length hTF has been produced in this system (Mizutani et al., 2010). Recent efforts using common baker's yeast, *Saccharomyces cerevisiae*, have produced homogenous non-glycosylated hTF at significantly higher yields (~1.5 g/l) than previously observed in any yeast system (Finnis et al., 2010).

Baculovirus-infected insect cells are another frequently used expression system. Although the production of hTF in suspended insect cells provides the convenience of fast growth rates (as in yeast and bacterial expression systems) and the ability to posttranslationally modify the expressed protein, maximum production of hTF in this system only reached ~20 mg/l (Ali, Joao, Csonga, Hammerschmid, & Steinkasserer, 1996; Retzer, Kabani, Button, Yu, & Schryvers, 1996).

By far, the most well-defined and characterized expression system used to produce recombinant hTF is the mammalian system developed in our laboratory (Funk, MacGillivray, Mason, Brown, & Woodworth, 1990; Mason, Funk, MacGillivray, & Woodworth, 1991; Mason et al., 1993, 2004). This system utilizes baby hamster kidney (BHK) cells transfected with the pNUT expression vector (Palmiter et al., 1987) containing the complementary DNA (cDNA) sequence for hTF. Importantly, the pNUT vector contains an ampicillin resistance gene that allows for propagation in

E. coli. The hTF gene expressed in the pNUT vector by our laboratory also possesses a number of key features, the first of which is the hTF 19-amino acid signal peptide. This sequence is critical to the natural secretion of hTF from the liver into the serum. Normally within the body, this signal peptide is cleaved. In order to assure cleavage of the signal peptide, the first four amino acids of hTF (Val-Pro-Asp-Lys) precede the N-terminal hexa-histidine tag (used for purification). Next, our hTF construct contains a factor Xa cleavage sequence (Ile-Glu-Gly-Arg) to allow for facile removal of the hexa-His tag. Finally, the plasmid contains the polynucleotide sequence coding for hTF (Yang et al., 1984). The resulting cDNA is then transfected into BHK cells using a standard calcium phosphate precipitation method (Mason et al., 1991). Of importance, the pNUT vector also includes a mutated dihydrofolate reductase enzyme that allows for rapid selection (1–2 weeks) with methotrexate of only those BHK cells containing the plasmid following transfection (Funk et al., 1990). The use of this recombinant system has provided the means to produce hTFs that are either incapable of binding (mutation of the two liganding Tyr residues to Phe precludes iron binding in one lobe to create authentic Fe_NhTF or Fe_ChTF constructs) (Mason et al., 2004) or releasing iron from one of the two lobes (mutation of residues in the dilysine trigger or C-lobe triad prevent iron removal to create $Lock_N$hTF or $Lock_C$hTF constructs) (Fig. 6) (Byrne, Chasteen, et al., 2010; Byrne & Mason, 2009; Halbrooks et al., 2003). Although production of the isolated N-lobe of hTF has been very successful and well documented (Funk et al., 1990; Mason et al., 1993), until recently, production of the isolated C-lobe has been far more problematic. Again, bacterial expression of the C-lobe is rendered nearly impossible by the need to correctly form 11 disulfide bonds (MacGillivray & Mason, 2002), although a low yield ($\sim$5%) of C-lobe with questionable conformation was reported by one laboratory (Hoefkens et al., 1996). Other attempts to produce the C-lobe of hTF using bacterial, yeast and mammalian systems have met with limited success and poor yields (Hoefkens et al., 1996; Steinlein et al., 1995; Steinlein & Ikeda, 1993). Even previous attempts to express the C-lobe as a recombinant entity using the BHK expression system, although successful, produced limited amounts of protein (Mason et al., 1997). Moreover, the C-lobe produced contained a complex glycosylation pattern at each of the two N-linked glycosylation sites in the C-lobe, resulting in a heterogeneous sample, further exacerbating the purification process. The Aisen laboratory produced a small amount of isolated C-lobe using the BHK cell system and an hTF construct with a factor Xa cleavage site in the bridge between

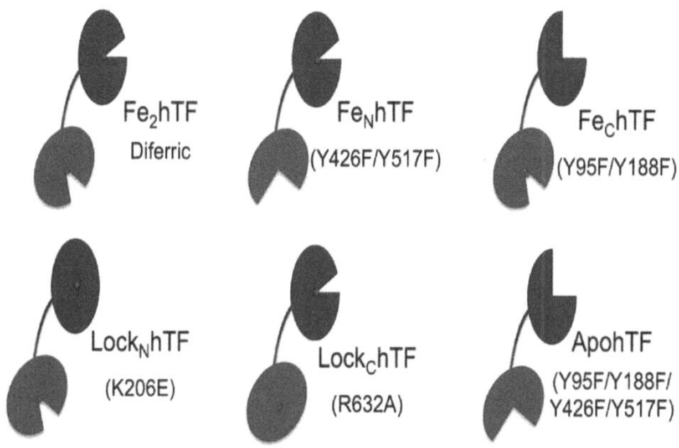

Figure 6 Recombinant hTF constructs. Schematic representing recombinant hTF constructs. Mutations introduced into hTF to prevent iron binding in a lobe (as in the case of the Fe_NhTF, Fe_ChTF and apohTF constructs) or prevent iron release from one lobe (as in the case of $Lock_N$hTF and $Lock_C$hTF) are indicated. All constructs containing an N-terminal hexa-histidine tag are non-glycosylated and have all been previously described (Byrne, Chasteen, et al., 2010; Mason et al., 2004). For color version of this figure, the reader is referred to the online version of this book.

the N- and C-lobes (Zak & Aisen, 2002). Using the BHK system, we followed a similar strategy in which the seven amino acids in the bridge were replaced by the tobacco etch virus (TEV) protease cleavage sequence, allowing utilization of the highly specific TEV protease to produce high yields of the isolated C-lobe (Steere, Roberts, et al., 2010). This particular C-lobe was recently crystallized (Noinaj et al., 2012).

Recently, recombinant production of human proteins in plants has become an appealing alternative expression system and the topic of a great deal of research. Plant sources provide a relatively inexpensive animal-free method to the production of recombinant proteins for biopharmaceutical applications. Plant-derived recombinant proteins eliminate any potential contamination by animal pathogens. Furthermore, edible transgenic plants provide an attractive possibility for therapeutic purposes through direct oral delivery. Thus far, attempts to produce recombinant hTF in tobacco (*Nicotiana tabacum*) have met with rather limited success (estimated 0.25% total soluble protein) (Brandsma et al., 2010), while the production of hTF in rice (*Oryza sativa*) is estimated to be ~40% of total soluble protein and is now available commercially under the trade name Optiferrin™ (Zhang et al., 2010). While Optiferrin™ appears to function almost identically to BHK-derived recombinant hTF in a number of aspects, the production of transferrins using this transgenic rice system requires ~1 to 2 years, making it somewhat impractical for the production of mutant transferrins (Steere, A.N., Bobst, C.E., Zhang, D., Pettit, S., Kaltashov, I.A., Huang, N., & Mason, A.B., Journal of Inorganic Biochemistry, In Press.). It seems imperative that all recombinant hTF and TFR samples undergo a rigorous evaluation of their ability to function in a physiologically relevant manner.

5.2. Recombinant Production and Purification of the sTFR

Initial studies involving the TFR utilized full-length TFR isolated from the placenta. Solubilization of the membrane-bound TFR required challenging purification techniques and the presence of detergent micelles, resulting in relatively low yields of TFR (Bali, Zak, & Aisen, 1991; Turkewitz, Amatruda, Borhani, Harrison, & Schwartz, 1988). The recombinant production of the sTFR (residues 121–706, Fig. 5) has helped overcome the challenges of working with the full-length placental-derived TFR (Byrne et al., 2006; Lawrence et al., 1999; Lebron & Bjorkman, 1999). The use of the sTFR eliminates the need for detergent required in the full-length TFR preparations. It is important to note that although two disulfide bonds in the stalk region (Cys89 and Cys98) are formed as a result of TFR dimerization, the homodimer is maintained even in their absence (as in the sTFR) due to the strong interaction between the helical domains of each monomer.

Production of the sTFR has been successful in a number of different expression systems including Chinese hamster ovary cells (Lawrence et al., 1999), baculovirus-infected insect cells (Giannetti et al., 2005; Giannetti, Snow, Zak, & Bjorkman, 2003; Lebron & Bjorkman, 1999; West et al., 2001) as well as our BHK cell system (Byrne et al., 2006). The expression in BHK cells and purification of the N-terminally hexa-His tagged sTFR in our laboratory is accomplished in much the same way as recombinant hTF and generally produces between 30 and 40 mg of protein per production run (Byrne et al., 2006).

6. KINETICS OF IRON RELEASE

Work from the Aisen laboratory provided some of the first mechanistic insights into the kinetics of iron release from hTF at pH 5.6 (Bali & Aisen, 1991, 1992). Specifically, the initial evidence that the TFR slows iron release from hTF at neutral pH while enhancing iron release from hTF at endosomal pH was provided by the polyethyleneglycol (PEG) precipitation studies of Bali and Aisen (Bali et al., 1991). By binding ^{59}Fe to apohTF, these experiments allowed iron release at various pH values to be monitored by PEG precipitation of the $^{59}Fe_2$hTF (the chelate–iron complex ^{59}Fe–PP$_i$ remains in the soluble fraction).

Over a number of years, the laboratory of el Hage Chahine has reported the results of stopped-flow absorbance and pH jump chemical relaxation studies monitoring iron release from serum-derived diferric hTF and monoferric C-lobe hTF in the presence of detergent-solubilized full-length placental-derived TFR (Chahine & Pakdaman, 1995; Hemadi & Ha-Duong, 2006). Discordant results have been obtained in this system in comparison to those of Aisen (Bali & Aisen, 1992) and our own laboratory, as recently discussed in some detail (Steere, Byrne, Chasteen, & Mason, 2012). Specifically, the detergent-solubilized TFR used by the el Hage Chahine laboratory promotes iron release preferentially from the N-lobe of hTF first. We provide evidence (below) that the sTFR promotes iron release preferentially from the C-lobe of hTF first. The numerous differences in the conditions used, including the absence of a chelator and pH values of <4 in the system of el Hage Chahine, may well account for the discrepancy. It appears that further work is required to completely resolve the substantial discrepancy in the findings. Of course, the use of any in vitro system to model what may be occurring in vivo is always subject to a good deal of uncertainty and questions of physiological relevance.

We suggest that, to date, our laboratory has provided the most comprehensive kinetic analysis of iron release from hTF in the absence and presence of the sTFR (Byrne, Chasteen, et al., 2010). Using an optimized stopped-flow iron removal assay, the increase in intrinsic tryptophan fluorescence (Ex. 280 nm) is measured as hTF, alone or in complex with the sTFR in 300 mM KCl is rapidly mixed with our standard iron removal buffer (200 mM MES (2-(N-morpholino)ethanesulfonic acid), pH 5.6 containing 300 mM KCl and 8 mM EDTA). In combination with the ability to produce recombinant hTF with specific properties of iron binding and/or release, the use of the stopped-flow spectrofluorometer has allowed a comprehensive determination of all kinetic steps involved in iron release from hTF in the presence and absence of the sTFR. With all the proper controls and equations, we can confidently extract accurate kinetic constants that provide important information with regard to the detailed mechanism of iron release from various hTF and TFR mutants. Understanding this complex system, especially with regard to the role of the TFR, is crucial to ultimately discovering how to manipulate it.

6.1. Kinetics of Iron Release from hTF in the Absence of the TFR

As mentioned, it is clear from decades of research that iron release from hTF is complex and involves a number of factors including pH, anion binding, chelator, lobe–lobe cooperativity and binding to the TFR. Under pseudo-first-order conditions, iron release from Fe_2hTF can occur via two pathways giving rise to four microscopic rate constants, (k_{1C}, k_{2N}, k_{1N}, and k_{2C}) (Bali & Aisen, 1992; Harris, Bali, & Crowley, 1992). The kinetic assessment of diferric locked constructs ($Lock_N$hTF and $Lock_C$hTF) allows the specific assignment of k_{1C} and k_{1N}, while the use of authentic monoferric constructs (Fe_NhTF and Fe_ChTF) allows the specific assignment of k_{2N} and k_{2C}, respectively. Curve fitting of Fe_2hTF shows that in the absence of the sTFR iron release from the N-lobe is very rapid (17.7 ± 2.2/min) and relatively slow from the C-lobe (0.65 ± 0.06/min). Thus, iron is released from Fe_2hTF almost exclusively ($\sim$96% of the time) via the $k_{1N} \rightarrow k_{2C}$ pathway, precluding the need to consider the negligible $k_{1C} \rightarrow k_{2N}$ pathway in the analysis of Fe_2hTF data. The validity of these values is further supported through analysis of the $Lock_C$hTF and Fe_ChTF constructs which provides independent measures of k_{1N} and k_{2C}.

6.1.1. Cooperativity between hTF Lobes

Given the bilobal nature of hTF, it is critical to examine whether cooperativity exists between the two iron-binding sites. Kinetic analysis of Fe_NhTF and the isolated N-lobe, along with k_{1N} values obtained from the Fe_2hTF and $Lock_C$hTF constructs, indicates that in the absence of the sTFR, the rate of iron release from the N-lobe (k_N) is impacted not only by the presence but also the iron occupancy of the C-lobe. However, the rate of iron release from the C-lobe of hTF (k_C) is clearly unaffected by the presence or iron status of the N-lobe. Kinetic cooperativity factors are used as a measure of the cooperative effect of iron release in one lobe on the other lobe (Bali & Aisen, 1992; Byrne, Chasteen, et al., 2010). Given as a ratio of k_2/k_1 of a respective lobe, a number greater than one indicates positive cooperativity between the two lobes, whereas a number equal to or less than one indicates a lack of cooperativity. Based on these determined kinetic values, the calculated kinetic cooperativity figures ($k_{2N}/k_{1N} = 24.8/17.9 = 1.4$ and $k_{2C}/k_{1C} = 0.79/0.72 = 1.1$) indicate that iron release from the N-lobe is accelerated by the absence of iron in the C-lobe as well as by the complete absence of the C-lobe, whereas iron occupancy of the N-lobe has little affect on the kinetics of iron release from the C-lobe.

6.1.2. Anion Binding

A complex relationship exists between anion binding to KISAB sites and iron release from hTF. At pH 5.6, anion concentration most dramatically affects the rate of iron release from the N-lobe (increasing it ca. four-fold from 50 to 600 mM [Cl^-]) in the absence of the TFR (Byrne, Chasteen, et al., 2010; Byrne, Steere, et al., 2010), while the rate of iron release from the C-lobe is relatively unaffected (Byrne, Chasteen, et al., 2010). The first authentic KISAB site in the N-lobe, residue Arg143, was not definitively identified until recently (Byrne, Steere, et al., 2010). Mutation of Arg143 to alanine completely eliminates iron release from the N-lobe of hTF at all tested chloride concentrations (50–600 mM) in the absence of the TFR. Thus, binding of anions to Arg143 may (at least in part) account for the salt sensitivity of iron release from the N-lobe of hTF. Additionally, mutation of C-lobe residue Lys569 in the Fe_2hTF background was previously found to slow iron release by 15- to 20-fold and inhibit chloride enhancement of iron release from the C-lobe (Zak, Tam, MacGillivray, & Aisen, 1997). Therefore, Lys569 was putatively identified as a KISAB site thought to be essential for iron release from the C-lobe of hTF. However, more recently we demonstrated that mutation of Lys569 to alanine (K569A) has little effect on the rate of iron release from Fe_2hTF in the absence or presence of the sTFR (Eckenroth et al., 2011). These data are consistent with the finding that the rate of iron release from the C-lobe is relatively unaffected by anion concentration. The discrepancy between more recent and previously published data again highlights the limitations of the steady-state tryptophan fluorescence technique used in the earlier work.

6.2. Kinetics of Iron Release from hTF in the Presence of the TFR

The TFR plays an integral role in physiological iron release from hTF. Therefore, it is absolutely essential that iron release from hTF also be monitored in the presence of the TFR. In accordance with previous findings (Bali & Aisen, 1991, 1992), under our standard conditions, the sTFR induces a switch in the order of iron release from hTF, such that iron is preferentially removed from the C-lobe first followed by the N-lobe ($k_{1C} \rightarrow k_{2N}$) (Byrne, Chasteen, et al., 2010). However, the fits of the kinetic data indicate that this is not the case 100% of the time. The $k_{1N} \rightarrow k_{2C}$ pathway is reduced from 96% in the absence of the sTFR to only 35% in the presence of the sTFR. It is clear that both pathways are physiologically relevant and must be included in the fits of Fe_2hTF/sTFR complex. While analyzing Fe_2hTF/sTFR complex kinetic data, in order to limit the number of variables, the values for the other pathway (k_{1N} and k_{2C}) are held constant based on kinetic values obtained from the $Lock_C$hTF/sTFR and Fe_ChTF/sTFR complexes. Again, the values obtained independently for the $k_{1C} \rightarrow k_{2N}$ pathway from the $Lock_N$hTF/sTFR and Fe_NhTF/sTFR complexes ($k_{1C} = 5.0 \pm 1.6$ and $k_{2N} = 1.7 \pm 0.6$) corroborate the validity of the Fe_2hTF/sTFR complex analysis.

Previous studies by the Aisen laboratory clearly demonstrated that the TFR enhances the rate of iron release from the C-lobe of hTF at pH 5.6 (Bali & Aisen, 1991; Zak & Aisen, 2003). Using our standard conditions, the sTFR enhances iron release from the C-lobe by ~7- to 11-fold ($k_{1C,\ complex}/k_{1C,\ alone}$ and $k_{2C,\ complex}/k_{2C,\ alone}$) and retards iron release from the N-lobe by 6- to 15-fold ($k_{1N,\ complex}/k_{1N,\ alone}$ and $k_{2N,\ complex}/k_{2N,\ alone}$) (Byrne, Chasteen, et al., 2010). Thus, not only does binding to the sTFR switch the order of iron release from hTF but also makes the rates of iron release from the two lobes more equivalent than in the absence of the sTFR. Consequently, the calculated cooperativity factors of each lobe are much different in the presence of the sTFR (N-$lobe_{alone} = 1.4$ versus N-$lobe_{complex} = 0.61$ and C-$lobe_{alone} = 1.1$ versus C-$lobe_{complex} = 1.5$) (Byrne, Chasteen, et al., 2010). Therefore, in the presence of the sTFR, the rate of iron release from the C-lobe is slowed by the presence of iron in the N-lobe, while for the N-lobe the reverse is true: iron occupancy in the C-lobe accelerates iron release from the N-lobe.

6.2.1. *pH Effects*

The endosomal pH (~5.6) is a critical component of the endocytic cycle and iron delivery by hTF. It is evident from a pH titration of the Fe_ChTF/sTFR complex that iron release from the hTF/TFR complex has an apparent pK_a of ~5.8 to 5.9 (Steere, Byrne, et al., 2010), a value similar to the pK_a of histidine residues. As previously mentioned, His349 in the C-lobe of hTF has been a residue of particular interest for some time (Giannetti et al., 2005). Interestingly, mutation of His349 to alanine or lysine completely abolishes the pH dependence of iron release from the Fe_ChTF/sTFR complex (Steere, Byrne, et al., 2010). This finding led to the proposal that His349 acts as a pH-inducible switch that accelerates iron release from the C-lobe of hTF. More recently, the effect of the H349A mutation in the Fe_2hTF background has been investigated. Curiously, the H349A Fe_2hTF/sTFR complex displays three kinetic rates (unlike the Fe_2hTF/sTFR control where only two rates, k_{1C} and k_{2N}, are obtained) (Eckenroth et al., 2011).

The first rate ($k_1 = 23.7 \pm 4.6$/min), ascribed to a rapid conformational change, is very similar to the conformational change observed in both monoferric complexes, as discussed above. However, unlike the Fe_2hTF/sTFR control complex, iron release from the H349A Fe_2hTF/sTFR complex follows only the $k_{1N} \rightarrow k_{2C}$ pathway. Interestingly, the rate of iron release from the N-lobe of the H349A Fe_2hTF/sTFR complex is nearly doubled ($k_{1N} = 6.7 \pm 0.3$/min) in comparison to the Fe_2hTF/sTFR complex ($k_{1N} = 2.8 \pm 0.8$/min) while iron release from the C-lobe of the H349A Fe_2hTF/sTFR complex is greatly retarded ($k_{2C} = 0.61 \pm 0.02$/min). Hence, the effect of receptor stimulation on iron release from the C-lobe is completely abrogated in the presence of the H349A mutation (Eckenroth et al., 2011), verifying its crucial role in the mechanism of iron release from the C-lobe of the hTF/TFR complex.

6.2.2. Anion Binding

In contrast to the findings of Egan et al. (1993), we find that at pH 5.6 the effect of anion concentration on iron release from hTF in the presence of the sTFR is significantly muted (Byrne, Chasteen, et al., 2010). Therefore, although a slight salt-induced acceleration in iron removal from the hTF/TFR complex is observed at pH 5.6, the overriding kinetic effect of the TFR appears to outweigh any kinetic anion effects. In accordance with these findings (Byrne, Chasteen, et al., 2010), the effect of the Arg143 mutation (described above) is largely abolished in the presence of the TFR (Byrne, Steere, et al., 2010).

7. STRUCTURE OF THE HTF/TFR COMPLEX

Given that the hTF/TFR interaction promotes conformational changes required for iron release, the molecular details of the interactions between the two proteins is clearly critical. Although transferrin has been studied extensively since the late 1940s and the existence of the TFR was confirmed in the 1970s, the precise structural details of the hTF/TFR interaction, as well as the functional implications of those interactions, have remained unclear. Therefore, identification of the contacts formed between the two proteins is critical. Studying the hTF/TFR interaction in the past was hindered by a lack of structural information on the complex. In 2004, the first model of the TF/sTFR complex was completed. A 7.5-Å resolution cryo-electron microscopy model of the TF/sTFR complex utilized the crystal structure of the unliganded sTFR and individually modeled in the structures of the isolated human N-lobe and rabbit C-lobe ($\sim$85% similar to the human C-lobe). The model indicated that the N-lobe is situated between the membrane and the TFR, while the C-lobe makes significant contacts with the helical domain of the TFR (PDB ID: 1SUV) (Cheng, Zak, Aisen, Harrison, & Walz, 2004). However, confidence in the model is limited by the relatively low resolution and the fact that placement of the N-lobe into the model required an $\sim$9 Å shift with respect to the C-lobe (Cheng et al., 2004; Sakajiri, Yamamura, Kikuchi, & Yajima, 2009). More recently, the inclusion of the available structure of the apohTF (Wally et al., 2006) in addition to consideration of various mutagenesis studies of hTF and the TFR led to a new in silico model of hTF bound to the TFR (Sakajiri et al., 2009). This computational model eliminated the need for the 9 Å gap between the N- and C-lobes of hTF and appeared to be consistent with all available mutagenesis data from both the TFR and the hTF (Giannetti et al., 2003; Mason et al., 2009).

The recently published high-resolution (3.22 Å) crystal structure of the Fe_NhTF/sTFR complex (Eckenroth et al., 2011) provided precise molecular details of the interaction between the two binding partners. Three distinct binding motifs between hTF and the TFR are observed: the hTF N1-TFR motif, the hTF N2-TFR motif and the hTF C1-TFR motif. Moreover, the structure clearly illustrates how binding to the TFR hinders iron release from the N-lobe while promoting iron release from the C-lobe. As in the structure of the hemochromatosis protein (HFE) and the TFR (Bennett, Lebron, & Bjorkman, 2000), the Fe_NhTF/sTFR complex structure definitively showed that conformational changes in the TFR occur as a result of ligand binding. Importantly, the insight gained from the recent crystal structure has provided new mutagenic targets in both the hTF and the TFR. By assessing various single-point mutants, we have been able to identify residues within both hTF and the sTFR that are critical to the receptor–ligand binding interaction and also influence the kinetics of iron release from hTF.

7.1. Features of the Fe_NhTF/sTFR Structure

An interesting feature of the Fe_NhTF/sTFR crystal structure was the presence of a Ca^{2+} ion bound between the apical and protease-like domains of the sTFR. Highlighting the structural stability provided by this Ca^{2+} ion, mutation of two of the coordinating amino acids (E465A/E468A) produces an unstable aggregated sTFR (Steere, Chasteen, et al., 2012). A number of intriguing questions remain with regard to this calcium ion in the sTFR. First, although the coordinated metal in sTFR produced in our BHK system was definitively identified as calcium by inductively coupled plasma mass spectrometry (Eckenroth et al., 2011), calcium may not be the physiologically relevant metal. It is possible that this metal-binding site in the sTFR may be occupied by some other metal ion (Zn^{2+} or Mg^{2+} for example) in vivo. Additionally, the identity of the physiological chelator of iron within the endosome remains unknown. We have previously shown that decreased pH (~5.6) alone is not sufficient to remove iron from hTF or the hTF/sTFR complex and that the presence of a chelator is indeed required to promote iron release in a physiologically relevant time frame (Byrne, Chasteen, et al., 2010). Although abundant small molecules such as citrate or PP_i have been suggested, to date, the endosomal chelator has not been definitively identified. Given the presence of another chelatable metal ion other than iron, the question of whether this other metal (putatively Ca^{2+}) is chelated within the endosome remains.

One especially interesting feature of the Fe_NhTF/sTFR structure was a large intersection formed between the apical and the protease-like domains of one TFR monomer, the helical domain of the other TFR monomer and the C1 subdomain of hTF (TFR–TFR′–C1 subdomain intersection) (Eckenroth et al., 2011). The significance of this intersection is emphasized by the large conformational changes in this region of the TFR that take place upon hTF binding. Furthermore, given the pH range experienced by hTF throughout the endocytic cycle (between ~7.5 and 5.5), it is not entirely surprising that histidine residues in both hTF (Eckenroth et al., 2011; Steere, Byrne, et al., 2010) and the sTFR (Steere, Chasteen, et al., 2012) within this TFR–TFR′–C1 subdomain intersection participate in pH-dependent receptor-stimulated iron release from the C-lobe of hTF. Through a series of interrelated pH-induced events involving His residues within the TFR–TFR′–C1 subdomain intersection iron release from the C-lobe of hTF is triggered. Specifically, protonation of sTFR residue His318 promotes a pH-inducible change in the sTFR that is communicated to the nearby C-terminus (residues 757–760) of the other TFR monomer (TFR′). Meanwhile, due to its protonation and a related conformational change in hTF, His349 (hTF) interacts with the C-terminus of the TFR′ and ultimately promotes iron release from the C-lobe of the hTF/sTFR complex (Steere, Chasteen, et al., 2012).

It is well established that highly proliferative cells have dramatically increased iron requirements and therefore an increased number of TFRs, making the TFR a prime candidate for targeted therapeutics (Daniels et al., 2011). Hence, the molecular details of how the TFR interacts with its ligand, hTF, are critical and could be used in order to develop small molecules or toxin conjugates to target the hTF/TFR pathway in cancer cells. Moreover, identification of specific residues that account for the pH-sensitive binding affinity of hTF throughout the endocytic cycle could be essential to delivery of therapeutics inside the cell. We are slowly identifying specific residues and/or structural elements within both the hTF and the TFR that directly correlate to events that occur during iron release.

In conclusion, after over a half century of research on transferrin, much has been learned (as detailed in the recent *Biochimica Biophysica Acta* Special Issue titled Transferrins: Molecular Mechanisms of Iron Transport and Disorders), but many interesting and important questions remain. Although the existence of the TFR was confirmed in the 1970s, the precise structural details of the hTF/TFR interaction, as well as the implications of those interactions on cellular iron delivery, have remained unclear. Collectively, these results help to establish a molecular basis for the pH-induced events that dictate efficient release of iron from each lobe within the endosome in a physiologically relevant time frame. We now enjoy a better understanding of how the TFR is able to influence iron release from each lobe of hTF at endosomal pH. Appreciation that the TFR is more than a passive binding partner and that it trumps other factors is emphasized by recent work.

ACKNOWLEDGMENTS

This work was supported by U.S. Public Service Grant R01 DK 21739 to A.B.M. and an American Heart Association Predoctoral Fellowship (10PRE4200010) to A.N.S. We wish to thank Drs N. D. Chasteen, Brian E. Eckenroth, Shaina L. Byrne and Nicholas G. James for their substantive contributions that helped to provide the basis for some of this work.

REFERENCES

Aisen, P. (2004). *The International Journal of Biochemistry and Cell Biology, 36*, 2137–2143.

Aisen, P., Enns, C., & Wessling-Resnick, M. (2001). *The International Journal of Biochemistry and Cell Biology, 33*, 940–959.

Aisen, P., Leibman, A., & Zweier, J. (1978). *The Journal of Biological Chemistry, 253*, 1930–1937.

Ali, S. A., Joao, H. C., Csonga, R., Hammerschmid, F., & Steinkasserer, A. (1996). *The Biochemical Journal, 319*(Pt 1), 191–195.

Anderson, B. F., Baker, H. M., Dodson, E. J., Norris, G. E., Rumball, S. V., Waters, J. M., et al. (1987). *Proceedings of the National Academy of Sciences of the United States of America, 84*, 1769–1773.

Andrews, N. C., & Schmidt, P. J. (2007). *Annual Review of Physiology, 69*, 69–85.

Arosio, P., & Levi, S. (2010). *Biochimica et Biophysica Acta, 1800*, 783–792.

Baker, E. N. (1994). *Advances in Inorganic Chemistry, 41*, 389–463.
Baldwin, D. A. (1980). *Biochimica et Biophysica Acta, 623*, 183–198.
Baldwin, D. A., & de Sousa, D. M. (1981). *Biochemical and Biophysical Research Communications, 99*, 1101–1107.
Bali, P. K., & Aisen, P. (1991). *Biochemistry, 30*, 9947–9952.
Bali, P. K., & Aisen, P. (1992). *Biochemistry, 31*, 3963–3967.
Bali, P. K., & Harris, W. R. (1990). *Archives of Biochemistry and Biophysics, 281*, 251–256.
Bali, P. K., Zak, O., & Aisen, P. (1991). *Biochemistry, 30*, 324–328.
Bennett, M. J., Lebron, J. A., & Bjorkman, P. J. (2000). *Nature, 403*, 46–53.
Bou-Abdallah, F., McNally, J., Liu, X. X., & Melman, A. (2011). *Chemical Communications, 47*, 731–733.
Brandsma, M. E., Diao, H., Wang, X., Kohalmi, S. E., Jevnikar, A. M., & Ma, S. (2010). *Plant Biotechnology Journal, 8*, 489–505.
Byrne, S. L., Chasteen, N. D., Steere, A. N., & Mason, A. B. (2010). *Journal of Molecular Biology, 396*, 130–140.
Byrne, S. L., Leverence, R., Klein, J. S., Giannetti, A. M., Smith, V. C., MacGillivray, R. T., et al. (2006). *Biochemistry, 45*, 6663–6673.
Byrne, S. L., & Mason, A. B. (2009). *Journal of Biological Inorganic Chemistry, 14*, 771–781.
Byrne, S. L., Steere, A. N., Chasteen, N. D., & Mason, A. B. (2010). *Biochemistry, 49*, 4200–4207.
Cannon, J. C., & Chasteen, N. D. (1975). *Biochemistry, 14*, 4573–4577.
Chahine, J. M. E. H., & Pakdaman, R. (1995). *European Journal of Biochemistry, 230*, 1102–1110.
Chasteen, N. D. (1983). *Advances in Inorganic Biochemistry, 5*, 201–233.
Cheng, Y., Zak, O., Aisen, P., Harrison, S. C., & Walz, T. (2004). *Cell, 116*, 565–576.
Collawn, J. F., Lai, A., Domingo, D., Fitch, M., Hatton, S., & Trowbridge, I. A. (1993). *The Journal of Biological Chemistry, 268*, 21686–21692.
Daniels, T. R., Bernabeu, E., Rodriguez, J. A., Patel, S., Kozman, M., Chiappetta, D. A., et al. (2011). *Biochimica et Biophysica Acta, 1820*, 291–317.
Daniels, T. R., Delgado, T., Rodriguez, J. A., Helguera, G., & Penichet, M. L. (2006). *Clinical Immunology, 121*, 144–158.
Delano, W. L. (2002). The PyMOL Molecular Graphics System.
de Smit, D. H., Hoefkens, P., de Jong, G., van Duin, J., van Knippenberg, P. H., & van Eijk, H. G. (1995). *The International Journal of Biochemistry and Cell Biology, 27*, 839–850.
Dewan, J. C., Mikami, B., Hirose, M., & Sacchettini, J. C. (1993). *Biochemistry, 32*, 11963–11968.
Dhungana, S., Taboy, C. H., Zak, O., Larvie, M., Crumbliss, A. L., & Aisen, P. (2004). *Biochemistry, 43*, 205–209.
Donovan, A., Brownlie, A., Zhou, Y., Shepard, J., Pratt, S. J., Moynihan, J., et al. (2000). *Nature, 403*, 776–781.
Eckenroth, B. E., Mason, A. B., McDevitt, M. E., Lambert, L. A., & Everse, S. J. (2010). *Protein Science, 19*, 1616–1626.
Eckenroth, B. E., Steere, A. N., Chasteen, N. D., Everse, S. J., & Mason, A. B. (2011). *Proceedings of the National Academy of Sciences of the United States of America, 108*, 13089–13094.
Egan, T. J., Ross, D. C., Purves, L. R., & Adams, P. A. (1992). *Inorganic Chemistry, 31*, 1994–1998.
Egan, T. J., Zak, O., & Aisen, P. (1993). *Biochemistry, 32*, 8162–8167.
Enns, C. A., & Sussman, H. H. (1981). *The Journal of Biological Chemistry, 256*, 9820–9823.
Fenton, H. J. H. (1876). *Chemical News, 33*, 190.
Fenton, H. J. H. (1893). *Chemical Society Proceedings, 9*, 113.

Finnis, C. J., Payne, T., Hay, J., Dodsworth, N., Wilkinson, D., Morton, P., et al. (2010). *Microbial Cell Factories, 9*, 87.
Fleming, M. D., Trenor, C. C., 3rd, Su, M. A., Foernzler, D., Beier, D. R., Di etrich, W. F., et al. (1997). *Nature Genetics, 16*, 383–386.
Folajtar, D. A., & Chasteen, N. D. (1982). *Journal of the American Chemical Society, 104*, 5775–5780.
Funk, W. D., MacGillivray, R. T., Mason, A. B., Brown, S. A., & Woodworth, R. C. (1990). *Biochemistry, 29*, 1654–1660.
Gaber, B. P., Miskowski, V., & Spiro, T. G. (1974). *Journal of the American Chemical Society, 96*, 6868–6873.
Giannetti, A. M., Halbrooks, P. J., Mason, A. B., Vogt, T. M., Enns, C. A., & Bjorkman, P. J. (2005). *Structure, 13*, 1613–1623.
Giannetti, A. M., Snow, P. M., Zak, O., & Bjorkman, P. J. (2003). *PLoS Biology, 1*, 341–350.
Grady, J. K., Mason, A. B., Woodworth, R. C., & Chasteen, N. D. (1995). *The Biochemical Journal, 309*(Pt 2), 403–410.
Grant, B. D., & Donaldson, J. G. (2009). *Nature Reviews Molecular Cell Biology, 10*, 597–608.
Greenberg, G. R., & Wintrobe, M. M. (1946). *The Journal of Biological Chemistry, 165*, 397.
Gubler, C. J., Lahey, M. E., Chase, M. S., Cartwright, G. E., & Wintrobe, M. M. (1952). *Blood, 7*, 1075–1092.
Gunshin, H., Fujiwara, Y., Custodio, A. O., Direnzo, C., Robine, S., & Andrews, N. C. (2005). *The Journal of Clinical Investigation, 115*, 1258–1266.
Gunshin, H., Mackenzie, B., Berger, U. V., Gunshin, Y., Romero, M. F., Boron, W. F., et al. (1997). *Nature, 388*, 482–488.
Halbrooks, P. J., He, Q. Y., Briggs, S. K., Everse, S. J., Smith, V. C., MacGillivray, R. T., et al. (2003). *Biochemistry, 42*, 3701–3707.
Hall, D. R., Hadden, J. M., Leonard, G. A., Bailey, S., Neu, M., Winn, M., et al. (2002). *Acta Crystallographica. Section D, Biological Crystallography, 58*, 70–80.
Hallberg, L., Bjorn-Rasmussen, E., Howard, L., & Rossander, L. (1979). *Scandinavian Journal of Gastroenterology, 14*, 769–779.
Harris, W. R. (1985). *Biochemistry, 24*, 7412–7418.
Harris, W. R., & Bali, P. K. (1988). *Inorganic Chemistry, 27*, 2687–2691.
Harris, W. R., Bali, P. K., & Crowley, M. M. (1992). *Inorganic Chemistry, 31*, 2700–2705.
Harris, W. R., Cafferty, A. M., Abdollahi, S., & Trankler, K. (1998). Binding of monovalent anions to human serum transferrin. *Biochimica et Biophysica Acta, 1383*, 197–210.
Harris, W. R., Nesset-Tollefson, D., Stenback, J. Z., & Mohamed-Hani, N. (1990). *Journal of Inorganic Biochemistry, 38*, 175–183.
Harrison, P. M., & Arosio, P. (1996). *Biochimica et Biophysica Acta, 1275*, 161–203.
Hayes, G. R., Williams, A. M., Lucas, J. J., & Enns, C. A. (1997). *Biochemistry, 36*, 5276–5284.
He, Q.-Y., & Mason, A. B. (2002). D. M. Templeton (Ed.), New York: Marcel Dekker, Inc. 95–124.
Hemadi, M., Ha-Duong, N. T., & el Hage Chahine, J. M. (2006). *Journal of Molecular Biology, 358*, 1125–1136.
Hershberger, C. L., Larson, J. L., Arnold, B., Rosteck, P. R., Jr., Williams, P., DeHoff, B., et al. (1991). *Annals of the New York Academy of Sciences, 646*, 140–154.
Hoefkens, P., de Smit, M. H., de Jeu-Jaspars, N. M., Huijskes-Heins, M. I., de Jong, G., & van Eijk, H. G. (1996). *The International Journal of Biochemistry and Cell Biology, 28*, 975–982.
Huebers, H. A., Josephson, B., Huebers, E., Csiba, E., & Finch, C. A. (1984). *Proceedings of the National Academy of Sciences of the United States of America, 81*, 4326–4330.

Ikeda, R. A., Bowman, B. H., Yang, F., & Lokey, L. K. (1992). *Gene, 117*, 265–269.
Jacobs, A. (1977). *Blood, 50*, 433–439.
Jandl, J. H., & Katz, J. H. (1963). *The Journal of Clinical Investigation, 42*, 314–326.
Jeffrey, P. D., Bewley, M. C., MacGillivray, R. T., Mason, A. B., Woodworth, R. C., & Baker, E. N. (1998). *Biochemistry, 37*, 13978–13986.
Kakhlon, O., & Cabantchik, Z. I. (2002). *Free Radical Biology and Medicine, 33*, 1037–1046.
Kaplan, J., & O'Halloran, T. V. (1996). *Science, 271*, 1510–1512.
Kretchmar, S. A., & Raymond, K. N. (1988). *Inorganic Chemistry, 27*, 1436–1441.
Kubal, G., Mason, A. B., Patel, S. U., Sadler, P. J., & Woodworth, R. C. (1993). *Biochemistry, 32*, 3387–3395.
Kuhn, L. C., McClelland, A., & Ruddle, F. H. (1984). *Cell, 37*, 95–103.
Lahey, M. E., Gubler, C. J., Chase, M. S., Cartwright, G. E., & Wintrobe, M. M. (1952). *Blood, 7*, 1053–1074.
Lambert, L. A., Perri, H., Halbrooks, P. J., & Mason, A. B. (2005). *Comparative Biochemistry and Physiology. Part B, Biochemistry and Molecular Biology, 142*, 129–141.
Laufberger, V. (1937). *Bulletin de la Société de Chimie Biologique*, 1575–1582.
Lawrence, C. M., Ray, S., Babyonyshev, M., Galluser, R., Borhani, D. W., & Harrison, S. C. (1999). *Science, 286*, 779–782.
Lebron, J. A., & Bjorkman, P. J. (1999). *Journal of Molecular Biology, 289*, 1109–1118.
Lehrer, S. S. (1969). *The Journal of Biological Chemistry, 244*, 3613–3617.
Leibman, A., & Aisen, P. (1977). *Biochemistry, 16*, 1268–1272.
Leibman, A., & Aisen, P. (1979). *Blood, 53*, 1058–1065.
Lestas, A. N. (1976). *British Journal of Haematology, 32*, 341–350.
Leverence, R., Mason, A. B., & Kaltashov, I. A. (2010). *Proceedings of the National Academy of Sciences of the United States of America, 107*, 8123–8128.
Levy, J. E., Jin, O., Fujiwara, Y., Kuo, F., & Andrews, N. C. (1999). *Nature Genetics, 21*, 396–399.
MacGillivray, R. T., Moore, S. A., Chen, J., Anderson, B. F., Baker, H., Luo, Y., et al. (1998). *Biochemistry, 37*, 7919–7928.
MacGillivray, R. T. A. & Mason, A. B. Templeton, D. M. (2002). New York: Marcel Dekker, Inc. 41–69.
Makey, D. G., & Seal, U. S. (1976). *Biochimica et Biophysica Acta, 453*, 250–256.
Marques, H. M., Egan, T. J., & Pattrick, G. (1990). *South African Journal of Science, 86*, 21–24.
Marques, H. M., Watson, D. L., & Egan, T. J. (1991). *Inorganic Chemistry, 30*, 3758–3762.
Mason, A. B., Byrne, S. L., Everse, S. J., Roberts, S. E., Chasteen, N. D., Smith, V. C., et al. (2009). *Journal of Molecular Recognition, 22*, 521–529.
Mason, A. B., Funk, W. D., MacGillivray, R. T., & Woodworth, R. C. (1991). *Protein Expression and Purification, 2*, 214–220.
Mason, A. B., Halbrooks, P. J., Larouche, J. R., Briggs, S. K., Moffett, M. L., Ramsey, J. E., et al. (2004). *Protein Expression and Purification, 36*, 318–326.
Mason, A. B., Miller, M. K., Funk, W. D., Banfield, D. K., Savage, K. J., Oliver, R. W., et al. (1993). *Biochemistry, 32*, 5472–5479.
Mason, A. B., Tam, B. M., Woodworth, R. C., Oliver, R. W., Green, B. N., Lin, L. N., et al. (1997). *The Biochemical Journal, 326*(Pt 1), 77–85.
Mason, A. B., Woodworth, R. C., Oliver, R. W., Green, B. N., Lin, L. N., Brandts, J. F., et al. (1996). *Protein Expression and Purification, 8*, 119–125.
McCance, R. A., & Widdowson, E. M. (1938). *Journal of Physiology, 94*, 148–154.
McClelland, A., Kuhn, L. C., & Ruddle, F. H. (1984). *Cell, 39*, 267–274.
Mizutani, K., Hashimoto, K., Takahashi, N., Hirose, M., Aibara, S., & Mikami, B. (2010). *Bioscience, Biotechnology and Biochemistry, 74*, 309–315.
Morgan, E. H. (1981). *Biochimica et Biophysica Acta, 642*, 119–134.

Morgan, E. H., & Appleton, T. C. (1969). *Nature, 223*, 1371–1372.
Noinaj, N., Easley, N. C., Oke, M., Mizuno, N., Gumbart, J., Boura, E., et al. (2012). *Nature, 483*, 53–58.
Ohgami, R. S., Campagna, D. R., Greer, E. L., Antiochos, B., McDonald, A., Chen, J., et al. (2005). *Nature Genetics, 37*, 1264–1269.
Palmiter, R. D., Behringer, R. R., Quaife, C. J., Maxwell, F., Maxwell, I. H., & Brinster, R. L. (1987). *Cell, 50*, 435–443.
Park, C. H., Bacon, B. R., Brittenham, G. M., & Tavill, A. S. (1987). *Laboratory Investigation, 57*, 555–563.
Park, I., Schaeffer, E., Sidoli, A., Baralle, F. E., Cohen, G. N., & Zakin, M. M. (1985). *Proceedings of the National Academy of Sciences of the United States of America, 82*, 3149–3153.
Patch, M. G., & Carrano, C. J. (1981). *Inorganica Chimica Acta, 56*, L71–L73.
Pecoraro, V. L., Harris, W. R., Carrano, C. J., & Raymond, K. N. (1981). *Biochemistry, 20*, 7033–7039.
Ponka, P., Borova, J., Neuwirt, J., & Fuchs, O. (1979). *FEBS Letters, 97*, 317–321.
Posey, J. E., & Gherardini, F. C. (2000). *Science, 288*, 1651–1653.
Price, E. M., & Gibson, J. F. (1972a). *The Journal of Biological Chemistry, 247*, 8031–8035.
Price, E. M., & Gibson, J. F. (1972b). *Biochemical and Biophysical Research Communications, 46*, 646–651.
Princiotto, J. V., & Zapolski, E. J. (1975). *Nature, 255*, 87–88.
Qiu, A., Jansen, M., Sakaris, A., Min, S. H., Chattopadhyay, S., Tsai, E., et al. (2006). *Cell, 127*, 917–928.
Raffin, S. B., Woo, C. H., Roost, K. T., Price, D. C., & Schmid, R. (1974). *The Journal of Clinical Investigation, 54*, 1344–1352.
Regoeczi, E., Bolyos, M., & Chindemi, P. A. (1989). *Archives of Biochemistry and Biophysics, 268*, 637–642.
Retzer, M. D., Kabani, A., Button, L. L., Yu, R. H., & Schryvers, A. B. (1996). *The Journal of Biological Chemistry, 271*, 1166–1173.
Sakajiri, T., Yamamura, T., Kikuchi, T., & Yajima, H. (2009). *Protein Journal, 28*, 407–414.
Schade, A. L., Reinhart, R. W., & Levy, H. (1949). *Archives of Biochemistry, 20*, 170–172.
Schlabach, M. R., & Bates, G. W. (1975). *The Journal of Biological Chemistry, 250*, 2182–2188.
Schneider, C., Owen, M. J., Banville, D., & Williams, J. G. (1984). *Nature, 311*, 675–678.
Seligman, P. A., Schleicher, R. B., & Allen, R. H. (1979). *The Journal of Biological Chemistry, 254*, 9943–9946.
Sharp, P., & Srai, S. K. (2007). *World Journal of Gastroenterology, 13*, 4716–4724.
Sharp, P. A. (2010). *International Journal for Vitamin and Nutrition Research, 80*, 231–242.
Shayeghi, M., Latunde-Dada, G. O., Oakhill, J. S., Laftah, A. H., Takeuchi, K., Halliday, N., et al. (2005). *Cell, 122*, 789–801.
Sheftel, A. D., Mason, A. B., & Ponka, P. (2012). *Biochimica et Biophysica Acta, 1820*, 161–187.
Steere, A. N., Byrne, S. L., Chasteen, N. D., & Mason, A. B. (2012). *Biochimica et Biophysica Acta, 1820*, 326–333.
Steere, A. N., Byrne, S. L., Chasteen, N. D., Smith, V. C., MacGillivray, R. T., & Mason, A. B. (2010). *Journal of Biological Inorganic Chemistry, 15*, 1341–1352.
Steere, A. N., Chasteen, N. D., Miller, B. F., Smith, V. C., Macgillivray, R. T., & Mason, A. B. (2012). *Biochemistry, 51*, 2113–2121.
Steere, A. N., Roberts, S. E., Byrne, S. L., Chasteen, N. D., Bobst, C. E., Kaltashov, I. A., et al. (2010). *Protein Expression and Purification, 72*, 32–41.
Steinlein, L. M., Graf, T. N., & Ikeda, R. A. (1995). *Protein Expression and Purification, 6*, 619–624.

Steinlein, L. M., & Ikeda, R. A. (1993). *Enzyme and Microbial Technology, 15*, 193–199.
Subramaniam, V. N., Summerville, L., & Wallace, D. F. (2002). *Cell Biochemistry and Biophysics, 36*, 235–239.
Sun, H., Li, H., & Sadler, P. J. (1999). *Chemical Review, 99*, 2817–2842.
Theil, E. C., & Goss, D. J. (2009). *Chemical Review, 109*, 4568–4579.
Thompson, C. P., McCarty, B. M., & Chasteen, N. D. (1986). *Biochimica et Biophysica Acta, 870*, 530–537.
Turkewitz, A. P., Amatruda, J. F., Borhani, D., Harrison, S. C., & Schwartz, A. L. (1988). *The Journal of Biological Chemistry, 263*, 8318–8325.
Vulpe, C. D., Kuo, Y. M., Murphy, T. L., Cowley, L., Askwith, C., Libina, N., et al. (1999). *Nature Genetics, 21*, 195–199.
Wally, J., Halbrooks, P. J., Vonrhein, C., Rould, M. A., Everse, S. J., Mason, A. B., et al. (2006). *The Journal of Biological Chemistry, 281*, 24934–24944.
Weaver, J., & Pollack, S. (1989). *The Biochemical Journal, 261*, 787–792.
West, A. P., Jr., Giannetti, A. M., Herr, A. B., Bennett, M. J., Nangiana, J. S., Pierce, J. R., et al. (2001). *Journal of Molecular Biology, 313*, 385–397.
Williams, J., Chasteen, N. D., & Moreton, K. (1982). *The Biochemical Journal, 201*, 527–532.
Williams, J., & Moreton, K. (1980). *The Biochemical Journal, 185*, 483–488.
Yang, F., Lum, J. B., McGill, J. R., Moore, C. M., Naylor, S. L., van Bragt, P. H., et al. (1984). *Proceedings of the National Academy of Sciences of the United States of America, 81*, 2752–2756.
Zak, O., & Aisen, P. (2002). *Biochemistry, 41*, 1647–1653.
Zak, O., & Aisen, P. (2003). *Biochemistry, 42*, 12330–12334.
Zak, O., Tam, B., MacGillivray, R. T., & Aisen, P. (1997). *Biochemistry, 36*, 11036–11043.
Zhang, D., Nandi, S., Bryan, P., Pettit, S., Nguyen, D., Santos, M. A., et al. (2010). *Protein Expression and Purification, 74*, 69–79.
Zuccola. (1993). *The crystal structure of monoferric human serum transferrin*. Atlanta, GA: Georgia Institute of Technology.

CHAPTER TWO

Structure, Function and Binding Selectivity and Stereoselectivity of Siderophore–Iron Outer Membrane Transporters

Isabelle J. Schalk[1], Gaëtan L.A. Mislin and Karl Brillet
UMR 7242, Université de Strasbourg-CNRS, ESBS, Boulevard Sébastien Brant, Strasbourg, France
[1]Corresponding author: E-mail: isabelle.schalk@unistra.fr

Contents

Abstract

To get access to iron, microorganisms produce and release into their environment small organic metal chelators called siderophores. In parallel, they produce siderophore–iron outer membrane transporters (also called TonB-Dependent Transporters or TBDT) embedded in the outer membrane; these proteins actively reabsorb the siderophore loaded with iron from the extracellular medium. This active uptake requires energy in the form of the proton motive force transferred from the inner membrane to the outer membrane transporter via the inner membrane TonB complex. Siderophores produced by microorganisms are structurally very diverse with molecular weights of 150 up to 2000 Da. Siderophore–iron uptake from the extracellular medium by TBDTs is a highly selective and sometimes even stereoselective process, with each siderophore

Current Topics in Membranes, Volume 69
ISSN 1063-5823,
http://dx.doi.org/10.1016/B978-0-12-394390-3.00002-1

having a specific TBDT. Unlike the siderophores, all TBDTs have similar structures and belong to the outer membrane β-barrel protein superfamily. The way in which the siderophore–iron complex passes through the TBDT is still unclear. In some bacteria, TBDTs are also partners of signaling cascades regulating the expression of proteins involved in siderophore biosynthesis and siderophore–iron acquisition.

1. INTRODUCTION

Iron is an essential element for almost all forms of life. The oxidation state of iron can change, reversibly, and it can coordinate and activate oxygen. These properties have resulted in iron being a cofactor of various enzymes involved in key metabolic processes, including ATP synthesis, nucleotide biosynthesis, and DNA replication, transcription and repair (Drakesmith & Prentice, 2008). The paradox of this essential metal is its extremely low solubility in aerobic and physiological environments. The presence of oxygen rapidly oxidizes iron(II) into insoluble ferric oxyhydroxides with a solubility product of 10^{-39}. In the host during infections, iron is also poorly bioavailable for pathogens, because it is always in complex with biological macromolecules, for example hemes, ferritins (iron storage proteins), transferin and lactoferrin (two proteins involved in iron transport). Thus, iron concentrations in various bacterial environments are estimated between 10^{-9} and 10^{-18} M (Raymond, Dertz, & Kim, 2003). To overcome this problem of iron bioavailability, microorganisms produce and use siderophores, small molecular weight iron chelators secreted into the environment. The function of siderophores is to allow iron uptake by microbial cells. Siderophores have a high affinity for iron(III) and are consequently able to solubilize iron from the bacterium's environment and to compete with the host's iron-chelating proteins. The siderophore–iron complexes formed are then transported back into the bacterial cytoplasm by specific transporters.

The Gram-negative bacterial envelope is composed of outer and inner membranes separated by a periplasm. Small molecules of molecular weight lower than 600 Da can cross the outer membrane (from the extracellular medium into the periplasm) by passive diffusion through porins (Nikaido, 2003). Siderophore–iron complexes are transported across the outer membrane by TBDTs (TonB-dependent transporters). These transporters are required by Gram-negative bacteria for uptake of compounds of molecular weight higher than 600 Da and/or present at low concentrations in the bacterial cell's environment. This uptake is energy dependent. The proton motive force of the inner membrane is used to activate the TBDTs

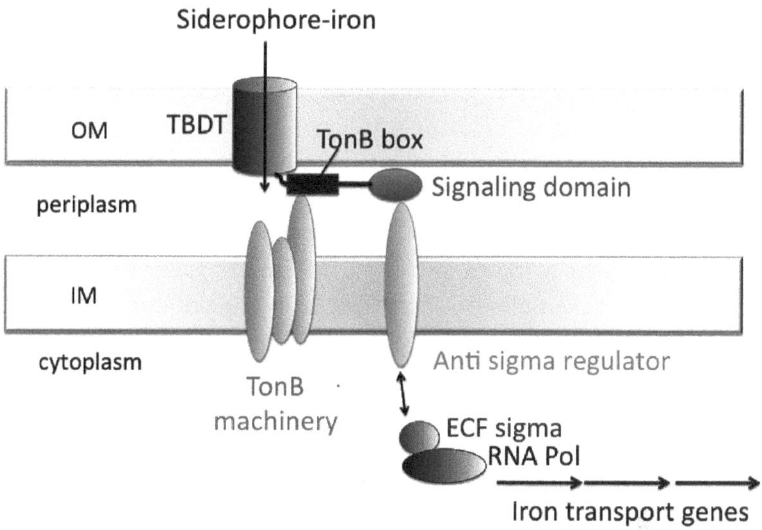

Figure 1 ***TBDTs are involved in siderophore–iron uptake and, in some cases, in signaling cascades regulating expression of genes involved in iron acquisition.*** Siderophore–iron complexes from the extracellular medium are recognized by their corresponding specific TBDT, which serves to transport the siderophore–iron complex into the periplasm. Some TBDTs are also involved in signaling cascades regulating the transcription of genes encoding proteins involved in iron uptake. This signaling cascade is initiated by the binding of the siderophore–iron complex to the TBDT, and involves several components: the N-terminal signaling domain of the TBDT, the inner membrane sigma regulator protein and the cytoplasmic extracytoplasmic function (ECF) sigma factor. Both transport and induction functions require energy transduction from the TonB machinery in the inner membrane. For color version of this figure, the reader is referred to the online version of this book.

via an inner membrane complex called the TonB machinery (Fig. 1) (Postle & Larsen, 2007). Once in the periplasm, the fate of siderophore–iron complexes depends on the siderophore and bacterium. Iron can be released from the siderophore in the periplasm, as for example in the pyoverdine pathway in *Pseudomonas aeruginosa* (Greenwald et al., 2007; Schalk, Abdallah, & Pattus, 2002; Yeterian, Martin, Lamont, & Schalk, 2010) or transported across the inner membrane into the cytoplasm by either ABC transporters or permeases (Cuiv, Clarke, Lynch, & O'Connell, 2004; Hannauer, Barda, Mislin, Shanzer, & Schalk, 2010; Koster, 2001). Iron release from the siderophore in the periplasm or cytoplasm involves the reduction of the iron and/or degradation or modification of the siderophore (Brickman & McIntosh, 1992; Hartman & Braun, 1980; Matzanke, Anemuller, Schunemann, Trautwein, & Hantke, 2004). Iron uptake has to be tightly controlled because an excess of free iron can catalyze the generation of highly reactive oxygen and nitrogen intermediates that damage biological macromolecules. Therefore, expression of the proteins involved in siderophore biosynthesis and in siderophore–iron uptake is tightly regulated by the intracellular concentration of iron (Braun & Braun, 2002; Fleischhacker & Kiley, 2011; Visca, Leoni, Wilson, & Lamont, 2002).

This review focuses only on the transport of siderophore–iron complexes from the extracellular medium into the bacterial periplasm by TBDTs. We first consider some examples of siderophores and describe the chemical and structural diversities of these compounds. Then, the structures of TBDTs solved to date are presented, and the specificity of the recognition between siderophore–iron complexes and TBDT-binding sites will be addressed. Mechanistic models of siderophore–iron translocation across the outer membrane by TBDT are discussed and the involvement of these proteins in signal cascades considered.

2. SIDEROPHORES

More than 500 different siderophores have now been identified and 270 are characterized structurally (for a review on siderophores, see Boukhalfa & Crumbliss, 2002; Hider & Kong, 2011). These compounds are organic chelators produced by microorganisms, fungi and plants to make iron accessible (Hider & Kong, 2011). They all have a low molecular weights (150–2000 Da), and present a high and specific affinity for iron(III).

The high affinity and selectivity of siderophores for Fe(III) is due to (i) stringent selection of metal-binding groups, (ii) the number of binding units, and (iii) their stereochemical arrangement. Oxygen atoms are the best donors for iron(III) coordination and the most common geometry for iron(III) coordination is octahedral with six heteroatoms arranged around the iron center (Boukhalfa & Crumbliss, 2002). This octahedral coordination favors the formation of the thermodynamically stable high-spin iron(III) species. The oxygen atoms of the siderophore are organized into bidentate functions, such as catechols, hydroxamates and α-hydroxycaboxylates (Fig. 2) (Hider & Kong, 2011). In some siderophores, nitrogen is also involved as donor atom, but in such structures, the affinity for iron(III) tends to be lower (Hider & Kong, 2011). The hexadentate structure is the most common siderophore form, with a mixture of coordination ligands (mostly hydroxamate and catechol functions) attached to either linear or cyclic

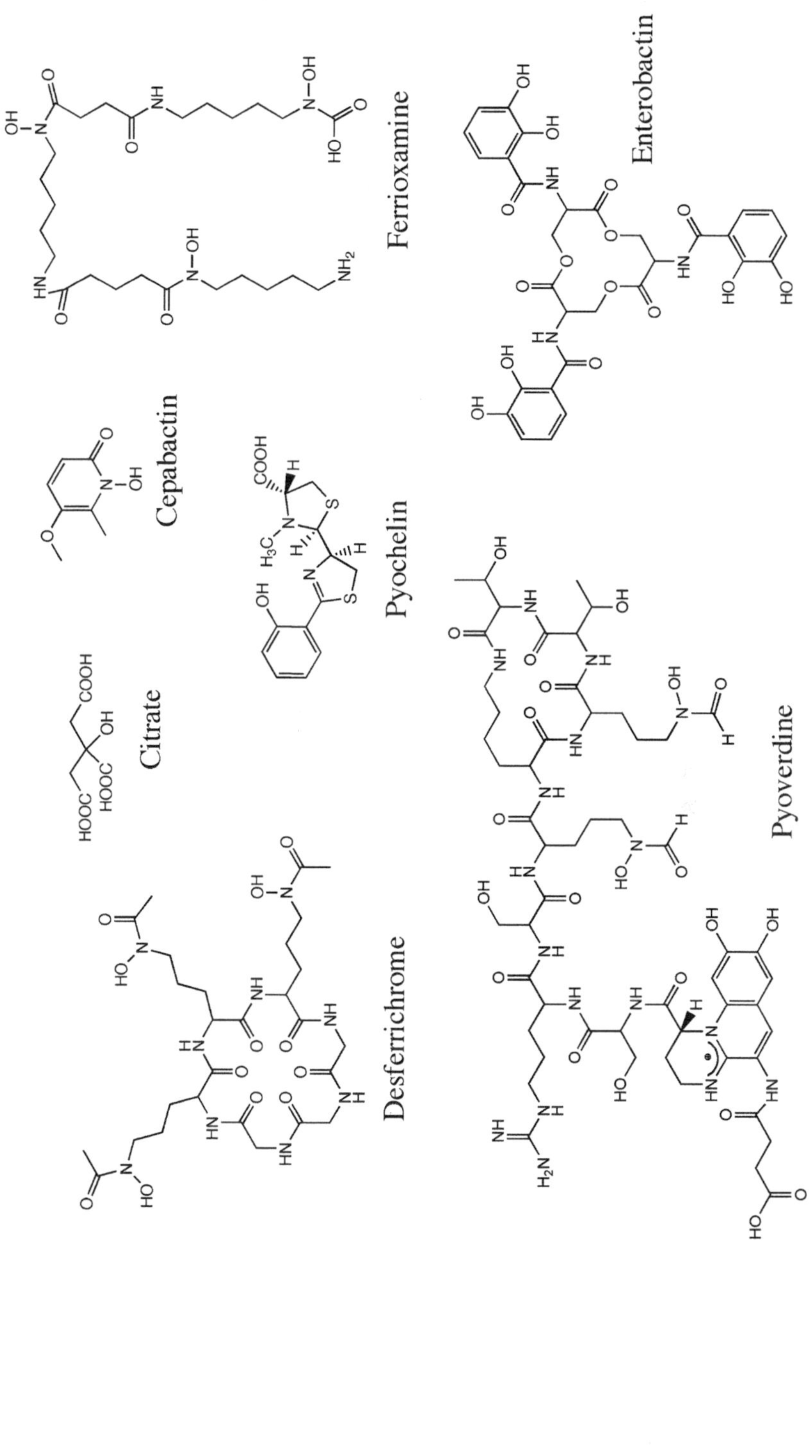

Figure 2 ***Structures of representative siderophores.***

structures (Fig. 2). Siderophores with lower denticity (tetradentate, tridentate and bidentate) are also produced by microorganisms, but these chelators cannot alone achieve full Fe(III) coordination and complexes with higher siderophore:Fe(III) stoichiometry have been described (Fig. 2) (Spasojevic, Boukhalfa, Stevens, & Crumbliss, 2001).

In general, hexadentate siderophores have a much higher affinity for Fe(III) than do tetradentate siderophores, which have a higher iron affinity than bidentate siderophores (Albrecht-Gary & Crumbliss, 1998). All siderophores possess a higher affinity for iron(III) than for iron(II). For example, desferrioxamine B, a hexadentate siderophore with three hydroxamate units, has an affinity for iron(III) that is 20 orders of magnitude higher than that for iron(II) (Helman & Lawrence, 1989).

For many siderophores, a large range of closely related structures have been reported. For example, more than 60 analogues of pyoverdines (Fuchs, Schafer, Geoffroy, & Meyer, 2001) have been reported, and 21 analogues of desferrichrome, 21 analogues of enterobactin and 20 analogues of ferrioxamines have been described (Hider & Kong, 2011). In siderophore structures, it is very common to find D-amino acids, unusual amino acids like ornithine, and N- and C-terminal blocked peptides, and these residues provide resistance to peptidase and to other lytic enzyme activities. Also a specific feature of siderophores produced by marine bacteria is the presence of one or two fatty acyl chains (Butler, 2005; Martinez et al., 2003; Xu, Martinez, Groves, & Butler, 2002); consequently, these chelators are surface active and tend to form molecular micelles (Butler, 2005; Martinez et al., 2003; Xu et al., 2002).

Most bacteria are able to produce several siderophores of different chemical structures and to use in addition heterologous siderophores (siderophores produced by other microorganisms). For example, *P. aeruginosa* produces two major siderophores, pyoverdine and pyochelin (Fig. 2), but is able to use at least five heterologous siderophores: cepabactin, ferrichrome, enterobactin, ferrioxamine and citrate (Hannauer et al., 2010; Llamas et al., 2006; Mislin et al., 2006; Schalk, 2008). Why bacteria synthesize and use multiple siderophores is still unclear. It does, however, incontestably illustrate the high importance of iron for microorganisms, and is presumably a mechanism driven by evolution. It has been proposed that this diversity results from the interplay between siderophore-producing and nonproducing microorganisms: when there are many non-producers exploiting a siderophore, it is beneficial to produce a siderophore unusable by the dominant population (Lee, van Baalen, & Jansen, 2012).

3. STRUCTURE OF SIDEROPHORE–IRON TBDTS

TBDTs are located in the bacterial outer membrane and their biological function is to import nutriments, and especially siderophore–iron complexes, from the extracellular medium. These transporters are characterized by a very stringent siderophore selectivity. This is associated with a strong correlation between the number of siderophores able to be used by a bacterium and the number of genes encoding iron-regulated TBDTs in its genome. Genome sequencing has revealed that many bacteria have numerous genes encoding these transporters, but not all are involved in iron acquisition (Blanvillain et al., 2007; Schauer, Gouget, Carriere, Labigne, & de Reuse, 2007). In the *P. aeruginosa* genome, there are 32 genes coding for TBDTs but only 11 seem to be involved in iron uptake (Llamas et al., 2006). *Escherichia coli* has seven genes coding for siderophore–iron TBDTs. The different siderophore–iron TBDTs share 20–45% sequence homology and two major groups can be identified. Members of the first group are composed of only two domains, a β-barrel domain and a plug domain (Fig. 3), and these transporters are only involved in siderophore–iron transport processes. The second group has an additional periplasmic domain called the signaling domain (Fig. 3); this domain is not involved in siderophore–iron uptake but plays a key role in the signaling cascade regulating the expression of proteins involved in siderophore biosynthesis and siderophore–iron uptake (Fig. 1) (Braun & Braun, 2002; Visca, 2004). An important conserved sequence of about seven residues, corresponding to the so-called TonB box, is found in the N-terminal part of the plug domain of proteins of group 1 and between the plug and the signaling domain of the proteins of group 2. This TonB box is necessary for TBDTs to become active: it is required for siderophore–iron complex transport from the extracellular medium into the periplasm and, for TBDTs of group 2, for activation of the signaling cascade (Shirley & Lamont, 2009). Indeed, deletion of this small region abolishes both transport and activation of the signaling cascade (Shirley & Lamont, 2009).

The first three structures of TBDTs were published at the end of 1998 and beginning of 1999: two structures of FhuA (Ferguson, Hofmann, Coulton, Diederichs, & Welte, 1998; Locher et al., 1998) and one of FepA (Buchanan et al., 1999), the *E. coli* ferrichrome and ferrienterobactin TBDTs, respectively. The structures of 12 more TBDTs have been solved

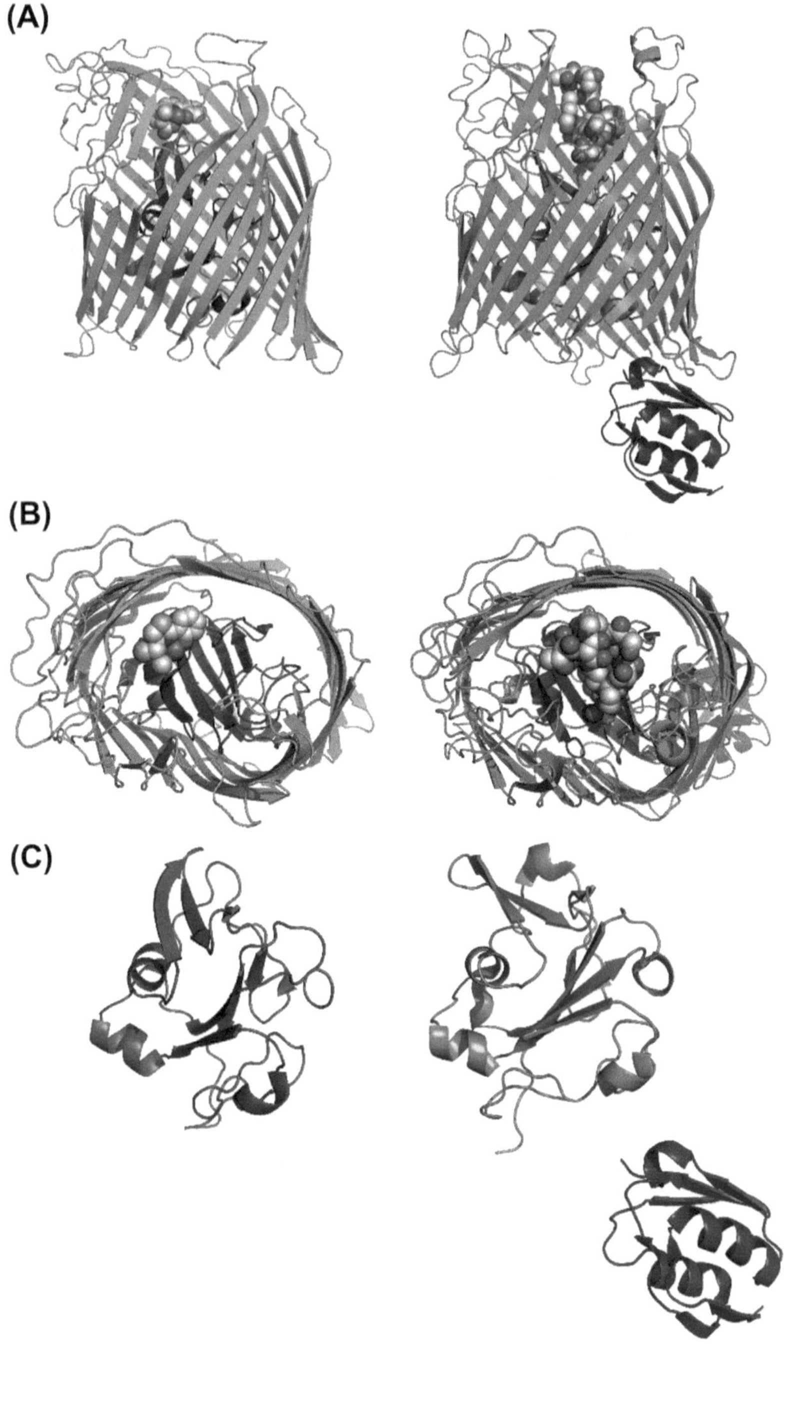
(A)
(B)
(C)

since then: six are siderophore–iron TBDTs (FecA, FpvA, FptA, FetA, Cir and FauA, Table 1), three are involved in the uptake of other compounds (vitamin B12 and heme) and one is apparently also involved in iron uptake (FrpB) but its siderophore has not been identified yet (Table 1). In addition, the structures have been solved of these TBDTs in various loading states, and for two of them bound to the periplasmic domain of TonB (FhuA and BtuB, the ferrichrome and vitamin B12 TBDTs in *E. coli*, respectively; Table 1). All these structures are composed of a 22-stranded antiparallel transmembrane β-barrel (Fig. 3). The lumen of the barrel is filled with the N-terminal globular domain (named plug, hatch or cork) and the binding site is always located on the extracellular face of the transporter, exposed to the solvent, and is constituted by residues of both the plug and the β-barrel domains.

The barrel domain of TBDTs is around 70 Å in height and most transmembrane strands extend beyond the outer leaflet of the bilayer. The tilt of the β-strands relative to the axis of the barrel is always around 45°. The β-strands of the transmembrane barrel are connected by 11 long and flexible extracellular loops and 10 periplasmic turns. The length and amino acid content of the extracellular loops differ between transporters and may play a role in ligand recognition and binding (see below).

The plug domains of TBDTs are mostly a mixed four-stranded β-sheet with surrounding loops and helices. They are globular in form and occlude the lumen of the β-barrel thereby preventing direct passage of the siderophore–iron complex across the barrel. One region of the plug domain is exposed to the periplasm and the other to the extracellular medium and participates in the ligand-binding site. The plug domain is kept in place in the β-barrel by 40–70 hydrogen bonds and two salt bridges involving four highly conserved residues: two Arg residues of the plug and two conserved Glu in the barrel domain (Chimento, Kadner, & Wiener, 2005; Wiener,

Figure 3 ***Structure of FptA (left) and FpvA (right), the* P. aeruginosa *pyochelin–iron and pyoverdine–iron TBDTs, respectively (Brillet et al., 2007; Wirth et al., 2007).*** Both transporters are loaded with their corresponding siderophore–iron complex. The β-barrel is colored in cyan in FptA and in green in FpvA, the plug domains are shown in purple (FptA) and in orange (FpvA) and the signaling domain in blue. The TonB boxes are disordered in both structures. The pyochelin- and pyoverdine–iron complexes are displayed in van der Waals spheres. (A) Overall side view of FptA and FpvA. (B) Overall top (extracellular) view. (C) Structure of the plug domains with the signaling domain for FpvA (side view). All the drawings in this review were prepared using PyMol. See the color plate.

Table 1 Known structures of TBDT involved in siderophore–iron transport

Origin	TBDT	Ligand or protein in complex	Resolution (Å)	PDB code	Reference
E. coli	FhuA	No ligand	2.50	2FCP	Ferguson et al. (1998)
E. coli	FhuA	No ligand	2.50	1QFG	Ferguson, Braun, et al. (2000)
E. coli	FhuA	No ligand	2.70	1BY3	Locher et al. (1998)
E. coli	FhuA	Ferrichrome	2.70	1QFF	Ferguson, Welte, et al. (2000)
E. coli	FhuA	Ferrichrome	2.60	1BY5	Locher et al. (1998)
E. coli	FhuA	Ferrichrome	2.70	1FCP	Ferguson et al. (1998)
E. coli	FhuA	Albomycin	3.10	1QKC	Ferguson, Braun, et al. (2000)
E. coli	FhuA	Rifamycin	2.90	1FI1	Ferguson et al. (2001)
E. coli	FhuA	Phenylferricrocin	2.95	1QJQ	Ferguson, Braun, et al. (2000)
E. coli	FhuA	TonB	3.30	2GRX	Pawelek et al. (2006)
E. coli	FecA	No ligand	2.50	1PNZ	Yue et al. (2003)
E. coli	FecA	No ligand	2.00	1KMO	Ferguson et al. (2002)
E. coli	FecA	Dicitrate	2.15	1PO0	Yue et al. (2003)
E. coli	FecA	Dicitrate–iron	2.50	1KMP	Ferguson et al. (2002)
E. coli	FecA	Dicitrate–iron	3.40	1PO3	Yue et al. (2003)
E. coli	FepA	No ligand	2.40	1FEP	Buchanan et al. (1999)
E. coli	Cir	No ligand	2.65	2HDF	Buchanan et al. (2007)

E. coli	Cir	Colicin Ia-R	2.50	2HDI	Buchanan et al. (2007)
P. aeruginosa	FpvA	No ligand	2.77	2O5P	Brillet et al. (2007)
P. aeruginosa	FpvA	No ligand	2.90	2W75	Greenwald et al. (2009)
P. aeruginosa	FpvA	Pyoverdine	3.60	1XKH	Cobessi, Célia, Folschweiller, et al. (2005)
P. aeruginosa	FpvA	Pyoverdine–iron	2.73	2IAH	Wirth et al. (2007)
P. aeruginosa	FpvA	Pyoverdine–iron	2.71	2W16	Greenwald et al. (2009)
P. aeruginosa	FpvA	$\text{Pyoverdine}_{\text{DSM50106}}$–iron	2.90	2W6T	Greenwald et al. (2009)
P. aeruginosa	FpvA	$\text{Pyoverdine}_{\text{G173}}$–iron	3.00	2W6U	Greenwald et al. (2009)
P. aeruginosa	FpvA	$\text{Pyoverdine}_{\text{Pa6}}$–iron	2.80	2W76	Greenwald et al. (2009)
P. aeruginosa	FpvA	$\text{Pyoverdine}_{\text{Pfl18.1}}$–iron	2.90	2W77	Greenwald et al. (2009)
P. aeruginosa	FpvA	$\text{Pyoverdine}_{\text{ATCC13535}}$–iron	3.00	2W78	Greenwald et al. (2009)
P. aeruginosa	FptA	Pyochelin–iron	2.00	1XKW	Greenwald et al. (2009)
P. aeruginosa	FetA	Enantio-pyochelin–iron	3.26	3QLB	Brillet et al. (2011)
N. meningitidis	FrpB	No ligand	2.40	/	Salem et al. (2012)
B. pertussis	FauA	No ligand	2.33	3EFM	Brillet, Meksem, Lauber, Reimmann, and Cobessi (2009)

E. coli, *Escherichia coli*; *P. aeruginosa*, *Pseudomonas aeruginosa*; *N. meningitidis*, *Neisseria meningitidis* ; *B. pertussis*, *Bordetella pertussis*; FhuA, TBDT of ferrichrome; FecA, TBDT of citrate–iron; FepA, TBDT of enterobactin–iron; Cir: TBDT of iron complex of dihydroxybenzoylserine; FpvA, TBDT of pyoverdine–iron; FptA, TBDT of pyochelin–iron; FetA, TBDT of enantio-pyochelin–iron; FrpB, TBDT of an unknown siderophore–iron complex; FauA, TBDT of alcaligin–iron.

2005). Several other conserved motifs are found in all TBDT structures (Chimento et al., 2005; Noinaj, Guillier, Barnard, & Buchanan, 2010; Wiener, 2005): TEE, PGV, IRG box, LIDG box, RP box and the Hβ4 motifs. In all TBDT structures, the plug is highly solvated inside the barrel pore (Chimento et al., 2005).

The conserved TonB box is located at the beginning of the plug domain on the periplasmic face of the TBDT and has either ordered or disordered conformations. However, no good correlation between the siderophore–iron loading status of the TBDT and the conformation of its TonB box has been demonstrated. (Fanucci, Lee, and Cafiso 2003) suggest that the conformation of this small domain may be very sensitive to its environment, and to solutes used during crystallization (Freed, Horanyi, Wiener, & Cafiso 2010).

Some TBDTs have an additional N-terminal domain, upstream from the plug domain and called the signaling domain. The signaling domain structure has only been observed in situ in FpvA (TBDT of *P. aeruginosa*) loaded or not loaded with pyoverdine–iron (Fig. 4) (Brillet et al., 2007; Wirth, Meyer-Klaucke, Pattus, & Cobessi, 2007). For FecA and PupA (citrate–iron and pseudobactin–iron TBDTs in *E. coli* and *Pseudomonas putida*, respectively), this domain was cloned without the rest of the protein and the 3D structure solved by Nuclear Magnetic Resonance (NMR) (Ferguson et al., 2007). The overall folding of this domain is well conserved and involves two α-helices sandwiched between two β-sheets (Fig. 4D and E). In FpvA, the signaling domain is linked to the plug by a long flexible loop called the connecting loop and containing the TonB box (Fig. 4A). In FpvA structures, the signaling domain moves between the two positions depending on the loading status of the transporter (Fig. 4A–C), suggesting that it is highly flexible probably because of the long connecting loop.

4. BINDING SITES OF SIDEROPHORE–IRON TBDTs

4.1. Siderophore Selectivity

The high siderophore selectivity of TBDTs was demonstrated in the 1990s by work based on siderophore–^{55}Fe uptake assays and mutagenesis. Each TBDT recognizes and transports a specific siderophore, in some cases a few structurally related siderophores, but never siderophores with different chemical structures. Over the last 10 years, the resolution of several TBDT

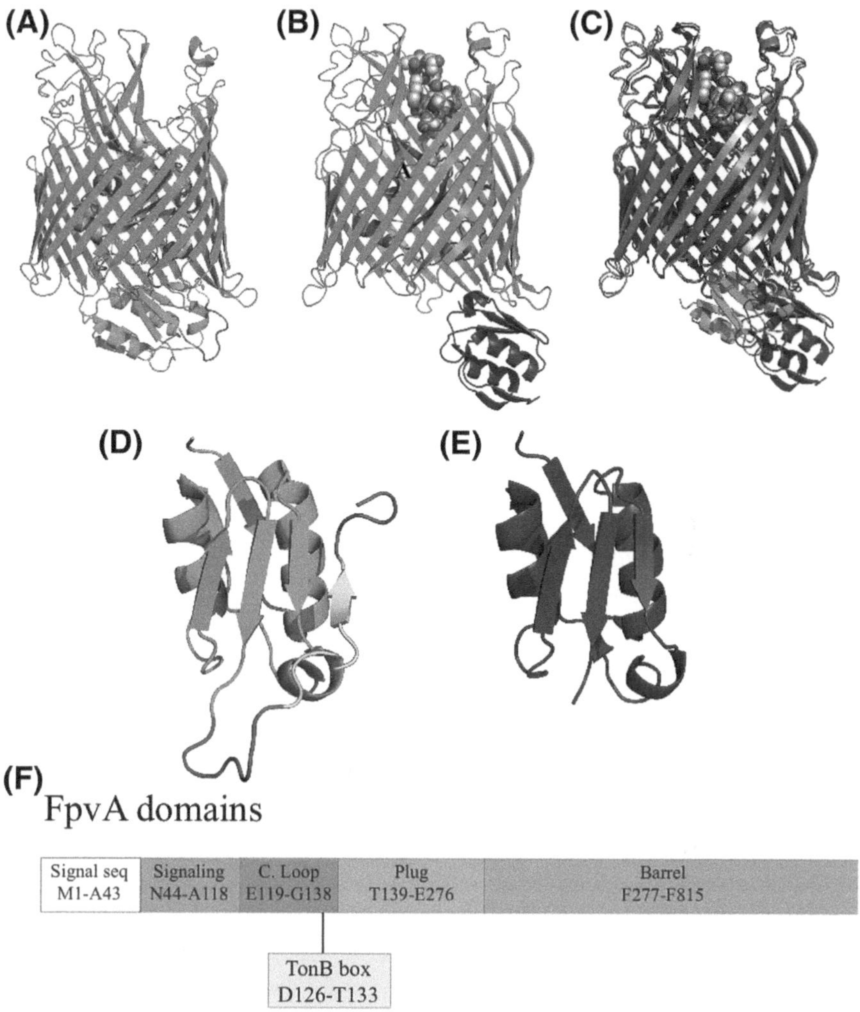

Figure 4 ***Conformational change in the signaling domain of FpvA upon binding of pyoverdine–iron (Brillet et al., 2007).*** The β-barrel and plug domains are colored in green and orange, respectively, in A and B. The signaling domain of apo FpvA and FpvA–pyoverdine–iron are cyan and blue, respectively. The connecting loop and TonB box are only visible in the apo FpvA structure and are in pink and in yellow, respectively. The pyoverdine–iron complexes are displayed in van der Waals spheres (A) Side view of apo FpvA. (B) Side view of FpvA–pyoverdine–iron. (C) Superimposition of apo FpvA and FpvA–pyoverdine–iron. (D) View of the signaling domain in FpvA. (E) View of the signaling domain in FpvA–Pvd–Fe. (F) The different FpvA domains: Signal seq., signal sequence; Signaling, signaling domain, C. loop, connecting loop; Plug, plug domain and barrel, barrel domain. See the color plate.

structures has revealed that the binding site is always located in the lumen on the extracellular face of the transporter and is composed of residues of the barrel and the plug domains. This rule applies to all known TBDT structures, whatever the siderophore recognized and transported. In all cases, the binding affinity is in the range of 0.1–5 nM (Clément, Mesini, Pattus, Abdallah, & Schalk, 2004; Hoegy et al., 2005; Payne et al., 1997; Schalk et al., 2001) and the binding kinetics are often biphasic (Clément et al., 2004; Payne et al., 1997). This binding is energy independent: the TBDT is able to bind its corresponding siderophore–iron complex even in the absence of any proton motive force and TonB expression, but no uptake occurs (Clément et al., 2004; Hoegy et al., 2005; Payne et al., 1997; Schalk et al., 2001). The nature of the amino acids involved in binding sites allowing specific siderophore recognition differs between TBDTs. For example, the citrate–iron recognition binding pocket of FecA is lined with positively charged residues (Ferguson et al., 2002; Yue, Grizot, & Buchanan, 2003), whereas the corresponding regions of FhuA and FptA (ferrichrome and enterobactin–iron TBDTs in *E. coli*) are predominantly lined with aromatic residues (Cobessi, Celia, & Pattus, 2005; Ferguson et al., 1998; Locher et al., 1998).

Uptake assays and TBDT structures indicate that some transporters are able to transport structurally related siderophores. The best examples of this are the pyoverdine–iron uptake pathways in fluorescent Pseudomonads. Each strain produces its own pyoverdine, composed of an aromatic chromophore derived from 2,3-diamino-6,7-dihydroquinoline, linked to a peptide of 8–12 residues. The peptide part is unique to each strain. Some are linear, others partially or entirely cyclic, and their composition and length are specific to the producer strain (Budzikiewicz, 2004). The production of a specific pyoverdine goes in Pseudomonads with the expression of a corresponding TBDT. Uptake assays using ^{55}Fe have shown that these different pyoverdine/TBDT pathways are highly specific with very little cross recognition or cross uptake (Meyer et al., 2002). The structure of the pyoverdine–TBDT, FpvA from *P. aeruginosa*, has been solved in complex with its own pyoverdine–iron complex and with non-cognate pyoverdines of high or low affinity (Greenwald et al., 2009). ^{55}Fe uptake by FpvA was found only with pyoverdines of high affinity (Greenwald et al., 2009). Surprisingly, all siderophores tested induced nearly the same conformational changes in FpvA and the chromophore; the iron and its chelating groups in the various pyoverdine–iron complexes have equivalent configurations in all structures solved (Fig. 5). Presumably, this is

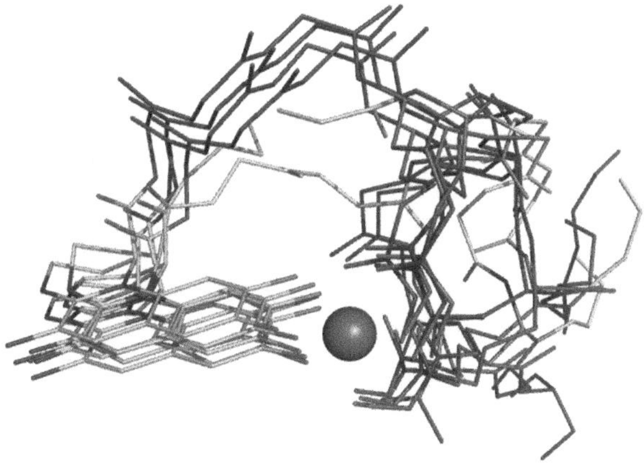

Figure 5 ***Overlay of the different pyoverdine–iron complexes bound to FpvA (Greenwald et al., 2009).*** The alignment was performed using only the FpvA backbone. For clarity, only the backbones of the pyoverdine peptides are represented. The chromophore ring (green) and the side chain of the modified amino acids involved in iron chelation (gray) are shown as sticks, with nitrogen in blue and oxygen in red. The iron atoms are represented as orange spheres. The color code is red for the high-affinity pyoverdines (PVDI, ATCC 13525, DSM 50106 and Pfl 18.1) and cyan for the low-affinity pyoverdines (G173 and Pa6). See the color plate.

because iron coordination obeys the same rules in all pyoverdine–iron complexes. A close relationship between transporter–siderophore–iron structures and the measured affinities was found not in the FpvA structures but in the different pyoverdine structures. The high- and low-affinity binders could be distinguished on the basis of the structure of the first amino acids of the pyoverdine peptide chain (Fig. 5), whereas the structure of the C-terminal portion of the siderophores, often cyclic, does not appear to make any large contribution to the interaction between the siderophore and the FpvA. According to these structures, the iron with its six dentates (involving the chromophore) seems to be the first element to position itself in the FpvA site, and then the peptide moiety positions according to its sequence and to its affinity for the binding site. FpvA has the same conformation in all seven pyoverdine–iron structures, so probably only the chromophore, the iron and its chelating groups cause the change of conformation observed in FpvA.

Some TBDTs are also able to bind and transport ligands other than siderophores, including in some cases antibiotics. The structures of FhuA (ferrichrome TBDT in *E. coli*) bound to albomycin and rifamycin have been solved (Ferguson, Braun, et al., 2000; Ferguson et al., 2001). These antibiotics bind to the ferrichrome binding pocket in FhuA. Albomycin is an antibiotic that consists of a hydroxamate-type iron-chelating siderophore, structurally analogous to ferrichrome, and an antibiotically active, thiribosyl pyrimidine moiety (Ferguson, Braun, et al., 2000). This similarity between ferrichrome and albomycin explains why this antibiotic is able to bind to FhuA binding pocket. FhuA also transports the semisynthetic rifamycin derivative CGP 4832, although the chemical structure of this antibiotic differs markedly from that of ferrichrome. The X-ray structure reveals that rifamycin CGP 4832 occupies the same ligand-binding site as ferrichrome and albomycin, thus demonstrating a surprising lack of selectivity of FhuA for this molecule.

TBDTs actively transport siderophores and siderophore–antibiotic conjugates like albomycin, but are also the targets of bacterial toxins including the colicins and bacteriophages. Colicins are large toxic peptides secreted by certain bacterial strains to kill other bacteria (for a recent review, see Cascales et al., 2007). The structure of *E. coli* TBDT Cir in complex with colicin I has been solved (Buchanan et al., 2007). Interestingly, the receptor-binding domain of colicin I (protein of 69 kDa) mimics siderophore binding by sitting in the binding pocket at approximately the same depth as metal chelates (Fig. 6) and like siderophore–iron complexes forms hydrogen bonds with residues of the plug domain. The binding of colicin I also induces a large unusual movement in extracellular loops L7 and L8: these loops move to open outward by 37° as compared with the same loops in the uncomplexed Cir structure.

4.2. Siderophore Stereoselectivity

Stereospecificity of siderophore-dependent iron uptake pathways has been reported for the following siderophores: pyochelin, parabactin, rhodotorulic acid, rhizoferrin, ferrichrome and enterobactin (Bergeron, Dionis, Elliott, & Kline, 1985; Bergeron, Xin, Weimar, Smith, & Wiegand, 2001; Drechsel, Jung, & Winkelmann, 1992; Muller, Isowa, & Raymond, 1985; Münzinger et al., 1999; Neilands, Erickson, & Rastetter, 1981; Winkelmann, 1979). The best example is pyochelin, a thiazoline-containing siderophore with three chiral centers, produced by *P. aeruginosa* (Hoegy et al., 2009). *Pseudomonas fluorescens* produces enantio-pyochelin, the same molecule but with the opposite configuration for the three asymmetric centers (Youard, Wenner, & Reimmann, 2011). Enantio-pyochelin promotes growth under iron limitation of *P. fluorescens*, but not that of *P.*

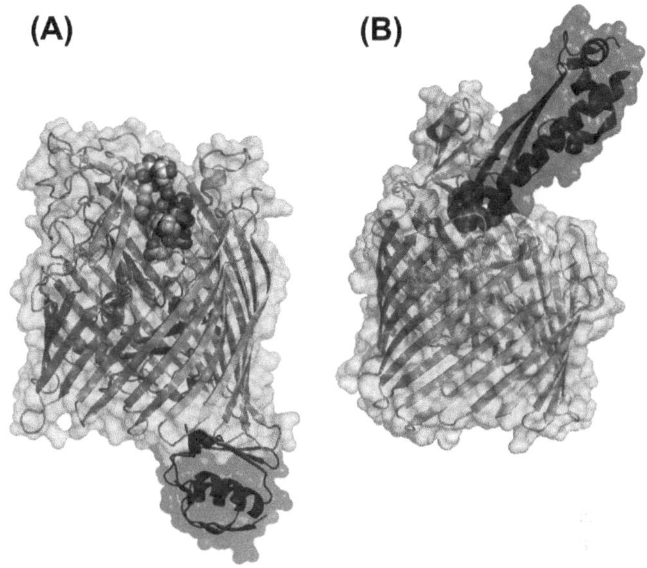

Figure 6 ***Pyoverdine–iron in its binding site on FpvA and interaction between colicin I and the Cir-binding site (Buchanan et al., 2007).*** FpvA is in green (barrel), orange (plug) and blue (signaling domain). Cir is colored in dark green (barrel) and yellow (plug) and colicin in red and the region of colicin interacting with Cir is represented in van der Waals spheres. The region of colicin I interacting with CirA mimics siderophore–iron binding and sits at approximately the same depth in the binding site as siderophore–iron in the structures of other TBDTs. For interpretation of the references to color in this figure legend, the reader is referred to the web version of this book.

aeruginosa (Hoegy et al., 2009). Reciprocally, pyochelin promotes growth of *P. aeruginosa* but is not functional for *P. fluorescens* (Hoegy et al., 2009). After iron chelation in the extracellular medium, pyochelin–Fe(III) and enantio-pyochelin–Fe(III) are recognized and transported by their specific outer membrane transporters: FptA in *P. aeruginosa* and FetA in *P. fluorescens* (Hoegy et al., 2009). The two TBDTs share little sequence identity but substantial structural similarities, and surprisingly the pyochelin and enantio-pyochelin binding pockets do not share any structural similarity but display similar physicochemical properties (Brillet et al., 2011; Cobessi, Celia, & Pattus, 2005). Structural analysis of FetA–enantio-pyochelin–Fe(III) and FptA–pyochelin–Fe(III) combined with mutagenesis and docking studies revealed that the stereospecific recognition of both enantiomers by their corresponding transporters is determined by the configuration of the siderophore's chiral C4″ center (Fig. 7) (Brillet et al., 2011; Cobessi, Celia, & Pattus, 2005; Hoegy et al., 2009). This recognition

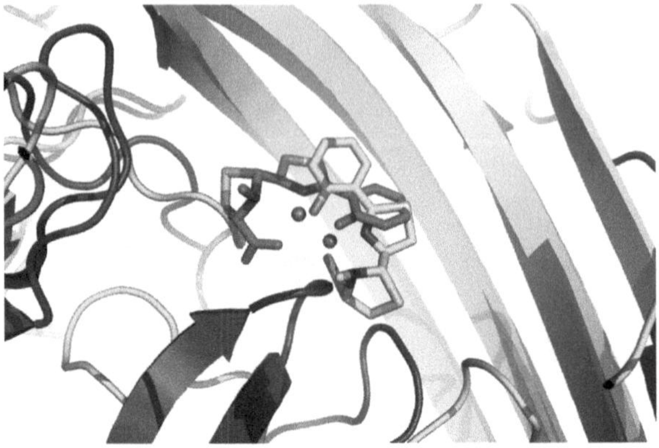

Figure 7 ***Superimposition of FptA–pyochelin–iron and FetA–enantio-pyochelin–iron showing the location and conformation of the siderophore–iron complexes in their corresponding binding sites (Brillet et al., 2011; Cobessi, Celia, & Pattus, 2005).*** FptA and FetA are colored purple and cyan, respectively. Pyochelin and enantio-pyochelin are in purple and cyan, respectively. The iron ion is orange. See the color plate.

involves specific hydrogen bonds between the Arg91 guanidinium group and enantio-pyochelin–Fe for FetA and between the Leu117–Leu116 main chain and pyochelin–Fe for FptA. FetA and FptA are the first TBDTs to be structurally described with opposing binding enantioselectivity for their ligands.

Stereochemistry also appears to be important for the interaction of pyoverdine–iron to its transporter, FpvA. The 3D structure of pyoverdine–Ga(III) was determined by NMR and reveals the presence in solution of two interchanging conformations of the siderophore upon binding of the gallium(III) ion (Wasielewski et al., 2008). Both conformers are present with roughly equivalent populations. The conformational interchange involves a stereoisomerization of the metal-binding coordination accompanied by a change in the global shape of the complex (Wasielewski et al., 2008). In the crystal structures of the various FpvA–pyoverdine–iron complexes determined (Greenwald et al., 2009; Wirth et al., 2007), only the Λ-4 configuration was observed, revealing a conformer selectivity of the binding site.

4.3. Metal Selectivity

There are extensive data showing that siderophores are able to chelate metal cations other than iron (Braud, Hoegy, Jezequel, Lebeau, & Schalk, 2009; Braud, Geoffroy, Hoegy, Mislin, & Schalk, 2010; Chen, Jurkewitch,

Bar-Ness, & Hadar, 1994; Hernlem, Vane, & Sayles, 1996; Neubauer, Nowack, Furrer, & Schulin, 2000). Binding experiments have shown that FpvA and FptA (pyoverdine–iron and pyochelin–iron TBDTs in *P. aeruginosa*) are able to bind their corresponding siderophore when complexed with metals other than iron (Braud, Hannauer, Mislin, & Schalk, 2009; Hannauer et al., 2012; Schalk et al., 2011). The affinities are highest for the ferric form of the siderophore and lower for complexes between siderophores and aluminum(III), manganese(II) or silver(I). In all TBDT structures solved, iron is always hexacoordinated by the siderophore and the transporter interacts only with the chelator, which forms like a shell around the metal. This absence of direct interaction between TBDTs and the metal chelated by siderophores probably explains the large metal specificity of this transporter. Uptake assays have shown that FpvA and FptA are both able to transport metals other than iron. Co^{2+}, Ga^{3+} and Ni^{2+} can accumulate via the pyochelin pathway in *P. aeruginosa* cells but with uptake rates 23- to 35-fold lower than for pyochelin–iron (Braud, Hannauer et al., 2009). Uptake assays with various mutants, monitored by inductively coupled plasma-atomic emission spectrometry for metal detection and by fluorescence for pyoverdine detection, show that FpvA has a broad metal specificity for both the binding and uptake of pyoverdine–metal complexes (Hannauer et al., 2012). The same approach shows that the metal selectivity of the pyoverdine/FpvA pathway is controlled by the PvdRT-OpmQ efflux pump, which exports all unwanted metals complexed with the siderophore from the bacterial periplasm (Hannauer et al., 2012). For all TBDTs with solved structures, the transporters interact only with the siderophore and not with the iron ion, so this wide metal permissiveness may be a common feature to transporters of this family.

4.4. Change of Conformations Induced by Siderophore–Iron Binding to TBDTs

Comparisons of TBDT structures, loaded or not loaded with their corresponding siderophore–iron complexes, have been possible only for FhuA, FecA and FpvA, the only TBDTs crystallized in different loading status (Table 1). Binding of siderophore–iron complexes to TBDTs induces conformational changes in the transporter: in the binding site itself, in some of the extracellular loops and in the N-terminal region.

Comparison of pyoverdine–gallium(III) in solution (structure solved by NMR) and pyoverdine–iron(III) bound to FpvA (X-ray structure) reveals a distinct change in the conformation of the siderophore peptide backbone upon binding to the TBDT and on regions neighboring the binding pocket of the receptor (Wasielewski et al., 2008; Wirth et al., 2007). This dynamic behavior suggests an induced fit mechanism during binding.

The crystal structures of FpvA bound to non-cognate pyoverdines with high or low affinities indicate that extracellular loop L7 moves, bringing a Trp and a Tyr closer to the high-affinity pyoverdines, but not to the low-affinity pyoverdines, for which little movement of L7 is observed (Greenwald et al., 2009). In the case of FecA, ligand binding leads to a major change in the conformation of extracellular loops L7 and L8: they close toward the ferricitrate binding pocket, thereby preventing the escape of the siderophore (Ferguson et al., 2002). Deletion of these loops inhibits citrate–iron binding to FecA (Sauter & Braun, 2004). Mutagenesis studies with FpvA implicate L4, L7 and L9 in pyoverdine–Fe uptake (Nader et al., 2007). Finally, no such loop closure has been detected for FhuA (Ferguson et al., 1998; Locher et al., 1998). However, molecular dynamics simulations suggest that L8 of FhuA closes the binding site following the binding of ferrichrome in a manner similar to that observed with FecA (Faraldo-Gomez, Smith, & Sansom, 2004), but deletion of this loop has little effect on ferrichrome transport (Endriss & Braun, 2004). Elimination of the L3 and L11 loops of FhuA abolishes ferrichrome binding and transport (Endriss & Braun, 2004), consistent with an important role for these loops in transport.

Ferrisiderophore binding also triggers additional structural changes at the periplasmic surface of the plug domain. FhuA and FecA contain a short helical region called the switch helix, which unwinds in the presence of bound substrate (Ferguson et al., 1998, 2001, 2002; Ferguson, Braun, et al., 2000; Locher et al., 1998; Yue et al., 2003), possibly leading to a movement of the TonB box required for interaction with TonB. The unwinding of the switch helices in crystal structures of ferrisiderophore-bound FhuA and FecA suggests that TonB may recognize a disordered conformation of the TonB box. However, as stated above, if we consider all TBDT structures, no real correlation can be established between the siderophore–iron loading status of the transporters and the conformation of the TonB box.

Moreover, in FpvA structures, the positioning of the signaling domain in the periplasm depends on the loading status of the transporter but the folding is not affected by loading (Brillet et al., 2007; Schalk, Lamont, & Cobessi, 2009; Wirth et al., 2007). The signaling domain of unloaded FpvA interacts with residues of its TonB box (Fig. 4) (Brillet et al., 2007). This interaction is very similar to those of the TonB boxes of BtuB and FhuA with TonB

(Pawelek et al., 2006; Shultis, Purdy, Banchs, & Wiener, 2006): the TonB box forms a four-stranded β-sheet with three β-strands of the signaling domain. When loaded with pyoverdine–Fe, the signaling domain moves, probably activating the signaling cascade and enabling the interaction of the TonB box with TonB. In FpvA–pyoverdine–Fe, the TonB box cannot be detected from electron density following pyoverdine–Fe binding suggesting that the flexibility of the connecting loop probably allows FpvA to interact with both FpvR, to regulate the expression of pyoverdine biosynthetic genes and *fpva*, and TonB for energy transduction and pyoverdine–Fe uptake.

5. SIDEROPHORE–IRON TRANSPORT BY TBDTs

5.1. Uptake by TBDTs Requires Energy

The pore through the barrel of TBDTs is completely occluded and does not allow diffusion of ligands without an energy input to open the pore. The outer membranes do not contain an energy source. The proton motive force of the inner membrane is the energy source for this transport and is transferred to the outer membrane TBDTs via a complex of three different inner membrane proteins: TonB, ExbB and ExbD. The N-terminal end of TonB is inserted in the cytoplasmic membrane and its C-proximal regions interact with the region called the TonB box of TBDTs. Bacteria have several 100–1000 copies of each TBDT and only a few 100 copies of TonB, so TonB is undoubtedly not permanently associated with TBDTs (Higgs, Larsen, & Postle, 2002). ExbB spans the inner membrane three times and most of the protein is in the cytoplasm. The organization of ExbD is similar to that of TonB.

The crystal structures of the periplasmic part of TonB in complex with FhuA loaded with ferrichrome or BtuB from *E. coli* loaded with vitamin B12 have been solved (Pawelek et al., 2006; Shultis et al., 2006). In these structures, the binding of TonB to the transporters causes only small changes in the structures of the TBDTs, except for the conformation of the TonB box, which is in a defined position. A four-stranded β-sheet is formed between a β-strand of the TonB box and the three-stranded β-sheet of TonB (Fig. 8). This interaction is very similar to that between the TonB box of FpvA and the three-stranded β-sheet of the signaling domain of this transporter (Brillet et al., 2007). The rest of the TBDT structure is not affected by the binding of TonB and looks likely to be an inactive form of the transporter, with the siderophore–iron complex still in its binding site on

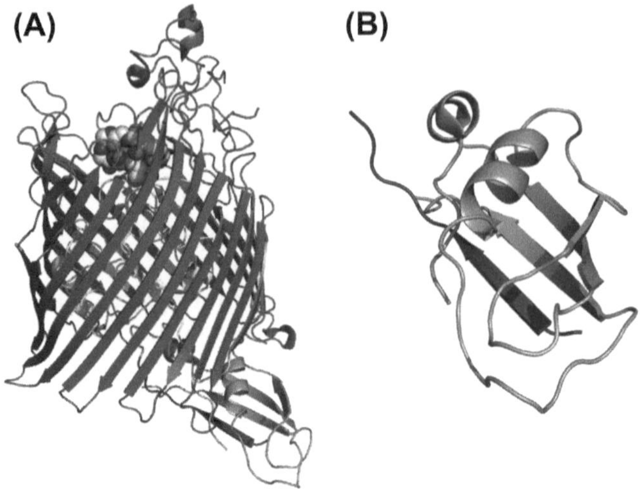

Figure 8 ***Structure of FhuA in complex with the periplasmic part of TonB (Pawelek et al., 2006).*** The β-barrel, plug domains and TonB box of FhuA are shown in blue, yellow and purple, respectively. TonB is in gray. The ferrichrome is displayed in van der Waals spheres. (A) Side view of TonB–FhuA–ferrichrome. (B) View of the interaction between the Ton box (purple) of FhuA and TonB (gray). See the color plate.

the extracellular face of the TBDT and the barrel domain still closed by the plug.

Many questions still remain unanswered about this TonB machinery and the way it activates TBDTs. It is still not known how the TonB–ExbB–ExbD complex responds to the proton motive force, what roles ExbB and ExbD play in this energy transfer, how this energy is transferred to the TBDT and how the TBDT responds to the energy input.

5.2. Mechanism of Siderophore–Iron Translocation across TBDT

The transport of molecules across the outer membrane needs formation of a channel; this channel needs to be able to handle siderophore-iron complexes up to 2000 Da and even compounds of 29–69 kDa like colicins (Cascales et al., 2007). No pore like those in porins for example or channel is found inside the receptor in any of the TonB-dependent structures. Several mechanisms have been proposed for channel formation, and in particular: (i) a major change in the conformation of the plug domain, (ii) the movement of this domain out of the β-barrel, and (iii) a combination of the two

(Eisenhauer, Shames, Pawelek, & Coulton, 2005; Endriss, Braun, Killmann, & Braun, 2003; Ma et al., 2007). Many different experimental approaches have been used to resolve this issue, but without success (Eisenhauer et al., 2005; Endriss et al., 2003; Ma et al., 2007). The data obtained are often contradictory between approaches used, making it impossible to conclude about the mechanism. Structure analyses of FhuA–TonB and BtuB–TonB led to the suggestion of a mechanical pulling model in which TonB induces total or partial unfolding of the plug domain, by exerting a very modest mechanical force on the transporter (Gumbart, Wiener, & Tajkhorshid, 2007). Numerous water molecules have been observed at the plug–barrel interface in the structures of TBDTs. These water molecules could reduce the activation energy barrier associated with any partial or total dissociation of the plug from the barrel thereby creating a channel between the barrel and the plug, allowing penetration of the siderophore–iron complex (Faraldo-Gomez et al., 2004). Note that for FpvA in *P. aeruginosa* (Nader, Journet, Meksem, Guillon, & Schalk, 2011) and FhuA in *E. coli* (Braun, Endriss, Killmann, & Braun, 2003), reconstituted TBDTs have been obtained by coexpression in the same strain of separately encoded plug and β-barrel domains, each endowed with a signal sequence. This indicates polypeptide complementation after secretion across the cytoplasmic membrane. The in vivo reconstituted FpvA binds pyoverdine–iron with the same affinity as wild-type FpvA, indicating that the plug is correctly and functionally inserted into the barrel despite the production of the plug and the β-barrel separately in vivo. For both reconstituted transporters (FpvA and FhuA), no passive diffusion of siderophore–iron complexes was observed; the siderophore–iron uptake was TonB dependent, but less efficient than that by the wild-type proteins. It is unknown whether this reconstitution occurs after β-barrel incorporation into the outer membrane or in the periplasm, with a periplasmic, partially folded β-barrel intermediate. Nevertheless, these data indicate that the barrel and the plug domains function independently and are able to associate. Clearly, further work is necessary to elucidate the mechanism of channel formation by TBDTs.

6. TBDT ARE INVOLVED IN SIGNALING CASCADES

In the case of the TonB-dependent receptors involved in signaling cascades, the signaling domain moves to allow interaction with the anti-sigma regulator of the inner membrane. This switch has been observed only in FpvA structures, both loaded and not loaded with pyoverdine–iron (Brillet et al., 2007; Wirth et al., 2007). In apo FpvA, the signaling domain is located below the plug domain and interacts with the TonB box forming a mixed four-stranded β-sheet composed of the three-stranded β-sheet of the signaling domain and one of the TonB box (Fig. 4). Binding of ferripyoverdine disrupts this interaction between the TonB box and the signaling domain, which moves to another position (Fig. 4). In this configuration, the TonB protein is able to interact with the TonB box and FpvR with the signaling domain. TonB is then able to induce the formation of a channel for siderophore–iron uptake and FpvR activates the signaling cascade.

7. CONCLUSION

Sequence alignment studies clearly show that all TBDTs, whatever the siderophore–iron complex transported, are organized as a β-barrel domain filled with a plug domain. The uptake process always involves the inner membrane proton motive force and a TonB protein. The binding site is highly specific for a single siderophore or for structurally related siderophores; it is always located on the extracellular face of the transporter and is composed of residues of both the barrel and the plug domains. The first structure of a TBDT was solved more than 14 years ago (1998) and today more than 14 TBDTs involved in siderophore–iron or other nutriment uptake have been crystallized and their structures, with different loading status, solved (a total of more than 45 different structures have been described). Despite all this structural information, the mechanism of ligand translocation across these transporters has not been clearly documented. The understanding of the different dynamic conformational changes necessary for translocation of the ligand across such structures remains an important challenge for the coming years.

ACKNOWLEDGMENTS

This work was partly funded by the Centre National de la Recherche Scientifique, by grants from the Centre International de Recherche au Frontière de la Chimie, from the ANR (Agence Nationale de Recherche, ANR-08-BLAN-0309-02) and from the Association Vaincre la Mucoviscidose.

REFERENCES

Albrecht-Gary, A. M., & Crumbliss, A. L. (1998). Coordination chemistry of siderophores: thermodynamics and kinetics of iron chelation and release. *Metal Ions in Biological Systems, 35*, 239–327.

Bergeron, R. J., Dionis, J. B., Elliott, G. T., & Kline, S. J. (1985). Mechanism and stereospecificity of the parabactin-mediated iron-transport system in *Paracoccus denitrificans*. *The Journal of Biological Chemistry, 260*(13), 7936–7944.

Bergeron, R. J., Xin, M. G., Weimar, W. R., Smith, R. E., & Wiegand, J. (2001). Significance of asymmetric sites in choosing siderophores as deferration agents. *Journal of Medicinal Chemistry, 44*(15), 2469–2478.

Blanvillain, S., Meyer, D., Boulanger, A., Lautier, M., Guynet, C., Denance, N., et al. (2007). Plant carbohydrate scavenging through *tonb*-dependent receptors: a feature shared by phytopathogenic and aquatic bacteria. *PLoS One, 2*(2), e224.

Boukhalfa, H., & Crumbliss, A. L. (2002). Chemical aspects of siderophore mediated iron transport. *Biometals, 15*(4), 325–339.

Braud, A., Geoffroy, V., Hoegy, F., Mislin, G. L. A., & Schalk, I. J. (2010). The siderophores pyoverdine and pyochelin are involved in *Pseudomonas aeruginosa* resistance against metals: another biological function of these two siderophores. *Environmental Microbiology Report, 2*(3), 419–425.

Braud, A., Hannauer, M., Mislin, G. L. A., & Schalk, I. J. (2009). The *Pseudomonas aeruginosa* pyochelin-iron uptake pathway and its metal specificity. *Journal of Bacteriology, 191*(11), 5317–5325.

Braud, A., Hoegy, F., Jezequel, K., Lebeau, T., & Schalk, I. J. (2009). New insights into the metal specificity of the *Pseudomonas aeruginosa* pyoverdine-iron uptake pathway. *Environmental Microbiology, 11*(5), 1079–1091.

Braun, M., Endriss, F., Killmann, H., & Braun, V. (2003). *In vivo* reconstitution of the FhuA transport protein of *Escherichia coli* K-12. *Journal of Bacteriology, 185*(18), 5508–5518.

Braun, V., & Braun, M. (2002). Iron transport and signaling in *Escherichia coli*. *FEBS Letters, 529*(1), 78–85.

Brickman, T. J., & McIntosh, M. A. (1992). Overexpression and purification of ferric enterobactin esterase from *Escherichia coli*. Demonstration of enzymatic hydrolysis of enterobactin and its iron complex. *The Journal of Biological Chemistry, 267*(17), 12350–12355.

Brillet, K., Journet, L., Celia, H., Paulus, L., Stahl, A., Pattus, F., et al. (2007). A β-strand lock-exchange for signal transduction in TonB-dependent transducers on the basis of a common structural motif. *Structure, 15*(11), 1383–1391.

Brillet, K., Meksem, A., Lauber, E., Reimmann, C., & Cobessi, D. (2009). Use of an in-house approach to study the three-dimensional structures of various outer membrane proteins: structure of the alcaligin outer membrane transporter FauA from *Bordetella pertussis*. *Acta Crystallographica. Section D, Biological Crystallography, 65*(Pt 4), 326–331.

Brillet, K., Reimmann, C., Mislin, G. L. A., Noël, S., Rognan, D., Schalk, I. J., et al. (2011). Pyochelin enantiomers and their outer membrane siderophore transporters in fluorescent *Pseudomonads*: structural bases of a unique enantiospecific recognition. *Journal of the American Chemical Society, 133*(41), 16503–16509.

Buchanan, S. K., Lukacik, P., Grizot, S., Ghirlando, R., Ali, M. M., Barnard, T. J., et al. (2007). Structure of colicin I receptor bound to the R-domain of colicin Ia: implications for protein import. *EMBO Journal, 26*(10), 2594–2604.

Buchanan, S. K., Smith, B. S., Venkatramani, L., Xia, D., Esser, L., Palnitkar, M., et al. (1999). Crystal structure of the outer membrane active transporter FepA from *Escherichia coli*. *Nature Structural Biology, 6*(1), 56–63.

Budzikiewicz, H. (2004). Siderophores of the *Pseudomonadaceae sensu stricto* (fluorescent and non-fluorescent *Pseudomonas* spp.). *Fortschritte der Chemie Organischer Naturstoffe, 87*, 81–237.
Butler, A. (2005). Marine siderophores and microbial iron mobilization. *Biometals, 18*(4), 369–374.
Cascales, E., Buchanan, S. K., Duche, D., Kleanthous, C., Lloubes, R., Postle, K., et al. (2007). Colicin biology. *Microbiology and Molecular Biology Review, 71*(1), 158–229.
Chen, Y., Jurkewitch, E., Bar-Ness, E., & Hadar, Y. (1994). Stability constants of pseudobactin complexes with transition metals. *Soil Science Society of American Journal, 58*, 390–396.
Chimento, D. P., Kadner, R. J., & Wiener, M. C. (2005). Comparative structural analysis of TonB-dependent outer membrane transporters: implications for the transport cycle. *Proteins, 59*(2), 240–251.
Clément, E., Mesini, P. J., Pattus, F., Abdallah, M. A., & Schalk, I. J. (2004). The binding mechanism of pyoverdin with the outer membrane receptor FpvA in *Pseudomonas aeruginosa* is dependent on its iron-loaded status. *Biochemistry, 43*(24), 7954–7965.
Cobessi, D., Célia, H., Folschweiller, N., Schalk, I. J., Abdallah, M. A., & Pattus, F. (2005). The crystal structure of the pyoverdin outer membrane receptor FpvA from *Pseudomonas aeruginosa* at 3.6 Å resolution. *Journal of Molecular Biology, 34*(1), 121–134.
Cobessi, D., Celia, H., & Pattus, F. (2005). Crystal structure at high resolution of ferric-pyochelin and its membrane receptor FptA from *Pseudomonas aeruginosa. Journal of Molecular Biology, 352*(4), 893–904.
Cuiv, P. O., Clarke, P., Lynch, D., & O'Connell, M. (2004). Identification of *rhtX* and *fptX*, novel genes encoding proteins that show homology and function in the utilization of the siderophores rhizobactin 1021 by *Sinorhizobium meliloti* and pyochelin by *Pseudomonas aeruginosa*, respectively. *Journal of Bacteriology, 186*(10), 2996–3005.
Drakesmith, H., & Prentice, A. (2008). Viral infection and iron metabolism. *Nature Reviews Microbiology, 6*(7), 541–552.
Drechsel, H., Jung, G., & Winkelmann, G. (1992). *Biometals, 5*, 141–148.
Eisenhauer, H. A., Shames, S., Pawelek, P. D., & Coulton, J. W. (2005). Siderophore transport through *Escherichia coli* outer membrane receptor FhuA with disulfide-tethered cork and barrel domains. *The Journal of Biological Chemistry, 280*(34), 30574–30580.
Endriss, F., Braun, M., Killmann, H., & Braun, V. (2003). Mutant analysis of the *Escherichia coli* FhuA protein reveals sites of FhuA activity. *Journal of Bacteriology, 185*(16), 4683–4692.
Endriss, F., & Braun, V. (2004). Loop deletions indicate regions important for FhuA transport and receptor functions in *Escherichia coli. Journal of Bacteriology, 186*(14), 4818–4823.
Fanucci, G. E., Lee, J. Y., & Cafiso, D. S. (2003). Spectroscopic evidence that osmolytes used in crystallization buffers inhibit a conformation change in a membrane protein. *Biochemistry, 42*(45), 13106–13112.
Faraldo-Gomez, J. D., Smith, G. R., & Sansom, M. S. P. (2004). Molecular dynamics simulations of the bacterial outer membrane protein FhuA: a comparative study of the ferrichrome-free and bound states. *Biophysical Journal, 85*, 1406–1420.
Ferguson, A. D., Amezcua, C. A., Halabi, N. M., Chelliah, Y., Rosen, M. K., Ranganathan, R., et al. (2007). Signal transduction pathway of TonB-dependent transporters. *Proceedings of the National Academy of Sciences of the United States of America, 104*(2), 513–518.
Ferguson, A. D., Braun, V., Fiedler, H. P., Coulton, J. W., Diederichs, K., & Welte, W. (2000). Crystal structure of the antibiotic albomycin in complex with the outer membrane transporter FhuA. *Protein Science, 9*(5), 956–963.

Ferguson, A. D., Chakraborty, R., Smith, B. S., Esser, L., van der Helm, D., & Deisenhofer, J. (2002). Structural basis of gating by the outer membrane transporter FecA. *Science, 295*(5560), 1715–1719.
Ferguson, A. D., Hofmann, E., Coulton, J. W., Diederichs, K., & Welte, W. (1998). Siderophore-mediated iron transport: crystal structure of FhuA with bound lipopolysaccharide. *Science, 282*(5397), 2215–2220.
Ferguson, A. D., Kodding, J., Walker, G., Bos, C., Coulton, J. W., Diederichs, K., et al. (2001). Active transport of an antibiotic rifamycin derivative by the outer-membrane protein FhuA. *Structure, 9*(8), 707–716.
Ferguson, A. D., Welte, W., Hofmann, E., Lindner, B., Holst, O., Coulton, J. W., et al. (2000). A conserved structural motif for lipopolysaccharide recognition by procaryotic and eucaryotic proteins. *Structure, 8*(6), 585–592.
Fleischhacker, A. S., & Kiley, P. J. (2011). Iron-containing transcription factors and their roles as sensors. *Current Opinion in Chemical Biology, 15*(2), 335–341.
Freed, D. M., Horanyi, P. S., Wiener, M. C., & Cafiso, D. S. (2010). Conformational exchange in a membrane transport protein is altered in protein crystals. *Biophysical Journal, 99*(5), 1604–1610.
Fuchs, R., Schafer, M., Geoffroy, V., & Meyer, J. M. (2001). Siderotyping—a powerful tool for the characterization of pyoverdines. *Current Topics of Medicinal Chemistry, 1*(1), 31–57.
Greenwald, J., Hoegy, F., Nader, M., Journet, L., Mislin, G. L. A., Graumann, P. L., et al. (2007). Real-time FRET visualization of ferric-pyoverdine uptake in *Pseudomonas aeruginosa*: a role for ferrous iron. *The Journal of Biological Chemistry, 282*, 2987–2995.
Greenwald, J., Nader, M., Celia, H., Gruffaz, C., Geoffroy, V., Meyer, J. M., et al. (2009). FpvA bound to non-cognate pyoverdines: molecular basis of siderophore recognition by an iron transporter. *Molecular Microbiology, 72*(5), 1246–1259.
Gumbart, J., Wiener, M. C., & Tajkhorshid, E. (2007). Mechanics of force propagation in TonB-dependent outer membrane transport. *Biophysical Journal, 93*(2), 496–504.
Hannauer, M., Barda, Y., Mislin, G. L., Shanzer, A., & Schalk, I. J. (2010). The ferrichrome uptake pathway in *Pseudomonas aeruginosa* involves an iron release mechanism with acylation of the siderophore and a recycling of the modified desferrichrome. *Journal of Bacteriology, 192*(5), 1212–1220.
Hannauer, M., Braud, A., Hoegy, F., Ronot, P., Boos, A., & Schalk, I. J. 2012. The PvdRT-OpmQ efflux pump controls the metal selectivity of the iron uptake pathway mediated by the siderophore pyoverdine in *Pseudomonas aeruginosa. Environmental Microbiology, 14*(7), 1696–1708.
Hartman, A., & Braun, V. (1980). Iron transport in *Escherichia coli*: uptake and modification of ferrichrome. *Journal of Bacteriology, 143*(1), 246–255.
Helman, R., & Lawrence, G. D. (1989). The increase in ferrioxamine B reduction potential with increasing acidity of the medium. *Journal Electroanalytical Chemistry (Bioelectrochemical Bioenergetics), 22*, 187–196.
Hernlem, B. J., Vane, L. M., & Sayles, G. D. (1996). Stability constants for complexes of the siderophore desferrioxamine B with selected heavy metal cations. *Inorganica Chimica Acta, 244*(2), 179–184.
Hider, R. C., & Kong, X. (2011). Chemistry and biology of siderophores. *Natural Product Report, 27*(5), 637–657.
Higgs, P. I., Larsen, R. A., & Postle, K. (2002). Quantification of known components of the *Escherichia coli* TonB energy transduction system: TonB, ExbB, ExbD and FepA. *Molecular Microbiology, 44*(1), 271–281.
Hoegy, F., Celia, H., Mislin, G. L., Vincent, M., Gallay, J., & Schalk, I. J. (2005). Binding of iron-free siderophore, a common feature of siderophore outer membrane transporters of

Escherichia coli and *Pseudomonas aeruginosa*. *The Journal of Biological Chemistry, 280*(21), 20222–20230.

Hoegy, F., Lee, X., Noël, S., Mislin, G. L., Rognan, D., Reimmann, C., et al. (2009). Stereospecificity of the siderophore pyochelin outer membrane transporters in fluorescent *Pseudomonads*. *The Journal of Biological Chemistry, 284*(22), 14949–14957.

Koster, W. (2001). ABC transporter-mediated uptake of iron, siderophores, heme and vitamin B12. *Research in Microbiology, 152*(3–4), 291–301.

Lee, W., van Baalen, M., & Jansen, V. A. (2012). An evolutionary mechanism for diversity in siderophore-producing bacteria. *Ecology Letters, 15*(2), 119–125.

Llamas, M. A., Sparrius, M., Kloet, R., Jimenez, C. R., Vandenbroucke-Grauls, C., & Bitter, W. (2006). The heterologous siderophores ferrioxamine B and ferrichrome activate signaling pathways in *Pseudomonas aeruginosa*. *Journal of Bacteriology, 188*(5), 1882–1891.

Locher, K. P., Rees, B., Koebnik, R., Mitschler, A., Moulinier, L., Rosenbusch, J. P., et al. (1998). Transmembrane signaling across the ligand-gated FhuA receptor: crystal structures of free and ferrichrome-bound states reveal allosteric changes. *Cell, 95*(6), 771–778.

Ma, L., Kaserer, W., Annamalai, R., Scott, D. C., Jin, B., Jiang, X., et al. (2007). Evidence of ball-and-chain transport of ferric enterobactin through FepA. *The Journal of Biological Chemistry, 282*(1), 397–406.

Martinez, J. S., Carter-Franklin, J. N., Mann, E. L., Martin, J. D., Haygood, M. G., & Butler, A. (2003). Structure and membrane affinity of a suite of amphiphilic siderophores produced by a marine bacterium. *Proceedings of the National Academy of Sciences of the United States of America, 100*(7), 3754–3759.

Matzanke, B. F., Anemuller, S., Schunemann, V., Trautwein, A. X., & Hantke, K. (2004). FhuF, part of a siderophore-reductase system. *Biochemistry, 43*(5), 1386–1392.

Meyer, J. M., Geoffroy, V. A., Baysse, C., Cornelis, P., Barelmann, I., Taraz, K., et al. (2002). Siderophore-mediated iron uptake in fluorescent Pseudomonas: characterization of the pyoverdine-receptor binding site of three cross-reacting pyoverdines. *Archives of Biochemistry and Biophysics, 397*(2), 179–183.

Mislin, G. L. A., Hoegy, F., Cobessi, D., Poole, K., Rognan, D., & Schalk, I. J. (2006). Binding properties of pyochelin and structurally related molecules to FptA of *Pseudomonas aeruginosa*. *The Journal of Molecular Biology, 357*(5), 1437–1448.

Muller, G., Isowa, Y., & Raymond, K. N. (1985). Stereospecificity of siderophore-mediated iron uptake in *Rhodotorula pilimanae* as probed by enantiorhodotorulic acid and isomers of chromic rhodotorulate. *The Journal of Biological Chemistry, 260*(26), 13921–13926.

Münzinger, M., Taraz, K., Budzikiewicz, H., Drechsel, H., Heymann, M., Winkelmann, G., et al. (1999). S, S-rhizoferrin (enantio-rhizoferrin)—a siderophore of *Ralstonia (Pseudomonas) pickettii* DSM 6297—the optical antipode of R, R-rhizoferrin isolated from fungi. *Biometals, 12*(2), 189–193.

Nader, M., Dobbelaere, W., Vincent, M., Journet, L., Adams, H., Cobessi, D., et al. (2007). Identification of residues of FpvA involved in the different steps of Pvd-Fe uptake in *Pseudomonas aeruginosa*. *Biochemistry, 46*(42), 11707–11717.

Nader, M., Journet, L., Meksem, A., Guillon, L., & Schalk, I. J. (2011). Mechanism of ferrisiderophore uptake by *Pseudomonas aeruginosa* outer membrane transporter FpvA: no diffusion channel formed at any time during ferrisiderophore uptake. *Biochemistry, 50*(13), 2530–2540.

Neilands, J. B., Erickson, T. J., & Rastetter, W. H. (1981). Stereospecificity of the ferric enterobactin receptor of *Escherichia coli* K-12. *The Journal of Biological Chemistry, 256*(8), 3831–3832.

Neubauer, U., Nowack, B., Furrer, G., & Schulin, R. (2000). Heavy metal sorption on clay minerals affected by the siderophore desferrioxamine B. *Environmental Science and Technology, 34*, 2749–2755.

Nikaido, H. (2003). Molecular basis of bacterial outer membrane permeability revisited. *Microbiology and Molecular Biology Reviews, 67*(4), 593–656.
Noinaj, N., Guillier, M., Barnard, T. J., & Buchanan, S. K. (2010). TonB-dependent transporters: regulation, structure, and function. *Annual Review of Microbiology, 64*, 43–60.
Pawelek, P. D., Croteau, N., Ng-Thow-Hing, C., Khursigara, C. M., Moiseeva, N., Allaire, M., et al. (2006). Structure of TonB in complex with FhuA, *E. coli* outer membrane receptor. *Science, 312*(5778), 1399–1402.
Payne, M. A., Igo, J. D., Cao, Z., Foster, S. B., Newton, S. M., & Klebba, P. E. (1997). Biphasic binding kinetics between FepA and its ligands. *The Journal of Biological Chemistry, 272*(35), 21950–21955.
Postle, K., & Larsen, R. A. (2007). TonB-dependent energy transduction between outer and cytoplasmic membranes. *Biometals, 20*(3–4), 453–465.
Raymond, K. N., Dertz, E. A., & Kim, S. S. (2003). Enterobactin: an archetype for microbial iron transport. *Proceedings of the National Academy of Sciences of the United States of America, 100*(7), 3584–3588.
Salem, M., Prince, S. M., Patel, H., Chan, H., Feavers, I. M., & Derrick, J. P. Refolding, purification and crystallization of the FrpB outer membrane iron transporter from *Neisseria meningitidis*. *Acta Crystallographica Section D, Biological Crystallography, 68*(Pt 2), 213–215.
Sauter, A., & Braun, V. (2004). Defined inactive FecA derivatives mutated in functional domains of the outer membrane transport and signaling protein of *Escherichia coli* K-12. *Journal of Bacteriology, 186*(16), 5303–5310.
Schalk, I. J. (2008). Metal trafficking via siderophores in Gram-negative bacteria: specificities and characteristics of the pyoverdine pathway. *Journal of Inorganic Biochemistry, 102*(5–6), 1159–1169.
Schalk, I. J., Abdallah, M. A., & Pattus, F. (2002). Recycling of pyoverdin on the FpvA receptor after ferric pyoverdin uptake and dissociation in *Pseudomonas aeruginosa*. *Biochemistry, 41*(5), 1663–1671.
Schalk, I. J., Hannauer, M., & Braud, A. (2011). New roles for bacterial siderophores in metal transport and tolerance. *Environmental Microbiology, 13*(11), 2844–2854.
Schalk, I. J., Hennard, C., Dugave, C., Poole, K., Abdallah, M. A., & Pattus, F. (2001). Iron-free pyoverdin binds to its outer membrane receptor FpvA in *Pseudomonas aeruginosa*: a new mechanism for membrane iron transport. *Molecular Microbiology, 39*(2), 351–360.
Schalk, I. J., Lamont, I. L., & Cobessi, D. (2009). Structure–function relationships in the bifunctional ferrisiderophore FpvA receptor from *Pseudomonas aeruginosa*. *Biometals, 22*(4), 671–678.
Schauer, K., Gouget, B., Carriere, M., Labigne, A., & de Reuse, H. (2007). Novel nickel transport mechanism across the bacterial outer membrane energized by the TonB/ExbB/ExbD machinery. *Molecular Microbiology, 63*(4), 1054–1068.
Shirley, M., & Lamont, I. L. (2009). Role of TonB1 in pyoverdine-mediated signaling in *Pseudomonas aeruginosa*. *Journal of Bacteriology, 191*(18), 5634–5640.
Shultis, D. D., Purdy, M. D., Banchs, C. N., & Wiener, M. C. (2006). Outer membrane active transport: structure of the BtuB: TonB complex. *Science, 312*(5778), 1396–1399.
Spasojevic, I., Boukhalfa, H., Stevens, R. D., & Crumbliss, A. L. (2001). Aqueous solution speciation of Fe(III) complexes with dihydroxamate siderophores alcaligin and rhodotorulic acid and synthetic analogues using electrospray ionization mass spectrometry. *Inorganic Chemistry, 40*(1), 49–58.
Visca, P. (2004). Iron regulation and siderophore signalling in virulence by *Pseudomonas aeruginosa*. In J.-L. Ramos (Ed.), Pseudomonas, *Vol. 2* (pp. 69–123). New York: Kluver Academic/Plenum Publishers.

Visca, P., Leoni, L., Wilson, M. J., & Lamont, I. L. (2002). Iron transport and regulation, cell signalling and genomics: lessons from *Escherichia coli* and *Pseudomonas*. *Molecular Microbiology, 45*(5), 1177–1190.

Wasielewski, E., Tzou, D. L., Dillmann, B., Czaplicki, J., Abdallah, M. A., Atkinson, R. A., et al. (2008). Multiple conformations of the metal-bound pyoverdine PvdI, a siderophore of *Pseudomonas aeruginosa*: a nuclear magnetic resonance study. *Biochemistry, 47*(11), 3397–3406.

Wiener, M. C. (2005). TonB-dependent outer membrane transport: going for Baroque? *Current Opinion in Structural Biology, 15*(4), 394–400.

Winkelmann, G. (1979). Evidence for stereospecific uptake of iron chelates in fungi. *FEBS Letters, 97*(7), 43–46.

Wirth, C., Meyer-Klaucke, W., Pattus, F., & Cobessi, D. (2007). From the periplasmic signaling domain to the extracellular face of an outer membrane signal transducer of *Pseudomonas aeruginosa*: crystal structure of the ferric pyoverdine outer membrane receptor. *Journal of Molecular Biology, 68*(2), 398–406.

Xu, G., Martinez, J. S., Groves, J. T., & Butler, A. (2002). Membrane affinity of the amphiphilic marinobactin siderophores. *Journal of the American Chemical Society, 124*(45), 13408–13415.

Yeterian, E., Martin, L. W., Lamont, I. L., & Schalk, I. J. (2010). An efflux pump is required for siderophore recycling by *Pseudomonas aeruginosa*. *Environmental Microbiology Report, 2*(3), 412–418.

Youard, Z. A., Wenner, N., & Reimmann, C. (2011). Iron acquisition with the natural siderophore enantiomers pyochelin and enantio-pyochelin in *Pseudomonas* species. *Biometals, 24*(3), 513–522.

Yue, W. W., Grizot, S., & Buchanan, S. K. (2003). Structural evidence for iron-free citrate and ferric citrate binding to the TonB-dependent outer membrane transporter FecA. *Journal of Molecular Biology, 332*(2), 353–368.

CHAPTER THREE

Iron Transport Machinery of Human Cells: Players and Their Interactions

Ningning Zhao and Caroline A. Enns[1]
Department of Cell and Developmental Biology, Oregon Health & Sciences University, Oregon, USA
[1]Corresponding author: E-mail: ennsca@ohsu.edu

Contents

Abstract

Organisms, like cells, maintain tight control of iron. In humans as well as other mammals, control is achieved through the regulation of iron uptake into the body rather than through the excretion of iron. The mechanisms by which humans and mice regulate both iron uptake and the distribution of iron within the body and cells are reviewed. Special emphasis is given to the iron transporters involved in this process.

Current Topics in Membranes, Volume 69
ISSN 1063-5823,
http://dx.doi.org/10.1016/B978-0-12-394390-3.00003-3

1. INTRODUCTION

Iron is an essential nutrient. Iron deficiency results in anemia, while overload leads to formation of reactive oxygen species, which causes cell and tissue damage. In a balanced state, about 0.5–2 mg of dietary iron is absorbed through intestinal enterocytes every day, and the same amount is lost in the urine, feces, sweat and sloughed cells (Andrews, 2000a; Smith, 1965). To keep the balance, iron homeostasis needs to be tightly controlled. How cells and organisms regulate their iron content, how various tissues coordinate iron distribution, and how dysregulated iron homeostasis leads to metabolic, hematological and neurodegenerative diseases are the hot topics of study in the field of iron biology (Hentze, Muckenthaler, & Andrews, 2004). In general, approximately 65% of the iron in the body is incorporated into hemoglobin (Hb) of red blood cells (RBCs) and 10% is present in myoglobin, other enzymes and cytochromes. The remaining body iron is stored in the liver, macrophages of the reticuloendothelial system and bone marrow (Andrews, 1999a; Munoz, Villar, & Garcia-Erce, 2009). The ability of iron to accept and donate electrons makes it an essential component of oxygen-binding molecules (e.g. Hb and myoglobin), cytochromes in the electron transport chain and as a cofactor in a variety of enzymes. Excess iron can be toxic by catalyzing the Fenton reaction, in which H_2O_2 is converted to the highly reactive hydroxyl radical (OH•). Hydroxyl radicals oxidize proteins, fatty acids, and nucleic acids, leading to cellular dysfunction (Aruoma, Halliwell, Laughton, Quinlan, & Gutteridge, 1989). To avoid the harmful effects of free iron during oxidative stress, iron is usually bound to specific proteins, such as intracellular ferritin for storage and circulating transferrin (Tf) for transport or utilization.

2. IRON IMPORT ACROSS THE PLASMA MEMBRANE

2.1. Tf and Transferrin Receptors

Tf is the major iron carrier protein in the blood. It is produced primarily in the liver as a glycoprotein. Two high-affinity Fe^{3+} binding sites are located in the amino- and carboxy-terminus (Crichton & Charloteaux-Wauters, 1987). The binding of Tf to iron atoms is associated with binding of an anion, typically bicarbonate (Aisen & Listowsky, 1980), and the Tf-bound iron (TBI) can be released by acidification (Nunez, Gaete, Watkins, & Glass, 1990). Mammalian cells take up TBI through receptor-mediated endocytosis.

There are two known transferrin receptors (TfR) in human body, TfR1 and TfR2. TfR1 is ubiquitously expressed. It functions as a glycosylated homodimer with the N-terminal 61 amino acids forming the cytoplasmic domain (Andrews, 2000b; Jing & Trowbridge, 1987; Kuhn, McClelland, & Ruddle, 1984). The cytoplasmic domain directs the rapid internalization of Tf–TfR1 complex through clathrin-coated vesicles (Fig. 1). Once internalized, the vesicles are acidified by proton pumps (Yamashiro, Tycko, Fluss, & Maxfield, 1984), decreasing the binding of Fe^{3+} to Tf. The resulting changes in both Tf and TfR1 facilitate the dissociation of iron from Tf (Sipe & Murphy, 1991). Fe^{3+} released from Tf is reduced by Steap (six-transmembrane epithelial antigen of the prostate) 3, the major ferrireductase in the erythroid Tf/TfR1 cycle (Ohgami et al., 2005; Sendamarai, Ohgami, Fleming, & Lawrence, 2008). After reduction, Fe^{2+} is transported into the cytosol by DMT1 (divalent metal transporter 1) or ZIP14 (RT, IRT_like protein) (Zhao, Gao, Enns, & Knutson, 2010). Both Tf and TfR1 return to the cell surface, where the iron-depleted Tf is released allowing TfR1 to bind the iron-loaded Tf for another round of internalization. This Tf/TfR1 cycle is the major pathway for iron uptake by erythrocytes.

Erythrocytes require iron incorporation into the heme of Hb to carry oxygen. Without adequate iron, the maturation of erythrocytes is impaired, leading to microcytic hypochromic anemia. In normal individuals, two-thirds of the total body iron is in developing erythroid precursors and mature RBCs. About 20–25 mg of iron is needed every day for Hb synthesis (Andrews & Schmidt, 2007). In the plasma, nearly all circulating iron is bound to Tf. The uptake of TBI into cells is dependent on the number of TfRs. The expression of TfR1 during erythroid maturation correlates with the changing rates of Hb production (Chan & Gerhardt, 1992). The importance of the Tf/TfR1-mediated iron uptake for erythropoiesis is demonstrated in the murine models. Hypotransferrinemic ($Trf^{hpx/hpx}$) mice carry a spontaneous mutation in the *Tf* gene (Bernstein, 1987). These mice have only about 1% of normal circulating Tf. They develop severe microcytic hypochromic anemia, indicating an essential role of Tf in iron delivery to developing erythroid precursors (Trenor, Campagna, Sellers, Andrews, & Fleming, 2000). $TfR1^{-/-}$ mice die in utero between embryonic day 9.5 and 11.5 as a result of severe anemia (Levy, Jin, Fujiwara, Kuo, & Andrews, 1999). The heterozygous mice lacking one copy of the *TfR1* gene also exhibit iron-deficient erythropoiesis.

TfR2 is a member of a TfR-like family. The amino acid sequence of the transmembrane and extracellular domains in TfR2 has 45% identity and 65%

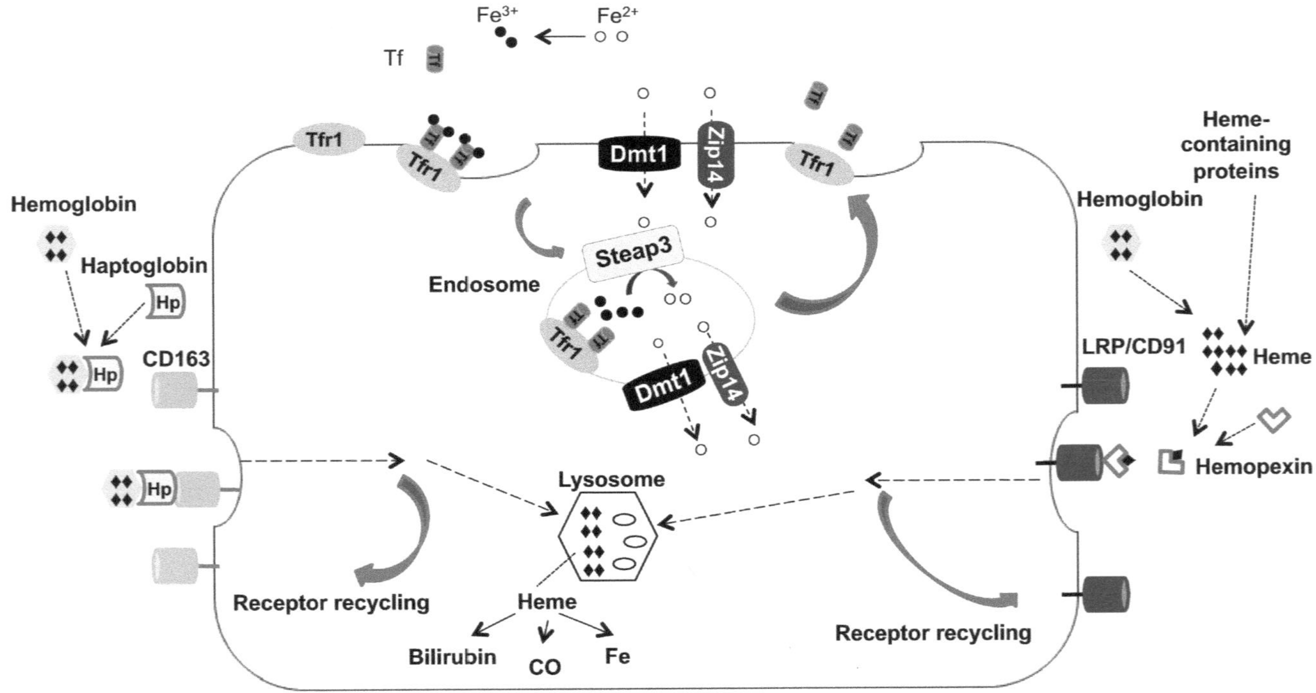

Figure 1 ***Model for the Tf/TfR1 cycle and overview of receptor-mediated pathways for endocytosis of extracellular heme and Hb.*** Holo-Tf binds TfR1 on the cell surface; the complex internalizes in endosomes, where acidification promotes iron release from Tf. Fe^{3+} is reduced to Fe^{2+} by Steap3, and then transported out of endosomes into the cytosol through DMT1 and Zip14. Both Tf and TfR1 return to the cell surface, where at neutral pH iron-depleted Tf is released from TfR1 and TfR1 is ready to bind more iron-laden Tf for another round of endocytosis. CD163 is a receptor for uptake of extracellular Hb bound to Hp. LRP/CD91 represents the pathway for uptake of heme–Hx complexes. Entocytosed heme can be further metabolized into bilirubin, Fe and carbon monoxide.

similarity to that of TfR1. Unlike the ubiquitously expressed TfR1, TfR2 is highly expressed in hepatocytes, the primary site of iron accumulation in the iron overload disease, hereditary hemochromatosis (HH). In untreated HH patients, the expression of TfR1 is undetectable in hepatocytes, but their livers accumulate iron, suggesting the existence of iron uptake pathways other than Tf/TfR1. HH type 3, is due to mutations in *TfR2* gene (Camaschella et al., 2000; Roetto et al., 2001). The function of TfR2 in iron metabolism is not clear. TfR2 is stabilized by diferric Tf in cell cultures (Johnson & Enns, 2004; Robb & Wessling-Resnick, 2004) and the cytoplasmic domain of TfR2 is responsible for its stabilization by diferric Tf (Chen & Enns, 2007). TfR2 is capable of binding and internalizing diferric Tf in vitro (Graham et al., 2008; Kawabata et al., 1999), However, cellular iron uptake may not be the major function of TfR2, because both human and mouse mutations, as well as the mouse knockout, lead to increased rather than decreased iron deposition in the liver (Camaschella et al., 2000; Fleming et al., 2002). Moreover, in mice lacking functional TfR2, Tf-mediated iron uptake by hepatocytes was only slightly decreased compared to wild-type mice with a similar amount of iron loading. Rather than playing a role in cellular iron uptake, TfR2 is more likely to be a modulator of hepcidin transcription. Hepcidin is a hormone secreted by the liver and binds to ferroportin1 (FPN1). FPN1 is the only known iron exporter on the surface of absorptive enterocytes and tissue macrophages, functioning in releasing dietary iron or iron recycled from RBCs into the blood circulation (Donovan et al., 2000; McKie et al., 2000). Upon hepcidin binding, FPN1 is internalized and degraded (Nemeth et al., 2004). HH causing mutations in *TfR2* reduce hepcidin expression, resulting in increased FPN and increased uptake of iron into the body.

2.2. Heme and Hb Receptors

The major source of plasma iron comes from macrophages that recycle iron from senescent or damaged erythrocytes rather than from intestinal iron absorption (Knutson & Wessling-Resnick, 2003). However, erythrophagocytosis is not the only way for macrophages to acquire iron. They can also obtain iron from circulating Tf via the Tf/TfR1 cycle and from the uptake of heme and Hb (Wilkins, Williams, & Cavill, 1977). Hb is the most abundant and functionally important protein in erythrocytes. But once released from RBCs, it becomes highly toxic because of the oxidative properties of heme (protoporphyrin IX and iron) via the Fenton reaction to produce reactive hydroxyl radical causing cell damage (Tolosano et al., 2002).

Normally, only a small amount of extravascular hemolysis occurs due to demolition of senescent erythrocytes, causing Hb release into plasma. Under various intravascular hemolysis-linked conditions, such as hemorrhage, hemoglobinopathies, ischemia reperfusion, sickle cell disease or malaria, large amounts of free Hb are released (Tolosano et al., 2002). Released plasma Hb is readily bound stoichiometrically by the liver-derived plasma protein haptoglobin (Hp), forming an Hb–Hp complex (Nielsen & Moestrup, 2009). CD163, an Hb scavenger receptor present on the surface of monocytes and macrophages in the liver and other tissues, mediates the endocytosis and subsequent degradation of Hb–Hp complex (Kristiansen et al., 2001). The Hb–Hp complex is degraded in lysosomes to release heme and other proteolytic products within the cells. The heme is then degraded by heme oxygenase 1 to release iron, carbon monoxide, and biliverdin. The free iron enters the same intracellular pool as iron taken up from other sources. Mice lacking functional Hp do not have obvious disorders in iron metabolism, suggesting that Hp does not play a major role in normal iron metabolism (Lim et al., 1998). When the binding capacity of Hp is exceeded, plasma Hb is quickly oxidized to ferrihemoglobin (Tolosano et al., 2002). Ferrihemoglobin dissociates into globin and ferriheme. Ferriheme binds to hemopexin (Hx), forming heme–Hx complex, which is internalized by receptor-mediated endocytosis into macrophages. The receptor for heme–Hx complex is LDL receptor-related protein (LRP)/CD91 (Hvidberg et al., 2005). Mice in which the *Hx* gene is inactivated do not exhibit disturbed iron metabolism, suggesting that the role of Hx in iron homeostasis is important only under pathological conditions (Tolosano et al., 1999). Hx levels in serum correlates with heme levels in the blood. A high Hx level indicates significant degradation of heme-containing compounds. Low Hx levels are one of the diagnostic features of hemolytic anemia. Extracellular heme can be cleared through these two receptor-mediated endocytic pathways, thus avoiding the strong oxidative features and proinflammatory effects of free heme.

2.3. Iron Transporters and their Implications in Human Diseases

2.3.1. Divalent Metal Transporter 1

DMT1 is the most extensively characterized iron transport protein. It was identified in 1995 from a screen for murine homologs of Nramp1 (natural resistance-associated macrophage protein 1), a protein involved in host immune defense. Therefore, DMT1 was first named Nramp2. Nramp1 is an integral membrane glycoprotein expressed in lysosomes of macrophages. It is targeted to the membrane of the phagosome after phagocytosis (Gruenheid, Pinner, Desjardins, & Gros, 1997). In contrast, DMT1 is ubiquitously expressed and does not seem particularly associated with organs or cells implicated in host defense. The *DMT1* gene is on chromosome 15, which is away from known host defense genes (Gruenheid, Cellier, Vidal, & Gros, 1995).

The *DMT1* gene encodes four different protein isoforms. The upstream 5′ exon, termed exon 1A, adds an in-frame translation initiation codon and extends the open reading frame (ORF) of the protein by 29–31 amino acids in different species (Hubert & Hentze, 2002). The sequences of the variants with or without exon 1A are identical after reaching exon 2 and until reaching the C-terminal sequences of the ORF. Two other DMT1 messenger RNA (mRNA) variants are produced by alternative splicing at the 3′ end. One contains the stem-loop structure, iron response element (IRE), and the other does not (Lee, Gelbart, West, Halloran, & Beutler, 1998). Four isoforms are produced by alternative splicing of the 5′ end exon 1A and the 3′ end IRE, yielding DMT-1A(+IRE), DMT-1B(+IRE), DMT-1A(−IRE), and DMT-1B(−IRE) isoforms (Hubert & Hentze, 2002). Like the IREs in the 3′ UTR of the *TfR1* mRNA, iron regulatory protein (IRP) binds under low iron conditions and stabilizes the *DMT1* mRNA, leading to increased DMT1 protein levels (Andrews, 1999b). However, in Caco2 cells, the 3′ end IRE is not necessary for iron regulation and the exon 1A itself is associated with iron regulation (Hubert & Hentze, 2002). The tissue distribution profile indicates that the exon 1B isoform is ubiquitously expressed, while the expression of exon 1A isoform is tissue specific and most abundant in the duodenum and the kidney. The same study also found that, in mice, DMT1 regulation in the kidney is associated with the presence of an IRE in the 3′ UTR, whereas in the duodenum, iron regulation is most strongly associated with the presence of exon 1A.

In Belgrade (*b*) rats and *mk* mice, rodent models of DMT1 dysfunction, a single nucleotide change in the *DMT1* gene results in a substitution of arginine for glycine at position 185 (G185R), leading to iron deficiency and anemia (Fleming et al., 1997, 1998). The *b* rat suffers from anemia accompanied by elevated plasma iron and iron-binding capacity, decreased stainable iron in tissues and decreased growth rate. (Oates and Morgan 1996) showed that the reduced uptake of both Fe^{3+} and Fe^{2+} iron in the *b* rat most likely involves a defective iron carrier associated with the apical membrane of the duodenum.

Using a positional cloning approach, (Fleming et al. 1998, 1997) identified *DMT1* as the defective gene for both *b* rats and *mk* mice. Injection of RNA synthesized from DMT1 complementary DNA in *Xenopus* oocytes promotes the uptake of iron as well as manganese, cobalt, and zinc (Gunshin et al., 1997). In Caco2 cells, DMT1 was shown to transport iron preferentially over other divalent cations (Tandy et al., 2000). Studies in other cultured mammalian cells have also demonstrated that DMT1 can transport a variety of divalent cations at the plasma membrane, including iron (Fleming et al., 1998; Zhang, Lee, Wang, & Soong, 2000). Uptake of metals by DMT1 is pH dependent. Iron uptake is most efficient in COS-7 and HEK 293 cells transfected with the DMT1(−IRE) isoform at pH 5.5–6.5, whereas iron uptake decreases to almost baseline level at pH 7.5 or above (Ludwiczek et al., 2007).

DMT1 expression is significantly induced in duodenal epithelial cells in rats fed iron-deficient diet (Gunshin et al., 1997). Dietary iron deficiency results in a dramatic upregulation of the DMT1(+IRE) form in proximal duodenum (Canonne-Hergaux, Gruenheid, Ponka, & Gros, 1999; Gunshin et al., 1997). In situ hybridization indicates that *DMT1* is highly expressed in enterocyte villus. An immunohistochemical study of human duodenum shows that DMT1 localizes to enterocytes, especially at the microvillus brush border membrane (Griffiths, Kelly, Smith, & Cox, 2000). Localization of DMT1 is in agreement with the known physiological site for iron absorption in the intestine, which is mostly restricted to the brush border of the proximal intestine (Muir & Hopfer, 1985). In Caco2 cells, preincubation with the DMT1-specific antibody significantly inhibits iron uptake at pH 5.5 (Tandy et al., 2000). Mice with specific inactivation of *DMT1* gene in the intestine are born alive, but they rapidly develop iron-deficiency anemia (Gunshin, Fujiwara, et al., 2005). Taken together, these results suggest that DMT1 is the major iron transporter in the small intestine.

In *b* rats, iron uptake from Tf by erythropoietic cells is reduced and globin synthesis is defective (Edwards, Sullivan, & Hoke, 1980; Sladic-Simic et al., 1966). A small decrease in endocytosis of Tf, associated with diminished iron uptake and increased iron release by exocytosis, indicates that the defect of iron uptake in *b* rats is after the Tf/TfR1 endocytosis (Bowen & Morgan, 1987). In fact, even though holo-Tf is taken up into reticulocytes, iron is poorly retained, and much is recycled to the extracellular space along with Tf in *b* rats, suggesting that their reticulocytes are unable to transport iron out of the vesicle after endocytosis (Garrick, Gniecko, Liu, Cohan, & Garrick, 1993). Given the evidence that both *b* rats and *mk* mice have the same G185R mutation in *DMT1* and endosomal iron release is not efficient in the *b* rat, DMT1 is hypothesized to function in iron transport out of endosomes (Fleming et al., 1998).

Several studies report cases of *DMT1* mutations in humans. (Mims et al., 2005) report the first case of a female with severe hypochromic microcytic anemia and iron overload. A G-to-C mutation in exon 12 (DMT1 1285 G → C) was found in this patient, resulting in a glutamic acid to aspartic acid (E399D) substitution. However, the predominant effect of this mutation is exon 12 skipping during processing of the mRNA present in erythroid cells. Removal of exon 12 deletes transmembrane domain 8, which may interfere with proper protein insertion into the membrane. E399D resides in the fourth predicted intracellular loop of DMT1 and forms part of a highly conserved transport signature motif among species (Lam-Yuk-Tseung, Mathieu, & Gros, 2005). The mutants E399D, E399Q, and E399A expressed in LLC-PK1 kidney cells are fully functional in terms of stability and targeting to the membrane as well as being transport competent, indicating that $DMT1^{G1285C}$ is not a complete loss of function. In healthy individuals, exon 12 skipping risk is 10%. However, G1285C mutation greatly increases the exon 12 skipping risk to 90%. Consequently, the amount of functional E399D produced in this patient is limited, and it may be adequate for iron absorption, but not sufficient for iron utilization in the erythrocytes. Therefore, the clinical phenotype appears to be due to exon 12 skipping rather than E399D substitution. Interestingly, a liver biopsy indicated that this patient has severe iron loading in both hepatocytes and Kupffer cells, suggesting the existence of another iron uptake pathway functioning in the liver.

Iolascon et al., (2006) reported an infant with hypochromic, microcytic anemia, and hepatic iron overload with increased serum iron, Tf saturation, and serum ferritin levels before treatment with recombinant erythropoietin (rEpo). Screening for *DMT1* identified two novel mutations: a 3-bp deletion in intron 4, causing a splicing abnormality, and a C to T transition in exon 13, resulting in the substitution of arginine with cysteine (R416C). This patient displayed severe iron loading in the liver despite low serum ferritin levels at age 5. The high iron stores were disproportionate to the iron from transfusions or bone marrow redistribution, indicating a significant increase of intestinal iron absorption. Administration of rEpo resulted in rapid improvement of anemia, suggesting the existence of a DMT1-independent pathway for iron utilization by the Tf/TfR1 cycle in erythrocytes.

(Beaumont et al., 2006) reported a case of congenital microcytic hypochromic anemia due to *DMT1* mutations in a 6-year-old French girl. This patient had two mutations in *DMT1*, including a deletion of a GTG codon in exon 5 and a glycine to valine (G212V) mutation in exon 8, resulting in the in-frame deletion of V114 (delV114) in TM2 and a G212V substitution in TM5, respectively. Different from two previous reported *DMT1* mutations, the anemia in this case was less severe, indicating partial loss of function.

(Bardou-Jacquet et al., 2011) reported a 27-year-old woman with hypochromic microcytic anemia and hepatic iron overload. This patient also had two mutations in *DMT1*, including the known G212V mutation and a novel N491S substitution. Overexpression of the mutant forms in Huh7 cells revealed that the N491S mutation led to disturbed protein trafficking with more DMT1 accumulating in endoplasmic reticulum.

(Barrios et al., 2012) and (Blanco, Kannengiesser, Grandchamp, Tasso, and Beaumont 2009) reported the same homozygous mutation in *DMT1* in a 7-year-old boy and a 10-year-old boy, respectively. The mutation is located in the first transmembrane domain and leads to a glycine to arginine substitution (G75R). Both patients had hypochromic microcytic anemia. The functional consequence of this mutation is unknown.

In summary, the identical mutation in *DMT1* of Belgrade rats and *mk* mice, together with reported cases of *DMT1* mutations in humans, emphasizes the role of DMT1 in erythropoiesis (Beaumont et al., 2006). In addition, DMT1 is expressed in bone marrow and optimally functions at acidic pH, such as in endosomes (Gunshin et al., 1997). These observations support the rationale that DMT1 is an endosomal iron transport protein, especially in erythroid cells.

2.3.2. Natural Resistance-Associated Macrophage Protein 1

Nramp1 was first identified as a gene responsible for intracellular protection against pathogens. It encodes a phagocyte-specific membrane protein (Malo et al., 1994; Soe-Lin, Sheftel, Wasyluk, & Ponka, 2008; Vidal, Malo, Vogan, Skamene, & Gros, 1993). Expression of *Nramp1* is highest in the reticuloendothelial organs (primarily the spleen and liver). Unlike *Nramp2*, *Nramp1* mutations have not been found to cause severe anemia in mice (Soe-Lin et al., 2009), but result in increased susceptibility to infection. Intracellular localization studies indicate that Nramp1 is localized to late endosomal/lysosomal compartments rather than plasma membrane of macrophages. After completion of phagocytosis, Nramp1 is rapidly recruited to the membrane of phagosomes (Gruenheid et al., 1997; Searle et al., 1998). By overexpressing an epitope-tagged Nramp1 in CHO cells and measuring radioisotope activities, (Forbes and Gros 2003) showed that Nramp1 can transport both Fe^{2+} and Mn^{2+}. Nramp1 transports divalent metals down the proton gradient, acting as a phagosomal efflux pump (Jabado et al., 2000). Thus, the antimicrobial effect of Nramp1 that leads to its identification appears to be due to the ability of Nramp1 to deplete bacteria-containing phagosomes of essential metal nutrients (Forbes & Gros, 2001).

A study by (Soe-Lin et al. 2009) demonstrated that *Nramp1*$^{-/-}$ mice have a higher splenic iron content and increased splenic macrophage iron deposits compared to wild-type animals, consistent with inefficient recycling of erythrophagocytosed iron and retention of iron within reticuloendothelial cells. The knockout mouse has increased Tf saturation, most likely due to compensatory increase in dietary iron absorption. Under the erythropoietic stress conditions induced by phenylhydrazine, *Nramp1*$^{-/-}$ mice show significantly decreased Tf saturation and elevated levels of nonheme iron in both the liver and the spleen compared to wild-type mice under the same conditions, indicating that *Nramp1*$^{-/-}$ mice is unable to use stored iron for efficient erythropoiesis. By using radiolabeled iron, they showed that iron could not be appropriately released from macrophages in *Nramp1*$^{-/-}$ mice. Overexpression of Nramp1 in RAW264.7 macrophages, which lack functional Nramp1, also results in an increased cellular iron release following phagocytosis of ^{59}Fe-labeled RBCs (Soe-Lin et al., 2008). Taken together, these results indicate that Nramp1 is involved in macrophage recycling of iron from the phagocytosis of senescent erythrocytes.

2.3.3. ZRT, IRT-Like Protein 14

ZIP14 (also known as SLC39A14, solute carrier 39 family, member 14) is a member of the ZIP[1] metal ion transporter superfamily of proteins. The ZIP family proteins take the name from ZRT, IRT-like protein, where ZRT represents zinc-regulated transporter and IRT stands for iron-regulated transporter (Guerinot, 2000). (Nomura et al., 1994) first cloned *Zip14*, which they named KIAA0062. Northern blot analysis detected ubiquitous expression of *Zip14*, with the highest expression in the liver and the lowest expression in the spleen, thymus, and peripheral blood leukocytes. A multiple tissue expression array also showed that *Zip14* was ubiquitously expressed with high expression in the liver, pancreas and heart (Taylor, Morgan, Johnson, & Nicholson, 2005).

(Taylor et al., 2005) first characterized the metal transport activity of ZIP14. They showed that ZIP14 overexpression in CHO cells stimulated

[1]Capital letters denote the human gene and lower case letters denote the the mouse gene.

the uptake of zinc. In 2005, (Liuzzi et al., 2005) also found that overexpression of mouse Zip14 in HEK 293 cells increased zinc uptake. In 2006, (Liuzzi, Aydemir, Nam, Knutson, and Cousins 2006) reported that overexpression of Zip14 in HEK 293 cells and Sf9 insect cells enhanced not only the uptake of zinc but also iron. The iron was presented to the cells as ferric citrate, the major form of non-Tf-bound iron (NTBI) that appears in plasma in conditions of iron overload (Grootveld et al., 1989). Zinc can inhibit iron uptake, suggesting that these two metals share the same transporter. Knockdown of endogenous Zip14 in AML12 cells (a mouse hepatocyte cell line) by small interfering RNA (siRNA) resulted in decreased uptake of iron from ferric citrate. Similarly, overexpression of Zip14 in *Xenopus* oocytes stimulates the uptake of ferrous iron into cells (Pinilla-Tenas et al., 2011). In 2008, Gao, Zhao, Knutson, and Enns (2008) showed that ZIP14 overexpression stimulated NTBI uptake in HeLa cells and that suppression of endogenous ZIP14 in the HepG2 cells decreased NTBI uptake. Moreover, Zip14 can transport iron efficiently at pH 7.4 (Liuzzi et al., 2006), the pH at the plasma membrane surface of hepatocytes. In contrast, DMT1, the first identified iron importer, transports iron optimally at pH 5.5 (Garrick et al., 2006; Gunshin et al., 1997) and is readily detected in endosomes but not on the plasma membrane of hepatocytes (Shindo et al., 2006). These observations provide strong support that ZIP14, rather than DMT1, mediates NTBI uptake in hepatocytes.

The majority of iron in the plasma is bound to Tf. When the amount of iron in the plasma exceeds the capacity of Tf, the level of NTBI increases, contributing to tissue iron loading. In iron overload disorders, such as HH, excess iron is mainly found in the liver, pancreas and heart (Hentze et al., 2004), where ZIP14 is highly expressed (Taylor et al., 2005). In HFE-associated HH, the most prevalent form of hemochromatosis (Olynyk et al., 1999), mutation of a single amino acid (C282Y) in the *HFE* gene leads to iron overload (Feder et al., 1996). Hepatocytes from $Hfe^{-/-}$ mice can take up more NTBI compared to wild-type animals (Chua, Olynyk, Leedman, & Trinder, 2004). The role of HFE in iron metabolism remains to be determined. Interestingly, overexpression of HFE in HepG2 cells decreases ZIP14 levels by decreasing the stability of ZIP14 (Gao et al., 2008). The reduced ZIP14 levels are associated with decreased uptake of both NTBI and TBI, suggesting that ZIP14 participates in both pathways of iron acquisition. Moreover, Zip14 localizes to the plasma membrane, as well as the Tf-containing endosomes (Zhao et al., 2010). Overexpression of Zip14 in HEK 293 cells increases TBI uptake, whereas depletion of endogenous

ZIP14 in HepG2 cells decreases TBI uptake. These results indicate that ZIP14 also participates in the iron uptake from Tf.

2.3.4. Transient Receptor Potential Cation Channel, Mucolipin Subfamily, Member 1

Most cells take up iron through receptor-mediated endocytosis. The endo-lysosome system consists of early endosomes, recycling endosomes, late endosomes, and lysosomes. DMT1 and Zip14 are candidates for endosomal iron release. How iron is released from lysosomes into the cytosol remains elusive. By measuring radioactive iron uptake, (Dong et al., 2008) showed that TRPML1 (transient receptor potential cation channel, mucolipin subfamily, member 1, also known as mucolipin1 MCOLN1or ML1) functions as a Fe^{2+}-permeable channel in late endosomes and lysosomes. In 2000, three groups cloned *TRPML1*, which is mutated in mucolipidosis type IV (ML IV) patients (Bargal et al., 2000; Bassi et al., 2000; Sun et al., 2000). TRPML1 belongs to the transient receptor potential (TRP) ion channel family and is primarily localized in late endosomes and lysosomes (Nilius, Owsianik, Voets, & Peters, 2007; Pryor, Reimann, Gribble, & Luzio, 2006). TRPML1 is ubiquitously expressed with the highest expression levels in the brain, kidney, spleen, liver, and heart (Sun et al., 2000). By comparing the amino acid sequence with known protein motifs, a TRP cation channel domain and an internal calcium and sodium channel pore region were identified.

Overexpression ML IV mutant forms of TRPML1 in HEK 293 cells demonstrated that ML IV mutations significantly impair the Fe^{2+} permeability of TRPML1 to various degrees, correlating well with disease severity. The $TRPML1^{-/-}$ ML IV human skin fibroblasts show an increase in lysosomal Fe^{2+} levels and a reduction in cytosolic Fe^{2+} levels compared to control cells. Moreover, some ML IV patients were reported to have iron deficiency or anemia (Altarescu et al., 2002). Based on these results, TRPML1 is proposed to mediate Fe^{2+} released from late endosomes and lysosomes. The hematological symptoms of ML IV patients may be due to impaired iron transport and the intralysosomal iron accumulation may also contribute to the neurodegenerative symptoms observed in these patients.

2.4. Ferrireductases

2.4.1. Ferrireductases for Iron Absorption

Dietary iron is found in two forms, either as heme iron, found in meat and meat products, or non heme iron, present in vegetables, fruits and beans.

Non-heme iron predominates in most diets, comprising 90–95% of total daily iron intake. The absorption of dietary iron occurs at the apical membrane of duodenal enterocytes. The primary form of nonheme iron in food is Fe^{3+}, which is insoluble and must be reduced to Fe^{2+} form before transporting across enterocytes (Fig. 2). The responsible enzyme is thought to be cytochrome b-like ferrireductase (Dcytb) (McKie et al., 2001). However, Gunshin, Starr, et al., (2005) demonstrated that the absence of Dcytb did not impair body iron stores when mice were fed normal chow diet, indicating that there was no major effect on intestinal absorption. They concluded that Dcytb is dispensable for intestinal iron absorption in mice. However, mice are capable of synthesizing ascorbic acid and may have less need for a duodenal surface ferric reductase (Sharp & Srai, 2007). In addition, direct iron absorption was not measured in the knockout study. The role of Dcytb in intestinal iron absorption remains unclear. The recently identified Steap2 protein might be another candidate for the ferrireductase role in the small intestine (Ohgami, Campagna, McDonald, & Fleming, 2006). Other dietary components, such as ascorbic acid, histidine and cysteine, are also able to reduce Fe^{3+} to Fe^{2+} (Glahn & Van Campen, 1997; Han, Failla, Hill, Morris, & Smith, 1995; Swain, Tabatabai, & Reddy, 2002).

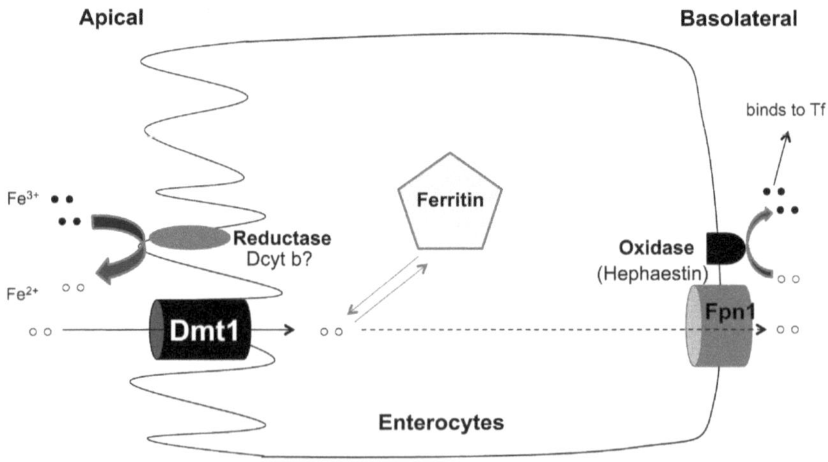

Figure 2 ***Schematic representation of intestinal iron absorption.*** The absorption of iron takes place predominantly in the duodenum. Dietary nonheme iron (mostly Fe^{3+}) is reduced to Fe^{2+} form by a ferric reductase. Fe^{2+} then enters enterocytes through DMT1. Depending upon the body's iron status, enterocyte iron can be stored in ferritin or exported into circulation through (FPN1). The exiting iron is reoxidized to Fe^{3+} through a ferrous oxidase (hephaestin) before loading onto Tf.

2.4.2. *Ferrireductase in Tf/TfR1 Cycle*

Most of the iron in blood is bound to Tf. Monoferric or diferric Tf is taken up into cells through Tf/TfR1 cycle (Fig. 1). This cycle begins with the binding of iron-bound Tf to TfR1 on the cell surface. This complex localizes to clathrin-coated pits, whose invagination initiates endocytosis. Once internalized, Fe^{3+} is dissociated from Tf due to endosomal acidification. An essential step before iron is exported into cytosol is the reduction of Fe^{3+} to Fe^{2+}. RBC precursors take up iron mostly through Tf cycle. A deficiency in iron acquisition by RBCs leads to hypochromic, microcytic anemia phenotype of the mutant mouse nm1054. By using positional cloning, (Ohgami et al., 2005) identified that *Steap3* gene is responsible for the iron deficiency anemia in these animals. They showed that Steap3 is highly expressed in hematopoietic tissues and colocalizes with the endosomal markers. Overexpression of Steap3 in HEK 293 cells stimulated the reduction of iron and facilitates TBI uptake. Moreover, by target deletion of *Steap3*, they found that mice lacking Steap3 were deficient in erythroid ferrireductase activity. Recently, a heterozygous nonsense mutation in exon 3 of the *Steap3* gene was identified in a family with three affected siblings (Grandchamp et al., 2011). Those individuals all have transfusion-dependent hypochromic anemia and iron overload. The mutation leads to a Cys 100 to stop codon substitution and causes very low expression level of this protein. Taken together, these results indicate that Steap3 is a ferrireductase required for efficient Tf-dependent iron uptake in erythrocytes.

3. IRON EXPORTERS AND THEIR IMPLICATIONS IN HUMAN DISEASES

3.1. Ferroportin 1

FPN1 also known as MTP1 and IREG1 (Abboud & Haile, 2000; Donovan et al., 2000; McKie et al., 2000) exports iron out of cells (Canonne-Hergaux, Donovan, Delaby, Wang, & Gros, 2006). Being the only known Fe^{2+} exporter in mammals, its regulation is key to both iron homeostasis as well as iron distribution in the body. Dietary iron absorbed by the intestine is exported out of enterocytes by FPN1 to be bound by Tf and transported through the blood. TBI is ultimately taken up into tissues expressing TfRs. Senescent RBCs are taken up by macrophages in the liver and spleen where the RBCs are broken down. Iron is scavenged from the Hb and the iron that is released by heme is transported out of the macrophage by FPN1.

FPN1 (SLC40A1) is controlled at the levels of transcription, translation, and posttranslation making it able to accommodate a large range of effectors. It is regulated transcriptionally by iron (Abboud & Haile, 2000; Donovan et al., 2000; Knutson & Wessling-Resnick, 2003; McKie et al., 2000). In addition FPN1 can transport both zinc and cadmium and is controlled transcriptionally by metal transcription factor-1 (MTF-1) through the ability of MTF-1 to bind zinc and cadmium (Troadec, Ward, Lo, Kaplan, & De Domenico, 2010). The binding of cytosolic zinc or cadmium to MTF-1 causes the translocation of MTF-1 from the cytoplasm to the nucleus where it activates FPN1 transcription thus increasing the transport of these metals out of the cell.

Most studies have concentrated on the translational and posttranslational mechanisms of FPN1 regulation. The iron-labile pool within the cell positively regulates the translation of FPN1 using a stem-loop, IRE, upstream of the translational start site (Abboud & Haile, 2000; McKie et al., 2000). Under conditions of low intracellular iron, iron-responsive proteins, IRP1 and IRP2, bind to the IRE and block the translation of FPN1 preventing further loss of iron from the cell. Under high iron conditions, the IRPs no longer bind to the IRE resulting in increased FPN1 translation. Increased levels of FPN1 help to lower intracellular iron levels. Tissue-specific alternate splicing of FPN1 that eliminates the IRE and creates an alternate transcriptional initiation site eliminates the control of FPN1 by intracellular iron (Zhang, Hughes, Ollivierre-Wilson, Ghosh, & Rouault, 2009). This form of FPN1 occurs in intestinal enterocytes and erythroid precursor cells. In the case of intestinal enterocytes, lack of intracellular iron regulation is desirable because otherwise low dietary iron would inhibit iron export from the enterocytes and high dietary iron would maximize iron export.

FPN1 levels are also regulated by hepcidin, iron transport, and ceruloplasmin through posttranslational mechanisms (De Domenico et al., 2007, 2011; Nemeth et al., 2004). The binding of hepcidin to FPN1 induces its phosphorylation by JAK2 kinase, internalization, ubiquitination, and subsequent degradation in lysosomes (De Domenico, Lo, Ward, & Kaplan, 2009, 2010). Thus, high levels of iron in the body result in the upregulation of hepcidin and the downregulation of FPN to limit further dietary iron uptake. Hepcidin expression is also positively regulated by inflammation (Nicolas et al., 2002). Limited dietary iron uptake and sequestration of iron in macrophages are part of an innate immune responses to lower circulation iron in the plasma that could be utilized by bacteria or contribute to inflammatory processes. These responses explain the anemia of chronic disease.

3.2. Feline Leukemia Virus-C Receptor and Heme Export

The feline leukemia virus-C receptor (FLVCR) was first identified by functional cloning and was predicted to be an organic anion carrier based on its sequence similarity to other members of this transporter family (Tailor, Willett, & Kabat, 1999). The observation that cats viremic with feline leukemia virus subgroup C were profoundly anemic and had defects in erythropoiesis leads (Quigley et al., 2000) to identify FLVCR as a heme exporter that is essential to prevent apoptosis of differentiating erythroid cells (Keel et al., 2008). *Flvcr* null mice die in midgestation from a lack of ability to undergo definitive erythropoiesis (Quigley et al., 2004). Mice with conditional deletions of *Flvcr* develop severe hyperchromic macrocytic anemia with a block in erythroid maturation at the proerythroblastic stage in addition to cardiomegaly and splenomegaly (Keel et al., 2008). Macrophages from *Flvcr*-deficient mice accumulate more intracellular iron in the form of ferritin than control mice. Further studies indicate that Flvcr facilitates the export of heme from macrophages. These studies lead to the proposal that FLVCR serves to function as a heme overflow valve if heme synthesis exceeds Hb synthesis. In addition, hepatocytes and duodenal enterocytes accumulate iron when Flvcr is downregulated implying a role for Flvcr in heme export in these cell types.

4. INTRACELLULAR IRON TRANSPORT AND IRON CHAPERONES

Whereas the uptake of iron via the Tf/TfR1, DMT1 pathway has been extensively studied, much less is known as to how iron is transported within the cell. Intracellular chaperones for both copper and zinc transport have been identified and reviewed (Robinson & Winge, 2010; Rosenzweig, 2002); however, the identification of iron chaperones was lacking until recently with the discovery that poly r(C)-binding proteins (PCBP) 1 and 2 facilitate the loading of iron into ferritin (Shi, Bencze, Stemmler, & Philpott, 2008). PCBP1 and 2 were originally identified as RNA-binding proteins. Further studies indicate that they also play a role in the transfer of iron to the iron-dependent prolyl hydroxylase 2 (PHD2) that negatively regulates the stability of hypoxia-inducible factor 1 α. Depletion of PCBP1 or PCBP2 in cells results in the loss of iron incorporation in PHD2 and loss of activity of this enzyme (Nandal et al., 2011).

Another iron chaperone protein, frataxin, is proposed to transfer iron to the Fe–S complex within mitochondria (Gerber, Muhlenhoff, & Lill, 2003; Muhlenhoff et al., 2003). Frataxin will be discussed further in the mitochondrial iron transport section.

5. MITOCHONDRIAL IRON IMPORT AND EXPORT

Extensive iron metabolism occurs in the mitochondrion. The incorporation of iron into protophorin IX to generate heme occurs in this organelle as well as most iron sulfur (Fe–S) biogenesis. Both heme and Fe–S-containing proteins are involved in the transfer of electrons in the oxidative phosphorylation pathway and in key enzymes in the tricarboxylic acid cycle within mitochondria. Mutations in the proteins involved in the transport of iron into and out of mitochondria result in neurodegenerative diseases, cancer, anemias and a host of other disorders.

Most of the information concerning mitochondrial iron transport is derived from mutations in the proteins involved in this process. The form of iron that is transported into mitochondria is unknown. Iron can be transferred directly from the endosome through DMT1 to the mitochondrion in erythroid precursors. How this is accomplished and the mitochondrial acceptor has not been identified (Ponka, Sheftel, & Zhang, 2002; Zhang, Sheftel, & Ponka, 2005) (Fig. 3). The cloning of the frascati mutation in zebra fish that causes anemia resulted in the identification of mitoferrin1 (SLS25A37), a transporter in the inner mitochondrial membrane. Mitoferrin1 is expressed in erythroid cells (Shaw et al., 2006). A paralog of mitoferrin1, mitoferrin2 (SLC25A28), is expressed ubiquitously (Shaw et al., 2006). Mitoferrin interacts with Abcb10, which also participates in this process (Chen et al., 2009). Both mitoferrin and Abcb10 are required for the synthesis of heme and Fe–S clusters suggesting that they may play a role in mitochondrial iron uptake (Chen et al., 2009). Frataxin is an iron-binding protein located in the mitochondria and plays an important part in Fe–S assembly by binding to scaffold proteins Isu1 and Isu2. Isu1 and 2 interact with Nfs1 and IsdII as part of the Fe–S assembly complex (Leidgens, De Smet, & Foury, 2010). Ferrochelatase is responsible for the final step of heme synthesis by loading protoporphyrin IX with iron. Ferrochelatase is an FeS cluster-containing protein, thereby linking heme and FeS biogenesis in the mitochondrion. Excess iron in the mitochondria of erythroid cells and testes is stored in mitoferritin, an H-ferritin-like protein (Levi et al., 2001).

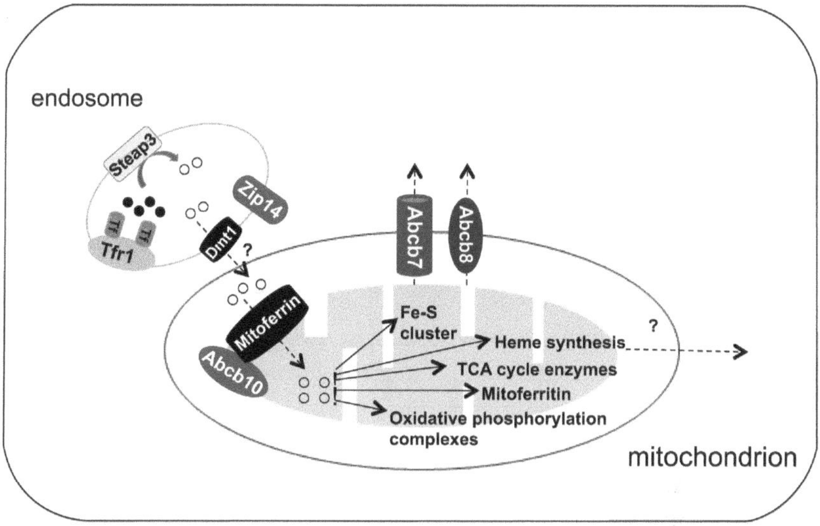

Figure 3 ***Schematic of mitochondrial iron transport.*** How iron crosses the outer membrane of mitochondrion is not clear. Evidence from erythroid precursors suggests that internalized Tf-containing endosomes are dock to mitochondria and iron is transported directly from the endosome into the mitochondrion. Mitoferrin and Abcb10 facilitate the transport of iron across the inner mitochondria. Iron in the mitochondrion can then be used for different metabolic processes, including heme synthesis and Fe/S cluster biosynthesis. Abcb7/8 facilitates the export of a component for the FeS assembly. The transporter responsible for the exit of heme from mitochondria remains to be determined.

Defects in heme biosynthesis or Fe–S clusters result in iron accumulation in mitochondria implying the involvement of specific transporters in the export of these molecules. Mutations in the ATP-binding cassette protein B7 (ABCB7) are associated with sideroblastic anemia and cerebellar ataxia (Allikmets et al., 1999; Bekri et al., 2000; Csere, Lill, & Kispal, 1998). In sideroblastic anemias, iron accumulates in mitochondria. Both ABCB7 and more recently ABCB8 appear to be involved in the export of iron out of mitochondria (Ichikawa et al., 2012). Whether these two proteins directly transport iron or stimulate another transporter remains unanswered.

6. FUTURE DIRECTIONS

With the recent discovery of a host of new membrane proteins that affect iron transport within the body, many issues remain to be resolved.

(1) How are the newly identified transporters affecting iron uptake and iron distribution in cells regulated at the transcriptional and post-transcriptional levels?

(2) Are there additional iron chaperones involved in the delivery of iron to different organelles and in the transcytosis of iron across epithelial and endothelial barriers?

(3) What forms of iron are transported across the mitochondrial membrane and what are their transporters?

(4) How is iron homeostasis achieved in individual organs such as the heart, brain, skeletal muscle, kidney, and pancreas?

(5) What are the best targets for altering iron homeostasis in the body in the case of severe iron overload in diseases like HH and β-thalassemia?

ACKNOWLEDGMENTS

We would like to thank our colleague, Dr An-Sheng Zhang, for his helpful comments and suggestions. This work was supported by NIH grants DK072166 and DK054488 (C.A.E.) and a Tartar Trust fellowship grant (N.Z.).

REFERENCES

Abboud, S., & Haile, D. J. (2000). A novel mammalian iron-regulated protein involved in intracellular iron metabolism. *Journal of Biological Chemistry, 275*, 19906–19912.

Aisen, P., & Listowsky, I. (1980). Iron transport and storage proteins. *Annual Review of Biochemistry, 49*, 357–393.

Allikmets, R., Raskind, W. H., Hutchinson, A., Schueck, N. D., Dean, M., & Koeller, D. M. (1999). Mutation of a putative mitochondrial iron transporter gene (ABC7) in X-linked sideroblastic anemia and ataxia (XLSA/A). *Human Molecular Genetics, 8*, 743–749.

Altarescu, G., Sun, M., Moore, D. F., Smith, J. A., Wiggs, E. A., Solomon, B. I., et al. (2002). The neurogenetics of mucolipidosis type IV. *Neurology, 59*, 306–313.

Andrews, N. C. (1999a). Disorders of iron metabolism. *The New England Journal of Medicine, 341*, 1986–1995.

Andrews, N. C. (1999b). The iron transporter DMT1. *International Journal of Biochemistry and Cell Biology, 31*, 991–994.

Andrews, N. C. (2000a). Iron homeostasis: insights from genetics and animal models. *Nature Reviews Genetics, 1*, 208–217.

Andrews, N. C. (2000b). Iron metabolism: iron deficiency and iron overload. *Annual Review of Genomics and Human Genetics, 1*, 75–98.

Andrews, N. C., & Schmidt, P. J. (2007). Iron homeostasis. *Annual Review of Physiology, 69*, 69–85.

Aruoma, O. I., Halliwell, B., Laughton, M. J., Quinlan, G. J., & Gutteridge, J. M. (1989). The mechanism of initiation of lipid peroxidation. Evidence against a requirement for an iron(II)-iron(III) complex. *Biochemical Journal, 258*, 617–620.

Bardou-Jacquet, E., Island, M. L., Jouanolle, A. M., Detivaud, L., Fatih, N., Ropert, M., et al. (2011). A novel N491S mutation in the human SLC11A2 gene impairs protein

trafficking and in association with the G212V mutation leads to microcytic anemia and liver iron overload. *Blood Cells, Molecules, and Diseases, 47*, 243–248.
Bargal, R., Avidan, N., Ben-Asher, E., Olender, Z., Zeigler, M., Frumkin, A., et al. (2000). Identification of the gene causing mucolipidosis type IV. *Nature Genetics, 26*, 118–123.
Barrios, M., Moreno-Carralero, M. I., Cuadrado-Grande, N., Baro, M., Vivanco, J. L., & Moran-Jimenez, M. J. (2012). The homozygous mutation G75R in the human SLC11A2 gene leads to microcytic anaemia and iron overload. *British Journal of Haematology, 157*, 514–516.
Bassi, M. T., Manzoni, M., Monti, E., Pizzo, M. T., Ballabio, A., & Borsani, G. (2000). Cloning of the gene encoding a novel integral membrane protein, mucolipidin- and identification of the two major founder mutations causing mucolipidosis type IV. *American Journal of Human Genetics, 67*, 1110–1120.
Beaumont, C., Delaunay, J., Hetet, G., Grandchamp, B., de Montalembert, M., & Tchernia, G. (2006). Two new human DMT1 gene mutations in a patient with microcytic anemia, low ferritinemia, and liver iron overload. *Blood, 107*, 4168–4170.
Bekri, S., Kispal, G., Lange, H., Fitzsimons, E., Tolmie, J., Lill, R., et al. (2000). Human ABC7 transporter: gene structure and mutation causing X-linked sideroblastic anemia with ataxia with disruption of cytosolic iron-sulfur protein maturation. *Blood, 96*, 3256–3264.
Bernstein, S. E. (1987). Hereditary hypotransferrinemia with hemosiderosis, a murine disorder resembling human atransferrinemia. *The Journal of Laboratory and Clinical Medicine, 110*, 690–705.
Blanco, E., Kannengiesser, C., Grandchamp, B., Tasso, M., & Beaumont, C. (2009). Not all DMT1 mutations lead to iron overload. *Blood Cells, Molecules, and Diseases, 43*, 199–201.
Bowen, B. J., & Morgan, E. H. (1987). Anemia of the Belgrade rat: evidence for defective membrane transport of iron. *Blood, 70*, 38–44.
Camaschella, C., Roetto, A., Cali, A., De Gobbi, M., Garozzo, G., Carella, M., et al. (2000). The gene TFR2 is mutated in a new type of haemochromatosis mapping to 7q22. *Nature Genetics, 25*, 14–15.
Canonne-Hergaux, F., Donovan, A., Delaby, C., Wang, H. J., & Gros, P. (2006). Comparative studies of duodenal and macrophage ferroportin proteins. *American Journal of Physiology. Gastrointestinal and Liver Physiology, 290*, G156–G163.
Canonne-Hergaux, F., Gruenheid, S., Ponka, P., & Gros, P. (1999). Cellular and subcellular localization of the Nramp2 iron transporter in the intestinal brush border and regulation by dietary iron. *Blood, 93*, 4406–4417.
Chan, L. N., & Gerhardt, E. M. (1992). Transferrin receptor gene is hyperexpressed and transcriptionally regulated in differentiating erythroid cells. *Journal of Biological Chemistry, 267*, 8254–8259.
Chen, J., & Enns, C. A. (2007). The cytoplasmic domain of transferrin receptor 2 dictates its stability and response to holo-transferrin in Hep3B cells. *Journal of Biological Chemistry, 282*, 6201–6209.
Chen, W., Paradkar, P. N., Li, L., Pierce, E. L., Langer, N. B., Takahashi-Makise, N., et al. (2009). Abcb10 physically interacts with mitoferrin-1 (Slc25a37) to enhance its stability and function in the erythroid mitochondria. *Proceedings of the National Academy of Sciences of the United States of America, 106*, 16263–16268.
Chua, A. C., Olynyk, J. K., Leedman, P. J., & Trinder, D. (2004). Nontransferrin-bound iron uptake by hepatocytes is increased in the Hfe knockout mouse model of hereditary hemochromatosis. *Blood, 104*, 1519–1525.
Crichton, R. R., & Charloteaux-Wauters, M. (1987). Iron transport and storage. *European Journal of Biochemistry, 164*, 485–506.
Csere, P., Lill, R., & Kispal, G. (1998). Identification of a human mitochondrial ABC transporter, the functional orthologue of yeast Atm1p. *FEBS Letters, 441*, 266–270.

De Domenico, I., Lo, E., Ward, D. M., & Kaplan, J. (2009). Hepcidin-induced internalization of ferroportin requires binding and cooperative interaction with Jak2. *Proceedings of the National Academy of Sciences of the United States of America, 106*, 3800–3805.

De Domenico, I., Lo, E., Ward, D. M., & Kaplan, J. (2010). Human mutation D157G in ferroportin leads to hepcidin-independent binding of Jak2 and ferroportin downregulation. *Blood, 115*, 2956–2959.

De Domenico, I., Lo, E., Yang, B., Korolnek, T., Hamza, I., Ward, D. M., et al. (2011). The role of ubiquitination in hepcidin-independent and hepcidin-dependent degradation of ferroportin. *Cell Metabolism, 14*, 635–646.

De Domenico, I., Ward, D. M., di Patti, M. C., Jeong, S. Y., David, S., Musci, G., et al. (2007). Ferroxidase activity is required for the stability of cell surface ferroportin in cells expressing GPI-ceruloplasmin. *The EMBO Journal, 26*, 2823–2831.

Dong, X. P., Cheng, X., Mills, E., Delling, M., Wang, F., Kurz, T., et al. (2008). The type IV mucolipidosis-associated protein TRPML1 is an endolysosomal iron release channel. *Nature, 455*, 992–996.

Donovan, A., Brownlie, A., Zhou, Y., Shepard, J., Pratt, S. J., Moynihan, J., et al. (2000). Positional cloning of zebrafish ferroportin1 identifies a conserved vertebrate iron exporter. *Nature, 403*, 776–781.

Edwards, J. A., Sullivan, A. L., & Hoke, J. E. (1980). Defective delivery of iron to the developing red cell of the Belgrade laboratory rat. *Blood, 55*, 645–648.

Feder, J. N., Gnirke, A., Thomas, W., Tsuchihashi, Z., Ruddy, D. A., Basava, A., et al. (1996). A novel MHC class I-like gene is mutated in patients with hereditary haemochromatosis. *Nature Genetics, 13*, 399–408.

Fleming, M. D., Romano, M. A., Su, M. A., Garrick, L. M., Garrick, M. D., & Andrews, N. C. (1998). Nramp2 is mutated in the anemic Belgrade (b) rat: evidence of a role for Nramp2 in endosomal iron transport. *Proceedings of the National Academy of Sciences of the United States of America, 95*, 1148–1153.

Fleming, M. D., Trenor, C. C., 3rd, Su, M. A., Foernzler, D., Beier, D. R., Dietrich, W. F., et al. (1997). Microcytic anaemia mice have a mutation in Nramp2, a candidate iron transporter gene. *Nature Genetics, 16*, 383–386.

Fleming, R. E., Ahmann, J. R., Migas, M. C., Waheed, A., Koeffler, H. P., Kawabata, H., et al. (2002). Targeted mutagenesis of the murine transferrin receptor-2 gene produces hemochromatosis. *Proceedings of the National Academy of Sciences of the United States of America, 99*, 10653–10658.

Forbes, J. R., & Gros, P. (2001). Divalent-metal transport by NRAMP proteins at the interface of host-pathogen interactions. *Trends in Microbiology, 9*, 397–403.

Forbes, J. R., & Gros, P. (2003). Iron, manganese, and cobalt transport by Nramp1 (Slc11a1) and Nramp2 (Slc11a2) expressed at the plasma membrane. *Blood, 102*, 1884–1892.

Gao, J., Zhao, N., Knutson, M. D., & Enns, C. A. (2008). The hereditary hemochromatosis protein, HFE, inhibits iron uptake via down-regulation of Zip14 in HepG2 cells. *Journal of Biological Chemistry, 283*, 21462–21468.

Garrick, M. D., Gniecko, K., Liu, Y., Cohan, D. S., & Garrick, L. M. (1993). Transferrin and the transferrin cycle in Belgrade rat reticulocytes. *Journal of Biological Chemistry, 268*, 14867–14874.

Garrick, M. D., Kuo, H. C., Vargas, F., Singleton, S., Zhao, L., Smith, J. J., et al. (2006). Comparison of mammalian cell lines expressing distinct isoforms of divalent metal transporter 1 in a tetracycline-regulated fashion. *Biochemical Journal, 398*, 539–546.

Gerber, J., Muhlenhoff, U., & Lill, R. (2003). An interaction between frataxin and Isu1/Nfs1 that is crucial for Fe/S cluster synthesis on Isu1. *EMBO Reports, 4*, 906–911.

Glahn, R. P., & Van Campen, D. R. (1997). Iron uptake is enhanced in Caco-2 cell monolayers by cysteine and reduced cysteinyl glycine. *Journal of Nutrition, 127*, 642–647.

Graham, R. M., Reutens, G. M., Herbison, C. E., Delima, R. D., Chua, A. C., Olynyk, J. K., et al. (2008). Transferrin receptor 2 mediates uptake of transferrin-bound and non-transferrin-bound iron. *Journal of Hepatology, 48*, 327–334.

Grandchamp, B., Hetet, G., Kannengiesser, C., Oudin, C., Beaumont, C., Rodrigues-Ferreira, S., et al. (2011). A novel type of congenital hypochromic anemia associated with a nonsense mutation in the STEAP3/TSAP6 gene. *Blood, 118*, 6660–6666.

Griffiths, W. J., Kelly, A. L., Smith, S. J., & Cox, T. M. (2000). Localization of iron transport and regulatory proteins in human cells. *The Quarterly Journal of Medicine, 93*, 575–587.

Grootveld, M., Bell, J. D., Halliwell, B., Aruoma, O. I., Bomford, A., & Sadler, P. J. (1989). Non-transferrin-bound iron in plasma or serum from patients with idiopathic hemochromatosis. Characterization by high performance liquid chromatography and nuclear magnetic resonance spectroscopy. *Journal of Biological Chemistry, 264*, 4417–4422.

Gruenheid, S., Cellier, M., Vidal, S., & Gros, P. (1995). Identification and characterization of a second mouse Nramp gene. *Genomics, 25*, 514–525.

Gruenheid, S., Pinner, E., Desjardins, M., & Gros, P. (1997). Natural resistance to infection with intracellular pathogens: the Nramp1 protein is recruited to the membrane of the phagosome. *Journal of Experimental Medicine, 185*, 717–730.

Guerinot, M. L. (2000). The ZIP family of metal transporters. *Biochimica et Biophysica Acta, 1465*, 190–198.

Gunshin, H., Fujiwara, Y., Custodio, A. O., Direnzo, C., Robine, S., & Andrews, N. C. (2005). Slc11a2 is required for intestinal iron absorption and erythropoiesis but dispensable in placenta and liver. *Journal of Clinical Investigation, 115*, 1258–1266.

Gunshin, H., Mackenzie, B., Berger, U. V., Gunshin, Y., Romero, M. F., Boron, W. F., et al. (1997). Cloning and characterization of a mammalian proton-coupled metal-ion transporter. *Nature, 388*, 482–488.

Gunshin, H., Starr, C. N., Direnzo, C., Fleming, M. D., Jin, J., Greer, E. L., et al. (2005). Cybrd1 (duodenal cytochrome b) is not necessary for dietary iron absorption in mice. *Blood, 106*, 2879–2883.

Han, O., Failla, M. L., Hill, A. D., Morris, E. R., & Smith, J. C., Jr. (1995). Reduction of Fe(III) is required for uptake of nonheme iron by Caco-2 cells. *Journal of Nutrition, 125*, 1291–1299.

Hentze, M. W., Muckenthaler, M. U., & Andrews, N. C. (2004). Balancing acts: molecular control of mammalian iron metabolism. *Cell, 117*, 285–297.

Hubert, N., & Hentze, M. W. (2002). Previously uncharacterized isoforms of divalent metal transporter (DMT)-1: implications for regulation and cellular function. *Proceedings of the National Academy of Sciences of the United States of America, 99*, 12345–12350.

Hvidberg, V., Maniecki, M. B., Jacobsen, C., Hojrup, P., Moller, H. J., & Moestrup, S. K. (2005). Identification of the receptor scavenging hemopexin-heme complexes. *Blood, 106*, 2572–2579.

Ichikawa, Y., Bayeva, M., Ghanefar, M., Potini, V., Sun, L., Mutharasan, R. K., et al. (2012). Disruption of ATP-binding cassette B8 in mice leads to cardiomyopathy through a decrease in mitochondrial iron export. *Proceedings of the National Academy of Sciences of the United States of America, 109*, 4152–4157.

Iolascon, A., d'Apolito, M., Servedio, V., Cimmino, F., Piga, A., & Camaschella, C. (2006). Microcytic anemia and hepatic iron overload in a child with compound heterozygous mutations in DMT1 (SCL11A2). *Blood, 107*, 349–354.

Jabado, N., Jankowski, A., Dougaparsad, S., Picard, V., Grinstein, S., & Gros, P. (2000). Natural resistance to intracellular infections: natural resistance-associated macrophage protein 1 (Nramp1) functions as a pH-dependent manganese transporter at the phagosomal membrane. *Journal of Experimental Medicine, 192*, 1237–1248.

Jing, S. Q., & Trowbridge, I. S. (1987). Identification of the intermolecular disulfide bonds of the human transferrin receptor and its lipid-attachment site. *The EMBO Journal, 6*, 327–331.

Johnson, M. B., & Enns, C. A. (2004). Diferric transferrin regulates transferrin receptor 2 protein stability. *Blood, 104*, 4287–4293.

Kawabata, H., Yang, R., Hirama, T., Vuong, P. T., Kawano, S., Gombart, A. F., et al. (1999). Molecular cloning of transferrin receptor 2. A new member of the transferrin receptor-like family. *Journal of Biological Chemistry, 274*, 20826–20832.

Keel, S. B., Doty, R. T., Yang, Z., Quigley, J. G., Chen, J., Knoblaugh, S., et al. (2008). A heme export protein is required for red blood cell differentiation and iron homeostasis. *Science, 319*, 825–828.

Knutson, M., & Wessling-Resnick, M. (2003). Iron metabolism in the reticuloendothelial system. *Critical Reviews in Biochemistry and Molecular Biology, 38*, 61–88.

Kristiansen, M., Graversen, J. H., Jacobsen, C., Sonne, O., Hoffman, H. J., Law, S. K., et al. (2001). Identification of the haemoglobin scavenger receptor. *Nature, 409*, 198–201.

Kuhn, L. C., McClelland, A., & Ruddle, F. H. (1984). Gene transfer, expression, and molecular cloning of the human transferrin receptor gene. *Cell, 37*, 95–103.

Lam-Yuk-Tseung, S., Mathieu, M., & Gros, P. (2005). Functional characterization of the E399D DMT1/NRAMP2/SLC11A2 protein produced by an exon 12 mutation in a patient with microcytic anemia and iron overload. *Blood Cells, Molecules and Diseases, 35*, 212–216.

Lee, P. L., Gelbart, T., West, C., Halloran, C., & Beutler, E. (1998). The human Nramp2 gene: characterization of the gene structure, alternative splicing, promoter region and polymorphisms. *Blood Cells, Molecules and Diseases, 24*, 199–215.

Leidgens, S., De Smet, S., & Foury, F. (2010). Frataxin interacts with Isu1 through a conserved tryptophan in its beta-sheet. *Human Molecular Genetics, 19*, 276–286.

Levi, S., Corsi, B., Bosisio, M., Invernizzi, R., Volz, A., Sanford, D., et al. (2001). A human mitochondrial ferritin encoded by an intronless gene. *Journal of Biological Chemistry, 276*, 24437–24440.

Levy, J. E., Jin, O., Fujiwara, Y., Kuo, F., & Andrews, N. C. (1999). Transferrin receptor is necessary for development of erythrocytes and the nervous system. *Nature Genetics, 21*, 396–399.

Lim, S. K., Kim, H., bin Ali, A., Lim, Y. K., Wang, Y., Chong, S. M., et al. (1998). Increased susceptibility in Hp knockout mice during acute hemolysis. *Blood, 92*, 1870–1877.

Liuzzi, J. P., Aydemir, F., Nam, H., Knutson, M. D., & Cousins, R. J. (2006). Zip14 (Slc39a14) mediates non-transferrin-bound iron uptake into cells. *Proceedings of the National Academy of Sciences of the United States of America, 103*, 13612–13617.

Liuzzi, J. P., Lichten, L. A., Rivera, S., Blanchard, R. K., Aydemir, T. B., Knutson, M. D., et al. (2005). Interleukin-6 regulates the zinc transporter Zip14 in liver and contributes to the hypozincemia of the acute-phase response. *Proceedings of the National Academy of Sciences of the United States of America, 102*, 6843–6848.

Ludwiczek, S., Theurl, I., Muckenthaler, M. U., Jakab, M., Mair, S. M., Theurl, M., et al. (2007). Ca2+ channel blockers reverse iron overload by a new mechanism via divalent metal transporter-1. *Nature Medicine, 13*, 448–454.

Malo, D., Vogan, K., Vidal, S., Hu, J., Cellier, M., Schurr, E., et al. (1994). Haplotype mapping and sequence analysis of the mouse Nramp gene predict susceptibility to infection with intracellular parasites. *Genomics, 23*, 51–61.

McKie, A. T., Barrow, D., Latunde-Dada, G. O., Rolfs, A., Sager, G., Mudaly, E., et al. (2001). An iron-regulated ferric reductase associated with the absorption of dietary iron. *Science, 291*, 1755–1759.

McKie, A. T., Marciani, P., Rolfs, A., Brennan, K., Wehr, K., Barrow, D., et al. (2000). A novel duodenal iron-regulated transporter, IREG1, implicated in the basolateral transfer of iron to the circulation. *Molecular Cell, 5*, 299–309.

Mims, M. P., Guan, Y., Pospisilova, D., Priwitzerova, M., Indrak, K., Ponka, P., et al. (2005). Identification of a human mutation of DMT1 in a patient with microcytic anemia and iron overload. *Blood, 105*, 1337–1342.

Muhlenhoff, U., Stadler, J. A., Richhardt, N., Seubert, A., Eickhorst, T., Schweyen, R. J., et al. (2003). A specific role of the yeast mitochondrial carriers MRS3/4p in mitochondrial iron acquisition under iron-limiting conditions. *The Journal of Biological Chemistry, 278*, 40612–40620.

Muir, A., & Hopfer, U. (1985). Regional specificity of iron uptake by small intestinal brush-border membranes from normal and iron-deficient mice. *American Journal of Physiology, 248*, G376–G379.

Munoz, M., Villar, I., & Garcia-Erce, J. A. (2009). An update on iron physiology. *World Journal of Gastroenterology, 15*, 4617–4626.

Nandal, A., Ruiz, J. C., Subramanian, P., Ghimire-Rijal, S., Sinnamon, R. A., Stemmler, T. L., et al. (2011). Activation of the HIF prolyl hydroxylase by the iron chaperones PCBP1 and PCBP2. *Cell Metabolism, 14*, 647–657.

Nemeth, E., Tuttle, M. S., Powelson, J., Vaughn, M. B., Donovan, A., Ward, D. M., et al. (2004). Hepcidin regulates cellular iron efflux by binding to ferroportin and inducing its internalization. *Science, 306*, 2090–2093.

Nicolas, G., Chauvet, C., Viatte, L., Danan, J. L., Bigard, X., Devaux, I., et al. (2002). The gene encoding the iron regulatory peptide hepcidin is regulated by anemia, hypoxia, and inflammation. *The Journal of Clinical Investigation, 110*, 1037–1044.

Nielsen, M. J., & Moestrup, S. K. (2009). Receptor targeting of hemoglobin mediated by the haptoglobins: roles beyond heme scavenging. *Blood, 114*, 764–771.

Nilius, B., Owsianik, G., Voets, T., & Peters, J. A. (2007). Transient receptor potential cation channels in disease. *Physiological Reviews, 87*, 165–217.

Nomura, N., Nagase, T., Miyajima, N., Sazuka, T., Tanaka, A., Sato, S., et al. (1994). Prediction of the coding sequences of unidentified human genes. II. The coding sequences of 40 new genes (KIAA0041-KIAA0080) deduced by analysis of cDNA clones from human cell line KG-1. *DNA Research, 1*, 223–229.

Nunez, M. T., Gaete, V., Watkins, J. A., & Glass, J. (1990). Mobilization of iron from endocytic vesicles. The effects of acidification and reduction. *Journal of Biological Chemistry, 265*, 6688–6692.

Oates, P. S., & Morgan, E. H. (1996). Defective iron uptake by the duodenum of Belgrade rats fed diets of different iron contents. *The American Journal of Physiology, 270*, G826–G832.

Ohgami, R. S., Campagna, D. R., Greer, E. L., Antiochos, B., McDonald, A., Chen, J., et al. (2005). Identification of a ferrireductase required for efficient transferrin-dependent iron uptake in erythroid cells. *Nature Genetics, 37*, 1264–1269.

Ohgami, R. S., Campagna, D. R., McDonald, A., & Fleming, M. D. (2006). The Steap proteins are metalloreductases. *Blood, 108*, 1388–1394.

Olynyk, J. K., Cullen, D. J., Aquilia, S., Rossi, E., Summerville, L., & Powell, L. W. (1999). A population-based study of the clinical expression of the hemochromatosis gene. *New England Journal of Medicine, 341*, 718–724.

Pinilla-Tenas, J. J., Sparkman, B. K., Shawki, A., Illing, A. C., Mitchell, C. J., Zhao, N., et al. (2011). Zip14 is a complex broad-scope metal-ion transporter whose functional properties support roles in the cellular uptake of zinc and nontransferrin-bound iron. *American Journal of Physiology. Cell Physiology, 301*, C862–C871.

Ponka, P., Sheftel, A. D., & Zhang, A. S. (2002). Iron targeting to mitochondria in erythroid cells. *Biochemical Society Transactions, 30*, 735–738.

Pryor, P. R., Reimann, F., Gribble, F. M., & Luzio, J. P. (2006). Mucolipin-1 is a lysosomal membrane protein required for intracellular lactosylceramide traffic. *Traffic, 7*, 1388–1398.

Quigley, J. G., Burns, C. C., Anderson, M. M., Lynch, E. D., Sabo, K. M., Overbaugh, J., et al. (2000). Cloning of the cellular receptor for feline leukemia virus subgroup C (FeLV-C), a retrovirus that induces red cell aplasia. *Blood, 95*, 1093–1099.

Quigley, J. G., Yang, Z., Worthington, M. T., Phillips, J. D., Sabo, K. M., Sabath, D. E., et al. (2004). Identification of a human heme exporter that is essential for erythropoiesis. *Cell, 118*, 757–766.

Robb, A., & Wessling-Resnick, M. (2004). Regulation of transferrin receptor 2 protein levels by transferrin. *Blood, 104*, 4294–4299.

Robinson, N. J., & Winge, D. R. (2010). Copper metallochaperones. *Annual Review of Biochemistry, 79*, 537–562.

Roetto, A., Totaro, A., Piperno, A., Piga, A., Longo, F., Garozzo, G., et al. (2001). New mutations inactivating transferrin receptor 2 in hemochromatosis type 3. *Blood, 97*, 2555–2560.

Rosenzweig, A. C. (2002). Metallochaperones: bind and deliver. *Chemistry and Biology, 9*, 673–677.

Searle, S., Bright, N. A., Roach, T. I., Atkinson, P. G., Barton, C. H., Meloen, R. H., et al. (1998). Localisation of Nramp1 in macrophages: modulation with activation and infection. *Journal of Cell Science, 111*(Pt 19), 2855–2866.

Sendamarai, A. K., Ohgami, R. S., Fleming, M. D., & Lawrence, C. M. (2008). Structure of the membrane proximal oxidoreductase domain of human Steap3, the dominant ferrireductase of the erythroid transferrin cycle. *Proceedings of the National Academy of Sciences of the United States of America, 105*, 7410–7415.

Sharp, P., & Srai, S. K. (2007). Molecular mechanisms involved in intestinal iron absorption. *World Journal of Gastroenterology, 13*, 4716–4724.

Shaw, G. C., Cope, J. J., Li, L., Corson, K., Hersey, C., Ackermann, G. E., et al. (2006). Mitoferrin is essential for erythroid iron assimilation. *Nature, 440*, 96–100.

Shi, H., Bencze, K. Z., Stemmler, T. L., & Philpott, C. C. (2008). A cytosolic iron chaperone that delivers iron to ferritin. *Science, 320*, 1207–1210.

Shindo, M., Torimoto, Y., Saito, H., Motomura, W., Ikuta, K., Sato, K., et al. (2006). Functional role of DMT1 in transferrin-independent iron uptake by human hepatocyte and hepatocellular carcinoma cell, HLF. *Hepatology Research, 35*, 152–162.

Sipe, D. M., & Murphy, R. F. (1991). Binding to cellular receptors results in increased iron release from transferrin at mildly acidic pH. *Journal of Biological Chemistry, 266*, 8002–8007.

Sladic-Simic, D., Zivkovic, N., Pavic, D., Marinkovic, D., Martinovic, J., & Martinovitch, P. N. (1966). Hereditary hypochromic microcytic anemia in the laboratory rat. *Genetics, 53*, 1079–1089.

Smith, R. S. (1965). Iron deficiency and iron overload. *Archives of Disease in Childhood, 40*, 343–363.

Soe-Lin, S., Apte, S. S., Andriopoulos, B., Jr., Andrews, M. C., Schranzhofer, M., Kahawita, T., et al. (2009). Nramp1 promotes efficient macrophage recycling of iron following erythrophagocytosis in vivo. *Proceedings of the National Academy of Sciences of the United States of America, 106*, 5960–5965.

Soe-Lin, S., Sheftel, A. D., Wasyluk, B., & Ponka, P. (2008). Nramp1 equips macrophages for efficient iron recycling. *Experimental Hematology, 36*, 929–937.

Sun, M., Goldin, E., Stahl, S., Falardeau, J. L., Kennedy, J. C., Acierno, J. S., Jr., et al. (2000). Mucolipidosis type IV is caused by mutations in a gene encoding a novel transient receptor potential channel. *Human Molecular Genetics, 9*, 2471–2478.

Swain, J. H., Tabatabai, L. B., & Reddy, M. B. (2002). Histidine content of low-molecular-weight beef proteins influences nonheme iron bioavailability in Caco-2 cells. *Journal of Nutrition, 132*, 245–251.

Tailor, C. S., Willett, B. J., & Kabat, D. (1999). A putative cell surface receptor for anemia-inducing feline leukemia virus subgroup C is a member of a transporter superfamily. *Journal of Virology, 73*, 6500–6505.

Tandy, S., Williams, M., Leggett, A., Lopez-Jimenez, M., Dedes, M., Ramesh, B., et al. (2000). Nramp2 expression is associated with pH-dependent iron uptake across the apical membrane of human intestinal Caco-2 cells. *Journal of Biological Chemistry, 275*, 1023–1029.

Taylor, K. M., Morgan, H. E., Johnson, A., & Nicholson, R. I. (2005). Structure-function analysis of a novel member of the LIV-1 subfamily of zinc transporters, ZIP14. *FEBS Letters, 579*, 427–432.

Tolosano, E., Fagoonee, S., Hirsch, E., Berger, F. G., Baumann, H., Silengo, L., et al. (2002). Enhanced splenomegaly and severe liver inflammation in haptoglobin/hemopexin double-null mice after acute hemolysis. *Blood, 100*, 4201–4208.

Tolosano, E., Hirsch, E., Patrucco, E., Camaschella, C., Navone, R., Silengo, L., et al. (1999). Defective recovery and severe renal damage after acute hemolysis in hemopexin-deficient mice. *Blood, 94*, 3906–3914.

Trenor, C. C., 3rd, Campagna, D. R., Sellers, V. M., Andrews, N. C., & Fleming, M. D. (2000). The molecular defect in hypotransferrinemic mice. *Blood, 96*, 1113–1118.

Troadec, M. B., Ward, D. M., Lo, E., Kaplan, J., & De Domenico, I. (2010). Induction of FPN1 transcription by MTF-1 reveals a role for ferroportin in transition metal efflux. *Blood, 116*, 4657–4664.

Vidal, S. M., Malo, D., Vogan, K., Skamene, E., & Gros, P. (1993). Natural resistance to infection with intracellular parasites: isolation of a candidate for Bcg. *Cell, 73*, 469–485.

Wilkins, M., Williams, P., & Cavill, I. (1977). Transferrin iron uptake by human synovium. *Annals of the Rheumatic Diseases, 36*, 474–475.

Yamashiro, D. J., Tycko, B., Fluss, S. R., & Maxfield, F. R. (1984). Segregation of transferrin to a mildly acidic (pH 6.5) para-Golgi compartment in the recycling pathway. *Cell, 37*, 789–800.

Zhang, A. S., Sheftel, A. D., & Ponka, P. (2005). Intracellular kinetics of iron in reticulocytes: evidence for endosome involvement in iron targeting to mitochondria. *Blood, 105*, 368–375.

Zhang, D. L., Hughes, R. M., Ollivierre-Wilson, H., Ghosh, M. C., & Rouault, T. A. (2009). A ferroportin transcript that lacks an iron-responsive element enables duodenal and erythroid precursor cells to evade translational repression. *Cell Metabolism, 9*, 461–473.

Zhang, L., Lee, T., Wang, Y., & Soong, T. W. (2000). Heterologous expression, functional characterization and localization of two isoforms of the monkey iron transporter Nramp2. *Biochemical Journal, 349*, 289–297.

Zhao, N., Gao, J., Enns, C. A., & Knutson, M. D. (2010). ZRT/IRT-like protein 14 (ZIP14) promotes the cellular assimilation of iron from transferrin. *Journal of Biological Chemistry, 285*, 32141–32150.

SECTION TWO

Ins and Outs of Copper Transport

CHAPTER FOUR

Structure and Function of Copper Uptake Transporters

Christopher R. Pope*,†, Adrian G. Flores*,†, Jack H. Kaplan‡ and Vinzenz M. Unger*,†,1

*Department of Molecular Biosciences, Northwestern University, Evanston, IL, USA
†Chemistry of Life Processes Institute, Northwestern University, Evanston, IL, USA
‡Department of Biochemistry and Molecular Genetics, University of Illinois College of Medicine, Chicago, IL, USA
[1]Corresponding author: E-mail: v-unger@northwestern.edu

Contents

Abstract

Owing to their redox and coordination chemistry copper ions play essential roles in cellular function. Research over the past 20 years has shed much light on the biochemistry of copper homeostasis, and the emergence of high-resolution crystal structures for many of the proteins that partake in cellular copper biology have began to provide insight into the molecular mechanisms by which cells handle this important metal. A notable gap in our understanding is related to the process by which cells acquire copper ions. This chapter describes recent progress in the structure determination of cellular copper uptake transporters and how the emerging structural information aids understanding of the molecular mechanisms that govern cellular copper acquisition and distribution.

1. INTRODUCTION

While commonly neglected, metal ions are a part of nearly every biological process that exists on earth. This dependency of life on metals, ranging from common ions such as sodium and potassium to more exotic elements like nickel and vanadium, is founded on the broad spectrum of

Current Topics in Membranes, Volume 69
ISSN 1063-5823,
http://dx.doi.org/10.1016/B978-0-12-394390-3.00004-5

chemical properties that allow metal ions to take on a large range of structural and catalytic roles. Here we summarize what is known about the molecular mechanisms by which cells acquire copper ions that due to the metal's coordination and redox chemistry have become an essential component in the chemical makeup of all eukaryotic cells.

The propensity of copper ions to engage sulfur, nitrogen and even oxygen atoms through weak coordinate bonds and the ability of copper ions to switch between mostly two oxidation states at potentials that are well within the physiological range are key properties that led nature to exploit this metal for vital cellular processes such as antioxidant defense, respiration, and iron homeostasis (Linder & Goode, 1991). The very same properties, however, also dictate that cells need to carefully regulate copper levels, distribution and coordination since an excess of copper ions or the presence of free copper ions are toxic and result in the uncontrolled formation of radicals that cause cellular damage (Bremner, 1998; Halliwell & Gutteridge, 1990). Accommodating both the good and the evil sides of copper, cells have evolved elaborate mechanisms to regulate both cellular flux and confinement of the metals' reactivity through coordination with accessory proteins, such as copper chaperones (O'Halloran & Culotta, 2000; Rosenzweig, 2002) and metallothioneins (Tottey, Harvie, & Robinson, 2005).

Much of what we currently know about copper homeostasis comes from work utilizing yeast and mammalian systems, which revealed that the molecular mechanisms of copper homeostasis are highly conserved (Kim, Nevitt, & Thiele, 2008; Maryon et al., 2007). In essence, cellular copper status is maintained by the interplay of copper transporters that transport copper in and out of the cells, copper chaperones that specifically deliver copper ions to its cellular targets, copper storage proteins that buffer cellular copper content, and copper sensing transcription factors that regulate the supply of copper handling proteins (Winge, 2002) and the cuproenzymes that require copper ions for catalytic function. Since 1994, a wealth of information has been gathered on the biochemistry of copper homeostasis. Adding to this, high-resolution structural information has emerged for many of the key players involved, paving the way for understanding copper metabolism at a near-atomic level (Abajian et al., 2004; Djinovic et al., 1991; Dmitriev et al., 2006; Lamb et al., 2000,2001; Rosenzweig et al., 1999). A notable exception to this growing body of knowledge is the structural biology of cellular copper uptake, which only recently came into being through the determination of a 7-Å structure of the major copper uptake transporter (CTRs) from humans (Aller & Unger, 2006; De Feo et al.,

2009). In the following sections, we will describe what this structure has taught us and how it ties in with functional studies that are focused on unraveling mechanistic details about cellular copper acquisition.

2. CTR—A CONSERVED FAMILY OF CTRS

Incidentally, CTRs were first identified in yeast during genetic screens investigating defects in iron homeostasis (Dancis et al., 1994a). This counterintuitive discovery fell into place when it became clear that iron acquisition in yeast requires copper-dependent reductases and oxidases that help mobilization of iron and prepare it for cellular uptake through iron transporters (Dancis et al., 1994b; Stearman et al., 1996). Since then, copper uptake transporters, have been identified in all eukaryotic cells but are missing from prokaryotes for which copper ions act as potent toxins (Dancis et al., 1994b; Kampfenkel et al., 1995; Lee et al., 2000; Zhou, Cadigan, & Thiele, 2003; Zhou & Gitschier, 1997; Zhou & Thiele, 2001).

Copper transporters are small integral membrane proteins that have three transmembrane (TM) helices, an extracellular N-terminus, and a C-terminus that is disposed on the intracellular side of the membrane (Dancis et al., 1994; De Feo et al., 2009; Eisses & Kaplan, 2002; Klomp et al., 2003; Zhou & Gitschier, 1997). Evolutionary distant CTRs (such as the human high-affinity transporter hCTR1 and the yeast high-affinity transporter yCTR3) are only poorly conserved at the sequence level. However, close inspection of the primary structure reveals that CTRs share an almost invariant $MX_4MX_{12}GX_4G$ sequence signature that is embedded within the second and third TM helices, which identifies these segments as a structurally and functionally important unit within the molecule (Fig. 1). Moreover, CTRs contain at least one putative metal-binding MXM sequence in the extracellular region (Jiang et al., 2005), and often feature potential metal-binding amino acids like cysteine and histidine at or close to the intracellular C-terminus. Besides these shared common features, there is little resemblance at the sequence level, giving rise to a highly diverse group of molecules whose molecular weights range from ~15 to 40 kDa. Much of the differences between CTRs are in the number and positioning of seemingly redundant, potentially metal-binding amino acid residues like histidine and methionine. The precise role of multiple redundant metal-binding sites remains a mystery because in genetic complementation studies, many of these sites are functionally dispensable (see below) (Lee et al., 2000; Zhou & Gitschier,

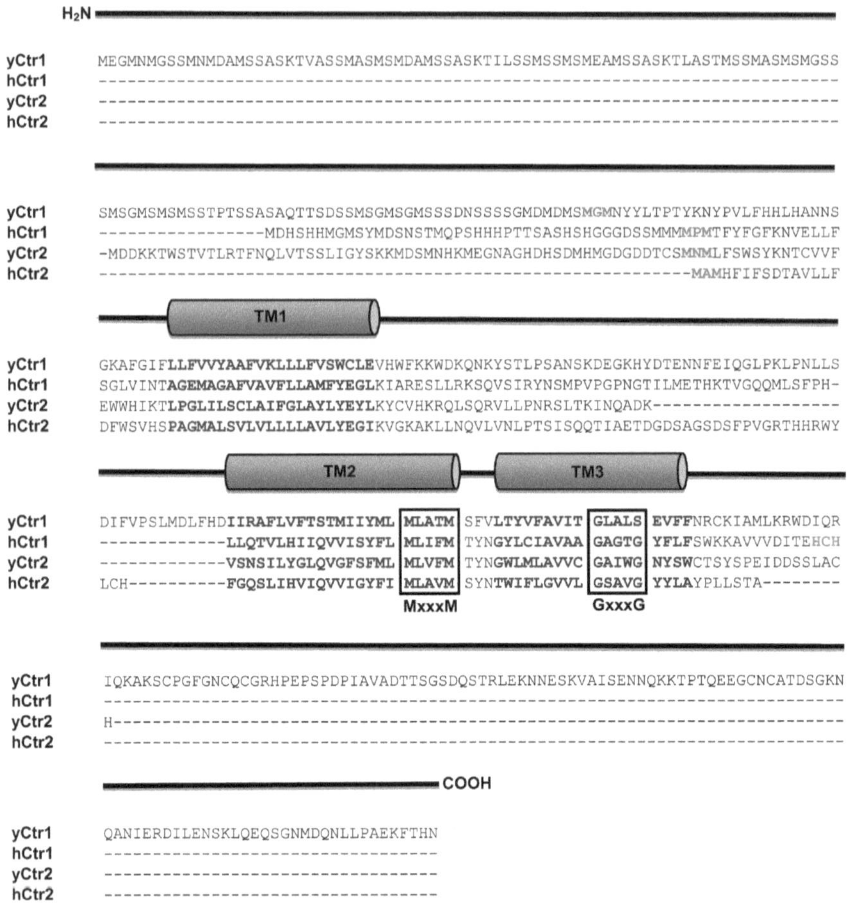

Figure 1 ***CTR Proteins Share Common Structural Signatures.*** Sequence alignments of CTRs 1 and 2 from *Sacchromyces cerevisiae* and *Homo sapiens*. The positions of the three TM helices of the four transporters were predicted using the TOPCON topology prediction algorithms from http://topcons.net/ (Bernsel et al., 2009). The protein sequences were aligned according to the three TM regions (demarcated by blue text and cylinders). Although the sequences between CTR1/2 from yeast to humans vary significantly, the overall architecture of these proteins is predicted to be highly similar, leading to the predictions that the other CTRs form similar oligomeric quaternary structures to that of hCTR1. At the sequence level, the most conserved elements are and MXM sequence in the extracellular N-terminus (highlighted in magenta), and the $MX_4MX_{12}GX_4G$ signature sequence that correlates the MxxM and GxxG motifs located in TM 2 and 3, respectively. Conservation of this signature sequence throughout the CTR-family gives rise to the prediction of a shared mechanism of copper transport across the bilayer. In the case of human CTR1, the C-terminal HCH residues (highlighted in orange) are thought to form a copper-binding site. Similar sequences are found in the majority of CTR proteins, but are absent from some, such as hCTR2. The functional and

1997; Zhou et al., 2003; Zhou & Thiele, 2001). Interestingly, N- and O-glycosylation add chemical complexity in the case of hCTR1 (Eisses & Kaplan, 2002; Maryon et al., 2007). O-glycosylation of hCTR1 was reported to protect hCTR1 from partial post translational proteolysis (Maryon et al., 2007), but the physiological and mechanistic roles of this protection remain unclear as does the role of N-glycosylation. Similarly, the functional significance of large intracellular domains like that found in the yeast high-affinity transporter CTR1 remains unknown, partially because the human homologue, which is missing an extensive C-terminus, can fully compensate for the lack of the native high-affinity copper uptake systems in yeast based on growth complementation assays (Zhou & Gitschier, 1997). However, there is some evidence that the C-terminal domain in yeast CTR1 may be involved in transferring copper from the transporter to at least one of the intracellular copper chaperones, Atx1 (Xiao & Wedd, 2002), hinting at the possibility that in addition to copper transport, CTR proteins may play further roles in intracellular copper distribution.

3. THE STRUCTURE OF HUMAN CTR1

The small size of the copper transporter membrane domain raised immediate mechanistic questions since three TM helices are not sufficient to generate a conduit for any solute, even as small as copper ions (Spencer & Rees, 2002). This puzzle was put into perspective by the independent observation from several laboratories that CTRs oligomerize (Dancis et al., 1994; Eisses & Kaplan, 2002; Klomp et al., 2003; Lee et al., 2002; Zhou & Thiele, 2001). Where studied, CTRs were observed to form homo-oligomers, with exception of yeast CTR4 and CTR5 that require hetero-oligomerization for function (Zhou & Thiele, 2001). While oligomerization was undisputed, the nature of the oligomer was under dispute until electron crystallography of human CTR1 directly visualized that the TM domain of CTR proteins forms radially symmetric trimers (Aller & Unger, 2006).

The initial view of hCTR1 was followed by a full 3D electron crystallographic reconstruction of the structure at 7 Å resolution (Fig. 2) (De Feo

mechanistic significance of these differences is unknown. Further adding to molecular and potentially mechanistic heterogeneity are large insertions/deletions at the termini. Although interesting, no data to date can explain the role of these large insertions or why they were not evolutionarily conserved. For interpretaton of the references to color in this figure legend, the reader is referred to the online version of this book.

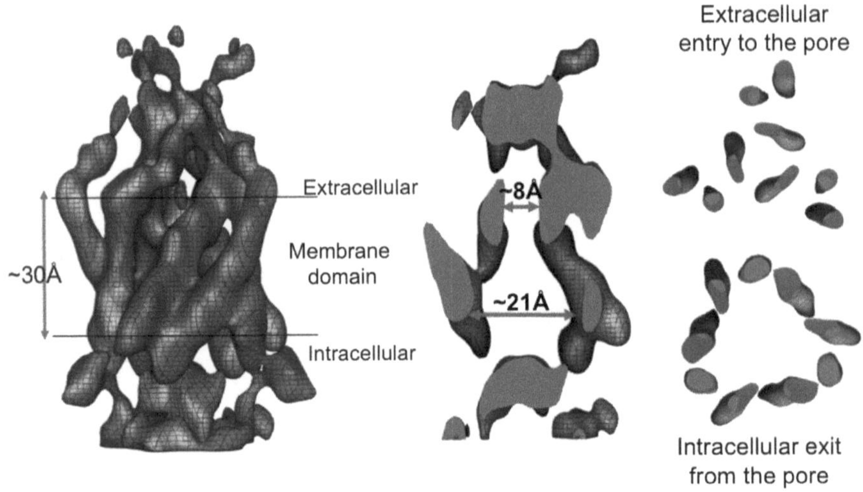

Figure 2 Structure of human CTR1 (hCTR1). This figure summarizes structural details of the 7-Å reconstruction of hCTR1 (adapted from Figure 5.1 in De Feo et al., 2009). Shown to the left is a view of the hCTR1 trimer seen parallel to the membrane. Rod-shaped density features, ~30 Å in length, represent TM alpha helices (contoured at 2σ). Thin black lines indicate the approximate boundaries of the hydrophobic core of the bilayer. The cellular topology of the reconstruction with respect to the membrane was established by immunogold-labeling experiments. A longitudinal section through the reconstruction (middle panel) reveals a funnel-shaped continuous TM conduit along the central three-fold axis of symmetry of the trimer. The pore is narrow at its extracellular end and widens from ~8 to 21 Å at the intracellular end (dimensions do not consider the contributions from side chains, which are not resolved at this resolution). The longitudinal section also reveals how two additional densities cap the TM pore on either side. These features are likely to contain contributions from the extracellular MXM sequence and the intracellular C-terminus, respectively. Moreover, on the intracellular side, some density is visible in addition to the density capping the pore. The nature of this additional feature is unresolved at this time. Two cross sections parallel to the membrane (right panel), taken close to the two ends of the pore, reveal the spatial organization of the TM helices within the trimer. Notably, only a single TM segment, believed to be the second TM helix, contributes to the pore at its narrow end. Closer to the extracellular side of the bilayer, all three TM helices line a cavity that based on its size has to be filled with water. see the color plate.

et al., 2009). While individual amino acid side chains were not resolved at this resolution, the structure nevertheless provided much insight into the design of CTRs and the potential mechanism by which these proteins move copper across bilayers. Specifically, the structure revealed that three CTR1 subunits in the trimer create a conduit that spans the entire width of the membrane. Located along the three-fold axis of the trimer, the putative copper pore is funnel shaped, measuring ~8 Å (not considering side chain contributions) at one end and widening to ~21 Å across (not considering side chain contributions), toward the other side of the membrane. Reminiscent of potassium channels, this molecular design creates a large aqueous pore in the middle of the bilayer, shortening the effective distance copper ions would have to travel and lowering the energetic barrier for copper translocation past the middle of the membrane (Doyle et al., 1998). Cross sections taken parallel to the membrane revealed structural signatures for nine helical segments within the trimer, confirming sequence-based predictions that CTR monomers contain three TM helices. Comparing cross sections taken at the narrowest and widest point of the pore showed that only a single helix from each subunit contributed to the narrowest part of the conduit, a second feature that CTR1 shared with traditional ion channels such as KcsA or mechanosensitive channels (Chang et al., 1998; Doyle et al., 1998).

In addition to the shape of the pore and the disposition of each subunit's TM helices along the TM conduit, the hCTR1 reconstruction revealed additional densities that were located close to the entrance and exit to the putative copper-permeable pore. The exact nature of these structural features still needs to be determined. However, the strategic placement of these densities, if correlated with the amino acid sequence, makes it likely that the highly conserved extracellular MXM sequence and the short hCTR1 C-terminus contribute to these structural features.

Using immunogold labeling of double-layered hCTR1 2D crystals, the relative topology of the structure with respect to the sequence was established, revealing that the narrow part of the central conduit is situated at the extracellular end of the putative copper-permeable pore (De Feo et al., 2009). Further clues in the structure, such as the apparent direct connectivity between two of the helices at the extracellular end, and biochemical studies allowed a tentative assignment of the helices in the structure to the helices in the amino acid sequence as well as a delineation of the subunit boundaries within the trimer (Fig. 3). Based on this assignment, the narrow constriction at the extracellular end of the pore is formed by the second TM segment from each of the three subunits. This design places the highly conserved MxxxM sequence on the second TM helix in a context where the sulfur atoms of the methionine could readily form copper coordination sites to support ion passage. This idea was supported by biochemical evidence showing that an M154C mutant of an otherwise cys-less hCTR1 was able to spontaneously form disulfide bridges (De Feo et al., 2009). Further support

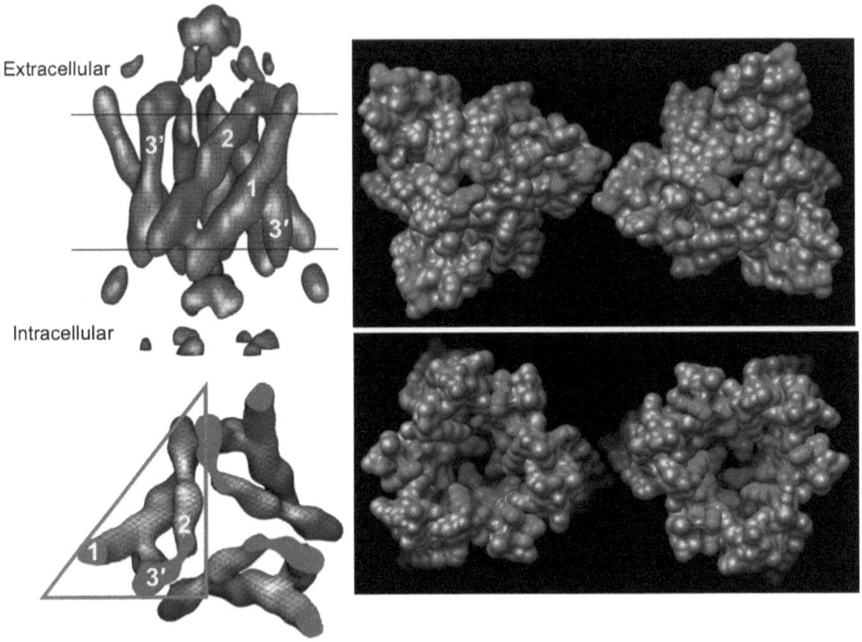

Figure 3 Interpretation of the hCTR1 structure. Immunogold labeling clues in the structure and biochemical studies suggested a putative assignment of the TM helices in the density map to the helices in the hCTR1 amino acid sequence (left panel; adapted from Figure 5.2 in De Feo et al., 2009). The top panel shows the explicit assignment for one of the subunits (1 = TM1, 2 = TM2, 3 = TM3, 3′ = TM3 from a neighboring subunit). This assignment is put into perspective in the bottom panel that outlines the putative subunit boundary of an hCTR1 monomer within the trimer. The assignment is consistent with the results of an extensive tryptophan scan of two evolutionary distant CTR proteins (hCTR1 and yCTR3), which revealed that introduction of steric bulk at the pore entry and along the wall of the aqueous cavity is detrimental to hCTR1 structure and/or function. The tryptophan scan also revealed that interactions between TM1 segments from adjacent trimers appear to be functionally significant. These observations are summarized in the right panels, showing views of a tentative structural model for hCTR1 seen from the extracellular side (top) and the intracellular side (bottom). For clarity, the pore-capping densities were removed. Replacement by tryptophan of amino acids shown in red caused structural and/or functional defects. Replacement of the residues shown in gray yielded hCTR1 variants that phenotypically behaved like wild type (figure adapted from Figure 5.3 in De Feo et al., 2010). The overall picture emerging from these studies strongly supports the hypothesis that copper transport occurs along the central pore within the hCTR1 trimer. See the color plate

for the helix assignment and the idea that the central interface between the three subunits is the active conduit came from an extensive tryptophan scan of two CTR proteins (hCTR1 and yCTR3) (Fig. 3) (Aller et al., 2004;

De Feo, Mootien, & Unger, 2010). These scanning experiments also suggested that the first TM helix functions as an adapter that stabilizes the monomer by holding the second and third TM in place and properly orients their $MX_4MX_{12}GX_4G$ signature sequence.

4. IMPLICATIONS FOR THE MECHANISM OF COPPER TRANSPORT

With the first structural description in place, what have we learned about the mechanisms of copper transport? This question will be discussed in more detail in the following subsections. However, it is noteworthy to point out that absorption spectroscopy and extended X-ray fine absorption spectroscopy (EXAFS) showed trimers of the N-terminally hemagglutanin-tagged N15Q-hCTR1 to bind two ions of Cu(I), each coordinated by three sulfur atoms (De Feo et al., 2009). Interestingly, the average bond length for the sulfur ligands was intermediate between the typical bond lengths of cysteine or methionine ligation, which suggested that both types of residues participate in copper binding. The experimental data could be most readily explained if one of the binding sites was located within the extracellular MXM motif, while the second binding site involved the C-terminal HCH sequence. This model, involving two distinct copper-binding sites, would be fully consistent with the observation that Cu(I)-binding sites facing an oxidizing environment prefer sulfur–ether ligation of copper, while the thermodynamically more stable ligation by sulfhydryls is preferred under reducing conditions of the cytosol (Rubino et al., 2011). In support for this current model are data showing that cysteine to alanine substitution of the penultimate residue in hCTR1 C-terminal HCH sequence results in a change of copper coordination in which three sulfur ligands are replaced by imidazole nitrogen atoms and the bond length of the remaining three sulfur ligands increases to the typical bond length observed for coordination by methionine residues (De Feo et al., 2009). While it cannot be ruled out at this point that these changes in the EXAFS spectra are caused by a global redistribution of copper ions within the trimer, the current data are consistent with the idea that hCTR1 has at least two tri-coordinate binding sites for Cu(I). Because of their likely location at the extracellular entry to the copper-permeable pore and its intracellular exit, it seems plausible that these binding sites regulate copper traffic across the bilayer. What needs to be shown is whether these sites communicate with each other about copper status and whether occupancy of the sites is a mandatory requirement for copper translocation. Equally intriguing, tri-coordinate binding of Cu(I) ions by CTR would mimic the transition state of copper exchange reactions between copper chaperones and their intracellular targets (Banci et al., 2006; Huffman & O'Halloran, 2001). That is, the trimeric nature of CTR1 proteins may not be due to size constraints of the substrate ion, but may have been kinetically selected to allow efficient copper transport at low concentrations and without a requirement for additional sources of energy. Clearly, more experimentation is required to address this critical mechanistic question. However, it becomes abundantly clear that the initial structural information has paved the way for a rational dissection of structure–function relationships in CTR protein and, as well be discussed next, allows many biochemical observations to be put into perspective.

5. COPPER TRANSPORT IN THE PHYSIOLOGICAL MILIEU

Many of the important details of copper transport in vivo remain unknown and prevent a full characterization of the overall process of copper homeostasis in higher organisms. Much of what has been learned about copper uptake has been based on complementation studies in yeast (Dancis et al., 1994; Lee et al., 2000; Zhou & Gitschier, 1997; Zhou & Thiele, 2001) or using tracer uptake studies in cell from higher organisms. The former has the advantage of ease in surveying large numbers of mutants, but lacks the ability to provide quantitative analysis of the transport process. In an initial characterization of tracer copper uptake in cells by hCTR1, it was suggested that extracellular K (and perhaps membrane potential) may influence the rate of copper uptake, and ascorbate provided a small enhancement in the rate of uptake, none of these factors has been systematically investigated (Eisses & Kaplan, 2002; Lee et al., 2002). In the serum, there are many proteins and perhaps small molecules that might serve to deliver copper to the extracellular surface of CTR1, or indeed compete with hCTR1 for any available copper. Although there have been several attempts to identify the copper donor, no clear answer has emerged. In the same way, it is well established that there are no free copper ions inside the cells (Rae et al., 1999). However, the identity of the initial intracellular copper acceptor that receives the metal ion from the transporter as it leaves the pore and enters the cell is similarly unknown. Since it is apparent that copper is transferred from chaperones to their target proteins via specific protein:protein interactions, it is tempting to believe that a similar mechanism may take place with the initial intracellular acceptor(s) and the transporter. Currently experimental proof of this hypothesis is lacking. The basic structural similarities between eukaryotic CTR molecules make it likely that there is a generally conserved mechanism of permeation. However, the structural differences that exist suggest that specific interactions with other cellular components may differ markedly among the members of the CTR family. The intracellular C-terminal tail in the yeast transporters has more than 100 amino acids and contains four or five cysteine residues (as potential copper coordination sites) with which to modulate interactions with intracellular components as well as the copper ion itself (Wu et al., 2009). The mammalian counterparts possess only a short tail of approximately 14 amino acids with the amino-terminal putative metal-binding HCH motif. This motif may be of importance in regulating pore function, but hCTR1 molecules lacking this motif are still able to mediate copper entry (Molloy, S.A., Maryon, E.B. & Kaplan, J.H. submitted).

The apparent lack of free copper in both the intracellular and the extracellular compartments complicates energetic (thermodynamic) considerations of the overall transport process. In the familiar transport of more abundant metal ions (such as Na, K and Ca), it is possible to determine whether or not the transport is passive (requiring no input of energy) or active (against electrochemical potential gradients and thus requiring energy). In the case of copper, the situation is less clear. There is no clear experimental evidence that the process of copper entry requires energy, either from ATP or via a coupled ion gradient. The current lack of an effective in vitro transport assay using recombinant or purified protein in a controlled system prevents clear resolution of this issue. If the activity of copper in physiological solutions is vanishingly small, then perhaps it is best treated as an "insoluble" substrate.

The selectivity of the CTR proteins for Cu(I) ions is striking. hCTR1 has been reported to transport only two other substrates. These are silver ions [which are of similar size and charge to Cu(I)] and the antitumor platinum drugs, typified by cis-Pt. It has been shown that Ag ions cause inhibition of copper uptake (Lee et al., 2002) and recently transport of Ag by hCTR1 was reported (Bertinato et al., 2010). The close similarity in size and charge density probably accounts for this activity. The potential involvement of hCTR1 in the uptake of Pt drugs is more surprising. Studies, predominantly in yeast deletant strains, first demonstrated that cells lacking hCTR1 were more resistant to cell killing by the Pt drug (Ishida et al., 2002). Later work led to the proposal that cis-Pt was transported by hCTR1 and handled in a similar way to copper (Howell et al., 2010). The structural studies, described above, provide a pore size that seems inadequate to accommodate the cis-dichloro,cis-diamino-platinum molecule. The issue is of some importance as the emergence of drug resistance to the Pt drugs severely limits their usefulness clinically, so that an understanding of their processing by cells has gained a higher priority. At the moment, the way in which the Pt anticancer drugs enter cells is not well understood and the way in which hCTR1 may be involved lacks clear resolution.

6. STRUCTURE–FUNCTION RELATIONS IN CTR PROTEINS

A series of studies carried out in yeast, mammalian systems or insect cells have utilized mutagenic analysis along with functional studies of Cu uptake rates or complementation studies to identify which residues may play essential roles in the transport process (Eisses & Kaplan, 2005; Puig et al., 2002). Among the residues that have been identified are the important methionine residues in the second and third TM segments referred to above and a tyrosine residue in the loop connecting the second and third TM segments. However, it appears that there are relatively few residues in the TM pore of the trimeric transporter that interact strongly with the metal ion and are essential for copper transport (Eisses & Kaplan, 2005). However, interactions within the pore, which might affect its structure and hence influence interactions with the permeating copper, do seem important. Mutation of histidine 139, close to the intracellular end of the second TM segment, to an arginine has a striking effect on transport. Both the V_{max} and the K_m for transport are increased substantially. The former by about 10-fold and the latter by about 7-fold (Eisses & Kaplan, 2005). This implies that interactions between the arginine residues in the trimer likely result in distortions in the pore structure or alterations in copper–protein interactions at the exit site that weaken CTR interactions with the copper ion and facilitate a higher rate of transport.

The precise functional roles of the intracellular C-terminal and of the extracellular N-terminal are less clear. Truncation of the first 53 amino acid residues that results in a molecule lacking the first two methionine-rich and histidine-rich sequences results in a substantial reduction in copper transport. Loss of the first 33 residues, which removes the first methionine-rich sequence but retains the second, has a much less significant effect on transport. The amino acid sequence of the amino-terminus is poorly conserved among higher eukaryotes and even mammals, so that it is difficult to identify specific and important residues in this region. However, if circulating donors of copper have to deliver the metal ion to the transporter, it seems most likely that interactions with the amino-terminus are important.

An examination of the functional consequences of two forms of post-translational modification provides contrasting and interesting insights into the transporter. hCTR1 has a single site of N-linked glycosylation at asparagine 15. Substitution of this residue to prevent glycosylation results in the predicted lower mass of the expressed transporter, but no significant effect on the copper transport efficiency (Eisses & Kaplan, 2002; Maryon et al., 2007) or protein trafficking (Klomp et al., 2003). hCTR1 also has O-linked glycosylation, at threonine 27 (Maryon et al., 2007). However, prevention of glycosylation of this residue, by mutagenesis or by using cells engineered to limit protein glycosylation, results in the appearance of a smaller version of hCTR1 at the plasma membrane of mammalian cells, having a mass of about 17 kDa. This protein lacking the first 30 residues of hCTR1 transports copper with a reduction of about 50% in its rate (Maryon et al., 2009). The role of this O-linked glycosylation is to protect the transporter from the actions of an intracellular protease that produces a copper transporter of reduced activity.

The overall picture that emerges on the mechanism of hCTR1-mediated copper transport is the following. Copper ions, carried on circulating molecules, are delivered to the entrance of the pore of hCTR1, presumably via interactions between the amino-terminal tail of hCTR1 and the copper donors. The copper is then steered, perhaps through interactions with the second set of methionine repeats, into the entry to the pore, formed by three hCTR1 monomers, where it interacts with a ring of three methionine 150 residues, is then transferred to the second ring of methionine 154 residues, further down the pore that brings the copper ion closer to the pore exit. The next step, exit from the pore, remains less well characterized. The copper ion that exits the pore finds its way to intracellular copper chaperones. Whether this occurs via direct interaction between the chaperone and the inner aspect of the transporter or via some other mechanism remains to be demonstrated. Currently although endogenous transport activity and the activity of mutant molecules can be readily compared via determinations of the V_{max} and K_m of the transport process, it is not yet possible to identify with confidence just where in the overall pathway the rate-limiting step for transport occurs or where are the key interactions between copper ions and the transporter that govern the substrate dependence (K_m).

REFERENCES

Abajian, C., Yatsunyk, L. A., Ramirez, B. E., & Rosenzweig, A. C. (2004). Yeast cox17 solution structure and copper(I) binding. *The Journal of Biological Chemistry, 279*(51), 53584–53592.

Aller, S. G, Eng, E. T., De Feo, C. J., & Unger, V. M. (2004). Eukaryotic CTR copper uptake transporters require two faces of the third transmembrane domain for helix packing, oligomerization, and function. *The Journal of Biological Chemistry, 279*(51), 53435–53441.

Aller, S. G., & Unger, V. M. (2006). Projection structure of the human copper transporter CTR1 at 6-A resolution reveals a compact trimer with a novel channel-like architecture. *Proceedings of the National Academy of Sciences of the United States of America, 103*(10), 3627–3632.

Banci, L., Bertini, I., Cantini, F., Felli, I. C., Gonnelli, L., Hadjiliadis, N., et al. (2006). The Atx1-Ccc2 complex is a metal-mediated protein-protein interaction. *Nature Chemical Biology, 2*(7), 367–368.

Bernsel, A., Viklund, H., Hennerdal, A., & Elofsson, A. (2009). TOPCONS: consensus prediction of membrane protein topology. *Nucleic Acids Research, 37*(Web Server issue), W465–W468.

Bertinato, J., Cheung, L., Hoque, R., & Plouffe, L. J. (2010). Ctr1 transports silver into mammalian cells. *Journal of Trace Elements in Medicine and Biology, 24*(3), 178–184.

Bremner, I. (1998). Manifestations of copper excess. *American Journal of Clinical Nutrition, 67*(Suppl. 5), 1069S–1073S.

Chang, G., Spencer, R. H., Lee, A. T., Barclay, M. T., & Rees, D. C. (1998). Structure of the MscL homolog from *Mycobacterium tuberculosis*: a gated mechanosensitive ion channel. *Science, 282*(5397), 2220–2226.

Dancis, A., Haile, D., Yuan, D. S., & Klausner, R. D. (1994). The *Saccharomyces cerevisiae* copper transport protein (Ctr1p). Biochemical characterization, regulation by copper, and physiologic role in copper uptake. *The Journal of Biological Chemistry, 269*(41), 25660–25667.

Dancis, A., Yuan, D. S., Haile, D., Askwith, C., Eide, D., Moehle, C., et al. (1994). Molecular characterization of a copper transport protein in *S. cerevisiae*: an unexpected role for copper in iron transport. *Cell, 76*(2), 393–402.

De Feo, C. J., Aller, S. G., Siluvai, G. S., Blackburn, N. J., & Unger, V. M. (2009). Three-dimensional structure of the human copper transporter hCTR1. *Proceedings of the National Academy of Sciences of the United States of America, 106*(11), 4237–4242.

De Feo, C. J., Mootien, S., & Unger, V. M. (2010). Tryptophan scanning analysis of the membrane domain of CTR-copper transporters. *The Journal of Membrane Biology, 234*(2), 113–123.

Djinovic, K., Gatti, G., Coda, A., Antolini, L., Pelosi, G., Desideri, A., et al. (1991). Structure solution and molecular dynamics refinement of the yeast Cu, Zn enzyme superoxide dismutase. *Acta Crystallographica. Section B, 47*(Pt 6), 918–927.

Dmitriev, O., Tsivkovskii, R., Abildgaard, F., Morgan, C. T., Markley, J. L., & Lutsenko, S. (2006). Solution structure of the N-domain of Wilson disease protein: distinct nucleotide-binding environment and effects of disease mutations. *Proceedings of the National Academy of Sciences of the United States of America, 103*(14), 5302–5307.

Doyle, D. A., Morais Cabral, J., Pfuetzner, R. A., Kuo, A., Gulbis, J. M., Cohen, S. L., et al. (1998). The structure of the potassium channel: molecular basis of K+ conduction and selectivity. *Science, 280*(5360), 69–77.

Eisses, J. F., & Kaplan, J. H. (2002). Molecular characterization of hCTR1, the human copper uptake protein. *The Journal of Biological Chemistry, 277*(32), 29162–29171.

Eisses, J. F., & Kaplan, J. H. (2005). The mechanism of copper uptake mediated by human CTR1: a mutational analysis. *The Journal of Biological Chemistry, 280*(44), 37159–37168.
Halliwell, B., & Gutteridge, J. M. (1990). Role of free radicals and catalytic metal ions in human disease: an overview. *Methods in Enzymology, 186*, 1–85.
Howell, S. B., Safaei, R., Larson, C. A., & Sailor, M. J. (2010). Copper transporters and the cellular pharmacology of the platinum-containing cancer drugs. *Molecular Pharmacology, 77*(6), 887–894.
Huffman, D. L., & O'Halloran, T. V. (2001). Function, structure, and mechanism of intracellular copper trafficking proteins. *Annual Review of Biochemistry, 70*, 677–701.
Ishida, S., Lee, J., Thiele, D. J., & Herskowitz, I. (2002). Uptake of the anticancer drug cisplatin mediated by the copper transporter Ctr1 in yeast and mammals. *Proceedings of the National Academy of Sciences of the United States of America, 99*(22), 14298–14302.
Jiang, J., Nadas, I. A., Kim, M. A., & Franz, K. J. (2005). A Mets motif peptide found in copper transport proteins selectively binds Cu(I) with methionine-only coordination. *Inorganic Chemistry, 44*(26), 9787–9794.
Kampfenkel, K., Kushnir, S., Babiychuk, E., Inze, D., & van Montagu, M. (1995). Molecular characterization of a putative *Arabidopsis thaliana* copper transporter and its yeast homologue. *The Journal of Biological Chemistry, 270*(47), 28479–28486.
Kim, B. E., Nevitt, T., & Thiele, D. J. (2008). Mechanisms for copper acquisition, distribution and regulation. *Nature Chemical Biology, 4*(3), 176–185.
Klomp, A. E., Juijn, J. A., van der Gun, L. T., van den Berg, I. E., Berger, R., & Klomp, L. W. (2003). The N-terminus of the human copper transporter 1 (hCTR1) is localized extracellularly, and interacts with itself. *Biochemical Journal, 370*(Pt 3), 881–889.
Lamb, A. L., Torres, A. S., O'Halloran, T. V., & Rosenzweig, A. C. (2000). Heterodimer formation between superoxide dismutase and its copper chaperone. *Biochemistry, 39*(48), 14720–14727.
Lamb, A. L., Torres, A. S., O'Halloran, T. V., & Rosenzweig, A. C. (2001). Heterodimeric structure of superoxide dismutase in complex with its metallochaperone. *Nature Structural Biology, 8*(9), 751–755.
Lee, J., Prohaska, J. R., Dagenais, S. L., Glover, T. W., & Thiele, D. J. (2000). Isolation of a murine copper transporter gene, tissue specific expression and functional complementation of a yeast copper transport mutant. *Gene, 254*(1–2), 87–96.
Lee, J., Pena, M. M., Nose, Y., & Thiele, D. J. (2002). Biochemical characterization of the human copper transporter Ctr1. *The Journal of Biological Chemistry, 277*(6), 4380–4387.
Linder, M. C., & Goode, C. A. (1991). *Biochemistry of copper. Biochemistry of the elements.* New York: Plenum Press. xiv, 525 p.
Maryon, E. B., Molloy, S. A., Zimnicka, A. M., & Kaplan, J. H. (2007). Copper entry into human cells: progress and unanswered questions. *Biometals, 20*(3–4), 355–364.
Maryon, E. B., Zhang, J., Jellison, J. W., & Kaplan, J. H. (2009). Human copper transporter 1 lacking O-linked glycosylation is proteolytically cleaved in a Rab9-positive endosomal compartment. *The Journal of Biological Chemistry, 284*(41), 28104–28114.
Maryon, E. B., Molloy, S. A., & Kaplan, J. H. (2007). O-linked glycosylation at threonine 27 protects the copper transporter hCTR1 from proteolytic cleavage in mammalian cells. *The Journal of Biological Chemistry, 282*(28), 20376–20387.
O'Halloran, T. V., & Culotta, V. C. (2000). Metallochaperones, an intracellular shuttle service for metal ions. *The Journal of Biological Chemistry, 275*(33), 25057–25060.
Puig, S., Lee, J., Lau, M., & Thiele, D. J. (2002). Biochemical and genetic analyses of yeast and human high affinity copper transporters suggest a conserved mechanism for copper uptake. *The Journal of Biological Chemistry, 277*(29), 26021–26030.
Rae, T. D., Schmidt, P. J., Pufahl, R. A., Culotta, V. C., & O'Halloran, T. V. (1999). Undetectable intracellular free copper: the requirement of a copper chaperone for superoxide dismutase. *Science, 284*(5415), 805–808.

Rosenzweig, A. C. (2002). Metallochaperones: bind and deliver. *Chemical Biology, 9*(6), 673–677.

Rosenzweig, A. C., Huffman, D. L., Hou, M. Y., Wernimont, A. K., Pufahl, R. A., & O'Halloran, T. V. (1999). Crystal structure of the Atx1 metallochaperone protein at 1.02 A resolution. *Structure,* 7(6), 605–617.

Rubino, J. T., Chenkin, M. P., Keller, M., Riggs-Gelasco, P., & Franz, K. J. (2011). A comparison of methionine, histidine and cysteine in copper(I)-binding peptides reveals differences relevant to copper uptake by organisms in diverse environments. *Metallomics, 3*(1), 61–73.

Spencer, R. H., & Rees, D. C. (2002). The alpha-helix and the organization and gating of channels. *Annual Review of Biophysics and Biomolecular Structure, 31*, 207–233.

Stearman, R., Yuan, D. S., Yamaguchi-Iwai, Y., Klausner, R. D., & Dancis, A. (1996). A permease-oxidase complex involved in high-affinity iron uptake in yeast. *Science, 271*(5255), 1552–1557.

Tottey, S., Harvie, D. R., & Robinson, N. J. (2005). Understanding how cells allocate metals using metal sensors and metallochaperones. *Accounts of Chemical Research, 38*(10), 775–783.

Winge, D. R. (2002). Copper metalloregulation of gene expression. *Advances in Protein Chemistry, 60*, 51–92.

Wu, X., Sinani, D., Kim, H., & Lee, J. (2009). Copper transport activity of yeast Ctr1 is down-regulated via its C terminus in response to excess copper. *The Journal of Biological Chemistry, 284*(7), 4112–4122.

Xiao, Z., & Wedd, A. G. (2002). A C-terminal domain of the membrane copper pump Ctr1 exchanges copper(I) with the copper chaperone Atx1. *Chemical Communications (Cambridge, England)*(6), 588–589.

Zhou, B., & Gitschier, J. (1997). hCTR1: a human gene for copper uptake identified by complementation in yeast. *Proceedings of the National Academy of Sciences of the United States of America, 94*(14), 7481–7486.

Zhou, H., Cadigan, K. M., & Thiele, D. J. (2003). A copper-regulated transporter required for copper acquisition, pigmentation, and specific stages of development in *Drosophila melanogaster. The Journal of Biological Chemistry, 278*(48), 48210–48218.

Zhou, H., & Thiele, D. J. (2001). Identification of a novel high affinity copper transport complex in the fission yeast *Schizosaccharomyces pombe. The Journal of Biological Chemistry, 276*(23), 20529–20535.

CHAPTER FIVE

Toward a Molecular Understanding of Metal Transport by P_{1B}-Type ATPases

Amy C. Rosenzweig[*,†,1] **and José M. Argüello**[‡]
[*]Department of Molecular Biosciences, Northwestern University, Evanston, IL, USA
[†]Department of Chemistry, Northwestern University, Evanston, IL, USA
[‡]Department of Chemistry and Biochemistry, Worcester Polytechnic Institute, Worcester, MA, USA
[1]Corresponding author: E-mail: amyr@northwestern.edu

Contents

Abstract

The P_{1B} family of P-type ATPases couples the transport of cytoplasmic transition metals across biological membranes to the hydrolysis of ATP. These ubiquitous transporters function in maintaining cytoplasmic metal quotas and in the assembly of metalloproteins, and have been classified into subfamilies (P_{1B-1}–P_{1B-5}) on the basis of their transported substrates (Cu^{+}, Zn^{2+}, Cu^{2+}, and Co^{2+}) and signature sequences in their transmembrane segments. In addition, each subgroup presents a characteristic membrane topology and specific regulatory cytoplasmic metal-binding domains. In recent years, significant major aspects of their transport mechanism have been described, including the stoichiometry of transport and the delivery of substrates to transport sites by metallochaperones. Toward understanding their structure, the metal

Current Topics in Membranes, Volume 69
ISSN 1063-5823,
http://dx.doi.org/10.1016/B978-0-12-394390-3.00005-7

coordination by transport sites has been characterized for Cu^+ and Zn^{2+}-ATPases. In addition, atomic resolution structures have been determined, providing key insight into the elements that enable transition metal transport. Because the Cu^+-transporting ATPases are found in humans and are linked to disease, this subfamily has been the focus of intense study. As a result, significant progress has been made toward understanding Cu^+-ATPase function on the molecular level, using both the human proteins and the bacterial homologs, most notably the CopA proteins from *Archaeoglobus fulgidus*, *Bacillus subtilis*, and *Thermotoga maritima*. This chapter thus focuses on the mechanistic and structural information obtained by studying these latter Cu^+-ATPases, with some consideration of how these aspects might differ for the other subfamilies of P_{1B}-ATPases.

1. INTRODUCTION

1.1. Overall Architecture

The P-type ATPases are a large family of integral membrane proteins that use the energy of ATP hydrolysis to transport cations and lipids across membranes (Bublitz, Morth, & Nissen, 2011; Palmgren & Nissen, 2011). In these enzymes, the hydrolysis of ATP and formation of a covalent phosphoenzyme intermediate are coupled to conformational changes in the transmembrane segments (TM) that allow ion translocation across the membrane (Palmgren & Nissen, 2011). The P-type ATPases are divided into subclasses on the basis of substrate specificity. These subfamilies include the P_{1A}-ATPases (K^+), the P_{1B}-ATPases (heavy metals), the P_2-ATPases (Ca^{2+}, Na^+/K^+, and H^+/K^+), the P_3-ATPases (H^+), the P_4-ATPases (phospholipids), and the P_5-ATPases (substrate unknown) (Lutsenko & Kaplan, 1995; Palmgren & Nissen, 2011). The P_2-ATPases include the well-characterized sarcoplasmic reticulum Ca^{2+}-ATPase (SERCA) and the Na^+/K^+-ATPase, both of which have been structurally characterized (Morth et al., 2007; Toyoshima, 2008). The P_3-ATPases are plasma membrane H^+ pumps and structurally resemble the P_2-ATPases (Pedersen, Buch-Pedersen, Morth, Palmgren, & Nissen, 2007).

The P_{1B}-ATPases, which transport heavy metal ions, are found in all kingdoms of life and are the prevalent P-type ATPases in bacteria and archaea (Palmgren & Nissen, 2011). While they share structural elements with P-type ATPases, there are several unique aspects to the architecture of the P_{1B}-ATPases. All P-type ATPases include two soluble cytoplasmic domains, the ATP-binding and actuator domains (ATPBD and A domain). The ATPBD comprises two distinct domains, the nucleotide-binding or N domain and the phosphorylation or P domain. All P-type ATPases also have a central core of six TM helices (denoted H1-H6) (Fig. 1). In most, but not all, P_{1B}-ATPases (Section 1.2), two additional TM helices are inserted at the N-terminal side of the core domain (denoted MA and MB) (Argüello, Eren, & González-Guerrero, 2007; Palmgren & Nissen, 2011). A second defining feature of P_{1B}-ATPases is the presence of various soluble MBDs at the N- and C-termini (Argüello et al., 2007; Argüello, González-Guerrero, & Raimunda, 2011).

1.2. Substrate Diversity

The P_{1B}-ATPases transport a wide array of substrates, including $Zn^{2+}/Cd^{2+}/Pb^{2+}$(Rensing, Sun, Mitra, & Rosen, 1998; Sharma, Rensing, Rosen, & Mitra, 2000), Cu^{2+}(Mana-Capelli, Mandal, & Argüello, 2003), Cu^{+}/Ag^{+}(Mandal, Cheung, & Argüello, 2002), and Co^{2+}(Rutherford, Cavet, & Robinson, 1999; Scherer & Nies, 2009). Residues in the last three TM helices (H4-H6) are proposed to form the metal-binding site(s) (MBS) and specificity is believed to derive from the identities and positions of residues within these helices (Argüello, 2003) (Fig. 1). In particular, mutagenesis data indicate that a three-residue, cysteine-containing sequence motif in H4 [CPC, CPH, SPC, (T/S) PCP, YPC, and APC] is important for metal transport activity (Argüello et al., 2007; Dutta, Liu, Hou, & Mitra, 2006; Fan & Rosen, 2002; Forbes & Cox, 1998; Lowe et al., 2004). These sequence motifs and limited experimental data

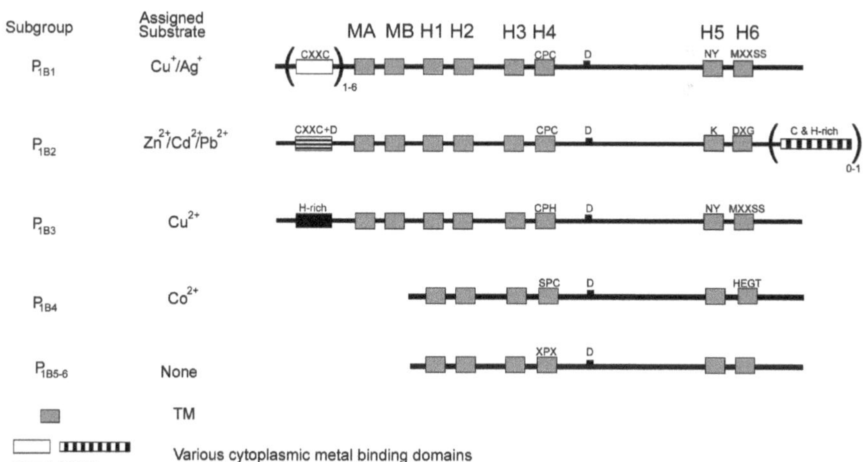

Figure 1 Structural features of P_{1B}-ATPase subgroups. Boxes represent TM and cytoplasmic metal-binding domains (MBDs). The aspartic acid residue phosphorylated during catalysis and the identified signature sequences involved in metal coordination are indicated.

linking various P_{1B}-ATPases to specific metal ions have been used to develop a subclassification scheme in which five families are designated P_{1B-1} through P_{1B-5} (Fig. 1) (Argüello, 2003; Argüello et al., 2007). The P_{1B-1}, P_{1B-2}, and P_{1B-3} families include the additional TM helices MA and MB at the N-terminus. The P_{1B-1}-ATPases are specific for Cu^{+} and Ag^{+}, P_{1B-2}-ATPases are $Zn^{2+}/Cd^{2+}/Pb^{2+}$transporters, P_{1B-3}-ATPases are Cu^{2+}-ATPases, and P_{1B-4}-ATPases are believed to transport primarily Co^{2+} and perhaps Zn^{2+} and Cd^{2+}. The P_{1B-5}-ATPase subgroup contains the (T/S)PCP sequence in H4 and QEXXD in H6, but their substrates have not been determined (Argüello, 2003; Traverso et al., 2010). Lastly, a small number of P_{1B}-ATPases with unknown substrates contain different signature sequences in their fourth TM (YPC, APC, etc.) and do not fit this classification (Argüello, 2003; Völlmecke, Lorenz Drees, Reimann, Albers, & Lübben, 2012).

The subfamilies are further differentiated on the basis of the type of MBDs present. Most bacterial P_{1B-1}-ATPases and P_{1B-2}-ATPases contain one or more MBDs (Fig. 1) (Argüello, 2003; Solioz & Vulpe, 1996). These resemble the Atx1-like family of Cu^{+} chaperones, small soluble proteins that bind Cu^{+} and can transfer it directly to partner P_{1B-1}-ATPases (Boal & Rosenzweig, 2009; Robinson & Winge, 2010; Rosenzweig, 2001) (Section 2.4). Eukaryote P_{1B-1}-ATPases contain up to six repeating N-terminal MBDs (Argüello, 2003; Solioz & Vulpe, 1996). A number of these MBDs have been characterized in vitro and bind a single Cu^{+} ion via a conserved CXXC motif (Boal & Rosenzweig, 2009; Singleton & LeBrun, 2007) (Section 3.1). The Atx1-like MBDs in the P_{1B-2}-ATPases contain, in addition to the two cysteines in the CXXC motif, conserved carboxylate residues that provide additional ligands for Zn^{2+} and Cd^{2+} (Banci et al., 2002; Eren, González-Guerrero, Kaufman, & Argüello, 2007). Some P_{1B-2}-ATPases, notably those in plants, contain other types of histidine- and cysteine-rich MBDs in their C-termini (Argüello, 2003; Eren et al., 2007; Eren, Kennedy, Maroney, & Argüello, 2006). The Cu^{2+}-specific P_{1B-3}-ATPases have a histidine-rich N-terminal MBD, which is necessary for maximal Cu^{2+} transport activity (Mana-Capelli et al., 2003). The two subfamilies that lack helices MA and MB also lack N-terminal MBDs. The P_{1B-4}-ATPases have no MBDs, but a number of P_{1B-5}-ATPases contain a novel C-terminal MBD, classified as a hemerythrin (Hr)-like domain and demonstrated to bind a diiron center (Traverso et al., 2010).

1.3. Biological Functions

The P_{1B}-ATPases have two important roles in both prokaryotes and eukaryotes (Lutsenko, Gupta, Burkhead, & Zuzel, 2008; Raimunda,

González-Guerrero, Leeber, & Argüello, 2011). First, they efflux cytoplasmic metal ions, protecting cells from the deleterious effects of elevated heavy metal concentrations. Second, they provide metal cofactors for biosynthesis of metalloproteins. In prokaryotes, there are many examples of systems in which genetic disruption of P_{1B}-ATPases leads to toxic metal accumulation, including *Escherichia coli* ZntA (Rensing, Mitra, & Rosen, 1997), *E. coli* CopA (Rensing, Fan, Sharma, Mitra, & Rosen, 2000), *Synechocystis* sp. PCC6803 PacS (Tottey, Rich, Rondet, & Robinson, 2001), *Enterococcus hirae* CopB (Odermatt, Suter, Krapf, & Solioz, 1993), and *Synechocystis* sp. PCC6803 CoaT (Rutherford et al., 1999). Prokaryotic P_{1B-1}-ATPases also play an important role in protein maturation, extruding Cu^+ ions to the periplasm for insertion into metalloenzymes. This dual functionality is consistent with the presence in many genomes of multiple genes encoding P_{1B-1}-ATPases (Raimunda et al., 2011). In *Pseudomonas aeruginosa*, the CopA1 Cu^+-ATPase provides copper tolerance, whereas disruption of a second Cu^+-ATPase, CopA2, results in reduced cytochrome oxidase activity and sensitivity to oxidative stress, suggesting a role in copper delivery to periplasmic chaperone proteins (González-Guerrero, Raimunda, Cheng, & Argüello, 2010). In the photosynthetic cyanobacteria *Synechocystis* sp. PCC6803, two P_{1B-1}-ATPases are present, CtaA in the plasma membrane and PacS in the thylakoid membrane, and are suggested to be involved in copper tolerance and copper delivery to *caa*$_3$ oxidase/plastocyanin, respectively (Tottey, Harvie, & Robinson, 2005).

The metal transport activity of P_{1B}-ATPases is also required for bacterial infection. Phagosomal copper and zinc overload has emerged as a potential bactericidal mechanism (Botella et al., 2011; Wagner et al., 2005; White, Lee, Kambe, Fritsche, & Petris, 2009) and bacterial Cu^+-ATPases responsible for maintaining low cytoplasmic Cu^+ levels appear essential for survival in the host (González-Guerrero et al., 2010; Schwan, Warrener, Keunz, Stover, & Folger, 2005; White et al., 2009; Zhang & Rainey, 2007). The Cu^+-ATPases associated with cytochrome *c* oxidase are also necessary for infection, apparently to cope with the phagosomal oxidative burst (González-Guerrero et al., 2010). It is likely that the observed induction of various P_{1B}-ATPases during infection might be linked to survival in the phagosomal noxious environment (Botella et al., 2011; Sassetti & Rubin, 2003).

The eukaryotic P_{1B}-ATPases are found in the trans-Golgi network, plasma membrane, vacuole, and chloroplast. In *Arabidopsis thaliana*, there are eight P_{1B}-ATPases (Williams & Mills, 2005). These include P_{1B-1}-, P_{1B-2}-

and P_{1B-4}-ATPases involved in both organismal metal distribution and metalloenzyme assembly (Williams & Mills, 2005). For instance, the participation of the Cu^{+}-ATPases HMA6, HMA7 in the assembly of plastocyanin and chloroplast Cu/Zn-SOD has been shown, as well as the role of HMA8 in the assembly of ethylene receptor (Abdel-Ghany, Muller-Moule, Niyogi, Pilon, & Shikanai, 2005; Hirayama et al., 1999). The function of the Zn^{2+}-ATPases HMA2 and HMA4 in controlling Zn^{2+} organismal distribution has also been described (Eren & Argüello, 2004; Mills, Krijger, Baccarini, Hall, & Williams, 2003). Fungi and animals contain primarily P_{1B-1}-ATPases. Among the earliest characterized P_{1B-1}-ATPases was the *Saccharomyces cerevisiae* Ccc2 ATPase that delivers Cu^{+} to the multicopper oxidase Fet3 (Yuan et al., 1995). In humans, there are two Cu^{+}-transporting P_{1B-1}-ATPases, ATP7A and ATP7B (Bull & Cox, 1994). ATP7A is primarily expressed in the intestine and the choroid plexus; ATP7B is primarily expressed in the liver and the brain (Lutsenko et al., 2008). These proteins cycle between the trans-Golgi network, endocytic vesicles, and the plasma membrane as a function of copper concentrations, and function both in Cu^{+} export and in biosynthesis of metalloenzymes, including tyrosinase, lysyl oxidase, peptidylglycine α−monooxygenase, and ceruloplasmin (Lutsenko, Barnes, Bartee, & Dmitriev, 2007; Lutsenko et al., 2008). In addition, the trafficking of ATP7A to the phagosome membrane during infection contributes to phagosomal copper accumulation and resultant bacterial killing (White et al., 2009).

Mutations in ATP7A and ATP7B lead to Menkes syndrome and Wilson disease, respectively (Bull & Cox, 1994). Menkes syndrome is caused by decreased Cu^{+} uptake across the small intestine, compounded by the lost ability to transfer copper from the cell cytosol into the lumen of the secretory pathway, as well as impaired copper distribution to a variety of tissues, including the brain (Kaler, 1998). The resultant deficiencies in copper-containing enzymes are manifested as severe impairment of neurological and connective tissue function (Mercer, 2001). By contrast, Wilson disease is characterized primarily by toxic copper overload in the liver due to reduced biliary excretion. The presence of yellow "Kayser–Fleischer" rings around the cornea of the eye, which signifies copper buildup in the brain, is sometimes the first symptom of Wilson disease (Das & Ray, 2006; Gitlin, 2003). Notably, increased expression of the Menkes and Wilson ATPases is also associated with resistance to the anticancer drug cisplatin (Katano et al., 2002; Safaei, Holzer, Katano, Samimi, & Howell, 2004).

2. MECHANISM

2.1. Catalytic Cycle

The Cu^+-ATPases follow the general steps of the classical Post–Albers cycle that describes the transport and catalysis of P-type ATPases with some differences associated with the chemistry and biological availability of their substrates (Argüello et al., 2011; Post, Hegyvary, & Kume, 1972) (Fig. 2). In this mechanism, enzymes assume two major conformational states, E1 and E2, presenting differential affinities for the transported substrates (Hatori, Majima, Tsuda, & Toyoshima, 2007; Mandal, Yang, Kertesz, & Argüello, 2004). In its E1 form, the enzyme binds nucleotides at the ATPBD and the metal substrate at the transmembrane transport sites from the cytoplasmic side with femtomolar affinity (González-Guerrero & Argüello, 2008). Although both nucleotide and metal can independently bind to the ATPase, the essential coupling of solute transport with ATP hydrolysis requires simultaneous ATP and metal binding to drive catalytic phosphorylation of the aspartic acid residue in the invariant DKTGT sequence and the subsequent enzyme turnover. Moreover, in the case of Cu^+-ATPases, which transport two Cu^+ ions per ATP, full occupancy of the transport sites is required for catalysis (González-Guerrero, Eren, Rawat, Stemmler, & Argüello, 2008; Mandal et al., 2004). Phosphorylation causes conformational changes leading to metal

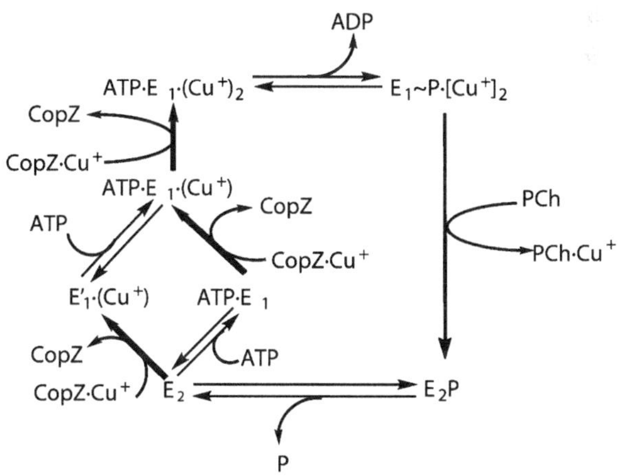

Figure 2 Catalytic and transport cycle of Cu^+-ATPases. E1 and E2 represent the two main enzyme conformations. CopZ is the bacterial cytoplasmic Cu^+ chaperone. PCh indicates a hypothetical periplasmic Cu^+ chaperone/acceptor. The thicker arrows represent largely irreversible partial reactions.

coordination rearrangements that force the release of metal into a luminal/periplasmic compartment as the enzyme transitions to a phosphorylated E2 state. The metal release appears to be a rate-limiting step in the catalytic cycle. This phenomenon explains the various transport rates observed with alternative substrates when the activity is measured in vitro (Mandal et al., 2002). It is significant that P_{1B}-ATPases turnover at relatively slow rates $<10/s$ (Hatori et al., 2007; Mandal et al., 2002; Sharma et al., 2000). These rates are much lower than those observed for P_2- and P_3-ATPases and highlight the differences between transition metal and alkali metal transport. Unlike Na^+, K^+, or Ca^{2+}, transition metal ions are not free/hydrated in the cell. Thus, although the free metal is transported, it binds the transport sites via a mechanism of metal exchange involving chaperone proteins and other sequestering molecules (see Section 2.4).

2.2. Transmembrane Metal-Binding Sites

The presence of distinct transmembrane metal-binding sites has been established for Cu^+- and Zn^{2+}-ATPases. A combination of metal-binding and mutagenesis experiments on *Archaeoglobus fulgidus* CopA clearly showed the presence of two TM sites with femtomolar affinity for Cu^+ (González-Guerrero & Argüello, 2008). This high affinity prevents release of the Cu^+ ions back into the cytoplasm prior to transfer. According to extended X-ray absorption fine structure (EXAFS) data, both sites exhibit trigonal planar coordination geometry with a mixture of sulfur and oxygen/nitrogen ligation. The first site involves Cys 380 and Cys 382 from the conserved CPC motif in H6 and Tyr 682 from H7 (González-Guerrero et al., 2008; Mandal et al., 2004). The second site comprises Asn 683 from H7 and residues Met 711 and Ser 715 from H8 (Fig. 3C). All these residues are required for Cu^+ binding and translocation. The high degree of conservation of these amino acids suggests that the binding of two Cu^+ ions and consequent transport stoichiometry is a general phenomenon in all Cu^+-ATPases. It should be noted, however, that the metal-binding stoichiometry may vary across the P_{1B}-ATPase subfamilies. For example, the *E. coli* Zn^{2+} P_{1B-2}-ATPase ZntA is reported to have a single TM Zn^{2+}-binding site (Liu, Dutta, Stemmler, & Mitra, 2006) in which the ion is coordinated in a tetragonal geometry by Cys 392 and Cys 394 in H4, an Asp 714 in H8 and likely Lys 693 in H5 (Dutta et al., 2006; Dutta, Liu, Stemmler, & Mitra, 2007; Okkeri & Haltia, 2006; Raimunda, Subramanian, Stemmler, & Argüello, 2012).

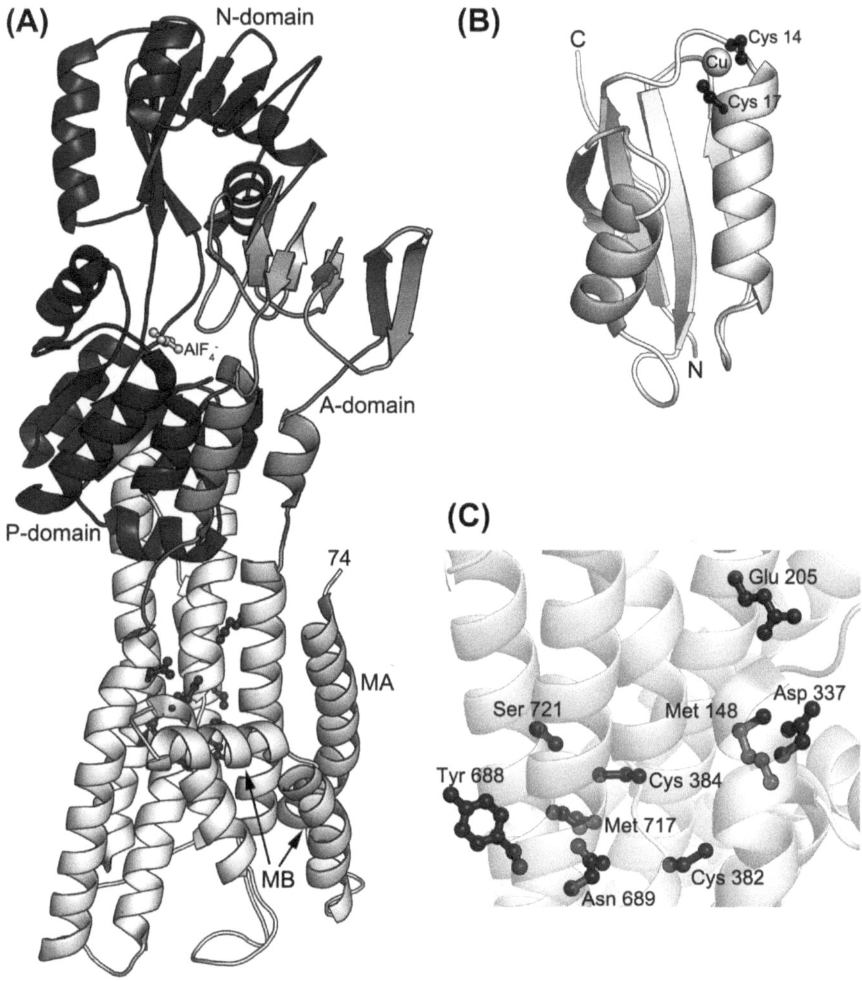

Figure 3 Structures of Cu^+-ATPases. (A) Crystal structure of *L. pneumophila* CopA (PDB accession code 3RFU). The TM regions are shown in white, the A domain is shown is gray, and the ATPBD is shown in black with the N and P domains labeled. The N-terminal MA and MB helices are labeled and platform and metal-binding residues, shown in more detail in panel C, are represented as ball and sticks. The model begins with residue 74; the N-terminal MBD is not visible in the electron density. (B) Solution structure of the N-terminal MBD of *B. subtilis* CopA with bound Cu^+ (PDB accession code 1KQK). (C) Close-up view of the platform and proposed metal-binding residues in *L. pneumophila* CopA (PDB accession code 3RFU).

2.3. Direction of Transport

The structural coupling of cytoplasmic substrate efflux to ATP hydrolysis and catalytic phosphorylation has been demonstrated in the P_2-ATPases by studies characterizing ion transport, ATP hydrolysis and various partial reactions (Palmgren & Nissen, 2011; Toyoshima & Inesi, 2004). Crystal structures of the SERCA Ca^{2+}-ATPase in various conformations have provided further support for this paradigm (Palmgren & Nissen, 2011; Toyoshima, 2008). As counterpart to this, when a second substrate is counter-transported (Na^+, K^+-ATPase, H^+, and K^+-ATPase), the inward transported ions bind the enzyme in an E2 form and consequently cannot drive catalytic phosphorylation. For the P_{1B}-ATPases, the efflux of cytoplasmic metals has been shown for Cu^+-, Zn^{2+}-, Cu^{2+}, and Co^{2+}-ATPases (Fan & Rosen, 2002; Mana-Capelli et al., 2003; Rensing et al., 1998; Scherer & Nies, 2009; Voskoboinik et al., 1999). In these cases, the direct measurement of transport was achieved by monitoring radioisotope uptake into everted vesicles obtained from systems heterologously expressing the target ATPases. However, phenotypes resulting from mutation of specific Cu^+-ATPases are not always consistent with the mechanism of transport described above. Whereas mutation of most bacterial Cu^+-ATPases leads to cytoplasmic copper overload, no changes or even a small decrease in the cytoplasmic copper content has been observed in some cases (González-Guerrero et al., 2010; Odermatt et al., 1993; Tottey et al., 2001). However, kinetic characterization combined with functional analysis indicates that Cu^+-ATPases involved in cuproprotein assembly do transport cytoplasmic Cu^+ to extracellular milieus, but at as low rate (González-Guerrero et al., 2010; Raimunda et al., 2011). This property, while compatible with their functional role, is not suitable to maintain cytoplasmic copper quotas, a task performed by homologous ATPases present in the genome.

2.4. Role of Metallochaperones

A unique feature of the Cu^+-ATPases is that, in the majority of the currently characterized systems, the Cu^+ substrate is delivered to the TM sites by soluble Cu^+ chaperone proteins. These ~70 residue proteins exhibit a conserved ferredoxin-like βαββαβ fold and bind Cu^+ ions via a conserved CXXC motif in a linear two-coordinate fashion (Boal & Rosenzweig, 2009; Robinson & Winge, 2010). The prototypical Cu^+ chaperone is *S. cerevisiae* Atx1. Disruption of Atx1 interferes with Cu^+ delivery from the Cu^+-ATPase Ccc2 to the multicopper oxidase Fet3, which is involved in iron transport (Lin, Pufahl, Dancis, O'Halloran, & Culotta, 1997). Atx1 and numerous homologs from humans (called Atox1), bacteria, and archaea (called CopZ) have been biochemically and structurally characterized (Robinson and Winge, 2010).

Observations that these chaperones are able to exchange Cu^+ with the N-terminal MBDs of their partner Cu^+-ATPases (Hamza, Schaefer, Klomp, & Gitlin, 1999; Walker et al., 2004; Walker, Tsivkovskii, & Lutsenko, 2002) initially led to a proposed mechanism in which substrate is routed from chaperone to MBDs and then from MBDs to the TM sites (Achila et al., 2006; Huffman & O'Halloran, 2000; O'Halloran & Culotta, 2000; Pufahl et al., 1997). In support of this pathway, the second MBD of ATP7B appears necessary for copper-dependent phosphorylation with Atox1 providing Cu^+ (Walker et al., 2004). However, in bacterial systems, removal of the MBDs or mutation of their metal-binding motifs does not affect copper-dependent ATPase activity (Fan & Rosen, 2002; Mandal & Argüello, 2003) and in the case of *A. fulgidus* CopA does not prevent activation by the copper-loaded chaperone CopZ (González-Guerrero & Argüello, 2008). Instead, the TM sites can be directly loaded by the chaperone and the MBDs cannot substitute for the chaperone in this role.

Metal transfer from CopZ to the TM-MBS in the ATPase is key to Cu^+ homeostasis because it allows the efflux of cytoplasmic Cu^+ pools while preventing the release of free ion into the cytoplasmic milieu. Similarly, this represents a departure from the well-characterized reversible interaction of alkali and alkali earth metals with their transporters (Argüello, Raimunda, & González-Guerrero, 2012). While the apo form of the chaperone apparently does not interact with the protein, the Cu^+-loaded chaperone probably docks with the "platform" region close to the TM site (Argüello et al., 2011; Gourdon et al., 2011) (Fig. 3A) (Section 3.2). The chaperone sequentially delivers Cu^+ to either TM site, and mutation of one site does not affect delivery to the other (González-Guerrero & Argüello, 2008). Once the Cu^+ is bound to a TM site, it cannot be exchanged with the apo chaperone. This, together with the high affinity of TM sites, leads to a practically irreversible Cu^+ delivery to TM sites (Fig. 2). Interestingly, once the first TM site is occupied with Cu^+, nucleotide binding is required for loading of the second TM site (González-Guerrero & Argüello, 2008).

2.5. Regulation

Although the MBDs are not required for activity of the bacterial P_{1B}-ATPases, they do affect the turnover rates (Fan & Rosen, 2002; Mana-Capelli et al., 2003; Mandal & Argüello, 2003; Mitra & Sharma, 2001). It is apparent that metal binding to the MBDs precludes their interaction with the ATPBD, allowing faster enzyme turnover (González-Guerrero, Hong, & Argüello, 2009; Hatori et al., 2007). The similar affinities of the MBDs and their cognate chaperones for Cu^+ are also consistent with a regulatory role (Huffman & O'Halloran, 2000; Walker et al., 2002; Wernimont, Yatsunyk, & Rosenzweig, 2004). Through these mechanisms, the MBDs can provide a direct readout of Cu^+ concentrations in the cell to the transporter.

In addition, the overall P_{1B}-ATPase cellular activity is controlled by a number of metal-sensing transcriptional regulators. Members of the MerR family of Cu^+ sensors (CueR like) are activators that respond to cellular Cu^+ excess and elicit Cu^+-ATPase expression (Ma, Jacobsen, & Giedroc, 2009; Osman & Cavet, 2008). CopY is a copper-responsive repressor present in *Lactobacillus lactis, Enterococcus hirae* and other gram-positive bacteria (Solioz, Abicht, Mermod, & Mancini, 2010). At low Cu^+, CopY remains bound to the promoter regions preventing transcription of Cu^+-ATPases. Finally, the recently described *Mycobacterium tuberculosis* CsoR is the founding member of a large family of Cu^+-sensing transcriptional repressors (Ma et al., 2009; Osman & Cavet, 2008).

The human Cu^+-ATPases are also regulated by Cu^+ binding to MBDs. Removal of the first five MBDs of ATP7B affects the rates of phosphorylation, and binding of Cu^+ to the MBDs prevents interaction with the ATPBD (Huster & Lutsenko, 2003; Tsivkovskii, MacArthur, & Lutsenko, 2001). Similar control of ATP7A Cu^+ transport has been described (Voskoboinik, Mar, Strausak, & Camakaris, 2001). In mammals, an additional layer of regulation derives from copper-dependent trafficking of the Cu^+-ATPases (Lutsenko et al., 2007). At basal copper concentrations, ATP7A and ATP7B are localized to the trans-Golgi network. At increased copper concentrations, both relocalize to the membrane for Cu^+ export, ATP7A to the basolateral membrane and ATP7B to the canalicular membrane to export Cu^+ into the bile.

3. STRUCTURE

3.1. Soluble Domains

Initial efforts to structurally characterize Cu^+-ATPases focused on the soluble domains, which are readily cloned, expressed, and purified in quantities sufficient for structure determination by X-ray crystallography or nuclear magnetic resonance (NMR). Numerous NMR structures of the N-MBDs have been reported, including single MBDs of yeast Ccc2, human ATP7A (all six), *Bacillus subtilis* CopA, and *Synechocystis* sp. PCC6803 PacS (Boal & Rosenzweig, 2009) (Fig. 3B). These structures have been determined both with Cu^+ bound and in the apo forms. The crystal structure of the apo C-terminal MBD from *A. fulgidus* CopA has also been reported (Agarwal et al., 2010). All MBDs exhibit the classic ferredoxin-like βαββαβ fold and bind Cu^+ with the conserved CXXC motif in a two-coordinate linear or approximately linear fashion (Fig. 3B). Structural alterations upon metal binding are typically confined to coordinating residues or residues adjacent to the metal-binding site. Other common features include stabilization of the metal-binding loop by hydrogen bonding interactions and hydrophobic interactions with residues in the protein core (Boal & Rosenzweig, 2009). Finally, analysis of the electrostatic surfaces of these domains has provided some insight into protein–protein interactions with Cu^+ chaperones (Banci, Bertini, McGreevy, et al., 2010; Huffman & O'Halloran, 2001).

Given that yeast and many bacterial Cu^+-ATPases have two sequential N-terminal MBDs and ATP7A and ATP7B have six N-terminal MBDs, it is also important to determine how these MBDs interact with one another on the molecular level and whether Cu^+ affects this interaction. Circular dichroism and EXAFS spectroscopic data suggest that Cu^+ binding affects the structure of the entire N-terminal soluble domain of ATP7B (DiDonato, Hsu, Narindrasorasak, Que, & Sarkar, 2000; Ralle, Lutsenko, & Blackburn, 2004). Structural studies of proteins containing multiple MBDs have been hampered by the presence of flexible linker sequences, but several NMR structures have been reported. In the structure of MBDs 5 and 6 from ATP7B, the two domains are oriented with the CXXC motifs at opposite sides of the molecule and the eight-residue linker in between the two domains (Achila et al., 2006). MBDs 3 and 4 of ATP7B have also been characterized. In this case, the two domains are connected by a 31-residue linker, which is disordered (Banci, Bertini, Cantini, Rosenzweig, & Yatsunyk, 2008). Interestingly, in all these structures of multiple MBDs, including a partial characterization of ATP7A domains 4–6, minimal perturbations are observed upon Cu^+ binding. Similar results were obtained from an NMR study of the two MBDs from *B. subtilis* CopA (Banci, Bertini, Ciofi-Baffoni, Gonnelli, & Su, 2003; Singleton et al., 2008).

Protein–protein interactions between the MBDs and the Cu^+ chaperone proteins have also been probed by NMR. These studies have focused on whether all the MBDs or just selected MBDs can interact with the chaperones and which residues specifically facilitate docking and recognition (Banci, Bertini, McGreevy, et al., 2010; Boal & Rosenzweig, 2009). Most of these studies involve chemical shift mapping, but NMR structures of complexes between *S. cerevisiae* Atx1 and MBD 1 of Ccc2 (Banci et al., 2006) and between Atox1 and domain 1 of ATP7A (Banci, Bertini, Calderone, et al., 2009) have also been determined. In these structures, a Cu^+ ion is primarily coordinated by two cysteine residues from the CXXC motif in the MBD and one cysteine from the chaperone CXXC motif. A nonphysiological Cd^{2+}-bridged complex between Atox1 and MBD 1 of ATP7A is also available. The complexation surfaces in these complexes include complementary electrostatic as well as hydrophobic interactions (Banci, Bertini, McGreevy, et al., 2010). Although these structures have provided insight into metal coordination and protein–protein recognition, limited conclusions can be drawn in the absence of the additional Cu^+-ATPase domains.

The other cytoplasmic domains of the Cu^+-ATPases, the ATPBD and the A domain, have also been characterized separately. Located between H4 and H5, the ATPBD comprises two subdomains, the N and P domains. The P domain contains an invariant DKTGT motif, of which the aspartic acid residue is phosphorylated during the catalytic cycle. The crystal structure of the ATPBD from *A. fulgidus* CopA reveals that the two domains are linked by a hinge region with the nucleotide-binding site in a cleft between domains (Sazinsky, Mandal, Argüello, & Rosenzweig, 2006). The overall fold of the P domain is a six-stranded β sheet flanked by α helices, and it structurally and sequentially resembles P domains from the other branches of the P-type ATPase family. The N domain also forms a mixed αβ structure, similar to those of other P-type ATPases, but its sequence is only conserved among members of the P_{1B}-ATPase family. Although bound nucleotide is not present in the *A. fulgidus* ATPBD structure, the two domains exhibit a closed conformation. However, the open (apo) and closed (nucleotide bound) forms of the ATPBD from *Thermotoga maritima* have also been crystallographically characterized, as well as the His462Gln variant (Tsuda & Toyoshima, 2009), which serves as a model for a frequent disease-causing mutation in ATP7B. These structures reveal that the N domain recognizes the adenine ring of ATP in a different fashion from that in other classes of P-type ATPases. The structure of the ATPBD from *Sulfolobus solfataricus* is also available and houses a sulfate ion at the phosphorylation site (Lübben et al., 2007). Finally, NMR structures of the N domains of both ATP7B and

ATP7A have been reported (Banci, Bertini, Cantini, et al., 2010; Dmitriev et al., 2006). The overall structures of the ATPBDs from prokaryotes and eukaryotes are quite similar, with the exception of some additional loops and structural elements present in the ATP7A and ATP7B N domain structures.

The A domain, located between H2 and H3, has also been structurally characterized. A conserved GE sequence in this domain required for dephosphorylation (Petris et al., 2002) likely interacts with the DKTGT motif in the P domain. An initial crystal structure of the *A. fulgidus* CopA A domain revealed 10 β strands with helices at the N- and C-termini leading into the membrane domains (Sazinsky, Agarwal, Argüello, & Rosenzweig, 2006). With the exception of several loop regions, the fold is very similar to that of the SERCA1 A domain. In SERCA1, the A domain rotates extensively during catalysis (Toyoshima & Inesi, 2004), and proteolytic mapping studies suggest similar conformational changes that occur in *T. maritima* CopA (Hatori et al., 2007). The solution structures of the A domains from ATP7A and ATP7B have also been reported and exhibit a similar overall fold to that of *A. fulgidus* CopA with an insertion between the first two β strands that are suggested to interact with the N-terminal MBDs (Banci, Bertini, Cantini, et al., 2009). The NMR data indicate that the GE loop is relatively rigid, suggesting that major rearrangements do not occur during the catalytic step.

3.2. Intact CopA

Although soluble domain structures have provided much insight into P_{1B}-ATPase function, a complete understanding requires structural characterization of a whole enzyme, including the TM regions. In the absence of suitable three-dimensional crystals of any P_{1B}-ATPase, the structures of intact Cu^{+}-ATPases were pursued by cryoelectron microscopy. Structures of *A. fulgidus* CopA lacking the C-terminal MBD and lacking both the N- and C-terminal MBDs were determined to ~17 Å resolution, and domains were assigned by docking the soluble domain crystal structures into the electron density (Wu, Rice, & Stokes, 2008). Comparison of the two resultant structures suggested several areas of conformational flexibility and, most importantly, identified a region between the ATPBD N domain and the A domain that might correspond to the N-terminal MBD. This location is consistent with cross-linking and proteolytic digestion studies (Hatori et al., 2007; Lübben et al., 2009). A 7Å projection map obtained using two-dimensional crystals of CtrA3 Cu^{2+}-ATPase from *Aquifex aeolicus* has also been reported and shows the arrangement of the eight TM helices (Chintalapati, Al Kurdi, van Scheltinga, & Kuhlbrandt, 2008). Both of these studies indicate a dimeric oligomerization state.

The first crystal structure of P_{1B}-ATPase, that of the Cu^{+}-ATPase CopA from *Legionella pneumophila*, was recently determined to 3.2 Å resolution (Gourdon et al., 2011). This structure has provided a unique scaffold to model other P_{1B}-ATPases and onto which Wilson and Menkes disease mutations can be mapped. In the structure, the six core TM helices are organized similarly to other P-type ATPases, but the N-terminal MA and MB helices have an unusual arrangement (Fig. 3A). The MA helix interacts with H2 and H6 and the MB helix interacts with H1 and H2. MB is divided into two segments, a short TM helix and an amphipathic helix at the membrane interface, linked by a kinked region. The structure was determined in the presence of AlF_4^- and is thus locked in a metal-free E2 conformation. Importantly, despite the absence of bound Cu^{+}, the structure reveals the locations of the ligands proposed on the basis of mutagenesis and spectroscopic studies of *A. fulgidus* CopA (González-Guerrero et al., 2008; Mandal et al., 2004). These residues, Cys 382 and Cys 384 from the M4 CPC motif, Tyr 688 and Asn 689 from M5, and Met 717 and Ser 721 of M6, occupy the same positions as the calcium-binding residues of SERCA1 in the calcium-free form (Olesen et al., 2007). Some rearrangement of these residues is necessary to form an appropriate Cu^{+}-binding site, and it remains to be answered how the atypical metal-binding residues Tyr 688, Asn 689, and Ser 721 might actually coordinate the metal ion. Although the entire CopA protein was crystallized, very little electron density for the N-MBD was observed, allowing only its tentative placement adjacent to the linker regions of the A domain. The proposed location is consistent with cryoelectron microscopy (Wu et al., 2008) and chemical cross-linking data (Hatori et al., 2007), however.

A key finding of the CopA structure is the identification of a "platform" formed by helix MB (Fig. 3A). Importantly, this "platform" contains positively charged residues, including Lys 135, Arg 136, and Lys 142, which may provide a complementary electrostatic surface patch for docking of a Cu^{+} chaperone and subsequent Cu^{+} delivery (Argüello et al., 2012; Gourdon et al., 2011). Given that the apo and Cu^{+}-loaded chaperones have similar surface properties, it is puzzling that Cu^{+} transfer experiments and calculations of the docking energetics suggest that the apo chaperone does not interact with the ATPase platform (Argüello et al., 2012; González-Guerrero & Argüello, 2008). It may be that the positive Cu^{+} charge is sufficient to transiently stabilize the interaction by interacting with negatively charged amino acids located proximal to the end of "platform". Three potential metal-binding residues, Met 148, Glu 205, and Asp 337, are proximal to the end of the "platform" and may represent the initial site of Cu^{+} coordination (Fig. 3A and C). These putative ligands are fully conserved in all Cu^{+}-ATPase sequences. Although this site is unable to stably bind Cu^{+}, a transport pathway might involve delivery of Cu^{+} to this transient-binding site on the platform followed by translocation to the TM sites. Conformational changes in helices H4, H5 and H6 might then facilitate release of Cu^{+} to the luminal/periplasmic compartment. It has been proposed that residues close to the external side of the membrane may form an exit site, which in the structure contains electron density modeled as a K^{+} ion (Gourdon et al., 2011). Several partially conserved residues, including Glu 189, Met 100, and Met 711, are located in this region. They may interact with sequestering molecules handling the metal upon release from the transporter.

4. FUTURE DIRECTIONS AND CHALLENGES

Although much is now known about the mechanism and structure of the Cu^{+}-ATPases, major questions remain to be addressed. From a mechanistic point of view, details of the metal release step need to be elucidated. For gram-negative bacteria, it is not clear whether specific periplasmic proteins receive the Cu^{+} and facilitate efflux or distribution to periplasmic or plasma membrane proteins. From a structural point of view, the location of the MBDs and the details of Cu^{+} coordination need to be addressed. Although specific amino acid ligands have been identified in Cu^{+}- and Zn^{2+}-ATPases, the changes in metal coordination and metrical/geometric parameters during transport will be of great interest. Finally, an overall structural and functional understanding requires structures of a Cu^{+}-ATPase in the different conformational states, with and without Cu^{+} substrate, and ideally, in complex with the Cu^{+} chaperone. Such studies will also provide new insight into the molecular basis for Wilson and Menkes diseases.

The study of the other P_{1B}-ATPase subfamilies has not reached the level of understanding that has been achieved for the Cu^{+}-ATPases. For example, transport of Co^{2+} (and alternative substrates) by the $P_{1B\text{-}4}$-ATPases has not been firmly established (Rutherford et al., 1999) and the substrate for the $P_{1B\text{-}5}$-ATPases remains unknown (Traverso et al., 2010). The functional diversity of these families is also relatively unexplored, and it is not known whether they play roles beyond efflux, perhaps in metalloprotein assembly. In addition, the details of metal binding by the TM regions have not been elucidated. Conserved residues have been identified, including the essential cysteine-containing motif in H4 (Fig. 1), but the stoichiometry of metal binding and the identities of specific ligands have barely been investigated. Similarly, the roles of the soluble MBDs have not been explored for these subfamilies. The histidine-rich N-terminal MBDs in the P_{1B-3}-ATPases are necessary for maximal Cu^{2+} transport activity (Mana-Capelli et al., 2003), but have not been characterized, and the function of the C-terminal Hr domain in the P_{1B-5}-ATPases (Traverso et al., 2010) is not known. Unraveling the coordination chemistry, mechanism, and structure of these ubiquitous metal ion transporters is critical to the field of metal ion homeostasis and has long-term implications for understanding the role of these enzymes in disease and virulence.

ACKNOWLEDGMENTS

Work in the Rosenzweig laboratory on P_{1B}-ATPases is supported by National Institutes of Health (NIH) grant GM058518. Work in the Argüello laboratory on P_{1B}-ATPases is supported by NIH grant 1R21AI082484-01, National Science Foundation (NSF) grant MCB-0743901, and United States Department of Agriculture-National Institute of Food and Agriculture (USDA-NIFA) grant 2010-65108-20606.

REFERENCES

Abdel-Ghany, S. E., Muller-Moule, P., Niyogi, K. K., Pilon, M., & Shikanai, T. (2005). Two P-type ATPases are required for copper delivery in *Arabidopsis thaliana* chloroplasts. *Plant Cell, 17*(4), 1233–1251.

Achila, D., Banci, L., Bertini, I., Bunce, J., Ciofi-Baffoni, S., & Huffman, D. L. (2006). Structure of human Wilson protein domains 5 and 6 and their interplay with domain 4 and the copper chaperone HAH1 in copper uptake. *Proceedings of the National Academy of Sciences of the United States of America, 103*(15), 5729–5734.

Agarwal, S., Hong, D. L., Desai, N. K., Sazinsky, M. H., Argüello, J. M., & Rosenzweig, A. C. (2010). Structure and interactions of the C-terminal metal binding domain of *Archaeoglobus fulgidus* CopA. *Proteins, 78*(11), 2450–2458.

Argüello, J. M. (2003). Identification of ion-selectivity determinants in heavy-metal transport P_{1B}-type ATPases. *The Journal of Membrane Biology, 195*, 93–108.

Argüello, J. M., Eren, E., & González-Guerrero, M. (2007). The structure and function of heavy metal transport P_{1B}-type ATPases. *Biometals, 20*(3–4), 233–248.

Argüello, J. M., González-Guerrero, M., & Raimunda, D. (2011). Bacterial transition metal P_{1B}-ATPases: transport mechanism and roles in virulence. *Biochemistry, 50*(46), 9940–9949.

Argüello, J. M., Raimunda, D., & Gonzalez-Guerrero, M. (2012). Metal transport across biomembranes: emerging models for a distinct chemistry. *The Journal of Biological Chemistry, 287*(17), 13510–13517.

Banci, L., Bertini, I., Calderone, V., Della-Malva, N., Felli, I. C., Neri, S., et al. (2009). Copper(I)-mediated protein–protein interactions result from suboptimal interaction surfaces. *The Biochemical Journal, 422*, 37–42.

Banci, L., Bertini, I., Cantini, F., Felli, I. C., Gonnelli, L., Hadjiliadis, N., et al. (2006). The Atx1-Ccc2 complex is a metal-mediated protein–protein interaction. *Nature Chemical Biology, 2*(7), 367–368.

Banci, L., Bertini, I., Cantini, F., Inagaki, S., Migliardi, M., & Rosato, A. (2010). The binding mode of ATP revealed by the solution structure of the N-domain of human ATP7A. *The Journal of Biological Chemistry, 285*(4), 2537–2544.

Banci, L., Bertini, I., Cantini, F., Migliardi, M., Natile, G., Nushi, F., et al. (2009). Solution structures of the actuator domain of ATP7A and ATP7B, the Menkes and Wilson disease proteins. *Biochemistry, 48*(33), 7849–7855.

Banci, L., Bertini, I., Cantini, F., Rosenzweig, A. C., & Yatsunyk, L. A. (2008). Metal binding domains 3 and 4 of the Wilson disease protein: solution structure and interaction with the copper(I) chaperone HAH1. *Biochemistry, 47*(28), 7423–7429.

Banci, L., Bertini, L., Ciofi-Baffoni, S., Finney, L. A., Outten, C. E., & O'Halloran, T. V. (2002). A new zinc-protein coordination site in intracellular metal trafficking: solution structure of the apo and Zn(II) forms of ZntA(46-118). *Journal of Molecular Biology, 323*(5), 883–897.

Banci, L., Bertini, I., Ciofi-Baffoni, S., Gonnelli, L., & Su, X. C. (2003). Structural basis for the function of the N-terminal domain of the ATPase CopA from *Bacillus subtilis*. *The Journal of Biological Chemistry, 278*(50), 50506–50513.

Banci, L., Bertini, I., McGreevy, K. S., & Rosato, A. (2010). Molecular recognition in copper trafficking. *Natural Products Reports, 27*(5), 695–710.

Boal, A. K., & Rosenzweig, A. C. (2009). Structural biology of copper trafficking. *Chemical Review, 109*(10), 4760–4779.

Botella, H., Peyron, P., Levillain, F., Poincloux, R., Poquet, Y., Brandli, I., et al. (2011). Mycobacterial P(1)-type ATPases mediate resistance to zinc poisoning in human macrophages. *Cell Host and Microbe, 10*(3), 248–259.

Bublitz, M., Morth, J. P., & Nissen, P. (2011). P-type ATPases at a glance. *Journal of Cellular Science, 124*(22), 2515–2519.

Bull, P. C., & Cox, D. W. (1994). Wilson disease and Menkes disease: new handles on heavy-metal transport. *Trends in Genetics, 10*(7), 246–252.

Chintalapati, S., Al Kurdi, R., van Scheltinga, A. C. T., & Kuhlbrandt, W. (2008). Membrane structure of CtrA3, a copper-transporting P-type-ATPase from *Aquifex aeolicus*. *Journal of Molecular Biology, 378*(3), 581–595.

Das, S. K., & Ray, K. (2006). Wilson's disease: an update. *Nature Clinical Practice. Neurology, 2*(9), 482–493.

DiDonato, M., Hsu, H.-F., Narindrasorasak, S., Que, L., Jr., & Sarkar, B. (2000). Copper-induced conformational changes in the N-terminal domain of the Wilson disease copper-transporting ATPase. *Biochemistry, 39*(7), 1890–1896.

Dmitriev, O., Tsivkovskii, R., Abildgaard, F., Morgan, C. T., Markley, J. L., & Lutsenko, S. (2006). Solution structure of the N-domain of Wilson disease protein: distinct nucleotide-binding environment and effects of disease mutations. *Proceedings of the National Academy of Sciences of the United States of America, 103*(14), 5302–5307.

Dutta, S. J., Liu, J. B., Hou, Z. J., & Mitra, B. (2006). Conserved aspartic acid 714 in transmembrane segment 8 of the ZntA subgroup of P_{1B}-type ATPases is a metal-binding residue. *Biochemistry, 45*(18), 5923–5931.

Dutta, S. J., Liu, J. B., Stemmler, A. J., & Mitra, B. (2007). Conservative and nonconservative mutations of the transmembrane CPC motif in ZntA: effect on metal selectivity and activity. *Biochemistry, 46*(12), 3692–3703.

Eren, E., & Argüello, J. M. (2004). Arabidopsis HMA2, a divalent heavy metal transporting PIB-type ATPase, is involved in cytoplasmic Zn2+ homeostasis. *Plant Physiology, 136*(3), 3712–3723.

Eren, E., González-Guerrero, M., Kaufman, B. M., & Argüello, J. M. (2007). Novel Zn^{2+} coordination by the regulatory N-terminus metal binding domain of *Arabidopsis thaliana* Zn^{2+}-ATPase HMA2. *Biochemistry, 46*(26), 7754–7764.

Eren, E., Kennedy, D. C., Maroney, M. J., & Argüello, J. M. (2006). A novel regulatory metal binding domain is present in the C terminus of *Arabidopsis* Zn^{2+}-ATPase HMA2. *The Journal of Biological Chemistry, 281*(45), 33881–33891.

Fan, B., & Rosen, B. P. (2002). Biochemical characterization of CopA, the *Escherichia coli* Cu(I)-translocating P-type ATPase. *The Journal of Biological Chemistry, 277*(49), 46987–46992.

Forbes, J. R., & Cox, D. W. (1998). Functional characterization of missense mutations in ATP7B: Wilson disease mutation or normal variant? *American Journal of Human Genetics, 63*(6), 1663–1674.

Gitlin, J. D. (2003). Wilson disease. *Gastroenterology, 125*(6), 1868–1877.

González-Guerrero, M., & Argüello, J. M. (2008). Mechanism of Cu^+-transporting ATPases: soluble Cu^+ chaperones directly transfer Cu^+ to transmembrane transport sites. *Proceedings of the National Academy of Sciences of the United States of America, 105*(16), 5992–5997.

González-Guerrero, M., Eren, E., Rawat, S., Stemmler, T. L., & Argüello, J. M. (2008). Structure of the two transmembrane Cu^+ transport sites of the Cu^+-ATPases. *The Journal of Biological Chemistry, 283*(44), 29753–29759.

González-Guerrero, M., Hong, D., & Argüello, J. M. (2009). Chaperone-mediated Cu^+ delivery to Cu^+ transport ATPases. Requirement of nucleotide binding. *The Journal of Biological Chemistry, 284*(31), 20804–20811.

González-Guerrero, M., Raimunda, D., Cheng, X., & Argüello, J. M. (2010). Distinct functional roles of homologous Cu^+ efflux ATPases in *Pseudomonas aeruginosa*. *Molecular Microbiology, 78*(5), 1246–1258.

Gourdon, P., Liu, X. Y., Skjørringe, T., Morth, J. P., Møller, L. B., Pedersen, B. P., et al. (2011). Crystal structure of a copper-transporting PIB-type ATPase. *Nature, 475*(7354), 59–74.

Hamza, I., Schaefer, M., Klomp, L. W. J., & Gitlin, J. D. (1999). Interaction of the copper chaperone HAH1 with the Wilson disease protein is essential for copper homeostasis. *Proceedings of the National Academy of Sciences of the United States of America, 96*(23), 13363–13368.

Hatori, Y., Majima, E., Tsuda, T., & Toyoshima, C. (2007). Domain organization and movements in heavy metal ion pumps—papain digestion of CopA, a Cu^+-transporting ATPase. *The Journal of Biological Chemistry, 282*(35), 25213–25221.

Hirayama, T., Kieber, J. J., Hirayama, N., Kogan, M., Guzman, P., Nourizadeh, S., et al. (1999). RESPONSIVE-TO-ANTAGONIST1, a Menkes/Wilson disease-related copper transporter, is required for ethylene signaling in Arabidopsis. *Cell, 97*(3), 383–393.

Huffman, D. L., & O'Halloran, T. V. (2000). Energetics of copper trafficking between the Atx1 metallochaperone and the intracellular copper-transporter, Ccc2. *The Journal of Biological Chemistry, 275*(25), 18611–18614.

Huffman, D. L., & O'Halloran, T. V. (2001). Function, structure, and mechanism of intracellular copper trafficking proteins. *Annual Review of Biochemistry, 70*, 677–701.

Huster, D., & Lutsenko, S. (2003). The distinct roles of the N-terminal copper-binding sites in regulation of catalytic activity of the Wilson's disease protein. *The Journal of Biological Chemistry, 278*(34), 32212–32218.

Kaler, S. G. (1998). Diagnosis and therapy of Menkes syndrome, a genetic form of copper deficiency. *American Journal of Clinical Nutrition, 67*(5), 1029S–1034S.

Katano, K., Kondo, A., Safaei, R., Holzer, A., Samimi, G., Mishima, M., et al. (2002). Acquisition of resistance to cisplatin is accompanied by changes in the cellular pharmacology of copper. *Cancer Research, 62*(22), 6559–6565.

Lin, S., Pufahl, R., Dancis, A., O'Halloran, T. V., & Culotta, V. C. (1997). A role for the *Saccharomyces cerevisiae* ATXI gene in copper trafficking and iron transport. *The Journal of Biological Chemistry, 272*(14), 9215–9220.

Liu, J. B., Dutta, S. J., Stemmler, A. J., & Mitra, B. (2006). Metal-binding affinity of the transmembrane site in ZntA: implications for metal selectivity. *Biochemistry, 45*(3), 763–772.

Lowe, J., Vieyra, A., Catty, P., Guillain, F., Mintz, E., & Cuille, M. (2004). A mutational study in the transmembrane domain of Ccc2p, the yeast Cu(I)-ATPase, shows different roles for each Cys-Pro-Cys cysteine. *The Journal of Biological Chemistry, 279*(25), 25986–25994.

Lübben, M., Güldenhaupt, J., Zoltner, M., Deigwelher, K., Haebel, P., Urbanke, C., et al. (2007). Sulfate acts as phosphate analog on the monomeric catalytic fragment of the CPx-ATPase CopB from *Sulfolobus solfataricus*. *Journal of Molecular Biology, 369*(2), 368–385.

Lübben, M., Portmann, R., Kock, G., Stoll, R., Young, M. M., & Solioz, M. (2009). Structural model of the CopA copper ATPase of *Enterococcus hirae* based on chemical cross-linking. *Biometals, 22*(2), 363–375.

Lutsenko, S., Barnes, N. L., Bartee, M. Y., & Dmitriev, O. Y. (2007). Function and regulation of human copper-transporting ATPases. *Physiological Review, 87*(3), 1011–1046.

Lutsenko, S., Gupta, A., Burkhead, J. L., & Zuzel, V. (2008). Cellular multitasking: the dual role of human Cu-ATPases in cofactor delivery and intracellular copper balance. *Archives of Biochemistry and Biophysics, 476*(1), 22–32.

Lutsenko, S., & Kaplan, J. H. (1995). Organization of P-type ATPases: significance of structural diversity. *Biochemistry, 34*(48), 15607–15613.

Ma, Z., Jacobsen, F. E., & Giedroc, D. P. (2009). Coordination chemistry of bacterial metal transport and sensing. *Chemical Reviews, 109*(10), 4644–4681.

Mana-Capelli, S., Mandal, A. K., & Argüello, J. M. (2003). *Archaeoglobus fulgidus* CopB is a thermophilic Cu^{2+}-ATPase. *The Journal of Biological Chemistry, 278*(42), 40534–40541.

Mandal, A. K., & Argüello, J. M. (2003). Functional roles of metal binding domains of the *Archaeoglobus fulgidus* Cu^{+}-ATPase CopA. *Biochemistry, 42*(37), 11040–11047.

Mandal, A. K., Cheung, W. D., & Argüello, J. M. (2002). Characterization of a thermophilic P-type Ag^{+}/Cu^{+}-ATPase from the extremophile *Archaeglobus fulgidus*. *The Journal of Biological Chemistry, 277*(9), 7201–7208.

Mandal, A. K., Yang, Y., Kertesz, T. M., & Argüello, J. M. (2004). Identification of the transmembrane metal binding site in Cu^{+}-transporting P_{IB}-type ATPases. *The Journal of Biological Chemistry, 279*(52), 54802–54807.

Mercer, J. F. B. (2001). The molecular basis of copper-transport diseases. *Trends in Molecular Medicine, 7*(2), 64–69.

Mills, R. F., Krijger, G. C., Baccarini, P. J., Hall, J. L., & Williams, L. E. (2003). Functional expression of AtHMA4, a P1B-type ATPase of the Zn/Co/Cd/Pb subclass. *Plant Journal, 35*(2), 164–176.

Mitra, B., & Sharma, R. (2001). The cysteine-rich amino-terminal domain of ZntA, a Pb(II)/Zn(II)/Cd(II)-translocating ATPase from *Escherichia coli*, is not essential for its function. *Biochemistry, 40*(25), 7694–7699.

Morth, J. P., Pedersen, B. P., Toustrup-Jensen, M. S., Sorensen, T. L. M., Petersen, J., Andersen, J. P., et al. (2007). Crystal structure of the sodium-potassium pump. *Nature, 450*(7172), 1043–1049.

O'Halloran, T. V., & Culotta, V. C. (2000). Metallochaperones, an intracellular shuttle service for metal ions. *The Journal of Biological Chemistry, 275*(33), 25057–25060.

Odermatt, A., Suter, H., Krapf, R., & Solioz, M. (1993). Primary structure of two P-type ATPases involved in copper homeostasis in *Enterococcus hirae*. *The Journal of Biological Chemistry, 268*(17), 12775–12779.

Okkeri, J., & Haltia, T. (2006). The metal-binding sites of the zinc-transporting P-type ATPase of *Escherichia coli*. Lys(693) and Asp(714) in the seventh and eighth transmembrane segments of ZntA contribute to the coupling of metal binding and ATPase activity. *Biochimica et Biophysica Acta, 1757*(11), 1485–1495.

Olesen, C., Picard, M., Winther, A. M. L., Gyrup, C., Morth, J. P., Oxvig, C., et al. (2007). The structural basis of calcium transport by the calcium pump. *Nature, 450*(7172), 1036–1042.

Osman, D., & Cavet, J. S. (2008). Copper Homeostasis in Bacteria. *Advances in Applied Microbiology, 65*, 217–247.

Palmgren, M. G., & Nissen, P. (2011). P-type ATPases. *Annual Review of Biophysics, 40*, 243–266.

Pedersen, B. P., Buch-Pedersen, M. J., Morth, J. P., Palmgren, M. G., & Nissen, P. (2007). Crystal structure of the plasma membrane proton pump. *Nature, 450*(7172), 1111–1114.

Petris, M. J., Voskoboinik, I., Cater, M., Smith, K., Kim, B.-E., Llanos, R. M., et al. (2002). Copper-regulated trafficking of the Menkes disease copper ATPase is associated with formation of a phosphorylated catalytic intermediate. *The Journal of Biological Chemistry, 277*(48), 46736–46742.

Post, R. L., Hegyvary, C., & Kume, S. (1972). Activation by adenosine triphosphate in the phosphorylation kinetics of sodium and potassium ion transport adenosine triphosphatase. *The Journal of Biological Chemistry, 247*(20), 6530–6540.

Pufahl, R. A., Singer, C. P., Peariso, K. L., Lin, S.-J., Schmidt, P., Culotta, V. C., et al. (1997). Metal ion chaperone function of the soluble Cu(I) receptor, Atx1. *Science, 278*(5339), 853–856.

Raimunda, D., González-Guerrero, M., Leeber, B. W., & Argüello, J. M. (2011). The transport mechanism of bacterial Cu^{+}-ATPases: distinct efflux rates adapted to different function. *Biometals, 24*(3), 467–475.

Raimunda, D., Subramanian, P., Stemmler, T., & Arguello, J. M. (2012). A tetrahedral coordination of Zinc during transmembrane transport by P-type Zn(2+)-ATPases. *Biochimica et Biophysica Acta, 1818*(5), 1374–1377.

Ralle, M., Lutsenko, S., & Blackburn, N. J. (2004). Copper transfer to the N-terminal domain of the Wilson disease protein (ATP7B): X-ray absorption spectroscopy of reconstituted and chaperone-loaded metal binding domains and their interaction with exogenous ligands. *The Journal of Biological Chemistry, 98*(5), 765–774.

Rensing, C., Fan, B., Sharma, R., Mitra, B., & Rosen, B. P. (2000). CopA: an *Escherichia coli* Cu(I)-translocating P-type ATPase. *Proceedings of the National Academy of Sciences of the United States of America, 97*(2), 652–656.

Rensing, C., Mitra, B., & Rosen, B. P. (1997). The *zntA* gene of *Escherichia coli* encodes a Zn(II)-translocating P-type ATPase. *Proceedings of the National Academy of Sciences of the United States of America, 94*(26), 14326–14331.

Rensing, C., Sun, Y., Mitra, B., & Rosen, B. P. (1998). Pb(II)-translocating P-type ATPases. *The Journal of Biological Chemistry, 273*(49), 32614–32617.

Robinson, N. J., & Winge, D. R. (2010). Copper metallochaperones. *Annual Review of Biochemistry, 79*, 537–562.

Rosenzweig, A. C. (2001). Copper delivery by metallochaperone proteins. *Accounts of Chemical Research, 34*(2), 119–128.

Rutherford, J. C., Cavet, J. S., & Robinson, N. J. (1999). Cobalt-dependent transcriptional switching by a dual-effector MerR-like protein regulates a cobalt-exporting variant CPx-type ATPase. *The Journal of Biological Chemistry, 274*(36), 25827–25832.

Safaei, R., Holzer, A. K., Katano, K., Samimi, G., & Howell, S. B. (2004). The role of copper transporters in the development of resistance to Pt drugs. *Journal of Inorganic Biochemistry, 98*(10), 1607–1613.

Sassetti, C. M., & Rubin, E. J. (2003). Genetic requirements for mycobacterial survival during infection. *Proceedings of the National Academy of Sciences of the United States of America, 100*(22), 12989–12994.

Sazinsky, M. H., Agarwal, S., Argüello, J. M., & Rosenzweig, A. C. (2006). Structure of the actuator domain from the *Archaeoglobus fulgidus* Cu^{+}-ATPase. *Biochemistry, 45*(33), 9949–9955.

Sazinsky, M. H., Mandal, A. K., Argüello, J. M., & Rosenzweig, A. C. (2006). Structure of the ATP binding domain from the *Archaeoglobus fulgidus* Cu^{+}-ATPase. *The Journal of Biological Chemistry, 281*(16), 11161–11166.

Scherer, J., & Nies, D. H. (2009). CzcP is a novel efflux system contributing to transition metal resistance in *Cupriavidus metallidurans* CH34. *Molecular Microbiology, 73*(4), 601–621.

Schwan, W. R., Warrener, P., Keunz, E., Stover, C. K., & Folger, K. R. (2005). Mutations in the cueA gene encoding a copper homeostasis P-type ATPase reduce the pathogenicity of Pseudomonas aeruginosa in mice. *International Journal of Medical Microbiology, 295*(4), 237–242.

Sharma, R., Rensing, C., Rosen, B. P., & Mitra, B. (2000). The ATP hydrolytic activity of purified ZntA, a Pb(II)/Cd(II)/Zn(II)-translocating ATPase from *Escherichia coli*. *The Journal of Biological Chemistry, 275*(6), 3873–3878.

Singleton, C., Banci, L., Ciofi-Baffoni, S., Tenori, L., Kihlken, M. A., Boetzel, R., et al. (2008). Structure and Cu(I)-binding properties of the N-terminal soluble domains of Bacillus subtilis CopA. *The Biochemical Journal, 411*(3), 571–579.

Singleton, C., & LeBrun, N. E. (2007). Atx1-like chaperones and their cognate P-type ATPases: copper-binding and transfer. *Biometals, 20*, 275–289.

Solioz, M., Abicht, H. K., Mermod, M., & Mancini, S. (2010). Response of Gram-positive bacteria to copper stress. *Journal of Biological Inorganic Chemistry, 15*(1), 3–14.

Solioz, M., & Vulpe, C. (1996). CPx-type ATPases: a class of P-type ATPases that pump heavy metals. *Trends in Biochemical Sciences, 21*(7), 237–241.

Tottey, S., Harvie, D. R., & Robinson, N. J. (2005). Understanding how cells allocate metals using metal sensors and metallochaperones. *Accounts of Chemical Research, 38*(10), 775–783.

Tottey, S., Rich, P. R., Rondet, S. A., & Robinson, N. J. (2001). Two Menkes-type ATPases supply copper for photosynthesis in *Synechocystis* PCC 6803. *The Journal of Biological Chemistry, 276*(23), 19999–20004.

Toyoshima, C. (2008). Structural aspects of ion pumping by Ca^{2+}-ATPase of sarcoplasmic reticulum. *Archives of Biochemistry and Biophysics, 476*(1), 3–11.

Toyoshima, C., & Inesi, G. (2004). Structural basis of ion pumping by Ca^{2+}-ATPase of sarcoplasmic reticulum. *Annual Review of Biochemistry, 73*, 269–292.

Traverso, M. E., Subramanian, P., Davydov, R., Hoffman, B. M., Stemmler, T. L., & Rosenzweig, A. C. (2010). Identification of a hemerythrin-like domain in a P_{1B}-type transport ATPase. *Biochemistry, 49*(33), 7060–7068.

Tsivkovskii, R., MacArthur, B. C., & Lutsenko, S. (2001). The Lys^{1010}-Lys^{1325} fragment of the Wilson's disease protein binds nucleotides and interacts with the N-terminal domain of this protein in a copper-dependent manner. *The Journal of Biological Chemistry, 276*(3), 2234–2242.

Tsuda, T., & Toyoshima, C. (2009). Nucleotide recognition by CopA, a Cu^{+}-transporting P-type ATPase. *EMBO Journal, 28*(12), 1782–1791.

Völlmecke, C., Lorenz Drees, S., Reimann, J., Albers, S. V., & Lübben, M. (2012). Both ATPases CopA and CopB contribute to copper resistance of the thermoacidophilic archaeon *Sulfolobus solfataricus*. *Microbiology, 158*, 1622–1633.

Voskoboinik, I., Mar, J., Strausak, D., & Camakaris, J. (2001). The regulation of catalytic activity of the Menkes copper-translocating P-type ATPase. *The Journal of Biological Chemistry, 276*(30), 28620–28627.

Voskoboinik, I., Strausak, D., Greenough, M., Brooks, H., Petris, M., Smith, S., et al. (1999). Functional analysis of the N-terminal CXXC metal-binding motifs in the human Menkes copper-transporting P-type ATPase expressed in cultured mammalian cells. *The Journal of Biological Chemistry, 274*(31), 22008–22012.

Wagner, D., Maser, J., Lai, B., Cai, Z. H., Barry, C. E., Bentrup, K. H. Z., et al. (2005). Elemental analysis of Mycobacterium avium-, Mycobacterium tuberculosis-, and Mycobacterium smegmatis-containing phagosomes indicates pathogen-induced microenvironments within the host cell's endosomal system. *Journal of Immunology, 174*(3), 1491–1500.

Walker, J. M., Huster, D., Ralle, M., Morgan, C. T., Blackburn, N. J., & Lutsenko, S. (2004). The N-terminal metal binding site 2 of the Wilson's disease protein plays a key role in the transfer of copper from Atox1. *The Journal of Biological Chemistry, 279*(15), 15376–15384.

Walker, J. M., Tsivkovskii, R., & Lutsenko, S. (2002). Metallochaperone Atox1 transfers copper to the NH_2-terminal domain of the Wilson's disease protein and regulates its catalytic activity. *The Journal of Biological Chemistry, 277*(31), 27953–27959.

Wernimont, A. K., Yatsunyk, L. A., & Rosenzweig, A. C. (2004). Binding of copper(I) to the Wilson disease protein and its copper chaperone. *The Journal of Biological Chemistry, 279*(13), 12269–12276.

White, C., Lee, J., Kambe, T., Fritsche, K., & Petris, M. J. (2009). A role for the ATP7A copper-transporting ATPase in macrophage bactericidal activity. *The Journal of Biological Chemistry, 284*(49), 33949–33956.

Williams, L. E., & Mills, R. F. (2005). P_{1B}-ATPases—an ancient family of transition metal pumps with diverse functions in plants. *Trends in Plant Science, 10*(10), 491–502.

Wu, C. C., Rice, W. J., & Stokes, D. L. (2008). Structure of a copper pump suggests a regulatory role for its metal-binding domain. *Structure, 16*(6), 976–985.

Yuan, D. S., Stearman, R., Dancis, A., Dunn, T., Beeler, T., & Klausner, R. D. (1995). The Menkes/Wilson disease gene homologue in yeast provides copper to a ceruloplasmin-like oxidase required for iron uptake. *Proceedings of the National Academy of Sciences of the United States of America, 92*(7), 2632–2636.

Zhang, X. X., & Rainey, P. B. (2007). The role of a P1-type ATPase from *Pseudomonas fluorescens* SBW25 in copper homeostasis and plant colonization. *Molecular Plant-Microbe Interactions, 20*(5), 581–588.

CHAPTER SIX

Regulation of Copper Transporters in Human Cells

Nesrin M. Hasan and Svetlana Lutsenko[1]
Department of Physiology, Johns Hopkins University, Baltimore, MD, USA
[1]Corresponding author: E-mail: lutsenko@jhmi.edu

Contents

Abstract

Copper is essential for normal growth and development of human organisms. The role of copper as a cofactor of important metabolic enzymes, such as cytochrome c oxidase, superoxide dismutase, lysyl oxidase, dopamine-β-hydroxylase, and many others, has been well established. In recent years, new regulatory roles of copper have emerged. Accumulating evidence points to the involvement of copper in lipid metabolism,

Current Topics in Membranes, Volume 69
ISSN 1063-5823,
http://dx.doi.org/10.1016/B978-0-12-394390-3.00006-9

antimicrobial defense, neuronal activity, resistance of tumor cells to platinum-based chemotherapeutic drugs, kinase-mediated signal transduction, and other essential cellular processes. For many of these processes, the precise mechanism of copper action remains to be established. Nevertheless, it is increasingly clear that many regulatory and signaling events are associated with changes in the intracellular localization and abundance of copper transporters, as well as distinct compartmentalization of copper itself. In this review, we discuss current data on regulation of the localization and abundance of copper transporters in response to metabolic and signaling events in human cells. Regulation by kinase-mediated phosphorylation will be addressed along with the emerging area of the redox-driven control of copper transport. We highlight mechanistic questions that await further testing.

1. COPPER AUTO REGULATES ITS CYTOPLASMIC LEVELS BY MODULATING THE AMOUNTS OF A HIGH-AFFINITY TRANSPORTER CTR1 AT THE PLASMA MEMBRANE

1.1. Acute Regulation of CTR1 through Copper-Dependent Trafficking

Copper enters human cells through at least two independent pathways. The high-affinity copper transporter CTR1 is responsible for the 60–70% of total copper influx, while the remaining uptake is mediated through a low-affinity transport mechanism (Lee, Petris, & Thiele, 2002). The identity of the low-affinity transporter(s) remains unknown. Several candidates have been proposed (Arredondo, Munoz, Mura, & Nunez, 2003; Zimnicka, Ivy, & Kaplan, 2011); the role of the divalent metal transporter DMT1 in copper acquisition in the brain has recently been questioned (Zheng, Chen, & Zheng, 2012). Human CTR1, hCTR1, is a homotrimer with a relatively long His/Met-rich N-terminal tail exposed at the extracellular side, three transmembrane (TM) segments, and a very short cytosolic C-terminus containing the His-Cys-His motif (Kaplan & Lutsenko, 2009; Puig et al., 2007). In the vast majority of cells, hCTR1 is responsible for acquisition of copper from the blood; in intestine, it plays an essential role in the dietary copper absorption (Nose, Kim, & Thiele, 2006). Copper uptake by hCTR1 is tightly controlled: the short-term regulation is mediated via copper-dependent changes in the abundance of hCTR1 at the plasma membrane (PM). Copper depletion increases the amount of hCTR1 at the PM through the recruitment from the intracellular pools (Kuo, Gybina, Pyatskowit, Gitschier, & Prohaska, 2006), whereas elevated copper induces rapid endocytosis of the transporter from the PM to vesicles (Guo, Smith, Lee, Thiele, & Petris, 2004) (Fig. 1).

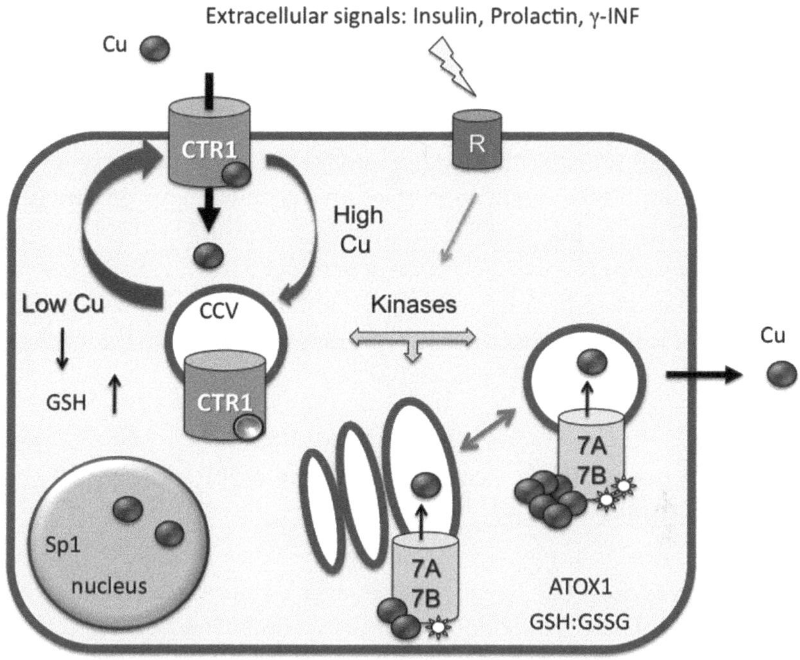

Figure 1 Main regulation modes of copper transporters in human cells. The high-affinity copper transporter hCTR1 is located at the PM, where it facilitates copper entry into the cell. It is currently unknown whether hCTR1 is functional in the intracellular compartments. The cycling between two compartments may be constitutive and involves the clathrin-dependent pathway (CCV, clathrin-coated vesicles). Copper binding to "sensor" sites favors the redistribution of hCTR1 toward intracellular vesicles, whereas copper depletion facilitates the delivery of hCTR1 to the PM. Copper-transporting ATPases 7A and 7B are located in the TGN under basal conditions where they have basal level of phosphorylation (indicated by a star). In response to copper elevation and hormonal signaling, the Cu-ATPases become hyperphosphorylated and traffic to vesicles, where they sequester excess copper. The return to the TGN is linked to the decrease in copper levels and return to a basal level of phosphorylation. Changes in glutathione (GSH) levels and/or alterations of the cellular redox balance GSH:GSSG influence the levels and activity of copper transporters.

The mechanisms through which copper triggers hCTR1 internalization and, reciprocally, the loss of copper initiates the return of hCTR1 to the PM are unclear. Biochemical and molecular modeling experiments revealed the presence of copper-binding sites at both the extracellular and cytosolic portions of hCTR1 (De Feo, Aller, Siluvai, Blackburn, & Unger, 2009; Haas, Putterman, White, Thiele, & Franz, 2011; Tsigelny et al., 2012). Mutations to Ala of residues at the entrance of the copper translocation pathway (Met150 and Met154 in the TM2) eliminate copper uptake. In contrast, mutations/deletions of Met, His, or Cys residues in the N-terminal or C-terminal tails of hCTR1, respectively, change the rate of transport (Du, Wang, Li, & Sun, 2012), but do not fully abolish the transport activity, pointing to the regulatory functions for the N-terminal and C-terminal copper-binding sites. The binding of copper to one or more of such regulatory sites and subsequent conformational changes in the protein (Eisses & Kaplan, 2002) are likely to be the first steps toward hCTR1 internalization. Further steps involve a clathrin-dependent pathway, as suggested by the co-localization of hCTR1 with the transferrin receptor and the inhibition of hCTR1 endocytosis by chlorpromazine and methyl-β-cyclodextrin (Petris, Smith, Lee, & Thiele, 2003).

1.2. Does Low or High Copper Serve as a Physiologically Relevant Signal for CTR1 Trafficking?

It should be noted that most of the current evidence for hCTR1 internalization and recycling has been obtained using the recombinant overexpressed protein. There is little doubt that the recombinant hCTR1 traffics to the intracellular compartments upon exposure to copper, whereas treatment with the copper chelator (or cisplatin) stabilizes protein levels at the PM (Guo et al., 2004). Given these observations, it might be worth considering why the results of early experiments on the endogenous hCTR1 are notably different from those with the overexpressed protein. In cells of various origin, the location of the endogenous hCTR1 was found to be predominantly intracellular. In addition to vesicles, the PM location was observed in HEK293 cells (Eisses et al., 2005), Caco-2 cells (Klomp, Tops, Van Denberg, Berger, & Klomp, 2002) and in polarized Jeg-3 cells (Hardman et al., 2006). High copper did not affect the PM localization of endogenous CTR1 in any of these cells, and no obvious internalization was observed. More recently, endocytosis of recombinant and endogenous hCTR1 was examined using surface biotinylation (Molloy & Kaplan, 2009). In these experiments, partial endocytosis (20–60%) was detected for both the recombinant and endogenous hCTR1 in some, but not all cell types. Internalization of endogenous CTR1 was seen in polarized MDCK cells and in HepG2 cells; no significant internalization of hCTR1 in response to extracellular copper was detected in HeLa and T47D cells (Molloy & Kaplan, 2009).

In mouse tissues, mCTR1 is also predominantly intracellular with a notable exception of the liver where it was immuno-detected at the PM (Hardman et al., 2006; Kuo et al., 2006; Ralle et al., 2010)].

In copper-deficient tissues, there is a distinct upregulation and recruitment of mCTR1 to the PM, especially in the intestine (Kuo et al., 2006). High copper has an opposite effect on protein localization, but not under all circumstances. The decreased levels of mCTR1 at the PM were observed in the liver tissue with a prolonged (20 weeks) intracellular accumulation of copper (Ralle et al., 2010), whereas high dietary copper did not seem to have such an effect (Araya et al., 2012). Altogether, these observations raise an interesting possibility that under physiologically relevant conditions the distribution of CTR1 between the intracellular and PM pools is controlled by the metabolic demands of a cell (i.e. by the need for copper for biosynthetic purposes) rather than by extracellular copper. This hypothesis would be consisted with a repeatedly observed increase in the CTR1 levels (total and at the PM) in response to copper deprivation. Hardman et al. (2006) have also found decreased intracellular staining of CTR1 in the insulin-treated Jeg-3 cells and proposed that insulin influenced the recruitment of hCTR1 to the PM through the mechanism resembling that of the glucose transporter GLUT4.

1.3. Copper Sensor Sites in CTR1 are Probably Cytosolic

How copper levels are sensed by CTR1 and then transferred into a signal for redistribution between cellular compartments is an interesting unresolved problem. The fact that the endogenous hCTR1 in HEK293 cells does not traffic in response to extracellular additions of copper, whereas a stably transfected hCTR1 does, suggests that the rate of copper entry and accumulation in the cytosol (which is higher in cells overexpressing hCTR1) may play an important role in triggering the internalization of the transporter. Consistent with this hypothesis, the recombinant CTR1, which lost its transport activity due to mutations of critical Met150 and Met154 (see above), does not traffic from the PM to vesicles despite the presence of the extracellular copper-binding sites at the N-terminal tail (Guo et al., 2004). Finally, since CTR1 internalization is a reversible process (i.e. copper depletion triggers the return of CTR1 to the PM) it seems that a "copper sensing" site is present at the cytosolic face of the molecule, where copper can be easily exchanged. Consistent with this idea, exposure of cells to higher copper delays the return of internalized CTR1 to the PM (Molloy & Kaplan, 2009), although this phenomenon could also be due to time needed to clear extra copper by the efflux systems.

1.4. Regulation of Copper Uptake by Modulation of Total Levels of CTR1

The long-term regulation of CTR1 is achieved through transcriptional/translational changes. The basal levels of CTR1 are influenced by the HIF2α transcription factor (Pourvali et al., 2012); however, which proteins directly control expression of CTR1 in response to changing copper levels or other metabolic needs is not entirely clear. Furthermore, similar to studies on CTR1 trafficking, there is some discrepancy in the literature on whether an elevated copper influences the levels of CTR1 messenger RNA (mRNA). Recent studies demonstrating that regulation of the CTR1 abundance could be affected by various metabolic factors offer potential explanation for such inconsistencies between the reports. For example, incubation of HepG2 cells in high copper results in decreased amounts of CTR1 mRNA and protein, whereas treatment with estradiol alleviates this copper effect (Arredondo et al., 2010). It is also interesting that overexpression of the exogenous CTR1 is associated with downregulation of the endogenous CTR1 mRNA indicating a tight homeostatic control of the total hCTR1 levels (Liang, Tsai, Lee, Savaraj, & Kuo, 2012).

The role of glutathione in regulating copper transport machinery is another intriguing finding. Overexpression of recombinant gamma-glutamyl-cysteine synthase γ-GCSht0090, a rate-limiting enzyme in the biosynthesis of glutathione, in human lung cancer cell line is associated with a two- to three-fold increase in the cellular levels of glutathione (Yamane et al., 1998). This, in turn, coincides with an increase in the CTR1 mRNA and protein level as well as increased copper uptake (Chen et al., 2008). In another set of experiments, the expression of CTR1 was shown to be upregulated upon copper depletion and downregulated under replete conditions. These changes in CTR1 levels paralleled changes in the levels of the transcription factor specificity protein 1 (Sp1) leading to the suggestion that Sp1 oscillations in response to copper availability regulate hCTR1 expression (Liang et al., 2012; Song et al., 2008).

Increases in the CTR1 mRNA and protein have also been observed in a cell model of inflammation. Treatment of BV-2 microglial cells with γ-interferon resulted in a two-fold increase of the CTR1 mRNA levels, increased rates of copper uptake, and higher copper accumulation (Zheng et al., 2010). Very similar phenomena, i.e. an increase in the CTR1 mRNA along with higher copper accumulation was observed in RAW264.7 cells exposed to hypoxic conditions (4% oxygen) (White, C., Kambe, T., Fulcher, Y.G., Sachdev, S.W., Bush, A.I., Fritsche, K., et al., 2009a). The mechanisms behind these interesting and potentially important regulatory events remain largely unexplored; however, the role of a kinase-mediated signaling appears increasingly likely. Recent study revealed an unexpected requirement for the CTR1-mediated copper uptake in signaling via the Ras/mitogen-activated protein (MAP) kinase pathway (Turski et al., 2012). In these studies, genetic inactivation of CTR1 in flies and in mouse cells as well as copper deficiency induced by metal chelation were shown to reduce the ability of the MAP kinase kinase Mek1 to phosphorylate the MAP kinase Extracellular signal regular Kinases (ERK1/2), whereas Akt kinase remained unaffected (Turski et al., 2012). Surprisingly, silver (which would be expected to mimic the effects of copper) had effect similar to copper chelation, i.e. it decreased cells' response to insulin, as evidenced by the diminished ERK1/2 phosphorylation. An important finding in this set of experiments was the observation that the addition of copper to the CTR1+/+ cells did not potentiate insulin signaling, whereas the same treatment of the copper-deficient CTR1−/− cells partially restores the ability of insulin to signal via the ERK1/2 phosphorylation pathway. These results seem to imply that the CTR1 function in insulin signaling is to maintain normal cell metabolism, rather than directly deliver copper to a copper-dependent kinase(s), although there is a convincing evidence for the direct Cu^{2+} binding to Mek1 kinase in vitro (Turski et al., 2012).

2. REGULATION OF COPPER EXPORT VIA PROTEIN TRAFFICKING AND KINASE-MEDIATED PHOSPHORYLATION

2.1. ATP7A and ATP7B are Copper-Transporting ATPases Important for Maintaining Cellular Copper Homeostasis

Human cells express two homologous copper-exporting ATPases, ATP7A and ATP7B, that transport copper across a membrane bilayer by using the energy derived from ATP hydrolysis. Both ATP7A and ATP7B belong to the P-type ATPases family, i.e. during their catalytic cycle they form a transient phosphorylated intermediate (catalytic phosphorylation) that is stimulated by binding of the exported ion, copper. ATP7A/B contains eight TM domains and several cytosolic domains (Fig. 2). The cytosolic N-terminal domain is composed of six metal-binding domains (MBDs) that are connected by flexible loops. Each MBD binds one copper (Lutsenko, Petrukhin, Cooper,

(A)

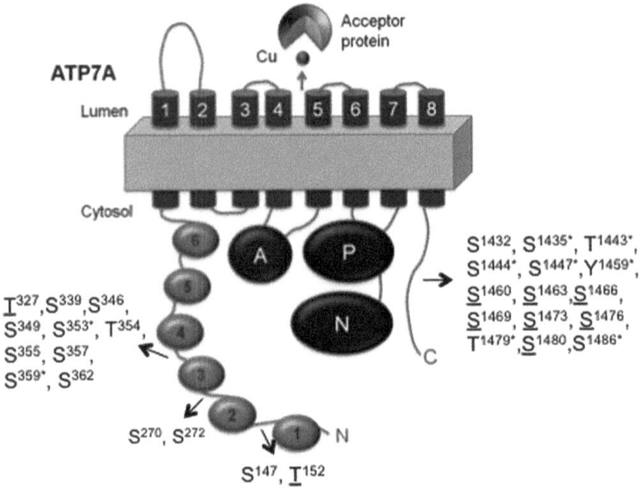

(B)

```
ATP7A_Homo_sapiens              345 VSITSEVESTSNSPSSSSLQKIPLNVVS ...
ATP7A_Rattus_norvegicus         345 VSISSEVESPTSSPSSSSLQKMPLNLVS ...
ATP7A_Mus_musculus              345 VSIASEVESTASSPSSSSLQKMPLNIVS ...

....1430 SPSEISVHVGIDDTSRNSPKLGLLDRIVNYSRASINSLLSDKRSLNSVVTSEPDKHS
....1422 SPSEISVHVGIDDTSRNSPRLGLLDRIVNYSRASINSLLSDKRSLNSVVTSEPDKHS
....1421 SPSEISVHVGIDDTSRNSPRLGLLDRIVNYSRASINSLLSDKRSLNSVVTSEPDKHS
```

(C)

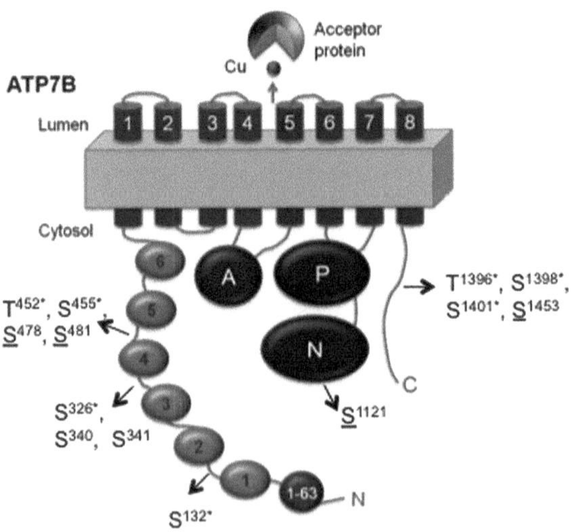

(D)

```
ATP7B_Homo_sapiens              130 AASWPSRSL ····
ATP7B_Rattus_norvegicus         128 AASWPSRSS ····
ATP7B_Mus_musculus              140 AASWPSRSS ····

450 DGTPTSVQEVAPHTGRLPANHAPDILAKSPQSTRAVAPQ ····· 1392 MKPLTASQVSVHIGMDDRWRDS
442 GDSPGSVQNMASDTRGLLTHQGPGYLSDSPPSPGGTASQ ····· 1379 MKPLSASQVSVHVGMDDRRRDS
452 GDIAGSVQKMAPDTRGLPTHQGPGHSSETPSSPGATASQ ····· 1389 MKPLSASQVSVHIGMDDRRRDS
```

Gilliam, & Kaplan, 1997); the binding induces conformational changes within the domain (DiDonato, Hsu, Narindrasorasak, Que, & Sarkar, 2000) that are accompanied by the rearrangements of the loops connecting MBDs (Bartee, Ralle, & Lutsenko, 2009). The nucleotide-binding domain (NBD), the phosphorylation (P) domain and the actuator (A) domain are central for the P-type ATPase function. The NBD and the P-domain are involved in ATP binding and hydrolysis, whereas the A-domain is required for conformational changes during ATP hydrolysis. Both the N-terminal domain and C-terminal tails of ATP7A/B are important for proper targeting of the protein. The 1–63 region of the N-terminal domain of ATP7B contains a region (F^{37}AFDNVGYE^{45}), which is involved in *trans*-Golgi network (TGN) retention/apical targeting (Braiterman et al., 2009), whereas the C-terminal tail contains residues important for endocytosis (Cater, La Fontaine, Shield, Deal, & Mercer, 2006; Francis et al., 1999; Stephenson, Dubach, Lim, Mercer, & La Fontaine, 2005) and copper-responsive trafficking to the apical membrane (Braiterman, Nyasae, Leves, & Hubbard, 2011).

2.2. Localization of Human Copper Efflux Transporters is Regulated

ATP7A and ATP7B have two main localizations within cells. One compartment is the TGN where these copper-ATPases supply the copper cofactor to cuproenzymes; the other compartment is vesicles-PM, to which copper-ATPases traffic in order to remove excess copper out of the cells (Gupta & Lutsenko, 2009). Copper regulates the distribution of ATP7A and ATP7B between these two locations. In basal or low copper, the TGN is the primary site for the localization and function of copper-ATPases; in high copper, the transporters exit the TGN and relocate to vesicles, which

Figure 2 The location of phosphorylated sites in ATP7A and ATP7B. (A, C) Schematic presentation of the domain organization of ATP7A (A) and ATP7B (C). Filled circles represent the N-terminal MBDs and are numbered 1–6. TM domains are numbered 1–8. The other domains are as follows: A: actuator domain; P: phosphorylation domain; N: NBD. The 1–63 region of ATP7B contains a region (F^{37}AFDNVGYE^{45}) important for TGN retention/apical targeting (Braiterman et al., 2009). The positions of identified phosphosites are shown. Underlined residues are putative copper-responsive phosphosites; the star (*) labels the putative phosphosites identified in other species and aligned with human ATP7A or ATP7B. (B,D) Sequence alignment of ATP7A (B) and ATP7B (D) orthologs in the regions with identified phosphosites. Sites for basal/constitutive phosphorylation identified by MS are in gray background. Copper-responsive phosphosites are in bold and underlined.

eventually fuse with the PM. This copper-regulated trafficking of ATP7A/7B is important for cellular copper homeostasis. Despite its central role, the exact mechanism of ATP7A7B trafficking between cellular compartments remains only partially characterized. It has been established that both ATP7A/7B undergo kinase-mediated phosphorylation and the level of phosphorylation is copper dependent and reversible. Therefore, the kinase-mediated phosphorylation of ATP7A/7B is likely to be an important regulatory step for trafficking of ATP7A/7B and/or for modulation of their copper transport function.

2.3. Both Copper-ATPases are Phosphorylated on Multiple Sites

Biochemical and mass spectrometry (MS) experiments suggest that both ATP7A and ATP7B are phosphorylated by kinases on serine residues (Vanderwerf, Cooper, Stetsenko, & Lutsenko, 2001; Vanderwerf & Lutsenko, 2002; Voskoboinik et al., 2003). The level of kinase-mediated phosphorylation is linked to the intracellular localization of copper-ATPases. Under basal conditions, while in TGN, ATP7A/B are basally phosphorylated. Upon copper elevation, the kinase-mediated phosphorylation increases, and this coincides with the trafficking of ATP7A/B to vesicles. Return to the basal media or copper depletion with chelators triggers trafficking back to the TGN. In parallel, the regulatory phosphorylation of ATP7A/7B reverts to its basal levels; suggesting that copper-dependent phosphorylation might be required for either exit from the TGN or retention in vesicles (Vanderwerf et al., 2001; Voskoboinik et al., 2003).

The identification of phosphorylation sites is necessary for determining the precise physiological role of ATP7A/7B phosphorylation. The in-cell and in vitro studies indicate that more than one residue is phosphorylated in either ATP7A or ATP7B (Pilankatta, Lewis, Adams, & Inesi, 2009; Vanderwerf et al., 2001; Voskoboinik et al., 2003). For ATP7B, it has been shown that the copper-responsive phosphorylation occurs at different site(s) than the basal phosphorylation site(s) and different regions of the protein are involved in these events (Vanderwerf et al., 2001). The C-terminal tail and the N-terminal domain of ATP7B are not required for basal phosphorylation, indicating that at least some of the basal phosphorylation site(s) are located in the Ser 796-Tyr 1384 region of ATP7B. On the other hand, the copper-induced phosphorylation requires the presence of the N-terminal domain (Vanderwerf et al., 2001), which is a target of phosphorylation both in vitro and in cells (Bartee et al., 2009). MS analysis of the in vitro phosphorylated N-terminal domain of ATP7B pointed to the loop connecting MBD3–MBD4 as a site for the kinase-mediated modification (Bartee et al., 2009). Other putative copper-responsive phosphorylation sites were identified by MS analysis of the full-length ATP7B expressed in mammalian cells (Table 1): these sites are located in the N-terminal domain (Ser 478 and Ser 481), the NBD (Ser 1121) and the C-terminal tail (Ser 1453) (Pilankatta et al., 2009; Pilankatta, Lewis, & Inesi, 2011).

Similar to ATP7B, ATP7A has multiple phosphorylation sites. Residues associated with the basal and copper-induced phosphorylation have been mapped to the N-terminal domain and the C-terminal tail of ATP7A (Veldhuis et al., 2009). Several other phosphosites/phosphopeptides have been identified by the MS analysis and the results are summarized in Table 2 (Fedjaev et al., 2012; Huttlin et al., 2010; Lundby et al., 2012; Monetti, Nagaraj, Sharma, & Mann, 2011). Despite these important advances, it remains unclear which, if any, of these sites represent primary targets of phosphorylation and how stoichiometry of phosphorylation is regulated in response to changing copper levels. Mutation of two putative phosphorylation sites revealed that the corresponding ATP7A variants, $ATP7A^{S1469A}$ and $ATP7A^{S1432A}$, have different intracellular localization, indicating that phosphorylation may have different consequences even when residues are located in the same domain (Veldhuis et al., 2009). At the same time, different signaling events may yield phosphorylation of the same region/sites triggering trafficking response similar to that induced by copper. For example, treatment of placental Jeg-3 cells with insulin stimulates ATP7A exit from the TGN and, interestingly, downregulates the ATP7B protein levels resulting in lower copper levels in cells (Hardman et al., 2006). In those earlier studies, the mechanism of insulin action was not explored. However, more recent studies using the large-scale MS identified numerous liver proteins phosphorylated following insulin signaling, including ATP7A and ATP7B. The residues phosphorylated in response to insulin were found in regions that were also known to be phosphorylated in response to changing copper level (Tables 1 and 2). How different signaling pathways intersect and interact with each other is a complex and fascinating question for further studies.

Several studies have been done to identify the kinases involved in the phosphorylation of ATP7A and ATP7B. Protein kinase D (PKD) (Pilankatta et al., 2011) and casein kinase II (Vanderwerf et al., 2001) were suggested to be involved in kinase-mediated phosphorylation of ATP7B. In contrast,

Table 1 Predicted phosphorylation sites in ATP7B

Phosphosites/phosphopeptides	Type of phosphorylation	Effect of mutation	Ref.
[Ser 478 (N-terminal), Ser 481(N-terminal), Ser 1121 (NBD), Ser 1453 (C-terminal)]^{h}	Copper responsive	Quadruple S > A mutant is not trafficking in response to copper.	(Pilankatta et al., 2009, 2011)
(340 SSSSHSPGSPPR)^{N,h}	Basal/constitutive	S340/341 mutants are localized in vesicles.	(Bartee et al., 2009)
(270 SCVLNIEENIGQLLGVQ SIQVSLENK)^{N,h} (352 NQVQGTCSTTL)^{N,h} (407 VISPEELR)^{N,h} (415 AIEDMGFEASVVSESCSTNP LGNHSAGN)^{N,h} (439 HSAGNSMV QTTDGTPTSVQEVAPHTGR)^{N,h} (444 SMVQTTDGTPTSVQEVA PHTGR)^{N,h} (496 GMTCASCVSNIER)^{N,h} (512 EAGVLSVLVALMAGK)^{N,h} (589 TNGITYASVALATSK)^{N,h} (595 ASVALATSK)^{N,h}	Basal/constitutive	ND	(Bartee et al., 2009)
(140 AAS^{142}WPSR)^{N,m}	Basal/constitutive	ND	(Huttlin et al., 2010)
(1389 MKPLS^{1393}AS^{1395}QVS^{1398} VHIGMDDR)^{*,C,m}	Basal/constitutive	ND	(Huttlin et al., 2010)

(1389 MKPLSAS^{1395}QVSVHI)^{C,m}	Insulin dependent	ND (putative kinase motif: NEK6)	(Monetti et al., 2011)
(458 GLLTHQGPGYLSDSPPS^{474} PGGTASQ)^{N,r}	Insulin dependent	ND	(Fedjaev et al., 2012)
(128 AAS^{130}WPSR)^{N,r}	Basal/constitutive	ND	(Lundby et al., 2012)
(323 VS^{324}LPDGLEK)^{N,r}	Basal/constitutive	ND	(Lundby et al., 2012)
(431 VSSGNSV PQAVGDS^{444}PGS^{447} VQNMASDTR)$^{*,\ N,r}$	Basal/constitutive	ND	(Lundby et al., 2012)
(457 GLLTHQGPGYLS^{468} DS^{470}PPS^{473}PGGT^{477}ASQK)$^{*,\ N,r}$	Basal/constitutive	ND	(Lundby et al., 2012)
(1381 PLS^{1383}AS^{1385}QVS^{1388} VHVGMDDR)$^{*,\ C,r}$	Basal/constitutive	ND	(Lundby et al., 2012)
(1439 WSLLLS^{1444}DR)$^{*,\ C,r}$	Basal/constitutive	ND	(Lundby et al., 2012)

Species: ^{m}-mouse; ^{r}-rat; ^{h}-human; ^{N}-N-terminal; ^{C}-C-terminal; ND: not determined.
* The identified phosphopeptide has different phosphorylated residue combinations identified by MS, and different phospho-occupancy.

Table 2 Predicted phosphorylation sites in ATP7A

Phosphosites/phosphopeptides	Type of phosphorylation	Effects of mutation	Ref.
(141 ELVPELS^{147}LDTGT^{152}LEKK)^{*,N,h}	Ser 147-Basal/constitutive Thr 152 copper responsive	ND	(Veldhuis et al., 2009)
(270 S^{270}PS^{272}YTNDSTATFIID)^{*,N,h}	Basal/constitutive	ND	(Veldhuis et al., 2009)
(321 YNASSVT^{327}PESLR)^{N,h}	Copper responsive	ND	(Veldhuis et al., 2009)
(334 AIEAVS^{339}PGLYR)^{N,h}	Basal/constitutive	ND	(Veldhuis et al., 2009)
(345 VS^{346}ITS^{349}EVEST^{354}S^{355}NS^{357} PSSSS^{362}LQK)^{*,N,h}	Basal/constitutive	ND	(Veldhuis et al., 2009)
(1430 SPS^{1432}EISVHVGIDDTSR)^{C,h}	Basal/constitutive	S1432A is mislocalized (vesicles in high Cu, proximal to basolateral PM)	(Veldhuis et al., 2009)
(1456 IVNYS^{1460}RAS^{1463}INS^{1466} LLS^{1469}DK)^{*,C,h}	Copper responsive	S1469A is mislocalized (vesicles in high Cu, but no PM localization)	(Veldhuis et al., 2009)
(1473 S^{1473}LNS^{1476}VVTS^{1480}EPDK)^{*,C,h}	Copper responsive	S1473A; basal localization and Cu-induced trafficking similar to WT.	(Veldhuis et al., 2009)
(345 VSIASEVESTAS^{356}S^{357}PS^{359} SSS^{362}LQK)^{*,N,m}	Basal/constitutive	ND	(Huttlin et al., 2010)
(1421 SPS^{1423}EIS^{1426}VHVGIDDT^{1434} S^{1435}RNS^{1438}PR)^{*,C,m}	Basal/constitutive	ND	(Huttlin et al., 2010)

(1447 IVNY^{1450}S^{1451}RAS^{1454}INS^{1457} LLS^{1460}DKR)^{*,C,m}	Basal/constitutive	ND	(Huttlin et al., 2010)
(1464 S^{1464}LNS^{1467}VVT^{1470}S^{1471}EPDKH S^{1477} LLVGDFEDDDTTL)^{*,C,m}	Basal/constitutive	ND	(Huttlin et al., 2010)
(1446 DRIVNYS^{1452}RAS^{1455}INS^{1458} LLS^{1461}DKRSLN)^{*,C,m}	Insulin dependent	ND; (putative kinase motifs: AURORA, AURORA-1, CK1, PKA, GSK3)	(Monetti et al., 2011)
(345 VS^{346}ISSEVES^{353}PTSS^{357} PSSSSLQK)^{*,C,r}	Basal/constitutive	ND	(Lundby et al., 2012)
(1454 AS^{1455}INS^{1458}LLSDK)^{*,C,r}	Basal/constitutive	ND	(Lundby et al., 2012)
(1465 S^{1465}LNSVVTSEPDK)^{C,r}	Basal/constitutive	ND	(Lundby et al., 2012)

Species: ^{m}-mouse; ^{r}-rat; ^{h}-human; ^{N}-N-terminal; ^{C}-C-terminal; ND: not determined.
* The identified phosphopeptide has different phosphorylated residue combinations identified by MS, and different phospho-occupancy.

PKD does not seem to affect ATP7A, whereas protein kinase A and Rho GTPase Cdc 42 were implicated in ATP7A anterograde trafficking (Cobbold, Ponnambalam, Francis, & Monaco, 2002).

2.4. Copper-Induced Trafficking of ATP7B Depends on the Conformational State of the Transporter

The cytosolic copper chaperone Atox1 was shown to transfer copper to the regulatory MBDs of the Cu-ATPases and regulate their occupancy (Walker et al., 2004). Current data indicate that copper binding to N-terminal domain of both ATP7A and ATP7B is necessary for their copper-responsive phosphorylation (Vanderwerf et al., 2001; Voskoboinik et al., 2003). Upon binding, copper induces rearrangements of the loops connecting MBDs in the N-terminal domain of ATP7B and also facilitates phosphorylation by a kinase, suggesting that conformational changes due to copper binding might be important in the exposure of additional phosphorylation sites (Bartee et al., 2009). An ATP7A mutant which cannot bind copper (due to mutation of CxxC in MBDs to SxxS) (Strausak et al., 1999) has a basal level of phosphorylation, but shows no increase in phosphorylation upon copper treatment and no trafficking (Voskoboinik et al., 2003).

The Wilson disease-causing mutant ATP7B^{G591D} is mislocalized to the endoplasmic reticulum (ER) under basal conditions (de Bie et al., 2007) and has a basal level of phosphorylation but no copper-induced phosphorylation (Vanderwerf et al., 2001). Similarly, the ATP7B^{R875} variant is also localized in ER under basal/copper-depleted conditions and is basally phosphorylated (Hasan et al., 2012). These findings indicate that the basal phosphorylation of ATP7B occurs early in the secretory pathway (presumably in the ER). The difference between ATP7B^{G591D} and ATP7B^{R875} is that ATP7B^{R875} binds copper and under elevated copper conditions traffics to vesicles, which is also coupled with the increase in the kinase-mediated phosphorylation (Hasan et al., 2012).

2.5. The Sequence of Events Associated with the TGN Exit/Trafficking of Copper-ATPases

It is not yet clear how hyperphosphorylation and trafficking are linked, or which occurs first. Copper binding to the regulatory N-terminal domain of ATP7A/7B might expose additional phosphorylation sites and the addition of an extra negative charge (hyperphosphorylation) may induce further structural changes necessary for the exit of copper-ATPases from the TGN.

Alternatively, copper binding to the ATPase and the subsequent structural changes may alone enable protein sorting within the TGN/interaction with the trafficking machinery, whereas hyperphosphorylation may be a result of the relocalization of ATP7A/7B to the correct vesicular compartment in which kinase(s) might be present.

Several sets of data support the important role of copper-induced structural changes in the initiation of ATP7B exit from the TGN. Copper binding not only influences the rearrangement of the N-terminal domain of ATP7B but also alters the domain–domain interactions within the entire ATP7B molecule. Relative to the copper-free form, copper-bound N-terminal domain interacts poorly with the NBD (Tsivkovskii, MacArthur, & Lutsenko, 2001), but strongly with the A-domain (Gupta et al., 2011). Since copper binding to the N-terminus of ATP7B alters the inter-domain contacts, the increased phosphorylation might be a result of the exposure of additional phosphorylation sites in more than one domain.

An alternative model is that the stabilization of ATP7B in a distinct conformational state is sufficient to cause its trafficking from TGN to vesicles. This model is supported by the observation that mutating residues that are not directly involved in copper binding may significantly alter the targeting and trafficking of ATP7B. The mutant ^{858}TGE^{860} > AAA, which is catalytically inactive, is found in vesicles under basal copper conditions (Cater, La Fontaine, & Mercer, 2007). This targeting is associated with the stabilization of this mutant protein in the so-called E2 conformational state, as was demonstrated by the high level of catalytic phosphorylation (whether this hyperphosphorylation caused by the loss of phosphatase activity by the mutant is also coupled to the increase in regulatory kinase-mediated phosphorylation has not been explored). Mutating the six N-terminal MBDs and/or the TM Cys-Proc-Cys(CPC) motif individually results in a non-trafficking ATP7B; however, simultaneous mutation of all sites produces protein targeted to vesicles even in basal conditions (Cater et al., 2007). The fact that the conformation acquired by the ATP7B$^{TGE>AAA}$ variant is sufficient to trigger its trafficking to vesicles suggests that trafficking is driven by structural changes in ATP7B, rather than interactions with other copper-binding proteins or phosphorylation by copper-dependent kinases.

Similar results were obtained by mutating S340/341 residues of the N-terminal domain of ATP7B. The ATP7B mutants with S340/341 mutations are found in vesicles under basal conditions. Metabolic labeling in cells with inorganic phosphate directly demonstrated that mutation of S340/341 residues significantly decreased the level of ATP7B phosphorylation, indicating that these residues were important for ATP7B modification by a kinase (Hasan et al., 2012). Therefore, vesicular localization of the $ATP7B^{Ser340/341Gly}$ protein was not due to mutation-induced hyperphosphorylation of some other residues, but rather due to conformational changes in ATP7B. Further studies revealed weakened interactions between the N-terminal and NBD domains in $ATP7B^{S340/341A}$ compared to control ATP7B (Hasan et al., 2012). Thus, the role of kinase-mediated phosphorylation might be to keep the N-terminal and NBD domains apart and stabilize the protein in a particular confirmation that can be recognized by the trafficking machinery, rather than acting as a specific recognition sequence for such machinery (Hasan et al., 2012). Although $^{858}TGE^{860}$ and $S^{340/341}$ residues are located in different domains of ATP7B, the mutations of these sites may have similar effects on the overall conformation of ATP7B and thus produce a similar trafficking response.

2.6. Copper Binding to the Regulatory N-terminal Sites and Catalytically Important TM Sites Has Different Effect on the Kinase-Mediated Phosphorylation of ATP7B

Mutating the N-terminal copper-binding site in MBD6 ($ATP7B^{C575/578A}$) interferes with a kinase-mediated phosphorylation, whereas mutating the intramembrane TM6 copper-binding site ($ATP7B^{C983/985A}$) does not have such an effect. Thus, copper binding to the MBD6 of N-terminal domain might be important for inducing the conformational changes within the domain so the phosphorylation sites can be accessed by the kinase. Neither mutant traffics in response to copper, i.e. the kinase-mediated phosphorylation of $ATP7B^{C983/985A}$ is not sufficient for triggering its relocalization. At the same time, the $ATP7B^{C983/985A}$ mutant has a lower phosphorylation level (or possibly a slower rate of phosphorylation) compared to the wild-type (WT) ATP7B, suggesting that the extent of phosphorylation might be important for ATP7B exit from the TGN or that this mutation indirectly altered other sites (Pilankatta et al., 2011).

2.7. The Catalytic and Kinase-Mediated Phosphorylation Occur through Different Mechanisms and are Independent Events

Both catalytic and kinase-mediated phosphorylation are stimulated by copper, but through different mechanisms (Pilankatta et al., 2011). Supporting the model that both phosphorylations are independent events, inhibition of the kinase-mediated phosphorylation using a kinase inhibitor does not affect the formation of catalytic phosphorylation by ATP7B (Pilankatta et al., 2011). Reciprocally, the catalytically inactive $ATP7B^{D1027A}$ and $ATP7B^{D1027N}$ mutants are phosphorylated indicating that the catalytic transfer of phosphate from ATP to D1027 is not necessary for their kinase-mediated phosphorylation (Bartee et al., 2009; Pilankatta et al., 2011). The same phenomenon was observed for the $ATP7B^{C983/985A}$ mutant, which had no catalytic phosphorylation, but showed kinase-mediated phosphorylation (Pilankatta et al., 2011). In the cases of $ATP7B^{D1027N}$ and $ATP7B^{C983/985A}$ mutants, the kinase-mediated phosphorylation level was lower (or the phosphorylation rate was slower compared to the WT), suggesting that the overall phosphorylation level of ATP7B might be altered as a result of catalytic impairment (Pilankatta et al., 2011).

In summary, both ATP7A and ATP7B undergo kinase-mediated phosphorylation, which is a regulatory posttranslational modification coupled to the intracellular localization of the protein. Copper binding to the N-terminal domain and/or trafficking to the post-Golgi compartments (vesicular/PM) is associated with hyperphosphorylation, but in some cases ($ATP7B^{S340/341G}$ or $ATP7B^{TGE>AAA}$) the conformational state attained by the protein determines its localization irrespective of copper binding. Several putative phosphorylation sites have been identified; however, it remains unclear whether the stoichiometry of phosphorylation or phosphorylation of distinct residues are important for the copper-responsive trafficking of ATP7A/B.

3. REDOX CONTROL OF COPPER UPTAKE AND EFFLUX FROM HUMAN CELLS

As described in Section 1, copper deficiency is associated with the increase of cellular glutathione levels, which in turn stimulate production of hCTR1 (Tennant, Stansfield, Yamaji, Srai, & Sharp, 2002). Recent studies indicate that the cellular glutathione status and the intracellular redox potential may also have a significant effect on the copper export activity of ATP7A and ATP7B. It has been suggested that under basal conditions, both copper-ATPases can be glutathionylated, presumably at their metal-binding sites and that the glutathionylation of ATP7A/7B and their interaction with glutaredoxin 1 are affected by changes in copper levels (Singleton et al., 2010). The downregulation of glutaredoxins by Small interfering RNA(siRNA) resulted in increased intracellular copper accumulation (Singleton et al., 2010), which was interpreted as evidence for direct regulation of copper export by glutaredoxin 1. While the data are compelling, it remains to be established more firmly whether the diminished copper export is due to a direct oxidation of copper-ATPases or due to the effect of experimental manipulation on a cellular redox environment and interference with the copper delivery system. Recent data suggest that copper export can be controlled at the stage of copper transfer from the cytosolic chaperone to the copper-ATPases.

The copper-ATPases receive copper from a copper chaperone Atox1. This small cytosolic protein has a characteristic copper-binding site CxxC, which is maintained in the reduced state by glutathione (Hatori, Clasen, Hasan, Barry, & Lutsenko, 2012). Decrease in the cellular ratio of reduced to oxidized glutathione (GSH:GSSG) results in Atox1 oxidation and a diminished delivery of copper to the secretory pathway, as evidenced by the decreased levels of secreted holo-ceruloplasmin (Hatori et al., 2012). The redox potential of Atox1 is estimated at approximately −220 mV and, in resting cells, Atox1 is expected to be fully reduced. However, upon cell proliferation and differentiation, the intracellular environment changes significantly (Dong-Yun, Yu-Ru, Shan-Lin, Ya-Dong, & Lian, 2003; Markovic et al., 2009; Ranjan et al., 2006). In Caco-2 cells, where both ATP7A and ATP7B are expressed, the redox state of cellular GSH/GSSG pair changes to be 40 mV more oxidizing, when cells progress from proliferation to contact inhibition and differentiation (Nkabyo, Ziegler, Gu, Watson, & Jones, 2002). This shift would be sufficient to produce a partial oxidation of Atox1 and presumably influence copper excretion. Direct testing of these predictions would greatly expand our understanding between copper transport and cellular redox balance.

Overall, the current results suggest that copper transport in and out of cells is likely to fluctuate depending on cellular metabolic status, the stage of cell division, or differentiation. An example of significant changes in the routes of copper distribution is provided by the effects of hypoxia in macrophages (White, C., Kambe, T., Fulcher, Y.G., Sachdev, S.W., Bush, A.I., Fritsche, K., et al., 2009a). Oxygen limitation in RAW264.7 macrophage cells stimulates copper uptake likely due to increased expression of hCTR1. Under these conditions, copper transfer to the secretory pathway is also enhanced, as inferred from the trafficking of ATP7A to vesicles. At the same time, the activities of cytosolic superoxide dismutase and mitochondria cytochrome c oxidase are reduced in response to hypoxia (White, C.,

Kambe, T., Fulcher, Y.G., Sachdev, S.W., Bush, A.I., Fritsche, K., et al., 2009a). Thus, under hypoxic conditions copper delivery to the secretory pathway becomes the primary pathway of copper distribution, presumably to facilitate copper incorporation into ceruloplasmin.

In conclusion, human copper transporters are subject of a complex multilayer regulation by various metabolic signals. Recent success in understanding the structure and biochemistry of these transporters (see other chapters in this volume) provides strong foundation for trying to link the molecular characteristics of the transporters to their behavior in cells.

ACKNOWLEDGMENTS

We thank Dr Arnab Gupta and Dr Jack H. Kaplan for stimulating discussions. This work was supported by the grants from the National Institute of Health P01GM01067166 and R01DK071865 to S.L.

REFERENCES

Araya, M., Nunez, H., Pavez, L., Arredondo, M., Mendez, M., Cisternas, F., et al. (2012). Administration of high doses of copper to capuchin monkeys does not cause liver damage but induces transcriptional activation of hepatic proliferative responses. *Journal of Nutrition, 142*(2), 233–237.

Arredondo, M., Nunez, H., Lopez, G., Pizarro, F., Ayala, M., & Araya, M. (2010). Influence of estrogens on copper indicators: in vivo and in vitro studies. *Biological Trace Element Research, 134*(3), 252–264.

Arredondo, M., Munoz, P., Mura, C. V., & Nunez, M. T. (2003). DMT1, a physiologically relevant apical Cu1+ transporter of intestinal cells. *American Journal of Physiology Cell Physiology, 284*(6), C1525–C1530.

Bartee, M. Y., Ralle, M., & Lutsenko, S. (2009). The loop connecting metal-binding domains 3 and 4 of ATP7B is a target of a kinase-mediated phosphorylation. *Biochemistry, 48*(24), 5573–5581.

Braiterman, L., Nyasae, L., Guo, Y., Bustos, R., Lutsenko, S., & Hubbard, A. (2009). Apical targeting and Golgi retention signals reside within a 9-amino acid sequence in the copper-ATPase, ATP7B. *American Journal of Physiology Gastrointestinal Liver Physiology, 296*(2), G433–G444.

Braiterman, L., Nyasae, L., Leves, F., & Hubbard, A. L. (2011). Critical roles for the COOH terminus of the Cu-ATPase ATP7B in protein stability, trans-Golgi network retention, copper sensing, and retrograde trafficking. *American Journal of Physiology Gastrointestinal Liver Physiology, 301*(1), G69–G81.

Cater, M. A., La Fontaine, S., & Mercer, J. F. (2007). Copper binding to the N-terminal metal-binding sites or the CPC motif is not essential for copper-induced trafficking of the human Wilson protein (ATP7B). *Biochemical Journal, 401*(1), 143–153.

Cater, M. A., La Fontaine, S., Shield, K., Deal, Y., & Mercer, J. F. (2006). ATP7B mediates vesicular sequestration of copper: insight into biliary copper excretion. *Gastroenterology, 130*(2), 493–506.

Chen, H. H., Song, I. S., Hossain, A., Choi, M. K., Yamane, Y., Liang, Z. D., et al. (2008). Elevated glutathione levels confer cellular sensitization to cisplatin toxicity by up-regulation of copper transporter hCtr1. *Molecular Pharmacology, 74*(3), 697–704.
Cobbold, C., Ponnambalam, S., Francis, M. J., & Monaco, A. P. (2002). Novel membrane traffic steps regulate the exocytosis of the Menkes disease ATPase. *Human Molecular Genetics, 11*(23), 2855–2866.
de Bie, P., van de Sluis, B., Burstein, E., van de Berghe, P. V., Muller, P., Berger, R., et al. (2007). Distinct Wilson's disease mutations in ATP7B are associated with enhanced binding to COMMD1 and reduced stability of ATP7B. *Gastroenterology, 133*(4), 1316–1326.
De Feo, C. J., Aller, S. G., Siluvai, G. S., Blackburn, N. J., & Unger, V. M. (2009). Three-dimensional structure of the human copper transporter hCTR1. *Proceedings of the National Academy of Sciences of the United States of Amercia, 106*(11), 4237–4242.
DiDonato, M., Hsu, H. F., Narindrasorasak, S., Que, L., Jr., & Sarkar, B. (2000). Copper-induced conformational changes in the N-terminal domain of the Wilson disease copper-transporting ATPase. *Biochemistry, 39*(7), 1890–1896.
Dong-Yun, S., Yu-Ru, D., Shan-Lin, L., Ya-Dong, Z., & Lian, W. (2003). Redox stress regulates cell proliferation and apoptosis of human hepatoma through Akt protein phosphorylation. *FEBS Letters, 542*(1–3), 60–64.
Du, X., Wang, X., Li, H., & Sun, H. (2012). Comparison between copper and cisplatin transport mediated by human copper transporter 1 (hCTR1). Metallomics. *4*(7), 679–685.
Eisses, J. F., Chi, Y., & Kaplan, J. H. (2005). Stable plasma membrane levels of hCTR1 mediate cellular copper uptake. *Journal of Biological Chemistry, 280*(10), 9635–9639.
Eisses, J. F., & Kaplan, J. H. (2002). Molecular characterization of hCTR1, the human copper uptake protein. *Journal of Biological Chemistry, 277*(32), 29162–29171.
Fedjaev, M., Parmar, A., Xu, Y., Vyetrogon, K., Difalco, M. R., Ashmarina, M., et al. (2012). Global analysis of protein phosphorylation networks in insulin signaling by sequential enrichment of phosphoproteins and phosphopeptides. *Molecular BioSystems, 8*(5), 1461–1471.
Francis, M. J., Jones, E. E., Levy, E. R., Martin, R. L., Ponnambalam, S., & Monaco, A. P. (1999). Identification of a di-leucine motif within the C terminus domain of the Menkes disease protein that mediates endocytosis from the plasma membrane. *Journal of Cell Science, 112*(Pt 11), 1721–1732.
Guo, Y., Smith, K., Lee, J., Thiele, D. J., & Petris, M. J. (2004). Identification of methionine-rich clusters that regulate copper-stimulated endocytosis of the human Ctr1 copper transporter. *Journal of Biological Chemistry, 279*(17), 17428–17433.
Gupta, A., Bhattacharjee, A., Dmitriev, O. Y., Nokhrin, S., Braiterman, L., Hubbard, A. L., et al. (2011). Cellular copper levels determine the phenotype of the Arg875 variant of ATP7B/Wilson disease protein. *Proceedings of the National Academy of Sciences of the United States of America, 108*(13), 5390–5395.
Gupta, A., & Lutsenko, S. (2009). Human copper transporters: mechanism, role in human diseases and therapeutic potential. *Future Medicinal Chem, 1*(6), 1125–1142.
Haas, K. L., Putterman, A. B., White, D. R., Thiele, D. J., & Franz, K. J. (2011). Model peptides provide new insights into the role of histidine residues as potential ligands in human cellular copper acquisition via Ctr1. *Journal of American Chemical Society, 133*(12), 4427–4437.
Hardman, B., Manuelpillai, U., Wallace, E. M., Monty, J. F., Kramer, D. R., Kuo, Y. M., et al. (2006). Expression, localisation and hormone regulation of the human copper transporter hCTR1 in placenta and choriocarcinoma Jeg-3 cells. *Placenta, 27*(9–10), 968–977.
Hasan, N. M., Gupta, A., Polishchuk, E., Yu, C. H., Polishchuk, R., Dmitriev, O. Y., et al. (2012). Molecular events initiating the exit of a copper-transporting ATPase ATP7B from the trans-Golgi network. *Journal of Biological Chemistry*.

Hatori, Y., Clasen, S., Hasan, N. M., Barry, A. N., & Lutsenko, S. (2012). Functional partnership of the copper export machinery and glutathione balance in human cells. *Journal of Biological Chemistry, 287*(32), 26678–26687.

Huttlin, E. L., Jedrychowski, M. P., Elias, J. E., Goswami, T., Rad, R., Beausoleil, S. A., et al. (2010). A tissue-specific atlas of mouse protein phosphorylation and expression. *Cell, 143*(7), 1174–1189.

Kaplan, J. H., & Lutsenko, S. (2009). Copper transport in mammalian cells: special care for a metal with special needs. *Journal of Biological Chemistry, 284*(38), 25461–25465.

Klomp, A. E., Tops, B. B., Van Denberg, I. E., Berger, R., & Klomp, L. W. (2002). Biochemical characterization and subcellular localization of human copper transporter 1 (hCTR1). *Biochemical Journal, 364*(Pt 2), 497–505.

Kuo, Y. M., Gybina, A. A., Pyatskowit, J. W., Gitschier, J., & Prohaska, J. R. (2006). Copper transport protein (Ctr1) levels in mice are tissue specific and dependent on copper status. *Journal of Nutrition, 136*(1), 21–26.

Lee, J., Petris, M. J., & Thiele, D. J. (2002). Characterization of mouse embryonic cells deficient in the ctr1 high affinity copper transporter. Identification of a Ctr1-independent copper transport system. *Journal of Biological Chemistry, 277*(43), 40253–40259.

Liang, Z. D., Tsai, W. B., Lee, M. Y., Savaraj, N., & Kuo, M. T. (2012). Specificity protein 1 (sp1) oscillation is involved in copper homeostasis maintenance by regulating human high-affinity copper transporter 1 expression. *Molecular Pharmacology, 81*(3), 455–464.

Lundby, A., Secher, A., Lage, K., Nordsborg, N. B., Dmytriyev, A., Lundby, C., et al. (2012). Quantitative maps of protein phosphorylation sites across 14 different rat organs and tissues. *Nature Communications, 3*, 876.

Lutsenko, S., Petrukhin, K., Cooper, M. J., Gilliam, C. T., & Kaplan, J. H. (1997). N-terminal domains of human copper-transporting adenosine triphosphatases (the Wilson's and Menkes disease proteins) bind copper selectively in vivo and in vitro with stoichiometry of one copper per metal-binding repeat. *Journal of Biological Chemistry, 272*(30), 18939–18944.

Markovic, J., Mora, N. J., Broseta, A. M., Gimeno, A., de-la-Concepcion, N., Vina, J., et al. (2009). The depletion of nuclear glutathione impairs cell proliferation in 3t3 fibroblasts. *PLoS One, 4*(7), e6413.

Molloy, S. A., & Kaplan, J. H. (2009). Copper-dependent recycling of hCTR1, the human high affinity copper transporter. *Journal of Biological Chemistry, 284*(43), 29704–29713.

Monetti, M., Nagaraj, N., Sharma, K., & Mann, M. (2011). Large-scale phosphosite quantification in tissues by a spike-in SILAC method. *Nature Methods, 8*(8), 655–658.

Nkabyo, Y. S., Ziegler, T. R., Gu, L. H., Watson, W. H., & Jones, D. P. (2002). Glutathione and thioredoxin redox during differentiation in human colon epithelial (Caco-2) cells. *American Journal of Physiology Gastrointestinal Liver Physiology, 283*(6), G1352–G1359.

Nose, Y., Kim, B. E., & Thiele, D. J. (2006). Ctr1 drives intestinal copper absorption and is essential for growth, iron metabolism, and neonatal cardiac function. *Cell Metabolism, 4*(3), 235–244.

Petris, M. J., Smith, K., Lee, J., & Thiele, D. J. (2003). Copper-stimulated endocytosis and degradation of the human copper transporter, hCtr1. *Journal of Biological Chemistry, 278*(11), 9639–9646.

Pilankatta, R., Lewis, D., Adams, C. M., & Inesi, G. (2009). High yield heterologous expression of wild-type and mutant Cu+-ATPase (ATP7B, Wilson disease protein) for functional characterization of catalytic activity and serine residues undergoing copper-dependent phosphorylation. *Journal of Biological Chemistry, 284*(32), 21307–21316.

Pilankatta, R., Lewis, D., & Inesi, G. (2011). Involvement of protein kinase D in expression and trafficking of ATP7B (copper ATPase). *Journal of Biological Chemistry, 286*(9), 7389–7396.

Pourvali, K., Matak, P., Latunde-Dada, G.O., Solomou, S., Mastrogiannaki, M., Peyssonnaux, C., et al. (2012). Basal expression of copper transporter 1 in intestinal epithelial cells is regulated by hypoxia-inducible factor 2alpha. *FEBS Letters, 586*(16): 2423–2427.

Puig, S., Mira, H., Dorcey, E., Sancenon, V., Andres-Colas, N., Garcia-Molina, A., et al. (2007). Higher plants possess two different types of ATX1-like copper chaperones. *Biochemical and Biophysical Research Communications, 354*(2), 385–390.

Ralle, M., Huster, D., Vogt, S., Schirrmeister, W., Burkhead, J. L., Capps, T. R., et al. (2010). Wilson disease at a single cell level: intracellular copper trafficking activates compartment-specific responses in hepatocytes. *Journal of Biological Chemistry, 285*(40), 30875–30883.

Ranjan, P., Anathy, V., Burch, P. M., Weirather, K., Lambeth, J. D., & Heintz, N. H. (2006). Redox-dependent expression of cyclin D1 and cell proliferation by Nox1 in mouse lung epithelial cells. *Antioxidants & Redox Signaling, 8*(9–10), 1447–1459.

Singleton, W. C., McInnes, K. T., Cater, M. A., Winnall, W. R., McKirdy, R., Yu, Y., et al. (2010). Role of glutaredoxin1 and glutathione in regulating the activity of the copper-transporting P-type ATPases, ATP7A and ATP7B. *Journal of Biological Chemistry, 285*(35), 27111–27121.

Song, I. S., Chen, H. H., Aiba, I., Hossain, A., Liang, Z. D., Klomp, L. W., et al. (2008). Transcription factor Sp1 plays an important role in the regulation of copper homeostasis in mammalian cells. *Molecular Pharmacology, 74*(3), 705–713.

Stephenson, S. E., Dubach, D., Lim, C. M., Mercer, J. F., & La Fontaine, S. (2005). A single PDZ domain protein interacts with the Menkes copper ATPase, ATP7A. A new protein implicated in copper homeostasis. *Journal of Biological Chemistry, 280*(39), 33270–33279.

Strausak, D., La Fontaine, S., Hill, J., Firth, S. D., Lockhart, P. J., & Mercer, J. F. (1999). The role of GMXCXXC metal binding sites in the copper-induced redistribution of the Menkes protein. *Journal of Biological Chemistry, 274*(16), 11170–11177.

Tennant, J., Stansfield, M., Yamaji, S., Srai, S. K., & Sharp, P. (2002). Effects of copper on the expression of metal transporters in human intestinal Caco-2 cells. *FEBS Letters, 527*(1–3), 239–244.

Tsigelny, I. F., Sharikov, Y., Greenberg, J. P., Miller, M. A., Kouznetsova, V. L., Larson, C. A., et al. (2012). An all-atom model of the structure of human copper transporter 1. *Cell Biochemistry and Biophysics, 63*(3), 223–234.

Tsivkovskii, R., MacArthur, B. C., & Lutsenko, S. (2001). The Lys1010-Lys1325 fragment of the Wilson's disease protein binds nucleotides and interacts with the N-terminal domain of this protein in a copper-dependent manner. *Journal of Biological Chemistry, 276*(3), 2234–2242.

Turski, M. L., Brady, D. C., Kim, H. J., Kim, B. E., Nose, Y., Counter, C. M., et al. (2012). A novel role for copper in Ras/mitogen-activated protein kinase signaling. *Molecular Cell Biology, 32*(7), 1284–1295.

Vanderwerf, S. M., Cooper, M. J., Stetsenko, I. V., & Lutsenko, S. (2001). Copper specifically regulates intracellular phosphorylation of the Wilson's disease protein, a human copper-transporting ATPase. *Journal of Biological Chemistry, 276*(39), 36289–36294.

Vanderwerf, S. M., & Lutsenko, S. (2002). The Wilson's disease protein expressed in Sf9 cells is phosphorylated. *Biochemical Society Transactions, 30*(4), 739–741.

Veldhuis, N. A., et al. (2009). Phosphorylation regulates copper-responsive trafficking of the Menkes copper transporting P-type ATPase. *International Journal of Biochemistry & Cell Biology, 41*(12), 2403–2412.

Voskoboinik, I., Fernando, R., Veldhuis, N., Hannan, K. M., Marmy-Conus, N., Pearson, R. B., et al. (2003). Protein kinase-dependent phosphorylation of the

Menkes copper P-type ATPase. *Biochemical and Biophysics Research Communications, 303*(1), 337–342.

Walker, J. M., Huster, D., Ralle, M., Morgan, C. T., Blackburn, N. J., & Lutsenko, S. (2004). The N-terminal metal-binding site 2 of the Wilson's Disease Protein plays a key role in the transfer of copper from Atox1. *Journal of Biological Chemistry, 279*(15), 15376–15384.

White, C., Kambe, T., Fulcher, Y. G., Sachdev, S. W., Bush, A. I., Fritsche, K., et al. (2009a). Copper transport into the secretory pathway is regulated by oxygen in macrophages. *Journal of Cell Science, 122*(Pt 9), 1315–1321.

Yamane, Y., Furuichi, M., Song, R., Van, N. T., Mulcahy, R. T., Ishikawa, T., et al. (1998). Expression of multidrug resistance protein/GS-X pump and gamma-glutamylcysteine synthetase genes is regulated by oxidative stress. *Journal of Biological Chemistry, 273*(47), 31075–31085.

Zheng, G., Chen, J., & Zheng, W. (2012). Relative contribution of CTR1 and DMT1 in copper transport by the blood-CSF barrier: implication in manganese-induced neurotoxicity. *Toxicology and Applied Pharmacology, 260*(3), 285–293.

Zheng, Z., White, C., Lee, J., Peterson, T. S., Bush, A. I., Sun, G. Y., et al. (2010). Altered microglial copper homeostasis in a mouse model of Alzheimer's disease. *Journal of Neurochemistry, 114*(6), 1630–1638.

Zimnicka, A. M., Ivy, K., & Kaplan, J. H. (2011). Acquisition of dietary copper: a role for anion transporters in intestinal apical copper uptake. *American Journal of Physiology Cell Physiology, 300*(3), C588–C599.

CHAPTER SEVEN

Metal Export by CusCFBA, the Periplasmic Cu(I)/Ag(I) Transport System of *Escherichia coli*

Tiffany D. Mealman*, Ninian J. Blackburn[†,1] and Megan M. McEvoy*

*Department of Chemistry and Biochemistry, University of Arizona, Tucson, AZ, USA

**Department of Science and Engineering, Oregon Health and Sciences University, Beaverton, OR, USA

[1]Corresponding author: E-mail: ninian@ebs.ogi.edu

Contents

Abstract

High levels of metal ions have the potential to cause cellular toxicity through a variety of mechanisms; therefore, cells have developed numerous systems that regulate their intracellular concentrations. The Cus resistance system aids in protection of *Escherichia coli* from high levels of Cu(I) and Ag(I) by actively transporting these metal ions to the extracellular environment. The Cus system forms a continuous complex, CusCBA, that spans the inner membrane, periplasm, and outer membrane of Gram-negative bacteria, together with a novel fourth component, the periplasmic metallochaperone, CusF. The metal-binding sites of CusA, CusB, and CusF are exquisitely tuned for Cu(I) and Ag(I), and thus effectively discriminate these ions for transport from other metals that may be required in the cell. Furthermore, direct transfer of metal from protein to protein within the Cus system during the transport process is likely to reduce the potential toxicity posed by the free metal ions. Here we review the wealth of structural, biochemical, and genetic information on the Cus system, which demonstrates the many intriguing aspects of function for metal-transporting efflux systems.

Current Topics in Membranes, Volume 69
ISSN 1063-5823,
http://dx.doi.org/10.1016/B978-0-12-394390-3.00007-0

1. INTRODUCTION

Microbial systems deal with a barrage of environmental challenges during their life cycle. Among these challenges is the requirement to obtain necessary levels of cofactors such as metals, while preventing toxic effects that an overabundance of these elements can cause. Copper ions present such a trial for living systems. Copper ions are required cofactors for numerous essential cellular reactions, such as electron transport, photosynthesis and respiration; however, excess copper leads to cell damage or death. Thus, to deal with these challenges, intricate and carefully regulated systems are in place to maintain metal homeostasis. Here we focus on one such system from Gram-negative bacteria that counter the challenges posed by high levels of copper and silver ions in the environment.

Metals in general exert their toxic effects through a variety of mechanisms. Copper surfaces, for example effectively kill bacterial cells, likely through the release of copper ions from metallic copper upon contact with the cells (Espirito Santo et al., 2011; Molteni, Abicht, & Solioz, 2010). The primary means for cellular harm by silver ions is probably a result of its high affinity for thiol groups, which can result in inappropriate binding to biomolecules and shut down essential functions such as the respiratory electron transport chain or DNA replication (Silver, Phung, & Silver, 2006). Through binding or displacement of a natural metal ion, enzyme activities or intermolecular interactions could be disrupted (Dibrov, Dzioba, Gosink, & Hase, 2002). In addition to ligand displacement, redox-labile metal ions such as copper can cause additional harm through the production of reactive oxygen species, which can lead to cellular damage (Halliwell & Gutteridge, 1984, 1990). The toxic properties of copper and silver ions have been exploited over the centuries by humans in efforts to limit microbial growth, since these metal ions are relatively nontoxic to humans at the concentrations that inhibit growth of many microbes. Some have suggested that the phrase "Born with a silver spoon in his mouth" refers to not only wealth but also the health that accompanies sanitary food consumption.

Metal homeostasis/resistance can be accomplished through a number of different strategies, including enzymatic detoxification, intra- or extracellular chelation, or transport systems (von Rozycki & Nies, 2009). In microorganisms, transport systems are prevalent and serve to limit metal concentrations through removal of metal from the cell (Nies, 2003). It is generally thought that the organism's specific living environment will reflect the specificities and levels of metal homeostasis/resistance systems in the organism. For example, *Escherichia coli* from the guts of pigs fed a copper supplemented diet have both chromosomally encoded and plasmid-borne copper resistance systems (Williams, Morgan, Rouch, Brown, & Lee, 1993). A strain of *Salmonella*, isolated from a burn ward at Massachusetts General Hospital in which seven people died of infections, was found to contain a plasmid bearing a silver resistance system likely selected in response to the silver treatments commonly used in burn wards (McHugh, Moellering, Hopkins, & Swartz, 1975). The most metal-resistant organisms, such as *Cupriavidus metallidurans*, seem to achieve a high level of metal resistance not through new systems that are not seen in more common organisms, but through having multiple variations of existing metal resistance systems (von Rozycki & Nies, 2009). *Escherichia coli* provides a powerful model system for analysis of a microorganism's response to metal stress, as it contains representatives of systems found in the most metal-resistant bacteria, such as *C. metallidurans*, yet lacks the duplicity of systems that would complicate efforts to understand specific mechanisms.

In *E. coli*, copper tolerance is mediated by two chromosomally encoded systems, the Cue and Cus systems, which are induced under conditions of excess copper (Rensing & Grass, 2003). The Cue system provides the first line of defense in copper resistance, as it is induced under low copper concentrations (Outten, Huffman, Hale, & O'Halloran, 2001). The Cus system is important for copper tolerance in anaerobic conditions and under aerobic conditions when levels of copper stress are extreme (Outten et al., 2001). The Cue system includes a multicopper oxidase, CueO, and a P-type ATPase, CopA. CopA pumps Cu(I) from the cytoplasm into the periplasm (Rensing, Fan, Sharma, Mitra, & Rosen, 2000). CueO resides in the periplasm, where it converts Cu(I) into the less toxic Cu(II) (Singh, Grass, Rensing, & Montfort, 2004). If copper concentrations continue to increase, the expression of the Cus system increases, as regulated by a two-component signaling system, CusR and CusS (Gudipaty, Larsen, Rensing, & McEvoy, 2012; Munson, Lam, Outten, & O'Halloran, 2000). The *cusCFBA* operon encoding the proteins CusC, CusF, CusB and CusA is induced by both copper and silver (Franke, Grass, & Nies, 2001; Franke, Grass, Rensing, & Nies, 2003; Munson et al., 2000). Genetic and phenotypic analyses strongly suggest that the products of *cusCFBA* form a CBA-type multimeric Cu(I) or Ag(I) efflux complex that spans from the cytoplasm to the extracellular space (Fig. 1) (Franke et al., 2003; Grass & Rensing, 2001; Munson et al., 2000; Outten et al., 2001).

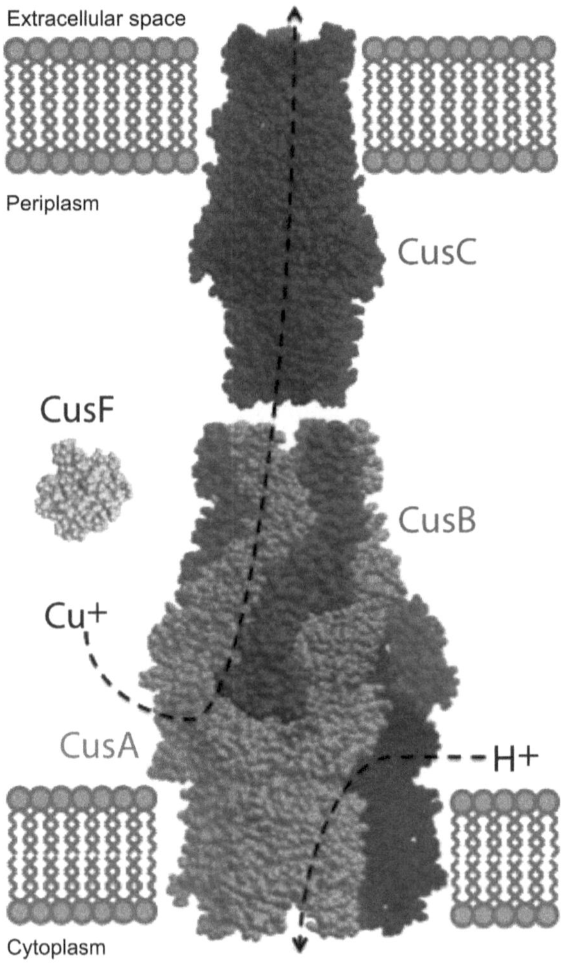

Figure 1 Cartoon representation of the Cus Cu(I)/Ag(I) export complex. The inner and outer membrane proteins, CusA and CusC, are trimers. There are six copies of the membrane fusion protein (MFP), CusB, shown in red and orange. Proton flow across the inner membrane is coupled with copper and silver export to the extracellular space. See the color plate.

CBA-type transporters are usually composed of three proteins and are involved in the export of metal ions, xenobiotics and drugs (Nies, 2003). The three conserved components of CBA transporters include a proton–substrate antiporter of the RND family (resistance, nodulation and cell division), an MFP (also called a periplasmic adaptor protein), and an outer membrane factor (OMF) (Nies, 2003; Zgurskaya & Nikaido, 1999, 2000).

Copper/silver efflux systems often have a fourth periplasmic component, such as CusF in the Cus system (Franke et al., 2003; Loftin et al., 2005), which can function as a metallochaperone (Bagai, Liu, Rensing, Blackburn, & McEvoy, 2008).

The RND superfamily of protein transporters are membrane-spanning proteins that drive substrate movement by coupling proton influx with substrate export (Goldberg, Pribyl, Juhnke, & Nies, 1999; Nies, 1995). RND transporters are involved in the efflux of a wide variety of substrates, including organic compounds, transition metals, and polypeptides (Saier, 2000; Tseng et al., 1999). While other types of efflux pumps (MATE, ABC, SMR, and MFS) of Gram-positive bacteria function individually (Piddock, 2006), the RND transporters of Gram-negative systems form tripartite protein complexes with a periplasmic MFP and an OMF protein, as described above for the Cus system in *E. coli*. RND proteins are fairly large in size; they contain more than 1000 amino acids and consist of two large hydrophobic domains and two large hydrophilic domains and are assembled into trimers (Goldberg et al., 1999).

Within the RND superfamily, seven families have been identified, of which two of these have been studied extensively. The first family is involved in the efflux of hydrophobic and amphiphilic substrates (HAE-RND) (Piddock, 2006), which consists of a range of compounds including antibiotics, bile salts, dyes, organic solvents, and peptides. A review of the structure and mechanism of multidrug efflux systems belonging to this RND protein family has been published recently (Nikaido, 2011). The second extensively characterized RND family consists of transporters that mediate heavy metal efflux (HME-RND) (Nies, 2003). *Escherichia coli* contains seven RND efflux transporters, six of which are multidrug efflux pumps of the HAE-RND family and include AcrB, AcrD, AcrF, MdtB, MdtC, and YhiV (Tseng et al., 1999). There is only one known HME-RND transporter in *E. coli*, the Cu(I)/Ag(I) efflux protein CusA (Franke et al., 2001, 2003). HME-RND systems from other organisms, such as SilA from *Salmonella enterica* serovar Typhimurium and CzcA from *C. metallidurans*, have also been identified and characterized (Rensing, Pribyl, & Nies, 1997; Silver, 2003).

In general, HAE-RND systems provide resistance to a wide variety of substrates and hence have been termed "periplasmic vacuum cleaners" (Lomovskaya, Zgurskaya, Totrov, & Watkins, 2007), while HME-RND systems demonstrate a high degree of substrate specificity, primarily transporting only metal ions with similar properties (Conroy, Kim, McEvoy, &

Rensing, 2010; Kittleson et al., 2006). The narrow substrate specificity of HME-RND systems is likely due to the metal-binding site(s) present in the RND protein itself, as well as metal-binding sites in the other proteins in the complex. In all Cus protein–metal ion complex structures determined, only the uncomplexed metal ions were observed bound to the proteins, and not metal ion–small molecule complexes (Loftin, Franke, Blackburn, & McEvoy, 2007; Long et al., 2010; Su et al., 2009; Xue et al., 2008). These metal-binding sites not only provide specificity but also are likely to serve a dual function in metal resistance by keeping the metal ion coordinated, thus preventing the cellular damage that can be caused by the free metal. As described below, the metal-binding sites and the transfer of metal ions from one protein to the next have been partially elucidated for the Cus system, providing valuable insight into the resistance mechanisms for this family of proteins.

To describe the molecular mechanisms of the Cus system, we will first begin with a summary of each of the protein components with respect to structure and metal-binding details. Then, where possible, we will discuss the interactions of the components and metal transfer events. Finally, we conclude with an overview of the system in which these pieces are assembled and discuss where further information is needed.

2. CUS SYSTEM COMPONENTS

2.1. CusA

The inner membrane RND proton/substrate antiporter CusA is the central component of the Cus system. If *cusA* is deleted, it leads to complete loss of *cus*-mediated copper resistance (Franke et al., 2003) and shows the same copper sensitivity as removal of the entire *cus* operon (Grass & Rensing, 2001; Grass & Rensing, 2001). Early sequence alignments of CusA with SilA (Gupta, Matsui, Lo, & Silver, 1999) and other putative RND transporters revealed the presence of five conserved methionines in one of the periplasmic domains of CusA (Franke et al., 2003). These residues are not conserved in divalent cation RND transport proteins, such as CzcA (Goldberg et al., 1999; Nies, Nies, Chu, & Silver, 1989). Individual mutations of three of these residues (M573, M623, and M672) to isoleucines resulted in the loss of copper resistance, suggesting their functional importance (Franke et al., 2003). Mutations of the other two conserved methionines in the first periplasmic domain, along with four additionally conserved methionines in the second periplasmic domain and a three-methionine cluster (MXXMXXM, a common copper-binding motif) adjacent to the second periplasmic domain, led to little or no effect on copper tolerance (Franke et al., 2003). Thus, while these residues are conserved and even include a common copper-binding motif, they do not appear to be essential for CusA function. From this work, the metal-binding site was proposed to consist of M573, M623, and M672 (Franke et al., 2003). These results, together with the identification of methionine-rich motifs in copper transport proteins of non-RND systems from *Schizosaccharomyces pombe* (Zhou & Thiele, 2001), support this proposed metal-binding site.

With the elucidation of the structure of the *E. coli* multidrug HAE-RND transporter AcrB in 2002, efforts were made to crystallize the HME-RND homolog CusA under similar conditions, but these efforts were unsuccessful. Sequence alignment of AcrB and CusA shows that these proteins share only 19% sequence identity and an extensive limited proteolysis study revealed that CusA is extremely flexible in solution compared to AcrB, suggesting an explanation for the difficulty observed in crystallization of CusA in similar conditions as AcrB (Deniaud, Goulielmakis, Coves, & Pebay-Peyroula, 2009). In 2010, the crystal structure of CusA was determined to 3.52 Å by (Long et al. 2010) making it the first RND transport protein of the HME subfamily to be determined. The structures of apo-CusA, Ag(I)-CusA, and Cu(I)-CusA have provided long-awaited structural information about HME-RND proteins. In these structures, CusA forms a homotrimer, as expected from the previously determined structures of HAE-RND proteins, AcrB (Murakami, Nakashima, Yamashita, & Yamaguchi, 2002) and MexB from *Pseudomonas aeruginosa* (Sennhauser, Bukowska, Briand, & Grutter, 2009), involved in multidrug efflux systems. The homotrimeric form is also in agreement with analytical ultracentrifugation experiments of CusA in solution (Stroebel et al., 2007).

CusA is a large protein comprised of 1047 amino acids and contains both periplasmic and transmembrane regions (Fig. 2A). Each CusA monomer has 12 transmembrane (TM) helices (TM1–TM12) and a large periplasmic domain formed by two large periplasmic loops, one between TM1 and TM2 and the other between TM7 and TM8. One difference between CusA and the structures of AcrB and MexB is the extension of four helices (TM4, TM5, TM10, and TM11) into the cytoplasm where they form a cytoplasmic domain. Additionally, TM2 and TM8 protrude into the periplasmic space and form portions of the periplasmic domain. Similar to the periplasmic

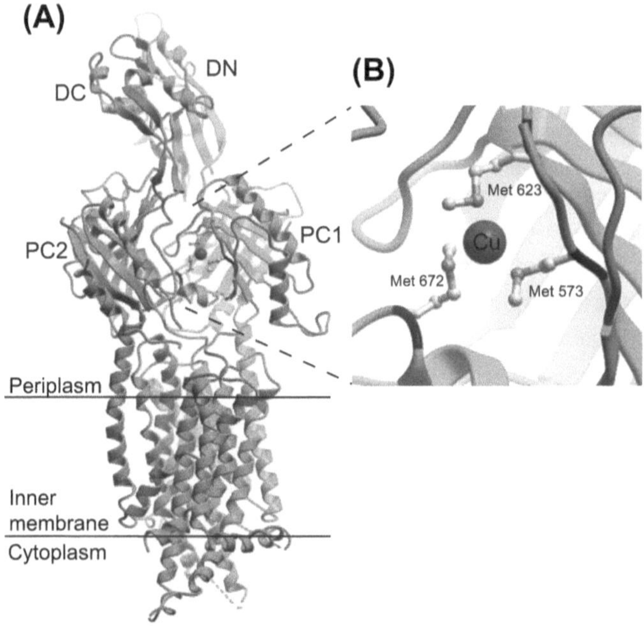

Figure 2 The inner membrane protein CusA. (A) Ribbon diagram of one subunit of CusA. Copper is shown in space-filling representation. (B) Close-up view of the three-coordinate metal-binding site consisting of Met 573, Met 623, and Met 672. Structures are based on PDB file 3KOI (Long et al., 2010). For color version of this figure, the reader is referred to the online version of this book.

domains of AcrB and MexB, the periplasmic portion of CusA is divided into six subdomains. Four subdomains constitute the pore domain (PN1, PN2, PC1, and PC2, in which N and C refer to N- or C-terminal) and two subdomains compose the docking domain (DN and DC), which is hypothesized to interact with the OMF CusC. Within the pore domain, PN1 is central and appears to stabilize the trimer, while the remaining subdomains, PN2, PC1, and PC2, are arranged facing out toward the periplasm. Subdomains PC1 and PC2 form an external periplasmic cleft that is closed in the apo-CusA structure by a single α-helix (residues 690–706) and three β-strands (residues 681–687, 711–716, and 821–827) that appear to be tilted into this cleft. A single horizontal α-helix, composed of residues 665–675, is present at the bottom of the cleft, roughly separating the periplasmic and transmembrane domains. The structural features of the periplasmic cleft are unique from that of AcrB and MexB, potentially governing the specificity of CusA for monovalent cations.

In the Cu(I)-CusA and Ag(I)-CusA structures (Long et al., 2010), metal ions are coordinated by a three-methionine cluster composed of M573, M623, and M672 (Fig. 2B), as was previously proposed (Franke et al., 2003). These residues are located above the horizontal helix at the base of the periplasmic cleft. Metal binding results in overall changes to the protein structure (Fig. 3A), such that superposition of Cu(I)-CusA and apo-CusA results in an rmsd of 3.9 Å over 1006 Cα atoms. The Cu(I)- and Ag(I)-bound structures are nearly identical. Though the apo and metal-bound structures of CusA were determined as symmetrical trimers, the structures of

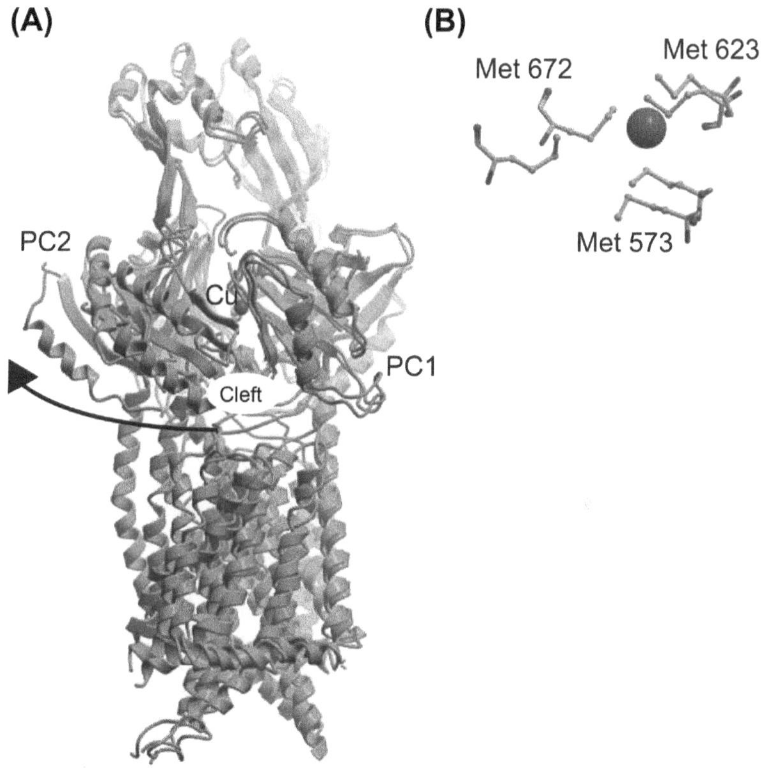

Figure 3 Movements in apo and metal-bound CusA. (A) Ribbon representation of one monomer of CusA in the apo state (cyan) and the Cu(I)-bound state (green). Movement of domain PC2 in the Cu(I)-bound state opens the cleft between PC1 and PC2. (B) Met 573 and Met 623 of CusA are in similar positions in both apo (cyan) and Cu(I)-bound (green) structures. The conformation of Met 672 changes significantly in the metal-bound state. Structures are based on PDB files 3KO7 and 3K0I (Long et al., 2010). For interpretation of the references to color in this figure legend, the reader is referred to the online version of this book.

the multidrug transporter AcrB showed asymmetry among the three subunits (Murakami, Nakashima, Yamashita, Matsumoto, & Yamaguchi, 2006; Seeger et al., 2006). The asymmetry is expected to reflect the different conformations that are needed to alternately allow binding, transport, and release of substrate in response to proton movement across the inner membrane. The apo and metal-bound conformations of CusA may represent two of these conformers, though further structural details are needed.

Metal binding to membrane transporters is expected to divide into two categories. First the metal may bind to a high-affinity site selective for the metal being transported. This event is designed to impart selectivity by inducing a conformational change which opens a pore or channel allowing access and transport of the metal ion, perhaps assisted by weaker (transient) binding sites along the channel pathway. The Cu(I)- and Ag(I)-bound CusA structures nicely underscore this paradigm. Here, the triad of methionines (M573, M623, and M672) is likely to bind copper and silver with high affinity (similar binding sites in CusB have been shown to have nanomolar or greater affinity; Bagai, Rensing, Blackburn, & McEvoy, 2007), but will discriminate against almost every other bioavailable cation. In addition, metal binding induces a large conformational change which is dictated by the positioning of the metal-binding Met residues in the apoprotein (Fig. 3B). Whereas M623 and M573 are well situated for Cu(I) binding, M672 is positioned too far for facile binding and must therefore move to facilitate Cu(I) binding by all three Met residues. M672 resides on the short helix at the base of the metal-binding site and metal coordination therefore causes upward movement of this helix, triggering a train of conformational changes. The result is the opening of the cleft formed between PC1 and PC2 which might provide a periplasmic entry point for metal ions and could be the site of interactions with the MFP CusB or the periplasmic copper chaperone CusF (Fig. 3A). It has been suggested that the three-methionine cluster could be involved in transient metal transfer with CusB or CusF, similar to that observed in the Atx1–Ccc2 complex (Banci et al., 2006).

As postulated above, transport of metal ions through the tripartite complex is expected to be facilitated by relays of more weakly binding residues which guide the metals along a kinetically labile vectorial pathway devoid of thermodynamic traps. The crystal structures may give clues as to the identity of such relays in the form of pairs of Met residues often described as "Met wires". The transmembrane domain of CusA contains six methionine residues that appear to form distinct pairs: M410/M501, M403/M486, and M391/M1009. All six methionine residues are conserved within HME-

RND-type Cu(I)/Ag(I) efflux pumps (when 30 of these pumps are aligned) (Long et al., 2010), though the overall sequences themselves are highly conserved. Thus, it has been suggested that these methionine pairs in the transmembrane domain of CusA could act as transient sites for metal transport from the cytoplasm. Two more residues (M271 and M755) in the periplasmic domain are close to each other and could potentially constitute another methionine pair although M755 is not completely conserved across the HME-RND pumps examined. Central to the hypothesis is the idea that the interaction of the Cu(I) or Ag(I) with the Met pairs should be weak with no thermodynamic barrier for exchange of metal from one pair to the next. To date although three- and four-coordinate Cu–S(Met)$_2X_n$ ($n=1$, 2) complexes are common [CusF: $X=$ His, Trp (Loftin et al., 2007), CusB: $X=$ Met (Bagai et al., 2007), CopK: $X=$ (Met)$_2$ (Sarret et al., 2010)], no examples of two-coordinate Cu(I)–S(Met)$_2$ have been described in either inorganic or biological systems. However, two-coordinate complexes with cysteine are well known and include CueR (Chen, Yuldasheva, Penner-Hahn, & O'Halloran, 2003), Atox1 (Ralle, Lutsenko, & Blackburn, 2003) and other Atx1-like N-terminal subdomains of copper-translocating ATPases such as mammalian ATP7A&B (Ralle, Cooper, Lutsenko, & Blackburn, 1998), (Achila et al., 2006) and yeast Ccc2 (Banci, Bertini, Ciofi-Baffoni, Huffman, & O'Halloran, 2001). Here the Cu–S distances are short at around 2.14 Å. This analysis suggests that an Met pair capable of a transient, kinetically labile interaction with Cu(I) should have sulfur to sulfur distances in excess of twice the linear two-coordinate Cu–S distance, i.e. in excess of 4.3 Å. Examination of distances in the proposed Met pairs in the CusA/CusB structure (Su et al., 2011) (see below) are M403/M486 = 5.5 Å, M410/M501 = 6.4 Å, M391/M1009 = 8.9 Å, and M271/M755 = 11 Å. The first two pairs with S–S distances of ~6 Å or less would thus appear to be candidates for a transient Cu(I)–(Met)$_2$ interaction, while the others appear to have S–S distances that would be too great to form transient sites without significant conformational change.

The functional importance of these Met wires has been investigated by mutagenesis. Whereas mutational analysis confirmed that individual mutations of the residues composing the metal-binding site (M573, M623, and M672) to isoleucines result in the complete loss of copper resistance (Long et al., 2010), as was previously demonstrated (Franke et al., 2003), mutation of residues involved in the methionine pairs (M391I, M410I, M486I, and M755I) showed a decrease in copper tolerance but not a complete loss of CusA function. This suggests that these methionine pairs contribute to CusA function and could potentially be involved in metal transport from the cytoplasm but do not appear to be essential for CusA function. The possibility that these mutants affect the structure of CusA and impair function of the pump cannot be ruled out.

These and other considerations raise questions as to whether the Met pairs are important for transport. Experiments with CusA-containing liposomes suggest that transport of Ag(I) occurs from the equivalent of the cytoplasm to the periplasm dependent on the proton gradient (Long et al., 2010). However, a genetic study demonstrated that CusA does not functionally substitute for the inner membrane transporter CopA, suggesting that the primary substrates for transport by the Cus system originate in the periplasm (Grass & Rensing, 2001). Lack of functional compensation may be due to different rates of transport for substrates in the cytoplasm or periplasm, though no transport rates have yet been measured. The distances between the methionine pairs/triad in the Cu(I)-bound structure (9.4, 13.0, 20.0 and 14.7 from the cytoplasmic side to the periplasmic side) do not support the idea that metal ions are transferred directly from one site to the next. Since the Cu(I)- or Ag(I)-bound structures do not show metal ions bound at these sites, it remains to be seen whether these sites indeed form a "Met wire" capable of transporting metal ions in a physiologically relevant manner.

Conformational changes associated with metal binding are also transmitted to TM8 whose movement could lead to signaling within the other transmembrane helices eventually initiating proton movement across the inner membrane (Long et al., 2010). Previous mutational analysis of conserved residues in or near TM4, D405, E412, and A399 indicate that these residues are essential for full CusA function and could be involved in the proton relay network (Franke et al., 2003). In the crystal structure, acidic D405 is found to interact with E939 of TM10 and K984 of TM11 (Long et al., 2010). Individual mutations of these three residues to alanine resulted in the loss of copper resistance, revealing that they are essential for transport function. Mutation of E412, which is located near the triad formed by D405, E939, and K984, did not result in the complete loss of copper resistance, but showed a significant decrease compared to the wild type. Together, it is suggested that these charged residues are involved in proton translocation across the inner membrane.

2.2. CusB

In addition to the inner membrane protein, the MFP (sometimes referred to as the periplasmic adaptor protein) (Dinh, Paulsen, & Saier, 1994; Saier,

Tam, Reizer, & Reizer, 1994) is an essential component of CBA-type transport systems (reviewed in Zgurskaya, Yamada, Tikhonova, Ge, & Krishnamoorthy, 2009). MFPs stabilize CBA transport complexes through association with both the inner and the outer membrane proteins and are required for tripartite complex assembly. Membrane association is a common feature of multidrug-resistant MFPs, which anchor to the inner membrane via an N-terminal transmembrane α-helix (N-TMS) or an N-terminal lipid modification (typically via cysteine) (Zgurskaya et al., 2009). The MFP in the Cus system, CusB, is essential for *cus*-mediated metal resistance in *E. coli* (Franke et al., 2003); however, CusB differs from other MFPs in that it purifies as a soluble protein and does not associate with either membrane (Bagai et al., 2007).

Initial CusB characterization in vitro indicates that this 379 amino acid protein is monomeric in solution and that the oligomeric state does not change upon metal binding (Bagai et al., 2007). CusB binds Ag(I) with a 1:1 stoichiometry and high affinity (K_d of 24.7 nM) by isothermal titration calorimetry (ITC) (Bagai et al., 2007). X-ray absorption spectroscopy (XAS) was utilized to probe the metal-binding site of CusB and determined that Cu(I) is bound in a three-coordinate environment by three sulfur atoms (Bagai et al., 2007). Mutational analyses demonstrated that M21I, M36I, or M38I (numbering is of the mature protein without the leader sequence) substitutions all had significant effects on metal binding in vitro. Furthermore, in vivo these mutants decreased the ability of *E. coli* cells to grow on copper-containing media and suggested that metal binding by CusB is functionally important for resistance (Bagai et al., 2007). Size exclusion chromatography demonstrated that CusB undergoes a conformational change upon metal binding, adopting a more compact structure; thus, it was proposed that CusB and other MFPs of RND-type transport systems play an active role in substrate efflux and are not simply adaptors (Bagai et al., 2007). Substrate-induced conformational changes have also recently been observed in ZneB, the MFP of the zinc efflux complex from *C. metallidurans* CH34 (De Angelis et al., 2010), supporting a similar role.

In 2009, the crystal structure of CusB was determined to 3.4 Å (Su et al., 2009) (Fig. 4). The structure contains 78% of the total protein sequence (residues 61–357, using numbering of the mature protein after processing of the signal peptide). Sixty residues from the N-terminus and 22 residues from the C-terminus were unable to be resolved (Su et al., 2009), suggesting that these regions could potentially be disordered, dynamic, or possibly proteolyzed during crystallization. Two molecules of CusB were in the

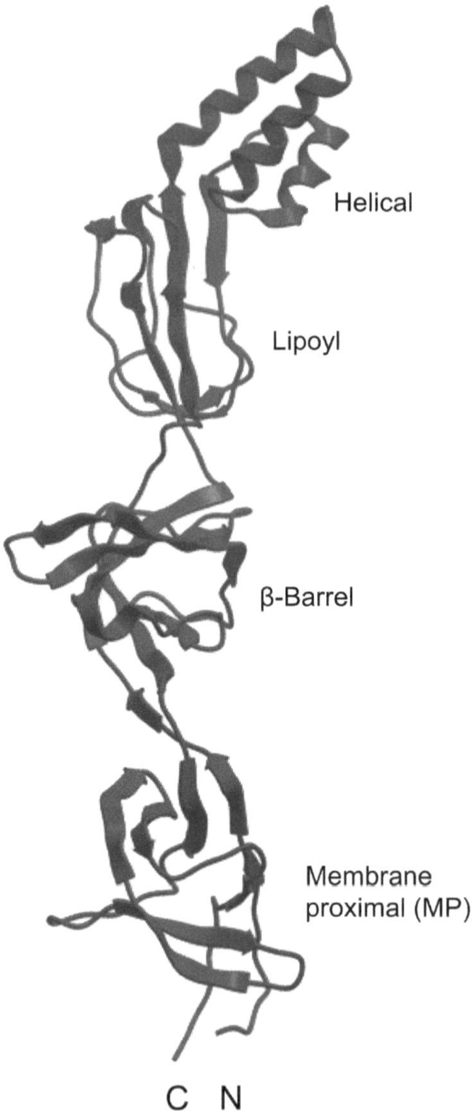

Figure 4 Ribbon representation of CusB, showing the four domains that are resolved in the crystal structure. Based on PDB file 3OOC (Su et al., 2009). For color version of this figure, the reader is referred to the online version of this book.

asymmetric unit of the crystal and both reveal an elongated protein structure (116–121 Å long and 37–40 Å wide); however, each molecule has a distinct conformation, which suggests that it is flexible in nature, as has been observed with other MFPs (Mikolosko, Bobyk, Zgurskaya, & Ghosh, 2006;

Vaccaro, Koronakis, & Sansom, 2006; Wang, Weng, Fan, & Wang, 2012). CusB consists of four distinct domains, three of which contain mostly β-strands and the fourth, closest to the outer membrane, is completely α-helical, similar to MFP structures from other systems. However, the N-terminal metal-binding region seems to be unique to CusB, as the MFPs of other systems lack a homologous region.

The first β-domain common to each MFP structure is called the membrane proximal (MP) domain because of its proximity to the inner membrane. This domain of CusB forms a β-barrel composed of six β-strands (one from the N-terminus and five from the C-terminus). The MP domain is predicted to interact with the RND transporter protein (De Angelis et al., 2010; Symmons, Bokma, Koronakis, Hughes, & Koronakis, 2009), which is supported by cross-linking data in the Cus system (Su et al., 2009). Further interactions between CusB and CusA have recently been elucidated with the CusBA co-crystal structure, showing direct interactions between these proteins (Su et al., 2011), described below.

Adjacent to the MP domain is the β-barrel domain, which, in one molecule of CusB, consists of six β-strands and one short α-helix and in the other molecule consists of six β-strands and two short α-helices. The N-terminal residues of CusB contribute one β-strand and the remaining secondary structural features are from the C-terminal end. The next globular β-domain is called the lipoyl domain and consists of a β-barrel formed by eight β-strands. These two domains are connected by a linker region composed of two small loops, which could potentially act as a flexible hinge for conformational changes necessary for CusB function and tripartite assembly. The presence of two distinct conformations of CusB within a single asymmetric unit, together with the presence of linkers between domains, points toward the conformational flexibility of the protein, also observed in AcrA (Mikolosko et al., 2006).

The most striking structural difference between CusB and other MFPs is in the all α-helical domain. While other MFPs contain an α-helical hairpin motif, the helical domain in CusB forms an antiparallel, three-helix bundle, a unique feature which has not been found in any other MFPs. The three-helix bundle comprised of a helix-turn-helix-turn-helix secondary structure is ~27 Å long, which makes it at least 20 Å shorter than the α-helical hairpin domains of MexA (Akama et al., 2004; Higgins, Bokma, Koronakis, Hughes, & Koronakis, 2004), AcrA (Mikolosko et al., 2006), and MacA (Yum et al., 2009). The α-helical domain of MFPs has been shown to be involved in direct interactions with the OMF protein (Tikhonova &

Zgurskaya, 2004; Touze, Eswaran, Bokma, Koronakis, Hughes, & Koronakis, 2004) and furthermore this interaction has proven to be essential for functional assembly of the tripartite complex (Bavro et al., 2008; Janganan et al., 2011; Kim et al., 2010; Lobedanz et al., 2007; Seeger et al., 2006). It is possible that the unique structure of the α-helical domain of CusB is essential for the recruitment of CusC and functional assembly of CusCBA.

By the addition of metal solutions to apo-CusB crystals, two putative metal-binding sites were proposed (Su et al., 2009). The first, composed of M296, F330, and R340, is located in the MP domain and the second, composed of W130, Q134, and M162, is within the three-helix bundle. Previous ITC and extended X-ray absorption fine structure (EXAFS) results demonstrated that full-length CusB binds a single metal ion in a three-sulfur coordination site consisting of conserved residues M21, M36, and M38 (Bagai et al., 2007). Mutation of these three methionines to isoleucines causes a loss of metal-binding activity and a significant decrease in copper resistance of *E. coli* cells in vivo. Mutation analysis of M162I and M296I shows no significant decrease in copper resistance (Kim E. H., Mealman T. D., Rensing C., McEvoy M. M., in preparation), thus suggesting that these residues are not physiologically relevant metal-binding sites, but rather a result of nonspecific metal association due to the soaking of CusB crystals in high metal concentrations. Furthermore, XAS analysis of the N-terminal region of CusB (residues 1–61) likewise shows copper coordinated to three methionine residues (Mealman et al., in preparation) indicating that this region of the full-length protein contains the metal-binding site. Functional analysis of this N-terminal region of CusB in vivo reveals that it is essential for full function of the CusCFBA efflux system (Mealman et al., in preparation). Structural information for this critical region of CusB is needed to more fully understand the function.

2.3. CusC

CusC is the outer membrane component of the CusCFBA efflux pump. Initial analysis of the *cusCFBA* operon revealed that deletion of *cusC* led to a strong decrease in copper resistance, but did not lead to complete loss of resistance, as was exhibited with deletion of *cusA* or *cusB* (Franke et al., 2003). Thus, it is likely that another OMF gene can complement the loss of CusC and confer slight resistance. It was hypothesized that TolC, the OMF from the *E. coli* multidrug efflux system AcrAB-TolC, may substitute this role; however, removal of the *tolC* gene (Δ*cueO*Δ*cusC*Δ*tolC*::Kan triple mutant strain) resulted in the same copper resistance as the Δ*cueO*Δ*cusC* double mutant strain, indicating that TolC cannot partially substitute for CusC. This is most likely because TolC is unable to form a functional assembly with CusBA. Potentially, one of the other OMF proteins in *E. coli* can substitute for CusC, but further investigation is needed.

The crystal structure of the OMF CusC from *E. coli* was solved in 2011 to 2.3 Å and was the first OMF of any HME-RND system to be determined (Kulathila, Indic, & van den Berg, 2011) (Fig. 5). Structures of the OMFs OprM (Akama et al., 2004) from the multidrug resistance system MexAB-OprM of *P. aeruginosa* and TolC (Koronakis, Sharff, Koronakis, Luisi, & Hughes, 2000) from *E. coli* had been determined previously. CusC shares 45% sequence identity with OprM; therefore, OprM was used as the model for molecular replacement in order to solve the structure of CusC. The final model contains the full sequence of the mature protein, except for residues 21–31 in the equatorial domain, which appear to be disordered. One

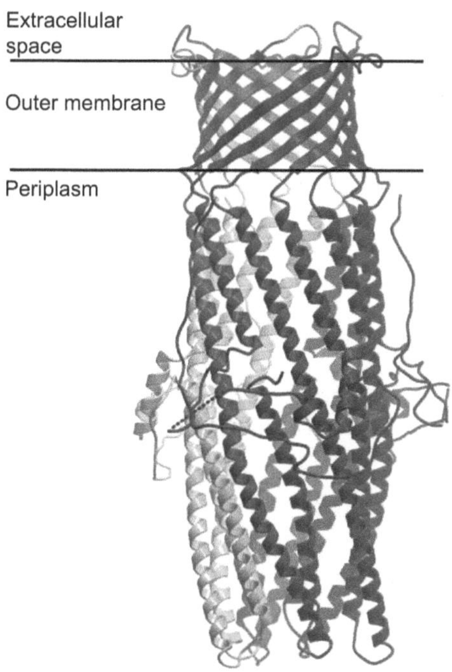

Figure 5 Ribbon representation of the outer membrane protein CusC. The three subunits of CusC are shown in three different colors. Based on PDB file 3PIK (Kulathila et al., 2011). See the color plate.

monomer (440 residues) of CusC is present in the asymmetric unit of the crystal. CusC forms a trimer within the crystal and its elution profile by gel filtration also matches that of a trimer in solution. This is in good agreement with the oligomeric behavior of all other OMF proteins; furthermore, the structure of CusC is very similar to OMF proteins solved previously. The overall trimeric structure of CusC is that of a cannon-shaped molecule formed by a long α/β-barrel, ~130 Å in length. The 12-stranded β-barrel of CusC is located within the outer membrane, in which four β-strands are contributed from each monomer. The remainder, and majority, of the protein forms an α-barrel in the periplasm, consisting of 12 long α-helices, 4 from each monomer of CusC. Two of these helices are not continuous, but are formed by two short separate helices that meet end to end. Together, the monomers form a trimer with a large water-filled internal cavity that is cylindrical in shape and has a volume of ~28,000 $Å^3$.

The charge distribution of CusC is similar to other OMF proteins, demonstrating that the outside surface of CusC does not contain extensive patches of charged residues. A space-filling view of the inside of the cavity, however, reveals that the interior is extremely electronegative. This feature raises the question as to how Cu(I)/Ag(I) ions are transported through CusC without getting trapped in the electronegative interior. It has been suggested that since the channel is ~25 Å in diameter, this area is sufficiently large enough for the monovalent cations to travel through the middle of the channel along with the bulk solvent without getting trapped by electrostatic interactions between the positively charged metal ions and the negatively charged residues lining the inside. Presumably, the relatively hydrophobic periplasmic end of the α-barrel interacts directly with the top of the periplasmic domain of the inner membrane protein CusA (Misra & Bavro, 2009; Symmons et al., 2009).

Similar to OprM, but unique from TolC, CusC has an N-terminal extension comprising 20 residues that extend ~25 Å between the β-barrel and the equatorial domain. Because the first amino acid of the N-terminus is a cysteine residue, CusC is predicted to be an OM lipoprotein. In the crystal structure of CusC, density is observed for three partial acyl chains, shown to attach to Cys1 of each CusC monomer, thus anchoring it into the outer membrane. It would appear that Cys1 of CusC, like the Braun lipoprotein from *E. coli* and other bacterial lipoproteins, is triacylated with *N*-acyl and *S*-diacylglycerol functionalities (Neilsen, Zimmerman, & McIntyre, 2001). The length of these acyl chains is unclear from the electron density; however, they are positioned in the same area as a region termed the aromatic belt. This region is formed by aromatic residues that surround the interface between the β-barrel and the α-helical barrel, corresponding to the inner leaflet of the outer membrane. To our knowledge, this is the first example of an OM lipoprotein that is anchored by a triacylated N-terminus. It is unclear why CusC is triacylated, as it also contains a hydrophobic OM-anchoring region of 20–25 Å. It has been hypothesized that either the presence of a serine residue in the middle of the OM region or the presence of the large periplasmic α-barrel could affect the OM channel insertion. However, because TolC is not acylated and still maintains a very similar structure to CusC and OprM, this suggests another reason still to be discovered for the additional membrane anchor in CusC.

Interestingly, even though the Cus system transports small substrates (Conroy et al., 2010), CusC has the largest extracellular opening observed for any OMF (~30 Å diameter). For comparison, the extracellular openings of TolC and OprM are ~12 Å in diameter, even though their substrates are much larger in size than monovalent cations. The observed extracellular opening of CusC, with loops splayed apart, is most likely a representation of the open state of the OMF. This is in contrast to the closed periplasmic α-barrel region that will require a substantial conformational change to open, as seen in other outer membrane proteins (Symmons et al., 2009). There are no known metal-binding motifs in CusC and it is very unlikely that the OMF protein contains a specific metal-binding site. The soaking of CusC crystals in high concentrations of Cu(I) and Ag(I) did not lead to any observable metal binding (Kulathila et al., 2011). Additionally, there are only three methionines located on the interior of each monomer of CusC, but they are located too far away from each other to constitute a three-coordinate binding site, as is exhibited by CusB and CusA. It is unlikely that CusF or CusB transfer metal to a binding site within CusC. Metal ions most likely enter the channel via CusA and are subsequently transported by diffusion with the bulk solvent.

2.4. CusF

CusF, the novel fourth component of the Cus efflux system, is only found in putative copper/silver resistance systems. It is a small periplasmic metal-lochaperone that delivers metal to the MFP CusB (Bagai et al., 2008). CusF is required for full copper resistance; however, deletion of *cusF* results in a copper-sensitive phenotype (Franke et al., 2003) but not a complete loss of *cus*-mediated copper tolerance, as is exhibited by deletion of either *cusA* or *cusB*. Under copper stress, the *cusF* gene transcript increases the most compared to all *E. coli* gene transcripts (Egler, Grosse, Grass, & Nies, 2005). Because CusF homologs are only found in putative copper/silver transport systems and not in well-characterized CBA-type transport systems, extensive investigations have been performed in order to determine the role of CusF in metal resistance.

The crystal structure of apo-CusF, which was determined in 2005 to a resolution of 1.5 Å, shows that CusF forms a small five-stranded β-barrel, consisting of two three-stranded antiparallel β-sheets that share the first β-strand in each sheet (Loftin et al., 2005) (Fig. 6). Interactions between β3 and β5 close the barrel and result in a topology typical of a domain with an OB-fold (Murzin, 1993). Proteins in the OB-fold family perform a variety of functions; however, many of them are oligonucleotide-binding proteins, from which the fold derives its name. To our knowledge, no other copper-binding protein has been identified with an OB-fold. Characterization of other copper metallochaperones, reviewed in (Robinson and Winge 2010), reveal distinctly different folds, such as the βαββαβ (ferredoxin)-fold (Arnesano, Banci, Bertini, Huffman, & O'Halloran, 2001; Banci, Bertini,

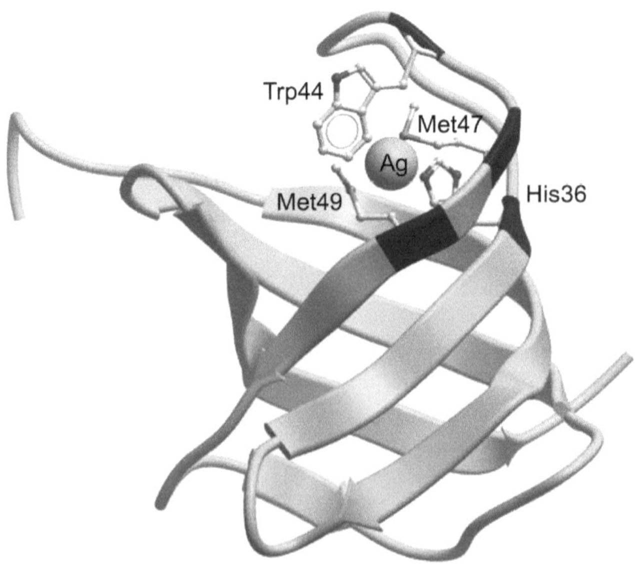

Figure 6 Ribbon representation of the periplasmic metallochaperone CusF. The major conformation of the metal-coordinating residues are shown as sticks with Ag(I) shown in space-filling representation. Based on PDB file 1QCP (Loftin et al., 2007). See the color plate.

Ciofi-Baffoni, Del Conte, & Gonnelli, 2003; Lamb et al., 1999; Stasser, Siluvai, Barry, & Blackburn, 2007), the cupredoxin-like-fold (Arnesano, Banci, Bertini, & Thompsett, 2002; Banci et al., 2005; Wernimont et al., 2003), or the helical hairpin motif (Abajian, Yatsunyk, Ramirez, & Rosenzweig, 2004).

In addition to its unique fold, CusF demonstrates a unique metal-binding site compared to other copper-binding proteins. A multiple sequence alignment of CusF homologs reveals three highly conserved residues in all instances: one histidine and two methionines (corresponding to H36, M47, and M49, using CusF numbering) (Loftin et al., 2005). In reducing environments, such as the cytoplasm, copper chaperones commonly utilize the sulfur atoms of cysteine residues for metal coordination (Lamb et al., 1999). Because the periplasm is an oxidizing environment and oxidation of cysteines would occur rapidly, methionine residues are commonly utilized by periplasmic copper-binding proteins (Davis & O'Halloran, 2008; Jiang, Nadas, Kim, & Franz, 2005; Roberts et al., 2002). The metal-binding site of CusF, initially mapped by nuclear magnetic resonance (NMR) chemical shift experiments (Kittleson et al., 2006; Loftin et al., 2005) and later confirmed by metal-bound crystal structures and XAS (Loftin et al., 2007; Xue et al., 2008), consists of the three conserved residues H36, M47, and M49, located within the second and third β-strands and clustered at one end of the protein. The side chains of the three ligands (histidine imidazole and two methionine thioesters) point toward the interior of the β-barrel.

The crystal structures of Cu(I) and Ag(I) complexes of CusF reveal an additional unique feature, viz the presence of a tryptophan residue that caps the binding site and appears to stabilize metal binding (Loftin et al., 2007; Xue et al., 2008). The metals are believed to form a π-cation interaction with the tryptophan aromatic ring, which, although within bonding distance, is too weak to be detected by EXAFS spectroscopy. The nature of this unique interaction and its possible functional role have been investigated by mutagenesis of W44 to Ala (Loftin, Blackburn, & McEvoy, 2009), and Met (Loftin et al., 2009; Xue et al., 2008) as well as by computational methods (Chakravorty, Wang, Ucisik, & Merz, 2011). These studies suggest that the Trp residue provides ~10 kcal of stabilization for the metal binding, in keeping with a modest decrease in affinity of the W44A mutant. On the other hand, mutation of the Trp residue to Met allows the Met at position 44 to bind, producing a four-coordinate Cu(I)-His(Met)$_3$ center, and increasing the affinity of the site for copper by about three orders of magnitude. Computational studies broadly confirm the structural data, but suggest that the metal center would pay a relatively large energetic penalty if the Tryp residue were to swing away during metal binding or transfer. Since many natural CusF homologues have Met residue at position 44, the significance of the Tryp residue is unclear. However, both EXAFS and DFT suggest that a major role maybe to protect the Cu(I) site from solvent access and subsequent oxidation to Cu(II). We note that this function would be important for copper but not for silver export. Clearly further work will be necessary to fully understand the role of this unique residue in copper binding and efflux.

3. CUS COMPLEX ASSEMBLY AND MECHANISM

Shortly after the X-ray crystal structures of the MFP CusB (Su et al., 2009) and the HME-RND transporter CusA (Long et al., 2010) were solved, Su et al. and Long et al. reported the co-crystal structure of the CusBA efflux complex (Su et al., 2011), resolved to a resolution of 2.90 Å. The CusBA complex structure reveals that two elongated molecules of CusB interact specifically with the periplasmic domain of one monomer of CusA (Fig. 7). Therefore, the CusA trimer makes direct connection with six CusB molecules, which together form a channel that is hypothesized to be involved in interaction with the OMF CusC. The overall conformation of CusA within the complex resembles that of the previously determined apo-CusA structure (Long et al., 2010). However, two distinct conformations of CusB were captured within the complex, resulting in a high rmsd of 7.5 Å (321 Cα atoms) when these two structures are superimposed. Comparisons of these two CusB molecules with the two conformations of the CusB molecules from the individual CusB crystal structure determination demonstrate four distinct conformations. The major differences reside in the MP domain, suggestive of the flexible nature of this domain, consistent with previous findings (De Angelis et al., 2010; Modali & Zgurskaya, 2011; Symmons et al., 2009). The MP and β-barrel domains of each CusB molecule form a concave surface, which establish a close fit with one CusA protomer. Previously, the N- and C-termini (residues 1–60 and 358–379) of the individual CusB crystal structure were hypothesized to be disordered, as these regions were missing from the electron density map of the isolated CusB structures (Su et al., 2009). However, in the co-crystal structure of CusBA, some of these residues form several short α-helices. C-terminal residues 364–371 of CusB molecule 1 form a short α-helix. Amino acids 1–50 are still absent from the structure and

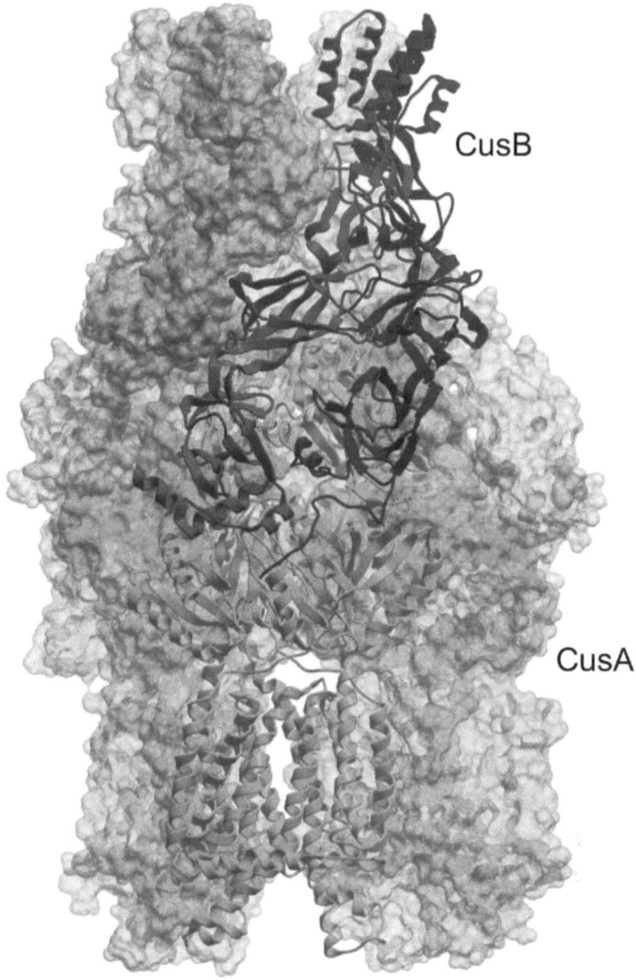

Figure 7 Structure of the CusB–CusA complex. CusB is shown in red and blue in ribbon representation, and in orange for the surface representation. CusA is shown in green in the ribbon representation, and in cyan for the surface representation. Based on PDB file 3NE5 (Su et al., 2011). See the color plate.

residues 51–67 compose a long random coil. In CusB molecule 2, residues 51–57 also form a random coil; however, residues 58–64 form a short α-helix. The C-terminal residues form two short α-helices instead of one, composed of residues 354–364 and 366–372.

Interactions between the six subunits of CusB, through side-by-side packing, result in a funnel-like structure and form an internal channel, which has recently been reported for multidrug tripartite efflux complexes (Xu et al., 2011). The MP domain and the lower half of the β-barrel domain mostly create a cap-like structure, through which they interact with the inner membrane protein CusA. The upper half of the β-barrel, the lipoyl, and the α-helical domains compose the central channel and contribute to the funnel structure. The inner surface of the channel is negatively charged, as was also previously determined for the inner surface of the channel formed by CusC (Kulathila et al., 2011). The close association of CusBA, which is expected to be contiguous with CusC, provides a protected environment for the movement of Cu(I) or Ag(I) to the extracellular environment.

Early investigations into the role of CusF hypothesized that CusF could either act as a metallochaperone, delivering metal to the CusCBA pump, or act as a metal-dependent regulatory protein, activating metal efflux of the CusCBA complex through metal binding and protein–protein interactions (Loftin et al., 2005). Based on positive results in yeast-two-hybrid experiments (Franke et al., 2003), in vitro experiments were performed to determine whether or not CusF could interact with and transfer metal to CusB. Through ITC and XAS experiments, specific and direct metal transfer between CusF and CusB was observed (Bagai et al., 2008). The first important result of these experiments is that the interaction is metal dependent. No interaction is observed in the absence of metal or when both proteins have metal bound. However, regardless of which protein starts with the metal [Ag(I)-CusF or Ag(I)-CusB] upon mixing with the apo form of the protein partner, an interaction is observed by ITC.

A direct transfer of metal between a presumed metallochaperone and a protein target is difficult to observe experimentally, and often relies on labeling one member of the chaperone–target pair with an affinity tag, followed by mixing and separation on an affinity chromatography column. This approach has two major drawbacks, first tagging may perturb protein–protein interactions critical for selectivity and second the process of mixing and separating the components takes minutes to hours resulting in the loss of kinetic information on the transfer process. With the Cus system, a novel XAS technique was developed to enable the spectroscopic measurement of metal transfer without the need for separation of the protein components (Bagai et al., 2008). In this approach, the metal-binding sites of CusF and CusB were distinguished from each other by replacing methionine residues with selenomethionine (SeM) in one of the two proteins, in this case CusF. This allows for the two methionine residues in CusF to be distinguished from the three methionine residues in CusB, in order to determine to which protein the metal ion is bound in protein mixtures. In these experiments, copper-loaded SeM-labeled CusF was mixed with apo *S*-methionine-labeled CusB and the mixtures frozen at set time points after mixing. EXAFS spectra of the mixtures were then recorded at the Cu and Se absorption edges. Initially, copper was bound to the Se atoms of the SeM-labeled CusF, but if transfer occurs it will migrate out of the Se environment of CusF into the non-selenium (S) environment of CusB. The extent of transfer can be determined from the decrease in the Se–Cu peak in the Fourier transform of the Se-edge data.

These XAS experiments demonstrated direct copper transfer between CusF and CusB. Additionally, transfer occurs in both directions and the metal is distributed about equally between the two proteins, which is probably due to their similar binding affinities in solution. Metal transfer is also very rapid, as transfer was observed in a sample that was allowed to interact for only 4 min. To test the specificity of this CusF/CusB interaction, similar experiments were performed with a close CusF homolog, SilF, which shares 51% sequence identity with CusF (Ref for SilF). SilF binds Ag(I) with a similar affinity as CusF. ITC showed no interaction between SilF and CusB (or SilF and CusF) in the presence (or absence) of silver. This further demonstrates that the interaction between CusF and CusB is specific and implies that recognition is necessary for the protein–protein interaction and direct metal transfer.

In order to identify the specific CusF/CusB interaction site, NMR spectroscopy and chemical cross-linking coupled to mass spectrometry were utilized (Mealman et al., 2011). NMR spectroscopic analysis of labeled CusF in the absence and presence of unlabeled CusB confirmed previous ITC results supporting a metal-dependent CusF/CusB interaction as no interaction was observed in the absence of metal or when both proteins were metal bound. The same observation was made by chemical cross-linking with the 7.7 Å cross-linker bis(sulfosuccinimidyl) glutarate (BS^2G); a cross-linked band corresponding to the mass of CusF cross-linked to CusB ($\sim$50 kDa band) was only observed in the presence of metal. NMR analysis of the ^1H-^{15}N chemical shift changes of labeled Ag(I)-CusF upon the addition of apo-CusB allowed the mapping of CusF residues involved in interaction with CusB. CusF residues L80, V42, N71, Q74, K31, S79, H35, T54, and T52 were significantly affected (listed in decreasing order of effect) upon addition of CusB and I32, H36, M47, T48, R50, F51, and Q75 were not observed in the spectrum due to line broadening which indicates an effect, although the magnitude is unable to be determined. The chemical shift changes can be mapped to the metal-binding face of CusF. Furthermore, the interaction was shown to be transient with NMR relaxation results revealing approximately 38% of CusF involved in complex with CusB, corresponding to a dissociation constant of ~160 μM. Chemical cross-linking in combination with high-resolution mass spectrometric analysis was further utilized to locate the region of CusB involved in the CusF/CusB interaction site, elucidating the resolution of this interaction down to the peptide/amino acid level. Two CusF/CusB cross-links were identified: CusF K31 to CusB K29 and CusF K58 to CusB K32, thus revealing that CusF interacts with the N-terminal region of CusB and the proteins interact in a transient, metal-dependent manner at their metal-binding sites.

Thus, CusF serves as a periplasmic metallochaperone that can deliver metal to CusB. Potentially this metal ion can be transferred from CusB to CusA for export from the cell. However, a further role of CusF may be in sequestration of periplasmic metal ions, which can prevent damage caused by the free metal. This role is supported by the observation that the *cusF* transcript is increased more than any other transcript in *E. coli* upon copper stress (Egler et al., 2005; Reffay et al., 2009). If there is a corresponding increase in protein levels, then an abundance of CusF would be expected in the periplasm. As the rates of transport by these types of efflux systems are expected to be rather slow (Lim and Nikaido, 2010), the storage of metal ions in CusF until their removal by the CusCBA complex takes place may be an important feature of CusF function.

How do metals pass through the Cus complex to the extracellular environment? The Cus system has a high degree of specificity for Cu(I) and Ag(I) (Conroy et al., 2010), particularly compared to the promiscuous multidrug resistance systems (Lomovskaya et al., 2007). Metal binding to multiple proteins and direct transfer could provide an additional check for the correct substrate and prevent toxicity of the free metal ion. The entry point of metal into the CusA metal-binding site has not yet been determined. It is possible that CusB transfers metal to CusA for subsequent export or that CusA acquires metal from either CusF or solution. These possibilities are discussed below.

Supporting the idea that CusB can transfer metal to CusA for export is the observation that in the co-crystal structure of CusBA, for both CusB molecules, the most N-terminal residues in the structure (starting at residue 51) are located near the periplasmic cleft of CusA formed by PC1 and PC2 (Fig. 8). The intriguing placement of these regions suggests that metal

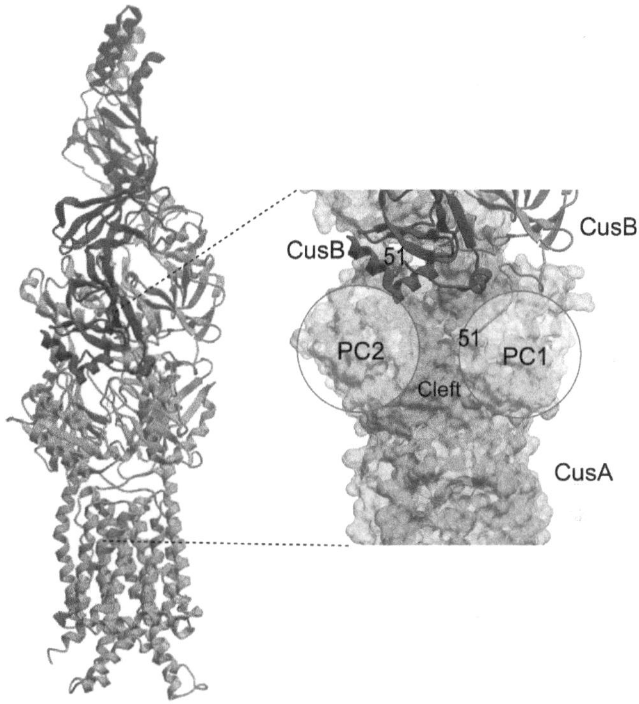

Figure 8 The N-terminal regions of CusB are near the cleft between PC1 and PC2 of CusA. Residues 51 of both CusB molecules are near the cleft leading to the metal-binding site of CusA, which suggests that the N-terminal metal-binding site of CusB (Met 21, Met 36 and Met 38) may also be nearby, though this region is missing from the crystal structure. Based on PDB file 3NE5 (Su et al., 2011). See the color plate.

transfer from the N-terminal metal-binding region of CusB to the three Met metal-binding site of CusA is possible, even though there is a mismatch in stoichiometry of the CusB to CusA molecules. The expected high degree of flexibility of the N-terminal region of CusB could contribute to dynamic interactions from both CusB subunits if transfer is to occur to a single CusA metal site. Knowledge of the precise location of the N-terminal regions of CusB, which contain the metal-binding site, would help resolve this issue.

If metal is not transferred from CusB to CusA, then what is the fate of the metal that is bound by CusB? Perhaps, the observed conformational change in CusB leads to activation of the Cus complex and functions in a regulatory role. As metal is removed from the periplasm by the Cus complex and metal concentrations decrease, metal could be transferred back to CusF, thus deactivating the Cus complex. The observed bidirectional transfer of metal between CusF and CusB supports this possibility (Bagai et al., 2008; Reffay et al., 2009).

If CusA acquires metal ions from a source other than CusB, from where is CusA most likely to obtain these metal ions? CusA may bind metal ions directly from the periplasm or by transfer from CusF. However, if metal ions are bound directly from the periplasm, this means they would be uncomplexed prior to entry into the Cus system, and thus potentially could cause cellular damage. We favor the possibility that CusA acquires metal from CusF or from CusB, which would maintain the metal ion in a sequestered environment until it is exported. Furthermore, metal binding by multiple proteins may provide a high degree of specificity for particular metal ions.

With the wealth of structural, functional, and genetic information available, the Cus system offers much valuable insight into the workings of RND-type efflux systems. Though many of the features of these systems are known, as we have summarized here, many interesting aspects of their function and mechanism still await discovery. An understanding of these systems can have a significant impact on human health as well as environmental issues, and therefore we look forward to future studies to clarify the questions still outstanding.

ACKNOWLEDGMENTS

The authors gratefully acknowledge support from the National Institutes of Health (GM079192 to M.M.M. and GM054803 to N.J.B.).

REFERENCES

Abajian, C., Yatsunyk, L. A., Ramirez, B. E., & Rosenzweig, A. C. (2004). Yeast cox17 solution structure and Copper(I) binding. *The Journal of Biological Chemistry, 279*(51), 53584–53592.

Achila, D., Banci, L., Bertini, I., Bunce, J., Ciofi-Baffoni, S., & Huffman, D. L. (2006). Structure of human Wilson protein domains 5 and 6 and their interplay with domain 4 and the copper chaperone HAM in copper uptake. *Proceedings of the National Academy of Sciences of the United States of America, 103*(15), 5729–5734.

Akama, H., Kanemaki, M., Yoshimura, M., Tsukihara, T., Kashiwagi, T., Yoneyama, H., et al. (2004). Crystal structure of the drug discharge outer membrane protein, OprM, of Pseudomonas aeruginosa: dual modes of membrane anchoring and occluded cavity end. *The Journal of Biological Chemistry, 279*(51), 52816–52819.

Akama, H., Matsuura, T., Kashiwagi, S., Yoneyama, H., Narita, S., Tsukihara, T., et al. (2004). Crystal structure of the membrane fusion protein, MexA, of the multidrug transporter in Pseudomonas aeruginosa. *The Journal of Biological Chemistry, 279*(25), 25939–25942.

Arnesano, F., Banci, L., Bertini, I., Huffman, D. L., & O'Halloran, T. V. (2001). Solution structure of the Cu(I) and Apo forms of the yeast metallochaperone, Atx1. *Biochemistry, 40*(6), 1528–1539.

Arnesano, F., Banci, L., Bertini, I., & Thompsett, A. R. (2002). Solution structure of CopC: a cupredoxin-like protein involved in copper homeostasis. *Structure, 10*(10), 1337–1347.

Bagai, I., Liu, W., Rensing, C., Blackburn, N. J., & McEvoy, M. M. (2007). Substrate-linked conformational change in the periplasmic component of a Cu(I)/Ag(I) efflux system. *The Journal of Biological Chemistry, 282*(49), 35695–35702.

Bagai, I., Rensing, C., Blackburn, N. J., & McEvoy, M. M. (2008). Direct metal transfer between periplasmic proteins identifies a bacterial copper chaperone. *Biochemistry, 47*(44), 11408–11414.

Banci, L., Bertini, I., Cantini, F., Felli, I. C., Gonnelli, L., Hadjiliadis, N., et al. (2006). The Atx1-Ccc2 complex is a metal-mediated protein–protein interaction. *Nature Chemical Biology, 2*(7), 367–368.

Banci, L., Bertini, I., Ciofi-Baffoni, S., Del Conte, R., & Gonnelli, L. (2003). Understanding copper trafficking in bacteria: interaction between the copper transport protein CopZ and the N-terminal domain of the copper ATPase CopA from Bacillus subtilis. *Biochemistry, 42*(7), 1939–1949.

Banci, L., Bertini, I., Ciofi-Baffoni, S., Huffman, D. L., & O'Halloran, T. V. (2001). Solution structure of the yeast copper transporter domain Ccc2a in the apo and Cu(I)-loaded states. *The Journal of Biological Chemistry, 276*(11), 8415–8426.

Banci, L., Bertini, I., Ciofi-Baffoni, S., Katsari, E., Katsaros, N., Kubicek, K., et al. (2005). A copper(I) protein possibly involved in the assembly of CuA center of bacterial cytochrome c oxidase. *Proceedings of the National Academy of Sciences of the United States of America, 102*(11), 3994–3999.

Bavro, V. N., Pietras, Z., Furnham, N., Perez-Cano, L., Fernandez-Recio, J., Pei, X. Y., et al. (2008). Assembly and channel opening in a bacterial drug efflux machine. *Molecular Cell, 30*(1), 114–121.

Chakravorty, D. K., Wang, B., Ucisik, M. N., & Merz, K. M., Jr. (2011). Insight into the cation–pi interaction at the metal binding site of the copper metallochaperone CusF. *Journal of the American Chemical Society, 133*(48), 19330–19333.

Chen, K., Yuldasheva, S., Penner-Hahn, J. E., & O'Halloran, T. V. (2003). An atypical linear Cu(I)-S2 center constitutes the high-affinity metal-sensing site in the CueR metalloregulatory protein. *Journal of the American Chemical Society, 125*(40), 12088–12089.

Conroy, O., Kim, E. H., McEvoy, M. M., & Rensing, C. (2010). Differing ability to transport nonmetal substrates by two RND-type metal exporters. *FEMS Microbiology Letters, 308*(2), 115–122.

Davis, A. V., & O'Halloran, T. V. (2008). A place for thioether chemistry in cellular copper ion recognition and trafficking. *Nature Chemical Biology, 4*(3), 148–151.

De Angelis, F., Lee, J. K., O'Connell, J. D., 3rd, Miercke, L. J., Verschueren, K. H., Srinivasan, V., et al. (2010). Metal-induced conformational changes in ZneB suggest an active role of membrane fusion proteins in efflux resistance systems. *Proceedings of the National Academy of Sciences of the United States of America, 107*(24), 11038–11043.

Deniaud, A., Goulielmakis, A., Coves, J., & Pebay-Peyroula, E. (2009). Differences between CusA and AcrB crystallisation highlighted by protein flexibility. *PLoS One, 4*(7), e6214.

Dibrov, P., Dzioba, J., Gosink, K. K., & Hase, C. C. (2002). Chemiosmotic mechanism of antimicrobial activity of Ag(+) in Vibrio cholerae. *Antimicrobial Agents Chemotherapy, 46*(8), 2668–2670.

Dinh, T., Paulsen, I. T., & Saier, M. H., Jr. (1994). A family of extracytoplasmic proteins that allow transport of large molecules across the outer membranes of gram-negative bacteria. *Journal of Bacteriology, 176*(13), 3825–3831.

Egler, M., Grosse, C., Grass, G., & Nies, D. H. (2005). Role of the extracytoplasmic function protein family sigma factor RpoE in metal resistance of *Escherichia coli*. *Journal of Bacteriology, 187*(7), 2297–2307.

Espirito Santo, C., Lam, E. W., Elowsky, C. G., Quaranta, D., Domaille, D. W., Chang, C. J., et al. (2011). Bacterial killing by dry metallic copper surfaces. *Applied Environmental Microbiology, 77*(3), 794–802.

Franke, S., Grass, G., & Nies, D. H. (2001). The product of the ybdE gene of the *Escherichia coli* chromosome is involved in detoxification of silver ions. *Microbiology, 147*(Pt 4), 965–972.

Franke, S., Grass, G., Rensing, C., & Nies, D. H. (2003). Molecular analysis of the copper-transporting efflux system CusCFBA of *Escherichia coli. Journal of Bacteriology, 185*(13), 3804–3812.

Goldberg, M., Pribyl, T., Juhnke, S., & Nies, D. H. (1999). Energetics and topology of CzcA, a cation/proton antiporter of the resistance-nodulation-cell division protein family. *The Journal of Biological Chemistry, 274*(37), 26065–26070.

Grass, G., & Rensing, C. (2001a). CueO is a multi-copper oxidase that confers copper tolerance in *Escherichia coli. Biochemistry and Biophysics Research Communications, 286*(5), 902–908.

Grass, G., & Rensing, C. (2001b). Genes involved in copper homeostasis in *Escherichia coli. Journal of Bacteriology, 183*(6), 2145–2147.

Gudipaty, S. A., Larsen, A. S., Rensing, C., & McEvoy, M. M. (2012). Regulation of Cu(I)/Ag(I) efflux genes in *Escherichia coli* by the sensor kinase CusS. *FEMS Microbiology Letters.*

Gupta, A., Matsui, K., Lo, J. F., & Silver, S. (1999). Molecular basis for resistance to silver cations in Salmonella. *Nature Medicine, 5*(2), 183–188.

Halliwell, B., & Gutteridge, J. M. C. (1984). Oxygen-toxicity, oxygen radicals, transition-metals and disease. *Biochemical Journal, 219*(1), 1–14.

Halliwell, B., & Gutteridge, J. M. C. (1990). Role of free-radicals and catalytic metal-ions in human-disease—an overview. *Methods in Enzymology, 186*, 1–85.

Higgins, M. K., Bokma, E., Koronakis, E., Hughes, C., & Koronakis, V. (2004). Structure of the periplasmic component of a bacterial drug efflux pump. *Proceedings of the National Academy of Sciences of the United States of America, 101*(27), 9994–9999.

Janganan, T. K., Zhang, L., Bavro, V. N., Matak-Vinkovic, D., Barrera, N. P., Burton, M. F., et al. (2011). Opening of the outer membrane protein channel in tripartite efflux pumps is induced by interaction with the membrane fusion partner. *The Journal of Biological Chemistry, 286*(7), 5484–5493.

Jiang, J., Nadas, I. A., Kim, M. A., & Franz, K. J. (2005). A Mets motif peptide found in copper transport proteins selectively binds Cu(I) with methionine-only coordination. *Inorganic Chemistry, 44*(26), 9787–9794.

Kim, H. M., Xu, Y., Lee, M., Piao, S., Sim, S. H., Ha, N. C., et al. (2010). Functional relationships between the AcrA hairpin tip region and the TolC aperture tip region for the formation of the bacterial tripartite efflux pump AcrAB-TolC. *Journal of Bacteriology, 192*(17), 4498–4503.

Kittleson, J. T., Loftin, I. R., Hausrath, A. C., Engelhardt, K. P., Rensing, C., & McEvoy, M. M. (2006). Periplasmic metal-resistance protein CusF exhibits high affinity and specificity for both CuI and AgI. *Biochemistry, 45*(37), 11096–11102.

Koronakis, V., Sharff, A., Koronakis, E., Luisi, B., & Hughes, C. (2000). Crystal structure of the bacterial membrane protein TolC central to multidrug efflux and protein export. *Nature, 405*(6789), 914–919.

Kulathila, R., Indic, M., & van den Berg, B. (2011). Crystal structure of *Escherichia coli* CusC, the outer membrane component of a heavy metal efflux pump. *PLoS One, 6*(1), e15610.

Lamb, A. L., Wernimont, A. K., Pufahl, R. A., Culotta, V. C., O'Halloran, T. V., & Rosenzweig, A. C. (1999). Crystal structure of the copper chaperone for superoxide dismutase. *Nature Structural Biology, 6*(8), 724–729.

Lim, S. P., & Nikaido, H. (2010). Kinetic parameters of efflux of penicillins by the multidrug efflux transporter AcrAB-TolC of *Escherichia coli. Antimicrobial Agents and Chemotherapy, 54*(5), 1800–1806.

Lobedanz, S., Bokma, E., Symmons, M. F., Koronakis, E., Hughes, C., & Koronakis, V. (2007). A periplasmic coiled-coil interface underlying TolC recruitment and the assembly of bacterial drug efflux pumps. *Proceedings of the National Academy of Sciences of the United States of America, 104*(11), 4612–4617.

Loftin, I. R., Blackburn, N. J., & McEvoy, M. M. (2009). Tryptophan Cu(I)-pi interaction fine-tunes the metal binding properties of the bacterial metallochaperone CusF. *Journal of Biological Inorganic Chemistry, 14*(6), 905–912.

Loftin, I. R., Franke, S., Blackburn, N. J., & McEvoy, M. M. (2007). Unusual Cu(I)/Ag(I) coordination of *Escherichia coli* CusF as revealed by atomic resolution crystallography and X-ray absorption spectroscopy. *Protein Science, 16*(10), 2287–2293.

Loftin, I. R., Franke, S., Roberts, S. A., Weichsel, A., Heroux, A., Montfort, W. R., et al. (2005). A novel copper-binding fold for the periplasmic copper resistance protein CusF. *Biochemistry, 44*(31), 10533–10540.

Lomovskaya, O., Zgurskaya, H. I., Totrov, M., & Watkins, W. J. (2007). Waltzing transporters and 'the dance macabre' between humans and bacteria. *Nature Reviews Drug Discovery, 6*(1), 56–65.

Long, F., Su, C. C., Zimmermann, M. T., Boyken, S. E., Rajashankar, K. R., Jernigan, R. L., et al. (2010). Crystal structures of the CusA efflux pump suggest methionine-mediated metal transport. *Nature, 467*(7314), 484–488.

McHugh, G. L., Moellering, R. C., Hopkins, C. C., & Swartz, M. N. (1975). Salmonella typhimurium resistant to silver nitrate, chloramphenicol, and ampicillin. *Lancet, 1*(7901), 235–240.

Mealman, T. D., Bagai, I., Singh, P., Goodlett, D. R., Rensing, C., Zhou, H., et al. (2011). Interactions between CusF and CusB identified by NMR spectroscopy and chemical cross-linking coupled to mass spectrometry. *Biochemistry, 50*(13), 2559–2566.

Mikolosko, J., Bobyk, K., Zgurskaya, H. I., & Ghosh, P. (2006). Conformational flexibility in the multidrug efflux system protein AcrA. *Structure, 14*(3), 577–587.

Misra, R., & Bavro, V. N. (2009). Assembly and transport mechanism of tripartite drug efflux systems. *Biochimica et Biophysica Acta, 1794*(5), 817–825.

Modali, S. D., & Zgurskaya, H. I. (2011). The periplasmic membrane proximal domain of MacA acts as a switch in stimulation of ATP hydrolysis by MacB transporter. *Molecular Microbiology, 81*(4), 937–951.

Molteni, C., Abicht, H. K., & Solioz, M. (2010). Killing of bacteria by copper surfaces involves dissolved copper. *Applied Environmental Microbiology, 76*(12), 4099–4101.

Munson, G. P., Lam, D. L., Outten, F. W., & O'Halloran, T. V. (2000). Identification of a copper-responsive two-component system on the chromosome of *Escherichia coli* K-12. *Journal of Bacteriology, 182*(20), 5864–5871.

Murakami, S., Nakashima, R., Yamashita, E., Matsumoto, T., & Yamaguchi, A. (2006). Crystal structures of a multidrug transporter reveal a functionally rotating mechanism. *Nature, 443*(7108), 173–179.

Murakami, S., Nakashima, R., Yamashita, E., & Yamaguchi, A. (2002). Crystal structure of bacterial multidrug efflux transporter AcrB. *Nature, 419*(6907), 587–593.

Murzin, A. G. (1993). OB (oligonucleotide/oligosaccharide binding)-fold: common structural and functional solution for non-homologous sequences. *EMBO Journal, 12*(3), 861–867.

Neilsen, P. O., Zimmerman, G. A., & McIntyre, T. M. (2001). *Escherichia coli* Braun lipoprotein induces a lipopolysaccharide-like endotoxic response from primary human endothelial cells. *Journal of Immunology, 167*(9), 5231–5239.

Nies, D. H. (1995). The cobalt, zinc, and cadmium efflux system CzcABC from Alcaligenes eutrophus functions as a cation-proton antiporter in *Escherichia coli*. *Journal of Bacteriology, 177*(10), 2707–2712.
Nies, D. H. (2003). Efflux-mediated heavy metal resistance in prokaryotes. *FEMS Microbiology Reviews, 27*(2–3), 313–339.
Nies, D. H., Nies, A., Chu, L., & Silver, S. (1989). Expression and nucleotide sequence of a plasmid-determined divalent cation efflux system from Alcaligenes eutrophus. *Proceedings of the National Academy of Sciences of the United States of America, 86*(19), 7351–7355.
Nikaido, H. (2011). Structure and mechanism of RND-type multidrug efflux pumps. *Advances in Enzymology and Related Areas of Molecular Biology, 77*, 1–60.
Outten, F. W., Huffman, D. L., Hale, J. A., & O'Halloran, T. V. (2001). The independent cue and cus systems confer copper tolerance during aerobic and anaerobic growth in *Escherichia coli*. *The Journal of Biological Chemistry, 276*(33), 30670–30677.
Piddock, L. J. (2006). Multidrug-resistance efflux pumps—not just for resistance. *Nature Reviews Microbiology, 4*(8), 629–636.
Ralle, M., Cooper, M. J., Lutsenko, S., & Blackburn, N. J. (1998). The Menkes disease protein binds copper via novel 2-coordinate Cu(I)-cysteinates in the N-terminal domain. *Journal of the American Chemical Society, 120*(51), 13525–13526.
Ralle, M., Lutsenko, S., & Blackburn, N. J. (2003). X-ray absorption spectroscopy of the copper chaperone HAH1 reveals a linear two-coordinate Cu(I) center capable of adduct formation with exogenous thiols and phosphines. *The Journal of Biological Chemistry, 278*(25), 23163–23170.
Reffay, M., Gambin, Y., Benabdelhak, H., Phan, G., Taulier, N., Ducruix, A., et al. (2009). Tracking membrane protein association in model membranes. *PLoS One, 4*(4), e5035.
Rensing, C., Fan, B., Sharma, R., Mitra, B., & Rosen, B. P. (2000). CopA: an *Escherichia coli* Cu(I)-translocating P-type ATPase. *Proceedings of the National Academy of Sciences of the United States of America, 97*(2), 652–656.
Rensing, C., & Grass, G. (2003). *Escherichia coli* mechanisms of copper homeostasis in a changing environment. *FEMS Microbiology Reviews, 27*(2–3), 197–213.
Rensing, C., Pribyl, T., & Nies, D. H. (1997). New functions for the three subunits of the CzcCBA cation-proton antiporter. *Journal of Bacteriology, 179*(22), 6871–6879.
Roberts, S. A., Weichsel, A., Grass, G., Thakali, K., Hazzard, J. T., Tollin, G., et al. (2002). Crystal structure and electron transfer kinetics of CueO, a multicopper oxidase required for copper homeostasis in *Escherichia coli*. *Proceedings of the National Academy of Sciences of the United States of America, 99*(5), 2766–2771.
Robinson, N. J., & Winge, D. R. (2010). Copper metallochaperones. *Annual Reviews of Biochemistry, 79*, 537–562.
Saier, M. H., Jr. (2000). A functional-phylogenetic classification system for transmembrane solute transporters. *Microbiology and Molecular Biology Reviews, 64*(2), 354–411.
Saier, M. H., Jr., Tam, R., Reizer, A., & Reizer, J. (1994). Two novel families of bacterial membrane proteins concerned with nodulation, cell division and transport. *Molecular Microbiology, 11*(5), 841–847.
Sarret, G., Favier, A., Coves, J., Hazemann, J. L., Mergeay, M., & Bersch, B. (2010). CopK from Cupriavidus metallidurans CH34 binds Cu(I) in a tetrathioether site: characterization by X-ray absorption and NMR spectroscopy. *Journal of the American Chemical Society, 132*(11), 3770–3777.
Seeger, M. A., Schiefner, A., Eicher, T., Verrey, F., Diederichs, K., & Pos, K. M. (2006). Structural asymmetry of AcrB trimer suggests a peristaltic pump mechanism. *Science, 313*(5791), 1295–1298.
Sennhauser, G., Bukowska, M. A., Briand, C., & Grutter, M. G. (2009). Crystal structure of the multidrug exporter MexB from Pseudomonas aeruginosa. *Journal of Molecular Biology, 389*(1), 134–145.

Silver, S. (2003). Bacterial silver resistance: molecular biology and uses and misuses of silver compounds. *FEMS Microbiology Reviews, 27*(2–3), 341–353.

Silver, S., Phung, L. T., & Silver, G. (2006). Silver as biocides in burn and wound dressings and bacterial resistance to silver compounds. *Journal of Industrial Microbiology and Biotechnology, 33*(7), 627–634.

Singh, S. K., Grass, G., Rensing, C., & Montfort, W. R. (2004). Cuprous oxidase activity of CueO from *Escherichia coli*. *Journal of Bacteriology, 186*(22), 7815–7817.

Stasser, J. P., Siluvai, G. S., Barry, A. N., & Blackburn, N. J. (2007). A multinuclear copper(I) cluster forms the dimerization interface in copper-loaded human copper chaperone for superoxide dismutase. *Biochemistry, 46*(42), 11845–11856.

Stroebel, D., Sendra, V., Cannella, D., Helbig, K., Nies, D. H., & Coves, J. (2007). Oligomeric behavior of the RND transporters CusA and AcrB in micellar solution of detergent. *Biochimica et Biophysica Acta, 1768*(6), 1567–1573.

Su, C. C., Long, F., Zimmermann, M. T., Rajashankar, K. R., Jernigan, R. L., & Yu, E. W. (2011). Crystal structure of the CusBA heavy-metal efflux complex of *Escherichia coli*. *Nature, 470*(7335), 558–562.

Su, C. C., Yang, F., Long, F., Reyon, D., Routh, M. D., Kuo, D. W., et al. (2009). Crystal structure of the membrane fusion protein CusB from *Escherichia coli*. *Journal of Molecular Biology, 393*(2), 342–355.

Symmons, M. F., Bokma, E., Koronakis, E., Hughes, C., & Koronakis, V. (2009). The assembled structure of a complete tripartite bacterial multidrug efflux pump. *Proceedings of the National Academy of Sciences of the United States of America, 106*(17), 7173–7178.

Tikhonova, E. B., & Zgurskaya, H. I. (2004). AcrA, AcrB, and TolC of *Escherichia coli* form a stable intermembrane multidrug efflux complex. *The Journal of Biological Chemistry, 279*(31), 32116–32124.

Touze, T., Eswaran, J., Bokma, E., Koronakis, E., Hughes, C., & Koronakis, V. (2004). Interactions underlying assembly of the *Escherichia coli* AcrAB-TolC multidrug efflux system. *Molecular Microbiology, 53*(2), 697–706.

Tseng, T. T., Gratwick, K. S., Kollman, J., Park, D., Nies, D. H., Goffeau, A., et al. (1999). The RND permease superfamily: an ancient, ubiquitous and diverse family that includes human disease and development proteins. *Journal of Molecular Microbiology and Biotechnology, 1*(1), 107–125.

Vaccaro, L., Koronakis, V., & Sansom, M. S. (2006). Flexibility in a drug transport accessory protein: molecular dynamics simulations of MexA. *Biophysical Journal, 91*(2), 558–564.

von Rozycki, T., & Nies, D. H. (2009). *Cupriavidus metallidurans*: evolution of a metal-resistant bacterium. *Antonie Van Leeuwenhoek, 96*(2), 115–139.

Wang, B., Weng, J., Fan, K., & Wang, W. (2012). Interdomain flexibility and pH-induced conformational changes of AcrA revealed by molecular dynamics simulations. *Journal of Physical Chemistry B*.

Wernimont, A. K., Huffman, D. L., Finney, L. A., Demeler, B., O'Halloran, T. V., & Rosenzweig, A. C. (2003). Crystal structure and dimerization equilibria of PcoC, a methionine-rich copper resistance protein from *Escherichia coli*. *Journal of Biological Inorganic Chemistry, 8*(1–2), 185–194.

Williams, J. R., Morgan, A. G., Rouch, D. A., Brown, N. L., & Lee, B. T. (1993). Copper-resistant enteric bacteria from United Kingdom and Australian piggeries. *Applied Environmental Microbiology, 59*(8), 2531–2537.

Xu, Y., Lee, M., Moeller, A., Song, S., Yoon, B. Y., Kim, H. M., et al. (2011). Funnel-like hexameric assembly of the periplasmic adapter protein in the tripartite multidrug efflux pump in Gram-negative bacteria. *The Journal of Biological Chemistry, 286*(20), 17910–17920.

Xue, Y., Davis, A. V., Balakrishnan, G., Stasser, J. P., Staehlin, B. M., Focia, P., et al. (2008). Cu(I) recognition via cation-pi and methionine interactions in CusF. *Nature Chemical Biology, 4*(2), 107–109.

Yum, S., Xu, Y., Piao, S., Sim, S. H., Kim, H. M., Jo, W. S., et al. (2009). Crystal structure of the periplasmic component of a tripartite macrolide-specific efflux pump. *Journal of Molecular Biology, 387*(5), 1286–1297.

Zgurskaya, H. I., & Nikaido, H. (1999). Bypassing the periplasm: reconstitution of the AcrAB multidrug efflux pump of *Escherichia coli*. *Proceedings of the National Academy of Sciences of the United States of America, 96*(13), 7190–7195.

Zgurskaya, H. I., & Nikaido, H. (2000). Multidrug resistance mechanisms: drug efflux across two membranes. *Molecular Microbiology, 37*(2), 219–225.

Zgurskaya, H. I., Yamada, Y., Tikhonova, E. B., Ge, Q., & Krishnamoorthy, G. (2009). Structural and functional diversity of bacterial membrane fusion proteins. *Biochimica et Biophysica Acta, 1794*(5), 794–807.

Zhou, H., & Thiele, D. J. (2001). Identification of a novel high affinity copper transport complex in the fission yeast *Schizosaccharomyces pombe*. *The Journal of Biological Chemistry, 276*(23), 20529–20535.

SECTION THREE

Zinc Transporters of Diverse Cellular Compartments

CHAPTER EIGHT

Molecular Architecture and Function of ZnT Transporters

Taiho Kambe
Division of Integrated Life Science, Graduate School of Biostudies, Kyoto University, Kyoto, Japan
Corresponding author: E-mail: kambe1@kais.kyoto-u.ac.jp

Contents

Abstract

Zinc transporters (ZnT) form an important class of zinc efflux proteins in mammals. Nine ZnT transporters, designated ZnT1–ZnT8 and ZnT10, have been identified to date. Recent reports have established that the ZnT transporters are localized to various intracellular compartments and the plasma membrane. They play essential roles in multiple biological processes and are involved in zinc-related human diseases. The ZnT transporters function as Zn^{2+}/H^{+} exchangers. Based on the X-ray structures of and sequence homology to bacterial homologues, they are thought to form dimers possessing one essential zinc-binding site within the transmembrane domains of each monomer, and a binuclear zinc-sensing and binding site in the cytoplasmic C-terminal region. This chapter summarizes the molecular characteristics of ZnT transporters and reviews our current knowledge of their structure, biological functions, and regulation, with the emphasis on the most recent advances. Molecular characterization of the ZnT homologues of bacteria, yeast, plant and model organisms are also described, because they have contributed to the major advances in our understanding of the ZnT protein function.

Current Topics in Membranes, Volume 69
ISSN 1063-5823,
http://dx.doi.org/10.1016/B978-0-12-394390-3.00008-2

1. INTRODUCTION

The zinc transporter (ZnT) family comprises the mammalian (and other vertebrate) members within the cation diffusion facilitator (CDF) family, which is found in all three phylogenetic domains (Archaea, Bacteria and Eukaryotes). The ZnT transporters are assigned to the SLC30A family (Palmiter & Huang, 2004) and are designated ZnT1–ZnT10 based on their order of characterization, although ZnT9 is a misnomer. ZnT9, which has stretches of sequence homology to the cation efflux domain (pfam01545), has been now renamed GAC63 because it acts as a component of the p160 coactivator complex in a signal transduction pathway (Chen, Kim, & Stallcup, 2005).

ZnT transporters facilitate efflux and compartmentalization of zinc from the cytosol into intracellular organelles and vesicles, in the opposite direction to ZIP (zinc-regulated transporter (ZRT), iron-regulated transporter (IRT)-like protein) transporters. The CDF transporters are classified into three groups (with 13 clusters), namely Zn-CDF, Zn/Fe-CDF and Mn-CDF, based on their phylogenetic relationships; ZnT proteins belong to the Zn-CDF group (Montanini, Blaudez, Jeandroz, Sanders, & Chalot, 2007). A more recent phylogenetic analysis using the maximum parsimony method classified CDF transporters into more than 13 groups (Gustin, Zanis, & Salt, 2011). Based on these analyses and sequence similarities, ZnT transporters can be divided into the following four groups: (1) ZnT1 and ZnT10; (2) ZnT2, ZnT3, ZnT4 and ZnT8; (3) ZnT5 and ZnT7; and (4) ZnT6 (Kambe, Suzuki, Nagao, & Yamaguchi-Iwai, 2006). ZnT1 and ZnT10 can also be divided into different groups, because a recent study showed that ZnT10 is involved in manganese homeostasis as a manganese transporter (Quadri et al., 2012; Tuschl et al., 2012). Although ZnT2, ZnT3 and ZnT8 show high sequence homology to one another (70–80% similarity in pfam01545), their physiological functions are very different. They are also expressed in a cell- and tissue-specific manner.

2. STRUCTURE AND TRANSPORT MECHANISM

2.1. Structure

The three-dimensional structure of eukaryotic ZnT transporters has not yet been determined, but they are predicted to have six transmembrane domains (TMDs) with the N- and C-terminal regions oriented toward the cytosol (Fig. 1). ZnT5 is unique, as it has an exceptionally long amino-terminal

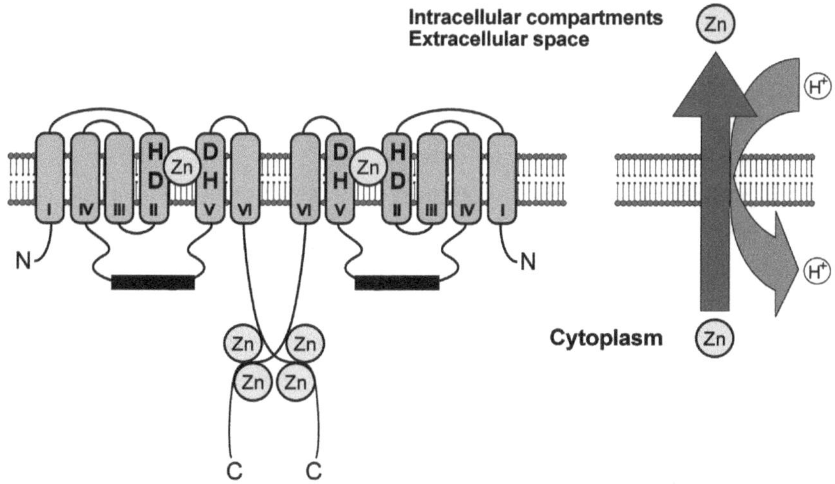

Figure 1 ***Predicted structure and zinc-binding sites of ZnT transporters.*** The predicted topologies of ZnT transporters are shown. ZnT transporters, except for ZnT5 and ZnT6, are thought to form homodimers, similar to the bacterial homologue YiiP. Based on the X-ray structure of YiiP and sequence similarities, ZnT transporters are predicted to have a zinc-binding site formed by two histidine (H) and two aspartic acid (D) residues within the TMDs and a binuclear zinc binding site in the cytoplasmic C-terminal portion (corresponding to Sites A and C in YiiP). The histidine-rich loop between TMDs IV and V, indicated by the thick black line, may help modulate zinc transport activity or form a zinc-binding site. The zinc-binding site corresponding to site B in YiiP is omitted because this site is not conserved. ZnT transporters are functional Zn^{2+}/H^{+} exchangers. The direction of zinc transport is shown on the right side. See the color plate.

region consisting of nine putative TMDs fused to the conserved six TMDs (Kambe et al., 2002). The function of this additional region has not been determined. Most ZnT transporters have a histidine-rich loop with an $HX_{1-6}H$ cluster (H is histidine, X is often serine or glycine) between TMDs IV and V that is exposed to the cytoplasm (Fukada & Kambe, 2011; Gaither & Eide, 2001; Kambe, Yamaguchi-Iwai, Sasaki, & Nagao, 2004), a common feature of members of the Zn-CDF group (Blindauer & Schmid, 2010). Posttranslational modification has been proposed for some ZnT transporters (Huang, Kirschke, & Gitschier, 2002), but glycosylation in ZnT2 is the only example of modification for which direct experimental evidence exists to date (Kelleher & Lonnerdal, 2003).

The X-ray structure of YiiP, a ZnT homologue in *Escherichia coli*, has been determined and improved our understanding of the structural and functional properties of the CDF family of proteins (Lu, Chai, & Fu, 2009;

Lu & Fu, 2007). YiiP forms a homodimer; each monomer has six TMDs and both the N- and C-termini located at the cytoplasmic side of the membrane, as previously predicted. Each YiiP protomer also contains four zinc-binding sites. The site within the TMDs (site A) is the primary zinc transport site; another site resides at the interface between the membrane and the cytoplasmic domain (site B). The remaining two sites are located in the cytoplasmic domain (site C), which has a metallochaperone-like structure with an αββα fold, despite lacking sequence homology to known metallochaperones (Lu et al., 2009; Lu & Fu, 2007). Sites A and C are highly conserved among ZnT transporters. Site A is formed by four conserved hydrophilic residues: three aspartic acid residues and one histidine residue in TMDs II and V (Figs 1 and 2). Site C functions as a binuclear zinc-sensing site in the cytoplasmic domain. Binding of zinc to this site is thought to induce a scissor-like movement, which affects the orientation of helices within the TMDs and modulates the coordination of zinc at site A. Thus, YiiP operates allosterically in the zinc-regulated zinc transport mechanism. Two additional X-ray structures of the cytoplasmic domain of bacterial ZnT homologues revealed that the cytoplasmic domain is sufficient for formation of a stable dimer, and that the structure of this domain is highly conserved despite a high degree of sequence variability (Cherezov et al., 2008; Higuchi, Hattori, Tanaka, Ishitani, & Nureki, 2009).

Name		TM II			TM V	
YiiP	42	ALVDSLVDIGASLTNLLVVRYSL	64	145	QAVRADMLHYQSDVM	159
ZnT1	40	DSFHMLSDVLALVVALVAERFAR	62	243	LNMRGVFLHVLGDAL	257
ZnT2	103	DAAHLLTDFASMLISLFSLWMSS	125	215	PSVRAAFIHVIGDFM	229
ZnT3	105	DAAHLLADVGSMMGSLFSLWLST	127	230	TSVRAAFVHVLGDLL	244
ZnT4	143	DALHMLTDLSAIILTLLALWLSS	165	269	LAVRAAFVHALGDLV	283
ZnT5	448	DGFHMLFDCSALVMGLFAALMSR	470	587	ANMRGVFLHVLADTL	601
ZnT6	63	YTYLTIFDLFSLMTCLISYWVTL	85	193	IFLPRMNPFVLIDLA	207
ZnT7	67	DSFHMFFDSTAILAGLAASVISK	89	232	QILQGVFLHILADTL	246
ZnT8	103	DAAHLLIDLTSFLLSLFSLWLSS	125	212	ASVRAAFVHALGDLF	226
ZnT10	40	DSFNMLSDLISLCVGLSAGYIAR	62	236	LNIRGVLLHVMGDAL	250

Figure 2 ***Multiple sequence alignment of TMDs II and V among ZnT transporters***. The sequences of TMDs II and V of YiiP and ZnT transporters are aligned. The conserved hydrophilic residues, shaded in black and gray, are predicted to contribute to the zinc-binding site within TMDs. The asparagine residue in TMD II of ZnT10 is indicated by the arrowhead. Non-conserved hydrophobic residues in ZnT6 are indicated by the open boxes. The indicated TMDs of ZnT5 correspond to TMDs XI and XIV. For color version of this figure, the reader is referred to the online version of this book.

Based on the structure of YiiP, structural models of ZnT3, ZnT5 (only the C-terminal six TMDs) and ZnT8 have been generated (Ohana et al., 2009; Salazar, Falcon-Perez, Harrison, & Faundez, 2009; Weijers, 2010). The models helped researchers to verify the location of a zinc-binding site and speculate on the zinc transport mechanism. Four hydrophilic residues in TMDs II and V, which form the zinc-binding site corresponding to site A in YiiP, are highly conserved, although one aspartic acid is replaced with histidine in ZnT transporters. Therefore, there are two aspartic acid and two histidine residues in their TMDs. The introduction of mutations into the conserved histidine and aspartic acid residues was found to abolish zinc transport activity (Ohana et al., 2009). This result confirmed that the identified site was essential for the zinc transport activity of ZnT transporters. The zinc-binding site in the cytoplasmic C-terminal portion, corresponding to site C in YiiP, has not yet been investigated. Therefore, it remains unknown how the ZnT transporters sense the cytoplasmic zinc concentrations. The cytoplasmic histidine-rich loop between TMDs IV and V is a characteristic feature of ZnT transporters that belong to the Zn-CDF group. This loop is missing in the X-ray structure of bacterial ZnT homologues, and the function of the loop is still unknown. However, the deletion and mutational analyses of this loop in the ZnT transporters and their homologues suggest that it may form a key zinc-binding site or may be involved in zinc delivery from the cytoplasm to the zinc-binding site within the TMDs (Kawachi, Kobae, Mimura, & Maeshima, 2008; Suzuki, Ishihara, Migaki, Nagao, et al., 2005), in concert with the cytoplasmic C-terminal region.

Oligomerization is crucial in enabling the zinc transport activity of ZnT transporters. Most ZnT transporters are thought to form homodimers, while ZnT5 and ZnT6 form heterodimers. In homodimers, each protomer contains a zinc-binding site within the TMDs. However, ZnT6 in the ZnT5–ZnT6 heterodimer does not form a zinc-binding site within the TMDs because of the lack of conserved hydrophilic residues (Fig. 2). Thus, ZnT6 is not directly involved in zinc transport across the cell membrane, but may function as a modulator (Fukunaka et al., 2009). A recent study showed that ZnT3 and ZnT10 could form a hetero-oligomer, although it is unknown whether they form functional complexes (Patrushev, Seidel-Rogol, & Salazar, 2012). These latter results raise the possibility that other ZnT transporters may also form hetero-complexes to modulate their functions.

The cytoplasmic C-terminal region of ZnT transporters is thought to be important for the regulation of complex formation because the C-terminal deletion of ZnT1 and ZnT5 results in a protein that is cytotoxic (Fukunaka et al., 2009; Palmiter & Findley, 1995), probably due to deposits of nonfunctional protein aggregates. Consistent with these findings, this region of ZnT5 participates in identifying ZnT6 as a partner molecule in the heterodimer (Fukunaka et al., 2009). The corresponding region of ZnT3 appears to regulate intracellular trafficking and zinc transport activity by forming a covalent dityrosine bond in response to oxidative stress (Salazar et al., 2009). The C-terminal region is also important for protein–protein interactions, as demonstrated by studies of interactions between the ZnT1 and the protein kinase Raf-1 (Jirakulaporn & Muslin, 2004).

The cytoplasmic N-terminal region seems to play an important role in zinc transport across the cell membrane. Deletion of the first TMD of ZnT1 results in a protein with dominant-negative properties (Palmiter & Findley, 1995). Some literature reports suggest that the splicing variants of ZnT transporters, which have different topologies due to missing TMDs or bearing an additional TMD, retain their zinc transport activity (Lopez & Kelleher, 2009; Valentine et al., 2007). The spliced variants may operate through the different modes of zinc transport compared to that of canonical ZnT transporters, or modulate their zinc transport activity by forming complexes with the latter. The structure–functional relationships of these noncanonical proteins need to be clarified to better understand the functional properties of ZnT transporters.

2.2. Mechanism of Zinc Transport

The zinc transport mechanism of ZnT transporters is still incompletely understood. However, direct experiments using fluorescence-based functional measurements have shown that ZnT transporters function as Zn^{2+}/H^{+} exchangers (Ohana et al., 2009). Several studies of the bacterial, yeast and plant ZnT homologues support this notion (Guffanti, Wei, Rood, & Krulwich, 2002; Kawachi et al., 2008; MacDiarmid, Milanick, & Eide, 2002). The stoichiometry of the Zn^{2+}/H^{+} exchange was calculated as 1:1 in a study using reconstituted proteoliposomes containing a bacterial ZnT homologue (Chao & Fu, 2004). Considering the vesicular localization of most ZnT transporters, this mode of zinc transport may provide some advantage, since the proton gradient between vesicle lumen and the cytosol can be used as a driving force. Although ZnT transporters belong to the CDF family (CDFs), the latter term is actually a misnomer because ZnT transporters and their homologues mobilize zinc against a concentration gradient as secondary active transporters. The specific transport mechanism of ZnT3 in neurons has been described. The zinc transport activity of ZnT3 is potentiated by a chloride channel, ClC-3, or vesicular glutamate transporter Vglut1 (Salazar, Craige, Love, Kalman, & Faundez, 2005; Salazar et al., 2004). Similar specific mechanisms may also operate in other ZnT transporters in a cell- or tissue-specific manner.

ZnT transporters, classified into the Zn-CDF group, are thought to specifically mobilize zinc under physiological conditions (Fukada & Kambe, 2011; Montanini et al., 2007). However, recent reports have suggested that ZnT10 plays a pivotal role as a manganese transporter (Quadri et al., 2012; Tuschl et al., 2012), which prompted researchers to reconsider the metal substrate specificity of ZnT transporters. In this regard, ZnT10 is unique because it contains an asparagine residue instead of histidine at the conserved position of TMD II (Fig. 2). This replacement may be critical for selectivity toward manganese, because this position appears to affect metal specificity, as recognized by conservation of hydrophilic residues in TMDs II and V (site A in YiiP). In fact, the corresponding position has an aspartic acid in the Mn-CDF group in plants (Montanini et al., 2007). Moreover, in ZnT10 the Arg and Lys clusters ($RX_{1-6}R$ and $KX_{1-6}K$, respectively) are found in the cytoplasmic N- and C-terminal regions and in the cytoplasmic loop of TMDs IV and V, instead of the histidine-rich loop in other ZnT transporters (Fukada & Kambe, 2011). Further studies are needed to determine whether these features contribute to the recognition of manganese by ZnT10.

Recent metal selectivity screening and mutational analyses using ZnT homologues from yeast and plants have identified residues important for the discrimination of zinc and other transition metals (Lin et al., 2009; Podar et al., 2012). Some of these residues are conserved in ZnT transporters and likely contribute to strict specificity of these transporters for their metal substrates.

3. EXPRESSION AND SUBCELLULAR LOCALIZATION

3.1. Regulation of Expression

The expression of ZnT transporters is transcriptionally and posttranslationally up- or downregulated in response to various stimuli, such as inflammation (Lang et al., 2007), lipopolysaccharides (LPS) (Kitamura et al., 2006), essential nutrients (Dubben, Honscheid, Winkler, Rink, & Haase, 2010; Suphioglu et al., 2010), glucose (Smidt et al., 2009), hormones (Patrushev et al., 2012), cytokines (El Muayed et al., 2010) and injury (Inoue et al., 2002) as well as by zinc contents, in a cell- or tissue-specific manner (reviewed in Devirgiliis, Zalewski, Perozzi, & Murgia, 2007; Kambe et al., 2004; Lichten & Cousins, 2009; Mocchegiani, Giacconi, & Malavolta, 2008; Sekler, Sensi, Hershfinkel, & Silverman, 2007). The ZnT expression is also finely regulated during differentiation (Kelleher et al., 2012; Ryu, Lichten, Liuzzi, & Cousins, 2008), development (Barila, Murgia, Nobili, Gaetani, & Perozzi, 1994; Kirschke & Huang, 2008; Liuzzi, Bobo, Cui, McMahon, & Cousins, 2003) and aging (Adlard, Parncutt, Finkelstein, & Bush, 2010; Iguchi et al., 2011), and in pathological conditions (Henshall et al., 2003; Smith, Xiong, Markesbery, & Lovell, 2006).

This section focuses on the regulation of ZnTs' expression at the molecular level. The transcription of *ZnT1* and *ZnT2* genes is induced in response to elevated zinc, which plays a role in cellular zinc homeostasis. This response is mediated by a zinc-sensing transcription factor metal response element-binding transcription factor-1 (MTF-1) through a metal response element similar to that in the *metallothionein* gene (Guo et al., 2010; Langmade, Ravindra, Daniels, & Andrews, 2000). Recent study has shown that a tumor suppressor phosphatase and tensin homolog deleted on chromosome 10 (PTEN) modulates the MTF-1-mediated expression of both *ZnT1* and *metallothionein* (Lin et al., 2012), which suggests an interesting relationship between tumorigenesis and zinc homeostasis. *ZnT2* expression increases in response to other stimuli. It is upregulated by prolactin in mammary cells via the JAK2/STAT5 signaling pathway and involves the STAT5 binding site in the *ZnT2* promoter (Qian, Lopez, Seo, & Kelleher, 2009). In pancreatic acinar cells, the STAT5 and glucocorticoid receptor interaction regulates *ZnT2* expression through the same response element (Guo et al., 2010).

The expression of *ZnT5* is transcriptionally upregulated by inducers of the unfolded protein response (UPR); the regulation is mediated by the transcription factor XBP1 through the conserved UPR element (5′-TGACGTGG-3′) in its promoter region (Ishihara et al., 2006). This response seems to be important as the elaborate feedback mechanism that maintains functions of the secretory pathway (Kambe, 2011). *ZnT5* expression is increased by LPS or interleukin-6 in the liver (Lichten & Cousins, 2009; Liuzzi et al., 2005). This latter effect is likely attributed to an endoplasmic reticulum (ER)-bound transcription factor, CREBH, because CREBH is activated in response to proinflammatory cytokines and LPS, and binds to the TGACGTGG sequence (Zhang et al., 2006). The islet-specific expression of *ZnT8* is regulated by the β cell-enriched transcription factor Pdx-1 through an intronic enhancer (Pound et al., 2011). Estrogen decreases ZnT3 expression in a non-transcriptional, posttranslational manner by down-regulating the expression of the δ-subunit adaptor protein complex (AP-3δ), which determines the level of ZnT3 (Lee et al., 2004), as described below.

3.2. Tissue Distribution and Subcelluar Localization

ZnT transporters change their cellular distribution according to extracellular zinc concentrations in a cell- or tissue-specific manner. The subcellular localization of the ZnT transporters is summarized in Fig. 3. ZnT1 is mainly localized to the plasma membrane (Palmiter & Findley, 1995), but studies of different cell types revealed some differences in cellular locations of ZnT1. Specifically, ZnT1 is localized at the basolateral membrane in enterocytes but at the apical membrane in pancreatic acinar cells (Cousins, Liuzzi, & Lichten, 2006; Cousins & McMahon, 2000; Liuzzi, Bobo, Lichten, Samuelson, & Cousins, 2004). More recently, ZnT1 was reported to be localized to the ER by forming complexes with EVER proteins in human keratinocytes (Lazarczyk et al., 2008).

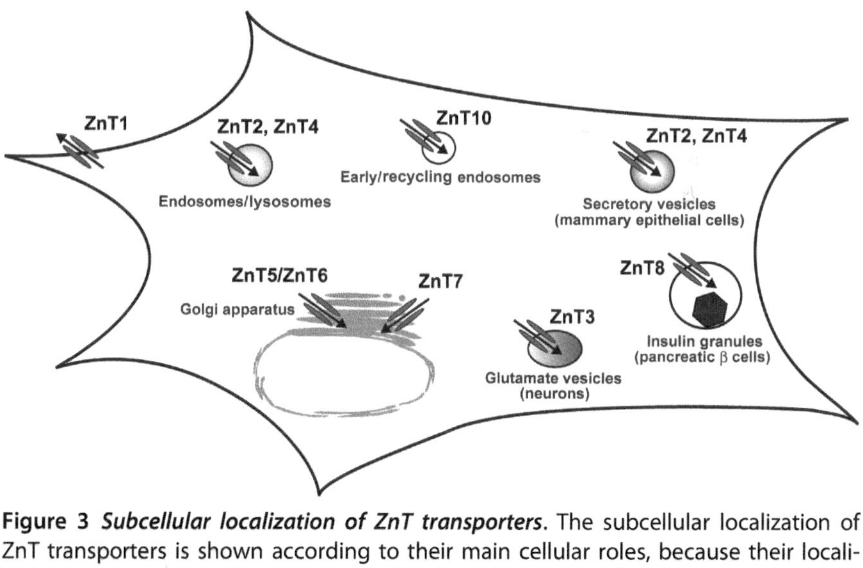

Figure 3 ***Subcellular localization of ZnT transporters.*** The subcellular localization of ZnT transporters is shown according to their main cellular roles, because their localization changes in cell-specific and developmentally regulated manners, and in response to various stimuli. ZnT1 is localized to the plasma membrane where it facilitates zinc efflux out of the cells, whereas the other members mobilize zinc into the indicated intracellular compartments For color version of this figure, the reader is referred to the online version of this book.

The subcellular localization of ZnT2 is also diverse. ZnT2 is localized to the secretory vesicles/granules in the highly specialized secretory tissues such as mammary glands, prostate and pancreas (Kelleher, McCormick, Velasquez, & Lopez, 2011), in addition to the endosomal/lysosomal compartments (Falcon-Perez & Dell'Angelica, 2007; Palmiter, Cole, & Findley, 1996). Mitochondrial localization of ZnT2, via its association with the inner membrane, has also been reported (Seo, Lopez, & Kelleher, 2011). Single nucleotide polymorphisms (SNPs) of *ZnT2* were found to alter its subcellular localization, which appears to affect cellular zinc metabolism (Seo & Kelleher, 2010). These results suggest that SNPs may have important implications for human health by affecting cellular functions of ZnT transporters.

In addition to ZnT2, the proteins ZnT3, ZnT4 and ZnT8 are localized to the vesicles in the regulated secretory pathway. ZnT3 is well known to be present in the synaptic vesicles of a subset of glutamatergic neurons in the hippocampus and neocortex (Sensi, Paoletti, Bush, & Sekler, 2009). The synaptic vesicle localization of ZnT3 is controlled by AP-3. Therefore, mice with a mutant form of AP-3 (*mocha* mouse) show disturbed localization or loss of ZnT3 (Kantheti et al., 1998). Recently, several studies have demonstrated that ZnT3 is expressed in various cells, including pancreatic β cells and retinal cells (Smidt & Rungby, 2012).

ZnT4 is localized to intracellular vesicles and the *trans*-Golgi network (TGN) as well as endosomal compartments (Falcon-Perez & Dell'Angelica, 2007; Ho et al., 2004; Palmiter & Huang, 2004). ZnT4 is also localized to the plasma membrane in some cells (Henshall et al., 2003; Overbeck, Uciechowski, Ackland, Ford, & Rink, 2008). ZnT4 and ZnT6 are redistributed from the perinuclear region to the cell periphery in response to high zinc conditions, probably to maintain intracellular zinc homeostasis (Huang et al., 2002). This trafficking is reminiscent of that of copper transporters ATP7A and 7B (Lutsenko, Barnes, Bartee, & Dmitriev, 2007), although it is unlikely to detoxify high zinc concentrations.

ZnT8 is predominantly localized to the membrane of insulin granules in pancreatic β cells, which has been a focus of research aimed at elucidating the etiology of diabetes mellitus. ZnT8 is also expressed in other tissues, including pancreatic α cells (Wijesekara et al., 2010). ZnT5, ZnT6 and ZnT7 are all localized to the early secretory pathway, including the ER, Golgi apparatus and cytoplasmic vesicles (Huang et al., 2002; Kambe et al., 2002; Kirschke & Huang, 2003). Although ZnT5 and ZnT6 are detected in the Golgi apparatus when overexpressed (Suzuki, Ishihara, Migaki, Nagao, et al., 2005), endogenous ZnT5 is concentrated in cytoplasmic vesicles, such as COPII-coated vesicles in many cells (Suzuki, Ishihara, Migaki, Matsuura, et al., 2005), and is associated with the insulin granule membrane in pancreatic β cells (Kambe et al., 2002). By contrast, endogenous ZnT6 is mainly localized to the TGN (Huang et al., 2002). These differences may suggest that each transporter has a specific function when they do not form heterodimers. ZnT7 is localized in the Golgi apparatus and cytoplasmic vesicles (Kirschke & Huang, 2003). Electron microscopy has shown that ZnT7 is predominantly localized in the *cis*-face or the forming face of the Golgi apparatus (Chi et al., 2006). Very recently, ZnT10 was reported to be localized in the early/recycling endosomes in vascular smooth muscle cells (Patrushev et al., 2012). The genes *ZnT1, ZnT4, ZnT5, ZnT6* and *ZnT7* are ubiquitously or widely transcribed, while *ZnT2, ZnT3, ZnT8* and *ZnT10* are expressed in a cell- or tissue-specific manner.

4. PHYSIOLOGICAL SIGNIFICANCE

4.1. Cellular and Physiological Functions

ZnT1 plays a protective role against zinc toxicity (Palmiter & Findley, 1995), which is particularly important in the brain because large amounts of zinc may suddenly accumulate in response to ischemia and seizures (Nolte et al., 2004; Tsuda et al., 1997). The functional properties and predominant localization at the plasma membrane suggest that ZnT1 facilitates efflux of cytosolic zinc out of the cells (neurons), although in some cases ZnT1 appears to have more complicated roles (Sekler et al., 2007). Moreover, ZnT1 could mediate zinc transfer into the circulation from enterocytes (McMahon & Cousins, 1998) and recover zinc from the glomerular filtrate in the kidney (Cousins & McMahon, 2000), because of its ability to efflux zinc out of cells and its localization on the basolateral membrane. In addition to regulating zinc distribution, ZnT1 appears to have multiple cellular effects by interacting with various proteins. For example, ZnT1 decreases cell surface expression of functional L-type calcium channels (LTCCs) by interacting with the LTCC β-subunit (Levy et al., 2009) or downregulates the transcription factors stimulated by MTF-1, c-Jun and Elk by forming hetero-complexes with EVER proteins (Lazarczyk et al., 2008). ZnT1 is also known to modulate Raf-1 activity through direct interactions (Jirakulaporn & Muslin, 2004). Since the binding of ZnT1 to Raf-1, which is inhibited by zinc, promotes enzymatic activity, ZnT1 possibly facilitates Raf-1 signaling by lowering cytosolic zinc concentrations (Jirakulaporn & Muslin, 2004).

A similar regulatory mechanism controlling the Ras–Raf–MEK–ERK pathway was found in the nematode ZnT1 orthologue (Bruinsma, Jirakulaporn, Muslin, & Kornfeld, 2002).

ZnT2 and ZnT4 both transport cytosolic zinc into acidic compartments, including endosomes/lysosomes and secretory vesicles/granules. Mobilization of zinc by these transporters into the endosome/lysosome compartments is particularly evident in high zinc conditions (Falcon-Perez & Dell'Angelica, 2007; Palmiter, Cole, & Findley, 1996; Palmiter & Huang, 2004), which suggests that they are involved in the homeostatic maintenance of intracellular zinc by sequestration. By contrast, zinc transport by these transporters into secretory vesicles/granules is important for exocrine function (Guo et al., 2010; Kelleher et al., 2011; Palmiter & Huang, 2004). ZnT3 is a critical transporter of zinc into synaptic vesicles of neurons in the brain (Palmiter, Cole, Quaife, & Findley, 1996). Synaptic vesicular zinc is depleted in *Znt3*-knockout mice (Cole, Wenzel, Kafer, Schwartzkroin, & Palmiter, 1999). Since zinc released into the synaptic cleft is an important modulator of neuronal activity (Sensi et al., 2009), as described below, ZnT3 has critical effects on numerous signaling pathways. For example, ZnT3 plays an essential role in presynaptic Erk1/2 signaling during hippocampus-dependent learning (Sindreu, Palmiter, & Storm, 2011). Besides its functions in the brain, ZnT3 was reported to regulate insulin production in pancreatic β cells (Smidt et al., 2009). In these cells, ZnT8 plays a critical role in zinc accumulation in the insulin granules, where insulin binds to zinc to form zinc–insulin crystals. Insulin granules in *Znt8*-knockout mice are immature with pale insulin "pro-granules" (Lemaire et al., 2009; Wijesekara et al., 2010). Meanwhile, overexpression of ZnT8 in insulinoma cells enhances glucose-stimulated insulin secretion (Chimienti et al., 2006).

ZnT5–ZnT6 heterodimers and ZnT7 homo-oligomers (probably homodimers) transport zinc into the lumen of compartments of the early secretory pathway. The enzymatic activity of tissue-nonspecific alkaline phosphatase (TNAP) is significantly decreased in the chicken DT40 cells lacking *ZnT5*, *ZnT6* and *ZnT7* (Suzuki, Ishihara, Migaki, Matsuura, et al., 2005; Suzuki, Ishihara, Migaki, Nagao, et al., 2005), indicating that the products of these genes are required for TNAP activation. Both complexes activate TNAP via a two-step mechanism, in which the first step regulates the stability of TNAP independently of its zinc transport activity and the second step supplies zinc to TNAP in the early secretory pathway (Fukunaka et al., 2011). Both complexes contribute to the homeostatic maintenance of the secretory pathway (Ishihara et al., 2006; Kambe, 2011). Since approximately one-third of all proteins in eukaryotes are targeted to the secretory pathway (Vembar & Brodsky, 2008), zinc homeostasis must be tightly regulated within this pathway.

ZnT5 is essential for activation of the nuclear factor-κB, which is mediated through the Fc epsilon receptor I (FcεRI)-induced translocation of the protein kinase C to the plasma membrane in mast cells (Nishida et al., 2009). This observation indicates that ZnT5 is also involved in important cellular events in the cytosol. ZnT10, in concert with ZnT3, regulates cellular senescence induced by angiotensin II in the vascular smooth muscle cells by reducing cytosolic zinc concentrations (Patrushev et al., 2012). ZnT10 plays a key role in the regulation of manganese homeostasis, although the mechanism has not yet been elucidated (Quadri et al., 2012; Tuschl et al., 2012).

As briefly described above, there is an increasing body of evidence to show that ZnT transporters fulfill their regulatory functions independently of their zinc transport activities through protein–protein interactions. This type of regulation by ZnT transporters must be clarified in more detail by analyzing their molecular properties.

4.2. Pathology of ZnT-Related Diseases and Phenotypes of the Knockout and Mutant Mice

Many reports have noted that several human genetic disorders are associated with changes in ZnT transporters. The nonsynonymous SNP of *ZNT8* (rs13266634), which introduces a substitution (arginine to tryptophan) at amino acid 325 at the dimer interface of the cytoplasmic C-terminal region, is associated with increased susceptibility for type 2 diabetes (Sladek et al., 2007). *ZNT8* has also received much interest as an autoimmunity marker for type 1 diabetes, because the autoantibodies to ZnT8 are present in approximately 60–80% of patients at the onset of clinical disease (Wenzlau et al., 2007). Moreover, the SNP rs13266634 encoding the Arg/Trp325 substitution determines specificity of anti-ZnT8 autoantibody in type 1 diabetes (Wenzlau et al., 2008).

The *ZNT2* gene is associated with a 75% reduction of zinc levels in milk, which causes transient zinc deficiency and dermatitis in infants (Chowanadisai, Lonnerdal, & Kelleher, 2006). More recently, *ZNT10* was reported to be involved in Parkinsonism by causing hypermanganesemia (Quadri et al., 2012; Tuschl et al., 2012). Although serum zinc concentrations are normal in individuals with a mutation in the *ZNT10* gene, the serum manganese concentration is significantly elevated, indicating that ZnT10 is a critical regulator of manganese homeostasis.

ZnT1 is implicated in epidermodysplasia verruciformis, a rare autosomal recessive genodermatosis associated with a high risk of developing skin carcinoma (Lazarczyk et al., 2008). Studies of ZnT3 revealed that its cortical levels decline with age in healthy humans and in people with Alzheimer disease (Adlard et al., 2010). This finding suggests that genetic ablation of *ZNT3* may phenocopy the deficits in learning and memory in Alzheimer disease. Altered expression of ZnT4 and ZnT6 in the brain has also been reported in Alzheimer disease (Smith et al., 2006), although presently there are no data to describe the underlying mechanisms.

Studies using the *Znt*-knockout mice have further demonstrated great biological significance of ZnT. Disruption of *Znt1* caused early embryonic death because of the impaired transfer of maternal zinc to the uterine and embryonic environments (Andrews, Wang, Dey, & Palmiter, 2004). Unfortunately, this property restricts our ability to investigate the physiological functions of ZnT1. Homozygous deletion of *Znt3* results in the lack of synaptic zinc (Cole et al., 1999), which is thought to be necessary for normal activity of neuronal ion channels, transporters and receptors in the synapse. Thus, ZnT3 and synaptic zinc transported by ZnT3 modulate synaptic transmission and plasticity (Sensi et al., 2009). Although the initial studies of *Znt3*-knockout mice failed to show impairments in spatial learning or memory (Cole, Martyanova, & Palmiter, 2001), recent studies have shown that *Znt3*-knockout mice show defects in fear memory (Martel, Hevi, Friebely, Baybutt, & Shumyatsky, 2010), age-dependent defects in learning and memory (over 6 months) (Adlard et al., 2010), and complete deficits in contextual discrimination and spatial working memory (Sindreu et al., 2011). In addition to these deficits in brain functions, *Znt3*-knockout mice showed mildly impaired glucose metabolism after streptozotocin treatment (Smidt et al., 2009).

Targeted deletion of *Znt5* results in poor growth, osteopenia, muscle weakness, low body fat and male-specific cardiac death (Inoue et al., 2002). *Znt5*-null mice exhibit marked defects in contact hypersensitivity and mast cell-mediated delayed allergic reactions, but not in the immediate allergic reactions (Nishida et al., 2009). Meanwhile, the *Znt7*-null mice show reduced absorption of dietary zinc and decreased body fat content (Huang, Yu, Kirschke, Gertz, & Lloyd, 2007). However, these mice had no apparent hair or skin abnormalities (Huang et al., 2007), which are typical symptoms of zinc deficiency.

Homozygous deletion of *Znt8* results in a decrease in the number of dense core insulin granules, and impairments in glucose tolerance and insulin secretion (Pound et al., 2009; Wijesekara et al., 2010), indicating that ZnT8 is necessary to maintain glycemic control. Recent analyses using global and pancreatic β cell-specific *Znt8*-knockout mice suggest that ZnT8 contributes to the risk of developing type 2 diabetes via β cell-specific and non-β cell-specific effects (Hardy et al., 2012).

A spontaneous mutation of *Znt4* in mice (lethal milk; *lm*) causes zinc deficiency, reducing milk zinc levels by 34% (Huang & Gitschier, 1997), similar to that in humans caused by a loss-of-function mutation in *ZNT2*. Zinc transport into milk via ZnT4 is likely to be specific to mice because mutations of *ZNT4* are not associated with low zinc milk concentrations in humans. The young *lm* mice fed a zinc-supplemented diet exhibit normal zinc homeostasis, but older mice develop alopecia and dermatitis, major symptoms of zinc deficiency (Erway & Grider, 1984), suggesting that ZnT4 participates in dietary zinc absorption. Moreover, *lm* mice have defects in balance during swimming because of a reduced number of otoliths (calcium carbonate crystals) in the inner ear.

Several reports have described the effects of breeding *Znt*-knockout mice with other transgenic mice, which have helped us to understand the roles of ZnT transporters in pathological conditions. For example, analyses of *hAPP+*:*ZnT3*-null mice have shown that synaptic zinc accumulated by ZnT3 may contribute to amyloid deposition. Meanwhile, analyses of TRAMP:*Znt7*-null mice (transgenic adenocarcinoma mouse prostate) suggest that the absence of *Znt7* may promote early onset of prostate tumors (Tepaamorndech, Huang, & Kirschke, 2011).

5. CONCLUSION

In the post-genomic era, it is becoming increasingly clear that ZnT transporters have important physiologic and pathogenic roles (Fukada & Kambe, 2011; Kambe, Weaver, & Andrews, 2008). Recent work has greatly improved our understanding of their molecular architectures and functions, but the mechanisms involved in zinc handling and the subsequent molecular events are still far from fully understood. One major limitation is that the X-ray structures of mammalian ZnT transporters have not been solved. Moreover, the regulatory mechanisms involved in posttranslational modification of ZnT transporters are still unknown, unlike for ZIP transporters, for which the kinase-mediated phosphorylation was shown to regulate zinc transport (Taylor, Hiscox, Nicholson, Hogstrand, & Kille, 2012). Defining the regulation of zinc transport by ZnT transporters is necessary to know the mechanism of fine-tuning the coordination of zinc to its target proteins (Maret & Li, 2009) and cellular signaling pathways (Fukada, Yamasaki, Nishida, Murakami, & Hirano, 2011). Further studies are needed to examine the interactions between ZnT transporters and other molecules. Molecular evidence from such studies will contribute to a comprehensive understanding of the physiologic and pathogenic roles of ZnT transporters.

ACKNOWLEDGMENTS

Support by grants-in-aid from the Ministry of Education, Culture, Sports, Science and Technology of Japan is gratefully acknowledged.

REFERENCES

Adlard, P. A., Parncutt, J. M., Finkelstein, D. I., & Bush, A. I. (2010). Cognitive loss in zinc transporter-3 knock-out mice: a phenocopy for the synaptic and memory deficits of Alzheimer's disease? *The Journal of Neuroscience, 30*, 1631–1636.

Andrews, G. K., Wang, H., Dey, S. K., & Palmiter, R. D. (2004). Mouse zinc transporter 1 gene provides an essential function during early embryonic development. *Genesis, 40*, 74–81.

Barila, D., Murgia, C., Nobili, F., Gaetani, S., & Perozzi, G. (1994). Subtractive hybridization cloning of novel genes differentially expressed during intestinal development. *European Journal of Biochemistry, 223*, 701–709.

Blindauer, C. A., & Schmid, R. (2010). Cytosolic metal handling in plants: determinants for zinc specificity in metal transporters and metallothioneins. *Metallomics, 2*, 510–529.

Bruinsma, J. J., Jirakulaporn, T., Muslin, A. J., & Kornfeld, K. (2002). Zinc ions and cation diffusion facilitator proteins regulate Ras-mediated signaling. *Developmental Cell, 2*, 567–578.

Chao, Y., & Fu, D. (2004). Kinetic study of the antiport mechanism of an Escherichia coli zinc transporter, ZitB. *The Journal of Biological Chemistry, 279*, 12043–12050.

Chen, Y. H., Kim, J. H., & Stallcup, M. R. (2005). GAC63, a GRIP1-dependent nuclear receptor coactivator. *Molecular and Cellular Biology, 25*, 5965–5972.

Cherezov, V., Hofer, N., Szebenyi, D. M., Kolaj, O., Wall, J. G., Gillilan, R., et al. (2008). Insights into the mode of action of a putative zinc transporter CzrB in Thermus thermophilus. *Structure, 16*, 1378–1388.

Chi, Z. H., Wang, X., Wang, Z. Y., Gao, H. L., Dahlstrom, A., & Huang, L. (2006). Zinc transporter 7 is located in the cis-Golgi apparatus of mouse choroid epithelial cells. *Neuroreport, 17*, 1807–1811.

Chimienti, F., Devergnas, S., Pattou, F., Schuit, F., Garcia-Cuenca, R., Vandewalle, B., et al. (2006). In vivo expression and functional characterization of the zinc transporter ZnT8 in glucose-induced insulin secretion. *Journal of Cell Science, 119*, 4199–4206.

Chowanadisai, W., Lonnerdal, B., & Kelleher, S. L. (2006). Identification of a mutation in SLC30A2 (ZnT-2) in women with low milk zinc concentration that results in transient neonatal zinc deficiency. *The Journal of Biological Chemistry, 281*, 39699–39707.

Cole, T. B., Martyanova, A., & Palmiter, R. D. (2001). Removing zinc from synaptic vesicles does not impair spatial learning, memory, or sensorimotor functions in the mouse. *Brain Research, 891*, 253–265.

Cole, T. B., Wenzel, H. J., Kafer, K. E., Schwartzkroin, P. A., & Palmiter, R. D. (1999). Elimination of zinc from synaptic vesicles in the intact mouse brain by disruption of the ZnT3 gene. *Proceedings of the National Academy of Sciences of the United States of America, 96*, 1716–1721.

Cousins, R. J., Liuzzi, J. P., & Lichten, L. A. (2006). Mammalian zinc transport, trafficking, and signals. *The Journal of Biological Chemistry, 281*, 24085–24089.

Cousins, R. J., & McMahon, R. J. (2000). Integrative aspects of zinc transporters. *The Journal of Nutrition, 130*, 1384S–1387S.

Devirgiliis, C., Zalewski, P. D., Perozzi, G., & Murgia, C. (2007). Zinc fluxes and zinc transporter genes in chronic diseases. *Mutation Research, 622*, 84–93.

Dubben, S., Honscheid, A., Winkler, K., Rink, L., & Haase, H. (2010). Cellular zinc homeostasis is a regulator in monocyte differentiation of HL-60 cells by 1 alpha,25-dihydroxyvitamin D3. *Journal of Leukocyte Biology, 87*, 833–844.

El Muayed, M., Billings, L. K., Raja, M. R., Zhang, X., Park, P. J., Newman, M. V., et al. (2010). Acute cytokine-mediated downregulation of the zinc transporter ZnT8 alters pancreatic beta-cell function. *Journal of Endocrinology, 206*, 159–169.

Erway, L. C., & Grider, A., Jr. (1984). Zinc metabolism in lethal-milk mice. Otolith, lactation, and aging effects. *Journal of Heredity, 75*, 480–484.

Falcon-Perez, J. M., & Dell'Angelica, E. C. (2007). Zinc transporter 2 (SLC30A2) can suppress the vesicular zinc defect of adaptor protein 3-depleted fibroblasts by promoting zinc accumulation in lysosomes. *Experimental Cell Research, 313*, 1473–1483.

Fukada, T., & Kambe, T. (2011). Molecular and genetic features of zinc transporters in physiology and pathogenesis. *Metallomics, 3*, 662–674.

Fukada, T., Yamasaki, S., Nishida, K., Murakami, M., & Hirano, T. (2011). Zinc homeostasis and signaling in health and diseases: zinc signaling. *Journal of Biological Inorganic Chemistry, 16*, 1123–1134.

Fukunaka, A., Kurokawa, Y., Teranishi, F., Sekler, I., Oda, K., Ackland, M. L., et al. (2011). Tissue nonspecific alkaline phosphatase is activated via a two-step mechanism by zinc transport complexes in the early secretory pathway. *The Journal of Biological Chemistry, 286*, 16363–16373.

Fukunaka, A., Suzuki, T., Kurokawa, Y., Yamazaki, T., Fujiwara, N., Ishihara, K., et al. (2009). Demonstration and characterization of the heterodimerization of ZnT5 and ZnT6 in the early secretory pathway. *The Journal of Biological Chemistry, 284*, 30798–30806.

Gaither, L. A., & Eide, D. J. (2001). Eukaryotic zinc transporters and their regulation. *Biometals, 14*, 251–270.

Guffanti, A. A., Wei, Y., Rood, S. V., & Krulwich, T. A. (2002). An antiport mechanism for a member of the cation diffusion facilitator family: divalent cations efflux in exchange for K+ and H+. *Molecular Microbiology, 45*, 145–153.

Guo, L., Lichten, L. A., Ryu, M. S., Liuzzi, J. P., Wang, F., & Cousins, R. J. (2010). STAT5-glucocorticoid receptor interaction and MTF-1 regulate the expression of ZnT2 (Slc30a2) in pancreatic acinar cells. *Proceedings of the National Academy of Sciences of the United States of America, 107*, 2818–2823.

Gustin, J. L., Zanis, M. J., & Salt, D. E. (2011). Structure and evolution of the plant cation diffusion facilitator family of ion transporters. *BMC Evolutionary Biology, 11*, 76.

Hardy, A., Wijesekara, N., Genkin, I., Prentice, K. J., Bhattacharjee, A., Kong, D., et al. (2012). Effects of high fat diet feeding on zinc transporter 8 (Znt8) null mice: differences between beta cell and global knockout of Znt8. *American Journal of Physiology—Endocrinology and Metabolism, 302*, E1084–E1096.

Henshall, S. M., Afar, D. E., Rasiah, K. K., Horvath, L. G., Gish, K., Caras, I., et al. (2003). Expression of the zinc transporter ZnT4 is decreased in the progression from early prostate disease to invasive prostate cancer. *Oncogene, 22*, 6005–6012.

Higuchi, T., Hattori, M., Tanaka, Y., Ishitani, R., & Nureki, O. (2009). Crystal structure of the cytosolic domain of the cation diffusion facilitator family protein. *Proteins, 76*, 768–771.

Ho, L. H., Ruffin, R. E., Murgia, C., Li, L., Krilis, S. A., & Zalewski, P. D. (2004). Labile zinc and zinc transporter ZnT4 in mast cell granules: role in regulation of caspase activation and NF-kappaB translocation. *The Journal of Immunology, 172*, 7750–7760.

Huang, L., & Gitschier, J. (1997). A novel gene involved in zinc transport is deficient in the lethal milk mouse. *Nature Genetics, 17*, 292–297.

Huang, L., Kirschke, C. P., & Gitschier, J. (2002). Functional characterization of a novel mammalian zinc transporter, ZnT6. *The Journal of Biological Chemistry, 277*, 26389–26395.

Huang, L., Yu, Y. Y., Kirschke, C. P., Gertz, E. R., & Lloyd, K. K. (2007). Znt7 (Slc30a7)-deficient mice display reduced body zinc status and body fat accumulation. *The Journal of Biological Chemistry, 282*, 37053–37063.

Iguchi, K., Morihara, N., Usui, S., Hayama, M., Sugimura, Y., & Hirano, K. (2011). Castration- and aging-induced changes in the expression of zinc transporter and metallothionein in rat prostate. *Journal of Andrology, 32*, 144–150.

Inoue, K., Matsuda, K., Itoh, M., Kawaguchi, H., Tomoike, H., Aoyagi, T., et al. (2002). Osteopenia and male-specific sudden cardiac death in mice lacking a zinc transporter gene, Znt5. *Human Molecular Genetics, 11*, 1775–1784.

Ishihara, K., Yamazaki, T., Ishida, Y., Suzuki, T., Oda, K., Nagao, M., et al. (2006). Zinc transport complexes contribute to the homeostatic maintenance of secretory pathway function in vertebrate cells. *The Journal of Biological Chemistry, 281*, 17743–17750.

Jirakulaporn, T., & Muslin, A. J. (2004). Cation diffusion facilitator proteins modulate Raf-1 activity. *The Journal of Biological Chemistry, 279*, 27807–27815.

Kambe, T. (2011). An overview of a wide range of functions of ZnT and Zip zinc transporters in the secretory pathway. *Bioscience, Biotechnology, and Biochemistry, 75*, 1036–1043.

Kambe, T., Narita, H., Yamaguchi-Iwai, Y., Hirose, J., Amano, T., Sugiura, N., et al. (2002). Cloning and characterization of a novel mammalian zinc transporter, zinc transporter 5, abundantly expressed in pancreatic beta cells. *The Journal of Biological Chemistry, 277*, 19049–19055.

Kambe, T., Suzuki, T., Nagao, M., & Yamaguchi-Iwai, Y. (2006). Sequence similarity and functional relationship among eukaryotic ZIP and CDF transporters. *Genomics Proteomics Bioinformatics, 4*, 1–9.

Kambe, T., Weaver, B. P., & Andrews, G. K. (2008). The genetics of essential metal homeostasis during development. *Genesis, 46*, 214–228.

Kambe, T., Yamaguchi-Iwai, Y., Sasaki, R., & Nagao, M. (2004). Overview of mammalian zinc transporters. *Cellular and Molecular Life Sciences, 61*, 49–68.

Kantheti, P., Qiao, X., Diaz, M. E., Peden, A. A., Meyer, G. E., Carskadon, S. L., et al. (1998). Mutation in AP-3 delta in the mocha mouse links endosomal transport to storage deficiency in platelets, melanosomes, and synaptic vesicles. *Neuron, 21*, 111–122.

Kawachi, M., Kobae, Y., Mimura, T., & Maeshima, M. (2008). Deletion of a histidine-rich loop of AtMTP1, a vacuolar Zn(2+)/H(+) antiporter of Arabidopsis thaliana, stimulates the transport activity. *The Journal of Biological Chemistry, 283*, 8374–8383.

Kelleher, S. L., & Lonnerdal, B. (2003). Zn transporter levels and localization change throughout lactation in rat mammary gland and are regulated by Zn in mammary cells. *The Journal of Nutrition, 133*, 3378–3385.

Kelleher, S. L., McCormick, N. H., Velasquez, V., & Lopez, V. (2011). Zinc in specialized secretory tissues: roles in the pancreas, prostate, and mammary gland. *Advances in Nutrition, 2*, 101–111.

Kelleher, S. L., Velasquez, V., Croxford, T. P., McCormick, N. H., Lopez, V., & Macdavid, J. (2012). Mapping the zinc-transporting system in mammary cells: molecular analysis reveals a phenotype-dependent zinc-transporting network during lactation. *Journal of Cellular Physiology, 227*, 1761–1770.

Kirschke, C. P., & Huang, L. (2003). ZnT7, a novel mammalian zinc transporter, accumulates zinc in the Golgi apparatus. *The Journal of Biological Chemistry, 278*, 4096–4102.

Kirschke, C. P., & Huang, L. (2008). Expression of the ZNT (SLC30) family members in the epithelium of the mouse prostate during sexual maturation. *Journal of Molecular Histology, 39*, 359–370.

Kitamura, H., Morikawa, H., Kamon, H., Iguchi, M., Hojyo, S., Fukada, T., et al. (2006). Toll-like receptor-mediated regulation of zinc homeostasis influences dendritic cell function. *Nature Immunology, 7*, 971–977.

Lang, C., Murgia, C., Leong, M., Tan, L. W., Perozzi, G., Knight, D., et al. (2007). Anti-inflammatory effects of zinc and alterations in zinc transporter mRNA in mouse models of allergic inflammation. *American Journal of Physiology - Lung Cellular and Molecular Physiology, 292*, L577–L584.

Langmade, S. J., Ravindra, R., Daniels, P. J., & Andrews, G. K. (2000). The transcription factor MTF-1 mediates metal regulation of the mouse ZnT1 gene. *The Journal of Biological Chemistry, 275*, 34803–34809.

Lazarczyk, M., Pons, C., Mendoza, J. A., Cassonnet, P., Jacob, Y., & Favre, M. (2008). Regulation of cellular zinc balance as a potential mechanism of EVER-mediated protection against pathogenesis by cutaneous oncogenic human papillomaviruses. *The Journal of Experimental Medicine, 205*, 35–42.

Lee, J. Y., Kim, J. H., Hong, S. H., Cherny, R. A., Bush, A. I., Palmiter, R. D., et al. (2004). Estrogen decreases zinc transporter 3 expression and synaptic vesicle zinc levels in mouse brain. *The Journal of Biological Chemistry, 279*, 8602–8607.

Lemaire, K., Ravier, M. A., Schraenen, A., Creemers, J. W., Van de Plas, R., Granvik, M., et al. (2009). Insulin crystallization depends on zinc transporter ZnT8 expression, but is not required for normal glucose homeostasis in mice. *Proceedings of the National Academy of Sciences of the United States of America, 106*, 14872–14877.

Levy, S., Beharier, O., Etzion, Y., Mor, M., Buzaglo, L., Shaltiel, L., et al. (2009). Molecular basis for zinc transporter 1 action as an endogenous inhibitor of L-type calcium channels. *The Journal of Biological Chemistry, 284*, 32434–32443.

Lichten, L. A., & Cousins, R. J. (2009). Mammalian zinc transporters: nutritional and physiologic regulation. *Annual Review of Nutrition, 29*, 153–176.

Lin, H., Burton, D., Li, L., Warner, D. E., Phillips, J. D., Ward, D. M., et al. (2009). Gain-of-function mutations identify amino acids within transmembrane domains of the yeast vacuolar transporter Zrc1 that determine metal specificity. *The Biochemical Journal, 422*, 273–283.

Lin, M. C., Liu, Y. C., Tam, M. F., Lu, Y. J., Hsieh, Y. T., & Lin, L. Y. (2012). PTEN interacts with metal-responsive transcription factor 1 and stimulates its transcriptional activity. *The Biochemical Journal, 441*, 367–377.

Liuzzi, J. P., Bobo, J. A., Cui, L., McMahon, R. J., & Cousins, R. J. (2003). Zinc transporters 1, 2 and 4 are differentially expressed and localized in rats during pregnancy and lactation. *The Journal of Nutrition, 133*, 342–351.

Liuzzi, J. P., Bobo, J. A., Lichten, L. A., Samuelson, D. A., & Cousins, R. J. (2004). Responsive transporter genes within the murine intestinal-pancreatic axis form a basis of zinc homeostasis. *Proceedings of the National Academy of Sciences of the United States of America, 101*, 14355–14360.

Liuzzi, J. P., Lichten, L. A., Rivera, S., Blanchard, R. K., Aydemir, T. B., Knutson, M. D., et al. (2005). Interleukin-6 regulates the zinc transporter Zip14 in liver and contributes to the hypozincemia of the acute-phase response. *Proceedings of the National Academy of Sciences of the United States of America, 102*, 6843–6848.

Lopez, V., & Kelleher, S. L. (2009). Zinc transporter-2 (ZnT2) variants are localized to distinct subcellular compartments and functionally transport zinc. *The Biochemical Journal, 422*, 43–52.

Lu, M., Chai, J., & Fu, D. (2009). Structural basis for autoregulation of the zinc transporter YiiP. *Nature Structural and Molecular Biology, 16*, 1063–1067.

Lu, M., & Fu, D. (2007). Structure of the zinc transporter YiiP. *Science, 317*, 1746–1748.

Lutsenko, S., Barnes, N. L., Bartee, M. Y., & Dmitriev, O. Y. (2007). Function and regulation of human copper-transporting ATPases. *Physiological Reviews, 87*, 1011–1046.

MacDiarmid, C. W., Milanick, M. A., & Eide, D. J. (2002). Biochemical properties of vacuolar zinc transport systems of Saccharomyces cerevisiae. *The Journal of Biological Chemistry, 277*, 39187–39194.

Maret, W., & Li, Y. (2009). Coordination dynamics of zinc in proteins. *Chemical Reviews, 109*, 4682–4707.

Martel, G., Hevi, C., Friebely, O., Baybutt, T., & Shumyatsky, G. P. (2010). Zinc transporter 3 is involved in learned fear and extinction, but not in innate fear. *Learning and Memory, 17*, 582–590.

McMahon, R. J., & Cousins, R. J. (1998). Regulation of the zinc transporter ZnT-1 by dietary zinc. *Proceedings of the National Academy of Sciences of the United States of America, 95*, 4841–4846.

Mocchegiani, E., Giacconi, R., & Malavolta, M. (2008). Zinc signalling and subcellular distribution: emerging targets in type 2 diabetes. *Trends in Molecular Medicine, 14*, 419–428.

Montanini, B., Blaudez, D., Jeandroz, S., Sanders, D., & Chalot, M. (2007). Phylogenetic and functional analysis of the cation diffusion facilitator (CDF) family: improved signature and prediction of substrate specificity. *BMC Genomics, 8*, 107.

Nishida, K., Hasegawa, A., Nakae, S., Oboki, K., Saito, H., Yamasaki, S., et al. (2009). Zinc transporter Znt5/Slc30a5 is required for the mast cell-mediated delayed-type allergic reaction but not the immediate-type reaction. *The Journal of Experimental Medicine, 206*, 1351–1364.

Nolte, C., Gore, A., Sekler, I., Kresse, W., Hershfinkel, M., Hoffmann, A., et al. (2004). ZnT-1 expression in astroglial cells protects against zinc toxicity and slows the accumulation of intracellular zinc. *Glia, 48*, 145–155.

Ohana, E., Hoch, E., Keasar, C., Kambe, T., Yifrach, O., Hershfinkel, M., et al. (2009). Identification of the Zn2+ binding site and mode of operation of a mammalian Zn2+ transporter. *The Journal of Biological Chemistry, 284*, 17677–17686.

Overbeck, S., Uciechowski, P., Ackland, M. L., Ford, D., & Rink, L. (2008). Intracellular zinc homeostasis in leukocyte subsets is regulated by different expression of zinc exporters ZnT-1 to ZnT-9. *Journal of Leukocyte Biology, 83*, 368–380.

Palmiter, R. D., Cole, T. B., & Findley, S. D. (1996). ZnT-2, a mammalian protein that confers resistance to zinc by facilitating vesicular sequestration. *The EMBO Journal, 15*, 1784–1791.

Palmiter, R. D., Cole, T. B., Quaife, C. J., & Findley, S. D. (1996). ZnT-3, a putative transporter of zinc into synaptic vesicles. *Proceedings of the National Academy of Sciences of the United States of America, 93*, 14934–14939.

Palmiter, R. D., & Findley, S. D. (1995). Cloning and functional characterization of a mammalian zinc transporter that confers resistance to zinc. *The EMBO Journal, 14*, 639–649.

Palmiter, R. D., & Huang, L. (2004). Efflux and compartmentalization of zinc by members of the SLC30 family of solute carriers. *Pflugers Archiv, 447*, 744–751.

Patrushev, N., Seidel-Rogol, B., & Salazar, G. (2012). Angiotensin II requires zinc and downregulation of the zinc transporters ZnT3 and ZnT10 to induce senescence of vascular smooth muscle cells. *PLoS One, 7*, e33211.

Podar, D., Scherer, J., Noordally, Z., Herzyk, P., Nies, D., & Sanders, D. (2012). Metal selectivity determinants in a family of transition metal transporters. *The Journal of Biological Chemistry, 287*, 3185–3196.

Pound, L. D., Hang, Y., Sarkar, S. A., Wang, Y., Milam, L. A., Oeser, J. K., et al. (2011). The pancreatic islet beta-cell-enriched transcription factor Pdx-1 regulates Slc30a8 gene transcription through an intronic enhancer. *The Biochemical Journal, 433*, 95–105.

Pound, L. D., Sarkar, S. A., Benninger, R. K., Wang, Y., Suwanichkul, A., Shadoan, M. K., et al. (2009). Deletion of the mouse Slc30a8 gene encoding zinc transporter-8 results in impaired insulin secretion. *The Biochemical Journal, 421*, 371–376.

Qian, L., Lopez, V., Seo, Y. A., & Kelleher, S. L. (2009). Prolactin regulates ZNT2 expression through the JAK2/STAT5 signaling pathway in mammary cells. *American Journal of Physiology—Cell Physiology, 297*, C369–C377.

Quadri, M., Federico, A., Zhao, T., Breedveld, G. J., Battisti, C., Delnooz, C., et al. (2012). Mutations in SLC30A10 cause Parkinsonism and dystonia with hypermanganesemia, polycythemia, and chronic liver disease. *The American Journal of Human Genetics, 90*, 467–477.

Ryu, M. S., Lichten, L. A., Liuzzi, J. P., & Cousins, R. J. (2008). Zinc transporters ZnT1 (Slc30a1), Zip8 (Slc39a8), and Zip10 (Slc39a10) in mouse red blood cells are differentially regulated during erythroid development and by dietary zinc deficiency. *The Journal of Nutrition, 138*, 2076–2083.

Salazar, G., Craige, B., Love, R., Kalman, D., & Faundez, V. (2005). Vglut1 and ZnT3 co-targeting mechanisms regulate vesicular zinc stores in PC12 cells. *Journal of Cell Science, 118*, 1911–1921.

Salazar, G., Falcon-Perez, J. M., Harrison, R., & Faundez, V. (2009). SLC30A3 (ZnT3) oligomerization by dityrosine bonds regulates its subcellular localization and metal transport capacity. *PLoS One, 4*, e5896.

Salazar, G., Love, R., Styers, M. L., Werner, E., Peden, A., Rodriguez, S., et al. (2004). AP-3-dependent mechanisms control the targeting of a chloride channel (ClC-3) in neuronal and non-neuronal cells. *The Journal of Biological Chemistry, 279*, 25430–25439.

Sekler, I., Sensi, S. L., Hershfinkel, M., & Silverman, W. F. (2007). Mechanism and regulation of cellular zinc transport. *Molecular Medicine, 13*, 337–343.

Sensi, S. L., Paoletti, P., Bush, A. I., & Sekler, I. (2009). Zinc in the physiology and pathology of the CNS. *Nature Reviews Neuroscience, 10*, 780–791.

Seo, Y. A., & Kelleher, S. L. (2010). Functional analysis of two single nucleotide polymorphisms in SLC30A2 (ZnT2): implications for mammary gland function and breast disease in women. *Physiological Genomics, 42A*, 219–227.

Seo, Y. A., Lopez, V., & Kelleher, S. L. (2011). A histidine-rich motif mediates mitochondrial localization of ZnT2 to modulate mitochondrial function. *American Journal of Physiology—Cell Physiology, 300*, C1479–C1489.

Sindreu, C., Palmiter, R. D., & Storm, D. R. (2011). Zinc transporter ZnT-3 regulates presynaptic Erk1/2 signaling and hippocampus-dependent memory. *Proceedings of the National Academy of Sciences of the United States of America, 108*, 3366–3370.

Sladek, R., Rocheleau, G., Rung, J., Dina, C., Shen, L., Serre, D., et al. (2007). A genome-wide association study identifies novel risk loci for type 2 diabetes. *Nature, 445*, 881–885.

Smidt, K., Jessen, N., Petersen, A. B., Larsen, A., Magnusson, N., Jeppesen, J. B., et al. (2009). SLC30A3 responds to glucose- and zinc variations in beta-cells and is critical for insulin production and in vivo glucose-metabolism during beta-cell stress. *PLoS One, 4*, e5684.

Smidt, K., & Rungby, J. (2012). ZnT3: a zinc transporter active in several organs. *Biometals, 25*, 1–8.

Smith, J. L., Xiong, S., Markesbery, W. R., & Lovell, M. A. (2006). Altered expression of zinc transporters-4 and -6 in mild cognitive impairment, early and late Alzheimer's disease brain. *Neuroscience, 140*, 879–888.

Suphioglu, C., De Mel, D., Kumar, L., Sadli, N., Freestone, D., Michalczyk, A., et al. (2010). The omega-3 fatty acid, DHA, decreases neuronal cell death in association with altered zinc transport. *FEBS Letter, 584*, 612–618.

Suzuki, T., Ishihara, K., Migaki, H., Matsuura, W., Kohda, A., Okumura, K., et al. (2005). Zinc transporters, ZnT5 and ZnT7, are required for the activation of alkaline phosphatases, zinc-requiring enzymes that are glycosylphosphatidylinositol-anchored to the cytoplasmic membrane. *The Journal of Biological Chemistry, 280*, 637–643.

Suzuki, T., Ishihara, K., Migaki, H., Nagao, M., Yamaguchi-Iwai, Y., & Kambe, T. (2005). Two different zinc transport complexes of cation diffusion facilitator proteins localized in the secretory pathway operate to activate alkaline phosphatases in vertebrate cells. *The Journal of Biological Chemistry, 280*, 30956–30962.

Taylor, K. M., Hiscox, S., Nicholson, R. I., Hogstrand, C., & Kille, P. (2012). Protein kinase CK2 triggers cytosolic zinc signaling pathways by phosphorylation of zinc channel ZIP7. *Science Signaling, 5*, ra11.

Tepaamorndech, S., Huang, L., & Kirschke, C. P. (2011). A null-mutation in the Znt7 gene accelerates prostate tumor formation in a transgenic adenocarcinoma mouse prostate model. *Cancer Letter, 308*, 33–42.

Tsuda, M., Imaizumi, K., Katayama, T., Kitagawa, K., Wanaka, A., Tohyama, M., et al. (1997). Expression of zinc transporter gene, ZnT-1, is induced after transient forebrain ischemia in the gerbil. *The Journal of Neuroscience, 17*, 6678–6684.

Tuschl, K., Clayton, P. T., Gospe, S. M., Jr., Gulab, S., Ibrahim, S., Singhi, P., et al. (2012). Syndrome of hepatic cirrhosis, dystonia, polycythemia, and hypermanganesemia caused by mutations in SLC30A10, a manganese transporter in man. *The American Journal of Human Genetics, 90*, 457–466.

Valentine, R. A., Jackson, K. A., Christie, G. R., Mathers, J. C., Taylor, P. M., & Ford, D. (2007). ZnT5 variant B is a bidirectional zinc transporter and mediates zinc uptake in human intestinal Caco-2 cells. *The Journal of Biological Chemistry, 282*, 14389–14393.

Vembar, S. S., & Brodsky, J. L. (2008). One step at a time: endoplasmic reticulum-associated degradation. *Nature Reviews Molecular Cell Biology, 9*, 944–957.

Weijers, R. N. (2010). Three-dimensional structure of beta-cell-specific zinc transporter, ZnT-8, predicted from the type 2 diabetes-associated gene variant SLC30A8 R325W. *Diabetology and Metabolic Syndrome, 2*, 33.

Wenzlau, J. M., Juhl, K., Yu, L., Moua, O., Sarkar, S. A., Gottlieb, P., et al. (2007). The cation efflux transporter ZnT8 (Slc30A8) is a major autoantigen in human type 1 diabetes. *Proceedings of the National Academy of Sciences of the United States of America, 104*, 17040–17045.

Wenzlau, J. M., Liu, Y., Yu, L., Moua, O., Fowler, K. T., Rangasamy, S., et al. (2008). A common nonsynonymous single nucleotide polymorphism in the SLC30A8 gene determines ZnT8 autoantibody specificity in type 1 diabetes. *Diabetes, 57*, 2693–2697.

Wijesekara, N., Dai, F. F., Hardy, A. B., Giglou, P. R., Bhattacharjee, A., Koshkin, V., et al. (2010). Beta cell-specific Znt8 deletion in mice causes marked defects in insulin processing, crystallisation and secretion. *Diabetologia, 53*, 1656–1668.

Zhang, K., Shen, X., Wu, J., Sakaki, K., Saunders, T., Rutkowski, D. T., et al. (2006). Endoplasmic reticulum stress activates cleavage of CREBH to induce a systemic inflammatory response. *Cell, 124*, 587–599.

CHAPTER NINE

The Cation Selectivity of the ZIP Transporters

Robert E. Dempski
Department of Chemistry and Biochemistry, Worcester Polytechnic Institute, Worcester, MA, USA
Corresponding author: E-mail: rdempski@wpi.edu

Contents

Abstract

Zinc is a required micronutrient for cellular homeostasis and is essential for the structure and/or function of 100s of biological processes. Despite the central importance of zinc in physiology, the mechanism by which this transition metal is transported into cells is not well understood. The first human zinc importer was identified in 2000. Since that time, a new family of proteins, named ZIP, for Zrt-, Irt-like proteins, has been shown to be expressed in a tissue-specific manner in humans. There are a total of 14 members of this family, which can be further divided into four subfamilies based on sequence similarities: ZIPI, ZIPII, gufA and LIV-1. It has been shown that these proteins are expressed on the plasma membrane as well as on intracellular organelles and each function or are proposed to function to increase the cytosolic concentration of zinc. While the subcellular localization of most of the ZIP family of proteins has been elucidated, the mechanistic details of these proteins including the driving force of cation translocation, residues essential for transport and the molecular determinants which define the differing cation selectivity between the ZIP family of proteins have not been well resolved. The objective of this review is to describe the current understanding of cation transport, mediated by the mammalian family of ZIP proteins and to present some molecular determinants that may contribute to the differing substrate specificity for this family of proteins.

1. INTRODUCTION

Zinc is the second most abundant transition metal in the body (after iron) and is essential for a variety of structural and functional requirements in

Current Topics in Membranes, Volume 69
ISSN 1063-5823,
http://dx.doi.org/10.1016/B978-0-12-394390-3.00009-4

cells. For example, recent data have shown that zinc modulates glycine and glutamate receptors and therefore is important in neuronal transmission (Sensi, Paoletti, Bush, & Sekler, 2009; Trombley, Blakemore, & Hill, 2011). Perhaps most prominently, zinc finger motifs are required for the function of many proteins. Zinc finger motifs encode both histidine and cysteine which directly coordinate zinc and are known to bind to specific DNA sequences and regulate DNA replication (Vallee & Auld, 1995). Recently, zinc fingers have been modified to cleave specific sequences of DNA. It has been proposed that these genetically engineered enzymes will be able to cleave specific regions of DNA, which are implicated in disease states, as a potential therapeutic treatment (Davis & Stokoe, 2010).

It has been recognized for a long time that zinc is mostly bound to protein and other biologically relevant molecules (Vallee, 1959). Because the free concentration of zinc is low, this has limited a molecular understanding of zinc transport into and out of cells *in vivo* as well as identification of biomolecules that are essential for chaperoning zinc within the cell. In order to further understand zinc cellular homeostasis, a number of studies have focused on determining the total cellular concentration of intracellular zinc. In general, the range of total cellular zinc content in mammalian cells is 0.1–1.0 mM (Costello, Fenselau, & Franklin, 2011). Considering that most zinc is coordinated to a variety of biomolecules which exchange zinc, it has been proposed that variations in the cellular concentration of free zinc may have a physiological effect (Costello, Guan, Franklin, & Feng, 2004). In order to address this hypothesis, zinc sensors have been developed to examine how small changes in the cellular concentration of free zinc affect the homeostasis of cells (Burdette et al., 2003; Qin, Dittmer, Park, Jansen, & Palmer, 2011). As a result of these and other experiments, it has been determined that the free concentration of cellular zinc is somewhere between 5 pM and 1 nM (Maret, 2009; Maret & Li, 2009).

In a more broader sense, it has been proposed that cellular zinc resides in one of three pools: 1) free zinc which is uncoordinated in solution, 2) loosely bound zinc which can associate and dissociate from biomolecules such as metallothionein, and 3) tightly bound zinc which cannot react with other biomolecules or be readily released (Costello et al., 2004, 2011; Krezel & Maret, 2006). It has been further proposed that the concentration of the loosely bound zinc pool is approximately 20 μM (Costello et al., 2011). When compared to the total concentration of cellular zinc, simple arithmetic demonstrates that the free zinc and loosely bound pools of zinc are less than 0.0004 and 8% of the total amount of cellular zinc in HT-29 cells, respectively

(Krezel & Maret, 2006). Considering that these two pools are such a small fraction of the total cellular zinc, the mechanism of cellular zinc influx and efflux has been difficult to decipher. While the values above calculate the total cellular concentrations of zinc, it is also important to note that blood plasma contains a substantial concentration of zinc (10–20 μM), where the concentration of free zinc is sub-nanomolar (0.2 nM) (Magneson, Puvathingal, & Ray, 1987). This adds further complexity to investigating zinc transport across the plasma membrane. For instance, whether free or liganded zinc is the physiological substrate is today mostly an unresolved question.

The first zinc import transporter, Zrt1 for zinc-regulated transporter, was identified in *Saccharomyces cerevisiae* in 1996 (Zhao & Eide, 1996a). In the same year, a second homologous protein from *S. cerevisiae* was characterized, Zrt2 (Zhao & Eide, 1996b). Both these proteins have eight transmembrane domains, which include a large soluble domain between transmembranes three and four which contains three histidine residues potentially involved in zinc coordination. In addition, the putative transmembrane domains two, four and five each contain one conserved histidine residue. Considering that histidine can coordinate zinc with high affinity, it was proposed that these residues are essential for transport. From a mechanistic standpoint, it was noteworthy that Zrt1 is a high-affinity zinc transporter (apparent $K_m = 0.6 \pm 0.1$ μM) while Zrt2 is a low-affinity zinc transporter (apparent $K_m = 10 \pm 1$ μM) (Zhao & Eide, 1996a, 1996b).

A subsequent screening of *Arabidopsis thaliana* complementary DNA to identify iron transport genes identified two genes, Irt1 and Irt2 for iron-regulated transporters, which restored iron-limited growth in *S. cerevisiae* (Eide, Broderius, Fett, & Guerinot, 1996). Irt1 is a low-affinity (apparent $K_m = 6 \pm 1$ μM) iron (II) transporter. In contrast, Irt2 is a high-affinity iron uptake protein, which is expressed intracellularly and is proposed to prevent iron toxicity by compartmentalizing iron into vesicles (Vert et al., 2009). Since the discovery of the Zrt and Irt proteins, it appears that it is not uncommon for two homologous proteins to exist in the same organism where one protein has a high affinity for a particular substrate while the second protein has a low affinity for the same substrate (Levy, Kafri, Carmi, & Barkai, 2011). It has recently been suggested that the presence of multiple transporters which have either high or low affinity for a particular substrate could protect these organisms from the loss of function of one of the two transporters, be the result of gene duplication or that the two proteins may function optimally under different environmental conditions (Eide, 2012; Raimunda, Gonzalez-Guerrero, Leeber, & Arguello, 2011).

Identification of zinc-regulated transporters and iron-regulated transporters in *S. cerevisiae* and *A. thaliana*, together with an expansion of the number of sequenced genomes, has heralded the discovery of 100s of proteins which are homologous to the Zrt and Irt proteins. Based on the Zrt and Irt proteins, a new family of proteins have been named the ZIP proteins, which stands for Zrt-, Irt-like proteins. These proteins are members of the solute carrier, SLC39 family of membrane transport proteins (Hediger et al., 2004). The ZIP family of proteins have been shown or are hypothesized to increase the cytosolic concentration of transition metals, most often zinc, but these proteins can also transport cations such as iron (II), copper (II), nickel, cadmium, and manganese. These transition metals are transported across the plasma membrane either from the extracellular space or from intracellular organelles into the cytosol. While BLAST searches can readily find many proteins with varying sequence homology to the Zrt and Irt proteins, the function of most ZIP member proteins is not known. Furthermore, many of these proteins are expressed in intracellular organelles, making a functional understanding of these proteins problematic. While consensus sequences among similar proteins have been examined, for example between Zrt1 and Irt1, less well understood is how subtle amino acid changes between more distant relatives of this protein family result in functional differences (Zhao & Eide, 1996a).

It is important to note that the ZIP family of proteins is not the only group of proteins to transport zinc. The Znt family of proteins (SLC30) act in a synergistic manner with the ZIP proteins to decrease the cytosolic levels of zinc by transporting zinc out of the cytosol, either into the extracellular space or into intracellular organelles (Palmiter & Huang, 2004). These six pass transmembrane proteins are expressed in all kingdoms and are divided into three subfamilies (Liuzzi & Cousins, 2004; Palmiter & Huang, 2004). Similar to the ZIP family of proteins, the Znt proteins encode a histidine-rich region between the fourth and fifth transmembrane domains which is a metal-binding domain (Liuzzi & Cousins, 2004; Lu & Fu, 2007). An analysis of a crystal structure of the bacterial Znt protein, YiiP, has provided important insights into the mechanism of this class of protein (Lu & Fu, 2007). In addition, there are a number of studies which describe tissue localization, the importance of these proteins in the secretory pathway, implications in disease states as well as mechanistic information for the Znt family of proteins (Chao & Fu, 2004; Devirgiliis, Zalewski, Perozzi, & Murgia, 2007; Kambe, 2011; Liuzzi & Cousins, 2004; Lu & Fu, 2007).

2. MAMMALIAN ZINC TRANSPORTER SUBFAMILIES

The mammalian ZIP family of proteins can be classified into four subgroups: ZIPI, ZIPII, LIV-1, and gufA (Gaither & Eide, 2001a). The LIV-1 subfamily has also been called the LZT family which stands for LIV-1 subfamily of ZIP zinc transporters (Taylor & Nicholson, 2003). The LIV-1 subfamily is the most populous subfamily in mammals (Table 1). LIV-1 was initially characterized as a protein which is expressed in breast cancer cells and is upregulated by estrogen treatment (Taylor, 2000; Taylor & Nicholson, 2003). Based on sequence similarities, there are nine human proteins which have been either directly shown or postulated to be divalent transition metal transporters. Each of the LIV-1 subfamily of proteins encodes a predicted metalloprotease motif within the transmembrane domain V. In addition, there is a conserved sequence (HSVFEGLAVGIQ) in most proteins within transmembrane segment four which has been proposed to be a part of the zinc translocation pathway (Dufner-Beattie, Langmade, et al., 2003). Finally, there is a CPALLY motif located immediately preceding the first transmembrane domain in most of the LIV-1 subfamily of proteins.

Many, but not all the LIV-1 subfamily, are localized to the plasma membrane. It is evident that many of these proteins, following a search on the National Center for Biotechnology Information website (http://www.ncbi.nlm.nih.gov/), have multiple isoforms. For example, hZIP4 (human ZIP4) has two major isoforms which contain 622 and 647 amino acids. Synthesis of the homologous shorter isoform in mice has been shown to be due to initiation of protein synthesis within the first intron of the messenger RNA (mRNA) (Dufner-Beattie et al., 2003). Whether there is a functional difference between these two proteins is unknown. On a broader level, it is not clear whether any of the isoforms for any individual ZIP protein has functional significance. However, a cursory glance between the number of isoforms and subcellular localization suggests a correlation where proteins which are localized to intracellular organelles have an increased number of isoforms when compared to those proteins which are localized to the plasma membrane (Table 1). Whether these isoforms create a greater pool of functional diversity for the ZIP family of proteins has not been addressed.

The second most populous subfamily is ZIPII which is comprised of ZIP1, ZIP2 and ZIP3. These were the first proteins identified of the ZIP family of proteins. This was due, in part, to a consensus sequence (HSVXXGL), within the fourth transmembrane domain, which is

Table 1 Listing of the hZIP proteins

Protein	Subfamily	Amino acids (isoforms in parenthesis)	Major location of protein expression	Subcellular localization
ZIP1	ZIPII	324	Prostrate, small intestine, kidney, liver, pancreas, prostrate (Franklin et al., 2003; Gaither & Eide, 2001b; Lioumi et al., 1999)	Plasma membrane, endoplasmic reticulum (Gaither & Eide, 2001b; Milon, Dhermy, Pountney, Bourgeois, & Beaumont, 2001)
ZIP2	ZIPII	309	Prostrate, uterine, epithelial cells, ovary liver, skin (Cao et al., 2001; Gaither & Eide, 2000)	Plasma membrane (Gaither & Eide, 2000; Peters et al., 2007)
ZIP3	ZIPII	314 (411)	Testes, pancreatic cells (Costello, Zou, Desouki, & Franklin, 2012)	Plasma membrane (Dufner-Beattie, Huang, Geiser, Xu, & Andrews, 2005)
ZIP4	LIV-1	647 (622)	Small intestine, stomach, colon, cecum, kidney, pancreas (Dufner-Beattie et al., 2003)	Plasma membrane (Wang et al., 2002)
ZIP5	LIV-1	539 (540)	Liver, kidney, spleen colon, pancreas (Wang, Kim, Petris, & Eide, 2004)	Plasma membrane (Wang, Kim, Petris, & Eide, 2004)
ZIP6	LIV-1	755 (433)	Testis (Croxford et al., 2011)	Plasma membrane (Croxford et al., 2011)
ZIP7	LIV-1	469	Brain, liver (Huang et al., 2005)	Golgi, endoplasmic reticulum (Hogstrand et al., 2009)
ZIP8	LIV-1	460 (476, 443, 393)	Pancreas, placenta, lung, liver, testis, thymus (Begum et al., 2002; Croxford et al., 2011)	Plasma membrane, lysosome, endosomes (Begum et al., 2002)

ZIP9	ZIPI	307 (305, 284, 232, 189	Prostrate (Wang et al., 2008)	Golgi (Matsuura et al., 2009)
ZIP10	LIV-1	831 (697)	Testis (Croxford et al., 2011)	Plasma membrane (Croxford et al., 2011; Kaler & Prasad, 2007)
ZIP11	gufA	342 (335, 227)	Mammary gland (Kelleher et al., 2012)	Golgi (Kelleher et al., 2012)
ZIP12	LIV-1	691 (690, 654, 404, 348, 309)	Lung, brain, testis	
ZIP13	LIV-1	371 (364, 325)		Endoplasmic reticulum (Bin et al., 2011)
ZIP14	LIV-1	492 (489, 481, 460, 393)	Liver, heart pancreas, placenta, lung, brain (Nomura et al., 1994)	Plasma membrane, mitochondria, endosome, lysosome (Liuzzi et al., 2006; Zhao et al., 2010)

conserved among Irt1, Zrt1, Zrt2, ZIP1, ZIP2 and ZIP3 (Dufner-Beattie et al., 2003). Each of the ZIPII subfamily is expressed on the plasma membrane of various cell types (Table 1).

The final two subfamilies contain one member each in humans. The ZIPI subfamily was initially proposed in 2000 to be comprised of proteins mostly of fungal and plant members; however, hZIP9 is a member of this family (Guerinot, 2000; Matsuura et al., 2009). Finally, ZIP11 is a member of the gufA subfamily, which is named after the gufA protein from *Myxococcus xanthus* (McGowan, Gorham, & Hodgson, 1993). A number of prokaryotic and eukaryotic proteins have similar sequences to gufA; however, the function of many of these proteins is unknown. The gufA protein is also homologous to Zrt3, a zinc transporter in *S. cerevisiae* which transports zinc from the lumen of vacuoles into the cytoplasm (MacDiarmid, Gaither, & Eide, 2000).

3. CATION SELECTIVITY OF THE ZIP PROTEINS

Despite the central role of zinc in cellular homeostasis, there is a surprising lack of information about the mechanism by which zinc is transported into cells. This is partly due to technical issues in investigating proteins which are localized to subcellular organelles. As an example of these challenges, both ZIP9 and ZIP11 are expressed in the Golgi (Table 1).

3.1. The ZIPII Subfamily of Proteins

The proteins ZIP1, ZIP2 and ZIP3 comprise the mammalian members of the ZIPII subfamily of proteins. In contrast to ZIP9 and ZIP11, these proteins are normally expressed on the plasma membrane (Table 1). hZIP1 is natively expressed in prostrate cells and has been shown to be a zinc transporter (Gaither & Eide, 2001b). Overexpression of hZIP1 in K562 erythroleukemia cells enabled the determination of the apparent K_m (3.5 μM) by measuring $^{65}Zn^{2+}$ uptake for this protein. While hZIP1-mediated zinc uptake is time, temperature, and concentration dependent, zinc uptake could be saturated at high concentrations. Furthermore, hZIP1-mediated zinc transport was not energy dependent nor was zinc transport stimulated by HCO_3^-. This is important because an earlier set of experiments suggested that there were two different mechanisms for zinc uptake in cells, one which was bicarbonate dependent and one independent of the concentration of bicarbonate (Vazquez & Grider, 1995).

In order to examine substrate specificity, zinc uptake by hZIP1 was measured in the presence of other divalent metals. Transport was inhibited by a 6.6-fold excess of divalent nickel, copper, iron and cadmium, but not by divalent manganese, magnesium, or cobalt (Gaither & Eide, 2001b). Considering that the free concentration of zinc is in the picomolar to nanomolar range, zinc transport was measured in the presence of complete culture medium with 10% fetal bovine serum (FBS), which is intended to mimic the concentrations of biomolecules (albumins and globulins) which coordinate zinc in the blood plasma. Since zinc was associated with K562 cells following incubation of radiolabeled zinc in the presence of complete culture medium with 10% fetal bovine serum and after washing with phosphate-buffered saline and ethylenediaminetetraacetic acid (EDTA), it was concluded that hZIP1 can transport zinc under physiological conditions, where there is very little free zinc in solution. Zinc transport by hZIP1 also appears to be pH dependent as decreasing the pH of the assay buffer from 8 to 6 substantially decreased the rate of zinc transport (Colvin et al., 2005). However, it is not clear whether K_m changes as a function of pH. An analysis of K_m as a function of pH may provide information as to whether biometal binding and/or permeation is pH dependent.

An analogous set of experiments in PC-3 cells (a prostate cancer cell line) showed that transfection of a mock CMV plasmid versus a CMV plasmid containing hZIP1 had no significant change in the apparent K_m of $^{65}Zn^{2+}$ uptake (6.6 ± 2.3 and 7.6 ± 2.3 μM, respectively) while V_{max} increased two-fold, leading the authors to propose an apparent K_m value around 7 μM for hZIP1 (Franklin et al., 2003). In addition, the authors were interested in knowing whether chelated zinc could be transported by hZIP1. Therefore, they added an excess of citrate or EDTA to 20 μM $ZnCl_2$ where the ligand to zinc ratio was 3:1. Under these conditions, there should be very little free zinc in solution. An analysis of the results of these experiments showed that addition of EDTA nearly abolished zinc uptake while addition of citrate had very little effect on zinc uptake. This could be due to the lowered binding affinity of citrate for zinc. Increasing the concentration of citrate decreased zinc uptake. Thus, it appears that hZIP1 can transport zinc when this cation is liganded to components of fetal bovine serum, but that zinc transport is inhibited if it is coordinated to EDTA or high concentrations of citrate. This could be due to differences in affinity between these molecules for zinc. How these data translate to *in vivo* conditions is to date unclear.

In order to obtain a molecular understanding of zinc transport, two histidine residues, H190 and H217, which are in the third and fourth transmembrane domain of hZIP1, respectively, and are conserved among the ZIPII subfamily of proteins, were mutated to alanine (Milon, Wu, Zou, Costello, & Franklin, 2006). Each construct (the wild-type protein and two single-mutant proteins) was fused on the N-terminus to cMYC and an antibody specific to cMYC was used to quantitate total cellular protein, where each construct had similar total cell expression and were localized to the plasma membrane. In order to examine the role of H190 and H217 in zinc transport, the amount of zinc uptake in PC-3 cells for the wild-type fusion protein as well as the two histidine to alanine mutant constructs were examined. An analysis of the experimental data demonstrated that only cMYC-hZIP1 had a significant increase ($p > 0.01$) in $^{65}Zn^{2+}$ uptake when compared to heterologous expression of cMYC alone. Since neither of the single-mutant constructs had significantly more $^{65}Zn^{2+}$ uptake when compared to heterologous expression of cMYC alone, the authors concluded from these data that H190A and H217A were important for zinc influx. However, these data should be considered in light of the experimental system presenting low signal to noise ratio as well as data dispersion.

A functional characterization of hZIP2 has shown that the apparent K_m of this protein for zinc is 3 μM following expression in K562 cells while mZIP2 had an apparent K_m of 1.6 μM following expression in HEK293 cells (Dufner-Beattie et al., 2003; Gaither & Eide, 2000). Zinc uptake for hZIP2 was strongly (>80%) inhibited by six-fold excess of divalent cadmium, copper and manganese, but not by either iron (II) or cobalt (Gaither & Eide, 2000). In contrast, zinc uptake for mZIP2 was only slightly (<50%) inhibited by 50-fold excess of copper (II), cadmium, cobalt, silver and nickel, but not by manganese or iron (II) (Dufner-Beattie et al., 2003). Despite the high homology between hZIP2 and mZIP2, it is noteworthy that there is a difference in results for inhibition of zinc transport between mZIP2 and hZIP2. A more complete molecular understanding of these differences has not yet been attempted but could be an interesting line of research to examine differences between the mouse and human orthologues. Finally, in contrast to ZIP1, addition of HCO_3^- increases the amount of zinc uptake for hZIP2 (Gaither & Eide, 2000).

The final member of the human ZIPII subfamily is ZIP3. mZIP3 is moderately (~50%) inhibited by a 50-fold excess of copper (II), cadmium, manganese, cobalt, silver, magnesium, and nickel, but not by iron (II) following overexpression in HEK293 cells (Dufner-Beattie et al., 2003). Furthermore, the apparent K_m of zinc transport for mZIP3 is 1.6 μM, which is similar to mZIP1 and mZIP2. Analogous to mZIP1, it has been shown that the cellular concentration of zinc regulates the surface expression of mZIP3 (Wang, Dufner-Beattie, et al., 2004).

As members of the ZIPII subfamily, ZIP1, ZIP2 and ZIP3 have both similarities and key differences which may provide important information as to why these proteins are expressed in different tissues. Each of the ZIPII subfamilies has been directly shown to transport zinc across the plasma membrane and each have similar apparent K_m values. While the most detailed experiments to date have been performed on ZIP1, it is interesting to note that even when the same experiments are performed by the same investigators for these three proteins, there are pronounced differences in the inhibition profiles of mZIP1, mZIP2 and mZIP3 (Dufner-Beattie et al., 2003). Furthermore, it has already been noted that ZIP1, ZIP2 and ZIP3 were the first zinc transporters identified, in part, due to a consensus sequence which is shared with Irt1, Zrt1, and Zrt2. However, considering that Irt1 is an iron transporter, while Zrt1 and Zrt2 are zinc transporters, these data suggest that subtle changes in the amino acids which encode these proteins and are outside of consensus sequences could have an important effect on the cation selectivity of the ZIPII subfamily of proteins.

3.2. The LIV-1 Subfamily of Proteins

The LIV-1 family is the largest subfamily of the hZIP proteins. The proteins which comprise this subfamily include ZIP4, ZIP5, ZIP6, ZIP7, ZIP8, ZIP10, ZIP12, ZIP13 and ZIP14. To date, information regarding ZIP6, ZIP7 or ZIP12 has been limited to tissue specificity of protein expression for ZIP6 and ZIP7 and in the case of ZIP7, a proposal that phosphorylation at specific residues has a functional effect (Chowanadisai, Kelleher, & Lonnerdal, 2005; Chowanadisai, Lonnerdal, & Kelleher, 2008; Hogstrand, Kille, Nicholson, & Taylor, 2009; Huang, Kirschke, Zhang, & Yu, 2005).

In contrast to ZIP6, ZIP7 and ZIP12, data for the remaining members of the LIV-1 subfamily outline the cation specificity for these proteins; however, a molecular understanding of function is lacking. hZIP4 was first identified as mutations in this protein result in *acrodermatitis enteropathica*, a zinc deficiency disease which results in abnormal development (Kury et al., 2002; Wang, Zhou, Kuo, Zemansky, & Gitschier, 2002). hZIP4 is expressed in the kidney, small intestines, stomach, colon, jejunum and duodenum; these are the main locations of zinc uptake and reabsorption (Table 1) (Liuzzi & Cousins, 2004; Schmitt et al., 2009; Wang et al., 2001; Weaver, Dufner-Beattie, Kambe, & Andrews, 2007).

Based on hydropathy analysis, the N- and C-termini of hZIP4 are located on the extracellular side of the plasma membrane, separated by eight predicted transmembrane domains. The transmembrane domains are organized into two blocks (1–3 and 4–8) which are separated by a stretch of 76 amino acids. Transmembranes four and five contain conserved histidine and aspartic acid residues that have been proposed to line a cavity through which zinc passes (Dufner-Beattie et al., 2003; Eide, 2004). In addition, there are two other histidine-rich regions of hZIP4 that may serve as zinc-binding sites. One is located in the loop between transmembranes three and four. The mechanistic significance of this region is unknown. However, mutations within this domain decrease ubiquitin-dependent degradation (Mao, Kim, Wang, Eide, & Petris, 2007). The second histidine-rich region is located in the N-terminal domain, proximal to the first transmembrane domain.

To date, the cation specificity of both the mouse and hZIP4 proteins have been examined (Antala & Dempski, 2012; Dufner-Beattie et al., 2003, 2004; Kim et al., 2004; Wang, Kim, Dufner-Beattie, et al., 2004; Wang et al., 2004a). The two proteins are about 70% homologous and the transmembrane domains are nearly identical (Dufner-Beattie et al., 2003). However, the N-terminal domains are 68% identical and the histidine-rich extracellular region within this domain is not conserved. Examination of mZIP4 following expression in HEK293 cells has shown that mZIP4 is a zinc-selective transporter as no divalent transition metal cations in 50-fold excess inhibit mZIP4-mediated transport (Dufner-Beattie et al., 2003).

More recently, it has been shown that zinc transport mediated by hZIP4, following expression of hZIP4 in *Xenopus laevis* oocytes, is substantially inhibited by copper (II) and nickel (Antala & Dempski, 2012). Subsequent experiments demonstrated that copper and nickel are transported by hZIP4. Zinc, copper (II) and nickel have apparent K_m values of 1.4 ± 0.3, 1.2 ± 0.1 and 2.9 ± 0.3 μM, respectively. A novel finding from these experiments was a nanomolar binding affinity where zinc, copper (II) and nickel have apparent K_m values of 76 ± 5, 7.9 ± 3 and 1.1 ± 0.1 nM, respectively (Antala & Dempski, 2012). While cooper (II) and zinc were transported in the nanomolar concentration range, nickel was bound but not transported. This could be due to differences in coordination geometry for nickel when compared to either zinc or copper. Alternatively, nickel could be binding to a non-transport site on hZIP4. Considering that nickel is a toxic trace element, which can result in asthma and skin irritation in those with nickel sensitivity as well as lung and larynx cancer following long-term exposure, it may be important to understand the role of hZIP4 in nickel transport (Sunderman, Dingle, Hopfer, & Swift, 1988). An analysis of additional experiments also demonstrated that labeling of histidine residues (with diethylpyrocarbonate (DEPC)) had a significant effect on zinc uptake, while no change in zinc uptake was observed following labeling with maleimide, a cysteine-directed reagent. It is important to note that the differences in inhibition profiles between the mouse and hZIP4 proteins, like some of the ZIPII subfamily of proteins, cannot be explained at this time.

The results from these experiments which demonstrated that free zinc at the nanomolar concentration level can be transported by hZIP4 revisit the question of whether free or liganded zinc is subsequently transported by the ZIP proteins. This is especially relevant as under normal physiological conditions free zinc is present in nanomolar quantities in the extracellular space. Alternatively, it could also be possible that both free and liganded zinc can be transferred to hZIP4 for transport. An interesting corollary to these results is the observation that the N-terminal domain of hZIP4 is cleaved from the holo-transporter during zinc deficiency (Kambe & Andrews, 2009). However, while it is not known how the N-terminal cleavage affects biometal transport, this could suggest that there are multiple mechanisms by which hZIP4 transports zinc into cells.

ZIP4 is expressed on the apical and basolateral side of polarized MDCK cells while ZIP5 is expressed on the basolateral side (Wang, Kim, Petris, & Eide, 2004). ZIP5 is also expressed in pancreatic ancular cells and embryonic visceral endoderm cells (Dufner-Beattie et al., 2004). The apparent K_m for ZIP5 is 1.7 ± 0.2 μM where 50-fold excess of divalent copper, cadmium, cobalt and silver moderately inhibit ZIP5-mediated zinc transport following expression in HEK293 cells. In contrast, manganese, magnesium and iron (II) have only small effects on ZIP5-mediated zinc transport (Wang, Kim, Petris, & Eide, 2004). Of additional interest is that the surface expression of mZIP5 is independent of the extracellular concentration of zinc. This is in contrast to mZIP4 where the surface expression of this protein is inversely proportional to the cytosolic concentration of zinc.

ZIP8 is expressed on the plasma membrane of cells such as the testis (Dalton et al., 2005). Functional expression of ZIP8 in MFF cells demonstrated that ZIP8 has an apparent K_m for Cd^{2+} of 620 nM (He et al., 2006). A comparable experiment examining the apparent K_m of ZIP8 following heterologous expression in *Xenopus laevis* oocytes resulted in apparent K_m values for Zn^{2+} and Cd^{2+} of 260 ± 90 and 480 ± 80 nM, respectively (Liu et al., 2008).

Cadmium uptake is a physiological concern because while eukaryotic organisms do not require this biometal for function, absorption of cadmium can result in lung cancer. Cadmium uptake is inhibited by Zn^{2+}, Cu^{2+}, Pb^{2+}, and Hg^{2+}, but not affected by Mn^{2+} or Fe^{2+} (Liu et al., 2008). Considering that ZIP8 is most similar to ZIP14, an HCO_3^--dependent transporter, the electrogenicity of ZIP8 was examined indirectly by varying the membrane potential of the cells by changing the extracellular solution (Girijashanker et al., 2008). In these experiments, the external concentration of potassium was varied to control the cellular membrane potential, which was calculated using the Nernst equation. Because the authors observe only a small decrease (from 30 to 15 fmol per oocyte) in cadmium uptake when the extracellular concentration of potassium is changed from 2 to 20 mM, which correlates to a change in the membrane potential from −53 to −114 mV, it was concluded that the cadmium transport must be electroneutral, which would correspond to two bicarbonate ions being translocated with each cadmium ion (Liu et al., 2008). Considering that the addition of extracellular potassium could modulate the function of endogenous proteins in oocytes, such as the Na^+,K^+-ATPase, there may be some inherent error in the calculated membrane potentials (Rakowski, 1993; Rakowski, Vasilets, LaTona, & Schwarz, 1991).

hZIP10 is expressed on the plasma membrane in the kidney and intestine in humans, while mRNA of mZIP10 was evident in the kidney, intestine, pancreas, brain, testis and liver (Croxford, McCormick, & Kelleher, 2011; Kaler & Prasad, 2007). A novel finding when compared to other members of the ZIP family is that the surface expression of mZIP10 decreases when mice are fed a zinc-deficient diet when compared to a normal diet (Croxford et al., 2011). In order to examine the cation specificity of ZIP10, the protein was expressed in LLC-PK1 cells (Kaler & Prasad, 2007). These studies showed that rZIP10 is a zinc transporter with an apparent K_m of 19.2 μM. Further studies with this expression system showed that cadmium, but not calcium, inhibited zinc uptake. Labeling of rZIP10 with iodoacetate, which reacts with cysteine residues, did not have a significant effect on zinc transport; however, addition of either β-mercaptoethanol or N,N′-dicyclohexylcarbodiimide, which labels -COOH groups both significantly, decreased zinc uptake.

It has recently been demonstrated that mutations in hZIP13 result in Ehlers–Danlos syndrome, a disease which is characterized by abnormalities of connective tissue (Giunta et al., 2008). hZIP13 has been proposed to be expressed in the Golgi where the N- and C-termini are on the luminal side of the membrane (Bin et al., 2011). Following expression in HEK293T cells, this protein was solubilized with NP-40 and using Western blot analysis it was concluded that hZIP13 is a dimer in the absence of β-mercaptoethanol, but is monomeric in the presence of β-mercaptoethanol (Bin et al., 2011). Protein purified from Sf9 cells, where the oligomeric state was analyzed by blue native electrophoresis and size exclusion chromatography, show that the hZIP13 was a mixture of monomeric and dimeric proteins (Bin et al., 2011). While it was demonstrated that increasing the expression of hZIP13 increased the intracellular concentration of Zn^{2+}, a systemic analysis of cation specificity for this protein has not been performed.

ZIP14, the final member of the LIV-1 subfamily, is a plasma membrane glycoprotein, expressed in the liver, pancreas, thyroid and heart (Taylor, Morgan, Johnson, & Nicholson, 2005). mZIP14 can transport zinc as well as iron (II) (Liuzzi, Aydemir, Nam, Knutson, & Cousins, 2006). Of note, Cd^{2+}, Mn^{2+} and Cu^{2+} inhibit ZIP14-mediated Zn^{2+} uptake where Cd^{2+} (apparent K_m of 1.10 ± 0.02 μM) and Mn^{2+} (apparent K_m of 18 ± 2 μM) can be transported by this protein (Girijashanker et al., 2008). Interestingly, it was observed that labeling with 4,4′-diisothiocyanatostilbene-2,2′-disulfonic acid (DIDS), a lysine-reactive probe, reduced $^{109}Cd^{2+}$ uptake in a concentration-dependent manner. DIDS has previously been used to inhibit anion transporters. However, the mechanism of inhibition by DIDS for other proteins, such as the Ca^{2+}-ATPase, is not well understood (Romero & Ortiz, 1988; Waisman, Smallwood, Lafreniere, & Rasmussen, 1982).

It was initially suggested that because ZIP14 can transport iron at pH 6.5, which is the pH where iron will dissociate from transferrin in endosomes, that ZIP14 could absorb iron from transferrin, a plasma protein which can ligand up to two Fe^{3+} cations (Zhao, Gao, Enns, & Knutson, 2010).

However, upon the observation that ZIP14 can transport ^{55}Fe in the presence of L-ascorbate, but not nitrilotriacetic acid, it has been more recently suggested that ZIP14 is specific for Fe^{2+} (Pinilla-Tenas et al., 2011). Similar to the homologous mouse ZIP14 protein, hZIP14 can transport Zn^{2+} (apparent K_m 1.9 ± 0.6 μM), Fe^{2+} (apparent K_m 2.3 ± 0.5 μM), Cd^{2+}, and Mn^{2+} (Pinilla-Tenas et al., 2011). It was also observed that while excess Zn^{2+} could completely inhibit $^{55}Fe^{2+}$ transport, Fe^{2+} only partially inhibited $^{65}Zn^{2+}$ transport. Together with the data showing that $^{55}Fe^{2+}$ transport was Ca^{2+} dependent, while $^{65}Zn^{2+}$ transport was not Ca^{2+} dependent, the above result led these authors to propose that there are two independent cation transport pathways where Fe^{2+} can translocate through one pathway, but that Zn^{2+} can translocate through both pathways. This proposal of two independent pathways may also be relevant for the two apparent K_m values observed for hZIP4 which have differing cation transport properties (Antala & Dempski, 2012). Interestingly, while ZIP14 may transport iron from transferrin, iron deficiency does not affect the expression of ZIP14 in mice, but does regulate the expression of ZIP5, ZIP6, ZIP7 and ZIP10 in hepatoma cell lines (Nam & Knutson, 2012).

As a subfamily of proteins, the basic outline of cation transport and inhibition profiles for the LIV-1 family has been defined. Apparent K_m values for these proteins range from low nanomolar to ~20 μM with one or two apparent K_m values which could have up to two independent pathways. While some members of the LIV-1 subfamily have not been studied, each of the remaining proteins transports zinc. It is also clear that biometals which inhibit these proteins as well as cations other than zinc which are transported by these proteins vary. Whether these variations are due to differences in experimental approaches including expression in human and mouse cell lines as well as *Xenopus laevis* oocytes in addition to characterization of purified proteins is unresolved. The LIV-1 subfamily has largely been defined by consensus sequences within the transmembrane domains and extracellular regions. However, the functional significance of these conserved regions is largely unknown. In addition, while consensus sequences have defined these member proteins, in a similar manner to the ZIPII subfamily, it is not unreasonable to propose that differences in the amino acid sequences outside of the consensus sequences may contribute to variations in transition metal specificity.

4. FUTURE PERSPECTIVES

The past decade has seen the identification of a novel class of proteins which transport zinc (and other transition metals) in a tissue-specific manner. While many of these proteins are expressed on the plasma membrane, the intracellular targeting of ZIP7, ZIP9, ZIP11 and ZIP13 has clearly hindered a molecular understanding of these proteins. Furthermore, while the tissue distribution has been studied for most of these proteins, many fundamental mechanistic questions are unresolved. These include the driving force of cation transport, the physiological substrate, whether free or liganded metals are recognized for transport, as well as identification of residues essential for transport.

Initial experiments, at the turn of the century, suggested that the ZIP family of proteins are zinc-selective transporters. However, a closer analysis within the past couple of years has demonstrated that these proteins have broader substrate specificity. In separate instances, ZIP4 and ZIP14 transport two metals, nickel and cadmium, respectively, which have a deleterious effect on health. This opens the question of whether this class of protein can also mediate harmful effects *in vivo*.

In order to examine the molecular determinants which define the ZIP family of proteins, it has been common to identify sequence motifs within these proteins. These motifs include a consensus sequence, HSVXXGL, within the fourth transmembrane sequence of the ZIPII subfamily of proteins as well as the HEXPHEXGD consensus sequence within the fifth transmembrane domain of the LIV-1 family of proteins. At this point, it has not been experimentally elucidated how the residues which comprise these consensus motifs contribute to cation selectivity and transport. Furthermore, differences in substrate specificity among these proteins could be accounted for by differences in the non-conserved regions of the transmembrane domains of these proteins. In order to further examine residues which are potentially important for function, we have performed a sequence alignment of all 14 hZIP proteins using ClustalW2 (Goujon et al., 2010; Larkin et al., 2007). It is our hypothesis that residues which are uniformly conserved among the entire ZIP family of proteins are essential for the structure and function of the ZIP family of proteins. Furthermore, we propose that amino acid differences within the non-conserved regions could contribute to the variations in substrate specificity as described above.

Sequence alignment of the 14 hZIP proteins shows that there are five universally conserved residues within the eight transmembrane domains (Fig. 1A). Four of these residues are glycine (in transmembrane domains two, four, seven and eight) and the remaining residue is alanine in transmembrane four. Considering that alanine and glycine are the third and fifth most abundant residues in transmembrane domains, this discovery is unsurprising (Liu, Engelman, & Gerstein, 2002). A closer inspection of residues proximal to each conserved glycine residue shows that each is located within a consensus sequence NxxxG, where N is predominately alanine, glycine or serine. The GxxxG has been proposed to be important for helix–helix interactions where glycine can be substituted with conservative mutations such as alanine and serine (Russ & Engelman, 2000).

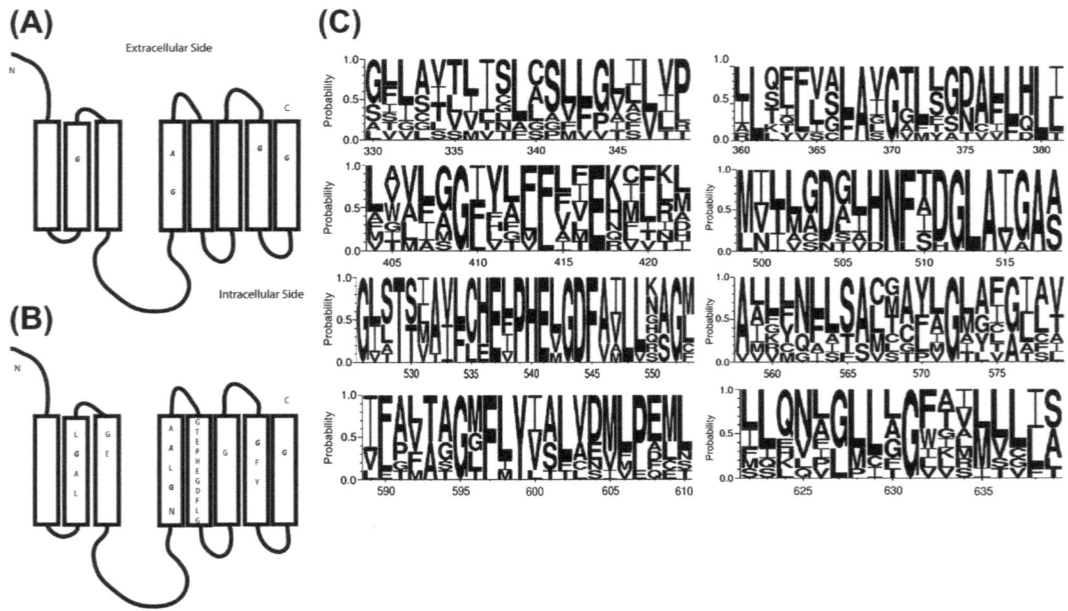

Figure 1 Schematic of conserved residues among the 14 hZIP proteins (A) and the LIV-1 subfamily of proteins (B). Intracellular and extracellular sides of the membrane are marked as well as the N- and C-termini. (C) Sequence alignment of the eight transmembrane domains for the human LIV-1 subfamily of proteins. Amino acid numbering is based on the sequence of 647 amino acid isoforms of hZIP4.

Therefore, it seems likely that these conserved residues are essential for the structural integrity for the entire ZIP family of proteins. The remaining conserved residue, alanine, is encoded within a near universally conserved region within the fourth transmembrane domain, $G_{14}L_{13}A_{14}I_{7}G_{13}$, of the 14 member proteins.

In order to identify residues of importance within the LIV-1 subfamily, a sequence alignment of these proteins was completed. An analysis of the sequence alignment demonstrates that most of the conserved residues within the transmembrane domains of LIV-1 are hydrophobic residues, such as leucine, alanine, and glycine which are often observed in transmembrane domains (Liu et al., 2002). A closer inspection of the conserved residues demonstrates that in addition to the five universally conserved residues described above for the entire hZIP family of proteins, most of the conserved residues within the LIV-1 subfamily are located in transmembrane five (Fig. 1B). A number of researchers have identified the potential significance of some of these residues in transmembrane five which include one histidine, two glutamates and an aspartic acid (Eide, 2004). However, direct experiments to elucidate the role of these residues in transition metal transport are incomplete. Furthermore, there is a conserved glutamate in transmembrane three which may be important for function. Again, a direct examination of the importance of this residue in transition metal transport has not yet been shown. A more complete sequence analysis of the transmembrane domains of the human LIV-1 subfamily of proteins reveals that there is considerable sequence diversity within these domains which are likely to be the major determinants of cation specificity (Fig. 1C).

This is an exciting time to investigate the mechanism of ion transport among the ZIP family of proteins. While the substrate specificity of some of these proteins have been examined, it should be clear that a mechanistic understanding of cation transport mediated by these proteins is unsolved. It is expected that the next couple of years should witness a more complete understanding of the importance of conserved residues in cation transport.

Furthermore, beyond these residues, it is not clear how small variations between otherwise homologous proteins result in functional differences. For example, while the LIV-1 subfamily has a highly conserved consensus sequence, HEXPHEXGD, within transmembrane five, ZIP14 codes for glutamate instead of histidine at the first position. In addition, while all members of the LIV-1 subfamily have the conserved consensus sequence DxxHNFxD in transmembrane four, ZIP13, the protein which is implicated in Ehlers–Danlos syndrome codes for NxxDNFxD. Finally, hZIP4, which has recently shown to have both a nanomolar and micromolar apparent K_m, has two cysteine residues (in transmembrane one and seven) as well as two histidine residues (in transmembrane five and eight) which are not present in any of the remaining subfamily members. The functional importance of these residues or whether these residues could contribute to the second apparent K_m observed for hZIP4 is unknown. Additional unresolved mechanistic questions include the role of bicarbonate in driving cations across the plasma membrane as well as whether free or liganded biometals are the physiological substrate for the ZIP family of proteins. The resolution of these and additional mechanistic questions in the upcoming years will bring a more complete understanding of transition metal homeostasis in mammalian cells.

REFERENCES

Antala, S., & Dempski, R. E. (2012). The human ZIP4 transporter has two distinct binding affinities and mediates transport of multiple transition metals. *Biochemistry, 5*, 963–973.

Begum, N. A., Kobayashi, M., Moriwaki, Y., Matsumoto, M., Toyoshima, K., & Seya, T. (2002). Mycobacterium bovis BCG cell wall and lipopolysaccharide induce a novel gene, BIGM103, encoding a 7-TM protein: identification of a new protein family having zn-transporter and zn-metalloprotease signatures. *Genomics, 6*, 630–645.

Bin, B. H., Fukada, T., Hosaka, T., Yamasaki, S., Ohashi, W., Hojyo, S., et al. (2011). Biochemical characterization of human ZIP13 protein: a homo-dimerized zinc transporter involved in the spondylocheiro dysplastic Ehlers-Danlos syndrome. *The Journal of Biological Chemistry, 46*, 40255–40265.

Burdette, S. C., Frederickson, C. J., Bu, W., & Lippard, S. J. (2003). ZP4, an improved neuronal Zn2+ sensor of the zinpyr family. *Journal of the American Chemical Society, 7*, 1778–1787.

Cao, J., Bobo, J. A., Liuzzi, J. P., & Cousins, R. J. (2001). Effects of intracellular zinc depletion on metallothionein and ZIP2 transporter expression and apoptosis. *Journal of Leukocyte Biology, 4*, 559–566.

Chao, Y., & Fu, D. (2004). Thermodynamic studies of the mechanism of metal binding to the Escherichia coli zinc transporter YiiP. *The Journal of Biological Chemistry, 17*, 17173–17180.

Chowanadisai, W., Kelleher, S. L., & Lonnerdal, B. (2005). Zinc deficiency is associated with increased brain zinc import and LIV-1 expression and decreased ZnT-1 expression in neonatal rats. *The Journal of Nutrition, 5*, 1002–1007.

Chowanadisai, W., Lonnerdal, B., & Kelleher, S. L. (2008). Zip6 (LIV-1) regulates zinc uptake in neuroblastoma cells under resting but not depolarizing conditions. *Brain Research*, 10–19.

Colvin, R. A., Fontaine, C. P., Thomas, D., Hirano, T., Nagano, T., & Kikuchi, K. (2005). Evidence for pH dependent Zn2+influx in K562 erythroleukemia cells: studies using ZnAF-2F fluorescence and 65Zn2+ uptake. *Archives of Biochemistry and Biophysics, 2*, 222–228.

Costello, L. C., Fenselau, C. C., & Franklin, R. B. (2011). Evidence for operation of the direct zinc ligand exchange mechanism for trafficking, transport, and reactivity of zinc in mammalian cells. *Journal of Inorganic Biochemistry, 5*, 589–599.

Costello, L. C., Guan, Z., Franklin, R. B., & Feng, P. (2004). Metallothionein can function as a chaperone for zinc uptake transport into prostate and liver mitochondria. *Journal of Inorganic Biochemistry, 4*, 664–666.

Costello, L. C., Zou, J., Desouki, M. M., & Franklin, R. B. (2012). Evidence for changes in RREB-1, ZIP3, and zinc in the early development of pancreatic adenocarcinoma. *Journal of Gastrointestinal Cancer.*

Croxford, T. P., McCormick, N. H., & Kelleher, S. L. (2011). Moderate zinc deficiency reduces testicular Zip6 and Zip10 abundance and impairs spermatogenesis in mice. *The Journal of Nutrition, 3*, 359–365.

Dalton, T. P., He, L., Wang, B., Miller, M. L., Jin, L., Stringer, K. F., et al. (2005). Identification of mouse SLC39A8 as the transporter responsible for cadmium-induced toxicity in the testis. *Proceedings of the National Academy of Sciences of the United States of America, 9*, 3401–3406.

Davis, D., & Stokoe, D. (2010). Zinc finger nucleases as tools to understand and treat human diseases. *BMC Medicine, 42.*

Devirgiliis, C., Zalewski, P. D., Perozzi, G., & Murgia, C. (2007). Zinc fluxes and zinc transporter genes in chronic diseases. *Mutation Research, 1–2*, 84–93.

Dufner-Beattie, J., Huang, Z. L., Geiser, J., Xu, W., & Andrews, G. K. (2005). Generation and characterization of mice lacking the zinc uptake transporter ZIP3. *Molecular and Cellular Biology, 13*, 5607–5615.

Dufner-Beattie, J., Kuo, Y. M., Gitschier, J., & Andrews, G. K. (2004). The adaptive response to dietary zinc in mice involves the differential cellular localization and zinc regulation of the zinc transporters ZIP4 and ZIP5. *The Journal of Biological Chemistry, 47*, 49082–49090.

Dufner-Beattie, J., Langmade, S. J., Wang, F., Eide, D., & Andrews, G. K. (2003). Structure, function, and regulation of a subfamily of mouse zinc transporter genes. *The Journal of Biological Chemistry, 50*, 50142–50150.

Dufner-Beattie, J., Wang, F., Kuo, Y. M., Gitschier, J., Eide, D., & Andrews, G. K. (2003). The acrodermatitis enteropathica gene ZIP4 encodes a tissue-specific, zinc regulated zinc transporter in mice. *The Journal of Biological Chemistry* 33474–33481.

Eide, D., Broderius, M., Fett, J., & Guerinot, M. L. (1996). A novel iron-regulated metal transporter from plants identified by functional expression in yeast. *Proceedings of the National Academy of Sciences of the United States of America, 11*, 5624–5628.

Eide, D. J. (2012). An "inordinate fondness for transporters" explained? *Science Signaling, 210*, pe5.

Eide, D. J. (2004). The SLC39 family of metal ion transporters. *Pflugers Archiv: European Journal of Physiology, 5*, 796–800.

Franklin, R. B., Ma, J., Zou, J., Guan, Z., Kukoyi, B. I., Feng, P., et al. (2003). Human ZIP1 is a major zinc uptake transporter for the accumulation of zinc in prostate cells. *Journal of Inorganic Biochemistry, 2–3*, 435–442.

Gaither, L. A., & Eide, D. J. (2001a). Eukaryotic zinc transporters and their regulation. *Biometals: An International Journal on the Role of Metal Ions in Biology, Biochemistry, and Medicine, 3–4*, 251–270.

Gaither, L. A., & Eide, D. J. (2000). Functional expression of the human hZIP2 zinc transporter. *The Journal of Biological Chemistry, 8*, 5560–5564.

Gaither, L. A., & Eide, D. J. (2001b). The human ZIP1 transporter mediates zinc uptake in human K562 erythroleukemia cells. *The Journal of Biological Chemistry, 25*, 22258–22264.

Girijashanker, K., He, L., Soleimani, M., Reed, J. M., Li, H., Liu, Z., et al. (2008). Slc39a14 gene encodes ZIP14, a metal/bicarbonate symporter: similarities to the ZIP8 transporter. *Molecular Pharmacology, 5*, 1413–1423.

Giunta, C., Elcioglu, N. H., Albrecht, B., Eich, G., Chambaz, C., Janecke,, A. R., et al. (2008). Spondylocheiro dysplastic form of the Ehlers-Danlos syndrome—an

autosomal-recessive entity caused by mutations in the zinc transporter gene SLC39A13. *American Journal of Human Genetics, 6*, 1290–1305.
Goujon, M., McWilliam, H., Li, W., Valentin, F., Squizzato, S., Paern, J., et al. (2010). A new bioinformatics analysis tools framework at EMBL-EBI. *Nucleic Acids Research* (Web Server Issue), W695–W699.
Guerinot, M. L. (2000). The ZIP family of metal transporters. *Biochimica et Biophysica Acta, 1–2*, 190–198.
He, L., Girijashanker, K., Dalton, T. P., Reed, J., Li, H., Soleimani, M., et al. (2006). ZIP8, member of the solute-carrier-39 (SLC39) metal-transporter family: characterization of transporter properties. *Molecular Pharmacology, 1*, 171–180.
Hediger, M. A., Romero, M. F., Peng, J. B., Rolfs, A., Takanaga, H., & Bruford, E. A. (2004). The ABCs of solute carriers: physiological, pathological and therapeutic implications of human membrane transport proteins. Introduction. *Pflugers Archiv: European Journal of Physiology, 5*, 465–468.
Hogstrand, C., Kille, P., Nicholson, R. I., & Taylor, K. M. (2009). Zinc transporters and cancer: a potential role for ZIP7 as a hub for tyrosine kinase activation. *Trends in Molecular Medicine, 3*, 101–111.
Huang, L., Kirschke, C. P., Zhang, Y., & Yu, Y. Y. (2005). The ZIP7 gene (Slc39a7) encodes a zinc transporter involved in zinc homeostasis of the golgi apparatus. *The Journal of Biological Chemistry, 15*, 15456–15463.
Kaler, P., & Prasad, R. (2007). Molecular cloning and functional characterization of novel zinc transporter rZip10 (Slc39a10) involved in zinc uptake across rat renal brush-border membrane. *American Journal of Physiology Renal Physiology*, F217–F229.
Kambe, T. (2011). An overview of a wide range of functions of ZnT and zip zinc transporters in the secretory pathway. *Bioscience, Biotechnology, and Biochemistry, 6*, 1036–1043.
Kambe, T., & Andrews, G. K. (2009). Novel proteolytic processing of the ectodomain of the zinc transporter ZIP4 (SLC39A4) during zinc deficiency is inhibited by acrodermatitis enteropathica mutations. *Molecular and Cellular Biology, 1*, 129–139.
Kelleher, S. L., Velasquez, V., Croxford, T. P., McCormick, N. H., Lopez, V., & MacDavid, J. (2012). Mapping the zinc-transporting system in mammary cells: molecular analysis reveals a phenotype-dependent zinc-transporting network during lactation. *Journal of Cellular Physiology, 4*, 1761–1770.
Kim, B. E., Wang, F., Dufner-Beattie, J., Andrews, G. K., Eide, D. J., & Petris, M. J. (2004). Zn2+-stimulated endocytosis of the mZIP4 zinc transporter regulates its location at the plasma membrane. *The Journal of Biological Chemistry, 6*, 4523–4530.
Krezel, A., & Maret, W. (2006). Zinc-buffering capacity of a eukaryotic cell at physiological pZn. *Journal of Biological Inorganic Chemistry: JBIC: A Publication of the Society of Biological Inorganic Chemistry, 8*, 1049–1062.
Kury, S., Dreno, B., Bezieau, S., Giraudet, S., Kharfi, M., Kamoun, R., et al. (2002). Identification of SLC39A4, a gene involved in acrodermatitis enteropathica. *Nature Genetics*, 239–240.
Larkin, M. A., Blackshields, G., Brown, N. P., Chenna, R., McGettigean, P. A., McWilliam, H., et al. (2007). Clustal W and clustal X version 2.0. *Bioinformatics* (Oxford, England), *21*, 2947–2948.
Levy, S., Kafri, M., Carmi, M., & Barkai, N. (2011). The competitive advantage of a dual-transporter system. *Science* (New York, N.Y.), *6061*, 1408–1412.
Lioumi, M., Ferguson, C. A., Sharpe, P. T., Freeman, T., Marenholz, I., Mischke, D., et al. (1999). Isolation and characterization of human and mouse ZIRTL, a member of the IRT1 family of transporters, mapping within the epidermal differentiation complex. *Genomics, 2*, 272–280.
Liu, Y., Engelman, D. M., & Gerstein, M. (2002). Genomic analysis of membrane protein families: abundance and conserved motifs. *Genome Biology, 10*. research0054.

Liu, Z., Li, H., Soleimani, M., Girijashanker, K., Reed, J. M., He, L., et al. (2008). Cd2+ versus Zn2+ uptake by the ZIP8 HCO3-dependent symporter: kinetics, electrogenicity and trafficking. *Biochemical and Biophysical Research Communications, 4*, 814–820.

Liuzzi, J. P., & Cousins, R. J. (2004). Mammalian zinc transporters. *Annual Review of Nutrition*, 151–172.

Liuzzi, J. P., Aydemir, F., Nam, H., Knutson, M. D., & Cousins, R. J. (2006). Zip14 (Slc39a14) mediates non-transferrin-bound iron uptake into cells. *Proceedings of the National Academy of Sciences of the United States of America, 37*, 13612–13617.

Lu, M., & Fu, D. (2007). Structure of the zinc transporter YiiP. *Science* (New York, N.Y.), *5845*, 1746–1748.

MacDiarmid, C. W., Gaither, L. A., & Eide, D. (2000). Zinc transporters that regulate vacuolar zinc storage in Saccharomyces cerevisiae. *The EMBO Journal, 12*, 2845–2855.

Magneson, G. R., Puvathingal, J. M., & Ray, W. J., Jr. (1987). The concentrations of free Mg2+ and free Zn2+ in equine blood plasma. *The Journal of Biological Chemistry, 23*, 11140–11148.

Mao, X., Kim, B. E., Wang, F., Eide, D. J., & Petris, M. J. (2007). A histidine-rich cluster mediates the ubiquitination and degradation of the human zinc transporter, hZIP4, and protects against zinc cytotoxicity. *The Journal of Biological Chemistry, 10*, 6992–7000.

Maret, W. (2009). Molecular aspects of human cellular zinc homeostasis: redox control of zinc potentials and zinc signals. *Biometals: An International Journal on the Role of Metal Ions in Biology, Biochemistry, and Medicine, 1*, 149–157.

Maret, W., & Li, Y. (2009). Coordination dynamics of zinc in proteins. *Chemical Reviews, 10*, 4682–4707.

Matsuura, W., Yamazaki, T., Yamaguchi-Iwai, Y., Masuda, S., Nagao, M., Andrews, G. K., et al. (2009). SLC39A9 (ZIP9) regulates zinc homeostasis in the secretory pathway: characterization of the ZIP subfamily I protein in vertebrate cells. *Bioscience, Biotechnology, and Biochemistry, 5*, 1142–1148.

McGowan, S. J., Gorham, H. C., & Hodgson, D. A. (1993). Light-induced carotenogenesis in Myxococcus xanthus: DNA sequence analysis of the carR region. *Molecular Microbiology, 4*, 713–735.

Milon, B., Dhermy, D., Pountney, D., Bourgeois, M., & Beaumont, C. (2001). Differential subcellular localization of hZip1 in adherent and non-adherent cells. *FEBS Letters, 3*, 241–246.

Milon, B., Wu, Q., Zou, J., Costello, L. C., & Franklin, R. B. (2006). Histidine residues in the region between transmembrane domains III and IV of hZip1 are required for zinc transport across the plasma membrane in PC-3 cells. *Biochimica et Biophysica Acta, 10*, 1696–1701.

Nam, H., & Knutson, M. D. (2012). Effect of dietary iron deficiency and overload on the expression of ZIP metal-ion transporters in rat liver. *Biometals: An International Journal on the Role of Metal Ions in Biology, Biochemistry, and Medicine, 1*, 115–124.

Nomura, N., Nagase, T., Miyajima, N., Sazuka, T., Tanaka, A., Sato, S., et al. (1994). Prediction of the coding sequences of unidentified human genes. II. The coding sequences of 40 new genes (KIAA0041-KIAA0080) deduced by analysis of cDNA clones from human cell line KG-1. *DNA Research: An International Journal for Rapid Publication of Reports on Genes and Genomes, 5*, 223–229.

Palmiter, R. D., & Huang, L. (2004). Efflux and compartmentalization of zinc by members of the SLC30 family of solute carriers. *Pflugers Archiv: European Journal of Physiology, 5*, 744–751.

Peters, J. L., Dufner-Beattie, J., Xu, W., Geiser, J., Lahner, B., Salt, D. E., et al. (2007). Targeting of the mouse Slc39a2 (Zip2) gene reveals highly cell-specific patterns of expression, and unique functions in zinc, iron, and calcium homeostasis. *Genesis* (New York, N.Y.: 2000), *6*, 339–352.

Pinilla-Tenas, J. J., Sparkman, B. K., Shawki, A., Illing, A. C., Mitchell, C. J., Zhao, N., et al. (2011). Zip14 is a complex broad-scope metal-ion transporter whose functional properties support roles in the cellular uptake of zinc and nontransferrin-bound iron. *American Journal of Physiology. Cell Physiology, 4*, C862–C871.

Qin, Y., Dittmer, P. J., Park, J. G., Jansen, K. B., & Palmer, A. E. (2011). Measuring steady-state and dynamic endoplasmic reticulum and golgi Zn2+ with genetically encoded sensors. *Proceedings of the National Academy of Sciences of the United States of America, 18*, 7351–7356.

Raimunda, D., Gonzalez-Guerrero, M., Leeber, B. W., 3rd, & Arguello, J. M. (2011). The transport mechanism of bacterial cu+-ATPases: distinct efflux rates adapted to different function. *Biometals: An International Journal on the Role of Metal Ions in Biology, Biochemistry, and Medicine, 3*, 467–475.

Rakowski, R. F. (1993). Charge movement by the Na/K pump in *xenopus* oocytes. *The Journal of General Physiology, 1*, 117–144.

Rakowski, R. F., Vasilets, L. A., LaTona, J., & Schwarz, W. (1991). A negative slope in the current-voltage relationship of the Na+/K+ pump in xenopus oocytes produced by reduction of external K+. *The Journal of Membrane Biology, 2*, 177–187.

Romero, P. J., & Ortiz, C. E. (1988). Electrogenic behavior of the human red cell Ca2+ pump revealed by disulfonic stilbenes. *The Journal of Membrane Biology, 3*, 237–246.

Russ, W. P., & Engelman, D. M. (2000). The GxxxG motif: a framework for transmembrane helix-helix association. *Journal of Molecular Biology, 3*, 911–919.

Schmitt, S., Kury, S., Giraud, M., Dreno, B., Kharfi, M., & Bezieau, S. (2009). An update on mutations of the SLC39A4 gene in acrodermatitis enteropathica. *Human Mutation, 6*, 926–933.

Sensi, S. L., Paoletti, P., Bush, A. I., & Sekler, I. (2009). Zinc in the physiology and pathology of the CNS. *Nature Reviews. Neuroscience, 11*, 780–791.

Sunderman, F. W., Jr., Dingle, B., Hopfer, S. M., & Swift, T. (1988). Acute nickel toxicity in electroplating workers who accidentally ingested a solution of nickel sulfate and nickel chloride. *American Journal of Industrial Medicine, 3*, 257–266.

Taylor, K. M. (2000). LIV-1 breast cancer protein belongs to new family of histidine-rich membrane proteins with potential to control intracellular Zn2+ homeostasis. *IUBMB Life, 4*, 249–253.

Taylor, K. M., Morgan, H. E., Johnson, A., & Nicholson, R. I. (2005). Structure-function analysis of a novel member of the LIV-1 subfamily of zinc transporters, ZIP14. *FEBS Letters, 2*, 427–432.

Taylor, K. M., & Nicholson, R. I. (2003). The LZT proteins; the LIV-1 subfamily of zinc transporters. *Biochimica et Biophysica Acta, 1–2*, 16–30.

Trombley, P. Q., Blakemore, L. J., & Hill, B. J. (2011). Zinc modulation of glycine receptors. *Neuroscience*, 32–38.

Vallee, B. L. (1959). Biochemistry, physiology and pathology of zinc. *Physiological Reviews, 3*, 443–490.

Vallee, B. L., & Auld, D. S. (1995). Zinc metallochemistry in biochemistry. *EXS*, 259–277.

Vazquez, F., & Grider, A. (1995). The effect of the acrodermatitis enteropathica mutation on zinc uptake in human fibroblasts. *Biological Trace Element Research, 2*, 109–117.

Vert, G., Barberon, M., Zelazny, E., Seguela, M., Briat, J. F., & Curie, C. (2009). Arabidopsis IRT2 cooperates with the high-affinity iron uptake system to maintain iron homeostasis in root epidermal cells. *Planta, 6*, 1171–1179.

Waisman, D. M., Smallwood, J., Lafreniere, D., & Rasmussen, H. (1982). The role of band III in calcium transport across the human erythrocyte membrane. *FEBS Letters, 2*, 337–340.

Wang, F., Dufner-Beattie, J., Kim, B. E., Petris, M. J., Andrews, G., & Eide, D. J. (2004). Zinc-stimulated endocytosis controls activity of the mouse ZIP1 and ZIP3 zinc uptake transporters. *The Journal of Biological Chemistry, 23*, 24631–24639.

Wang, F., Kim, B. E., Dufner-Beattie, J., Petris, M. J., Andrews, G., & Eide, D. J. (2004). Acrodermatitis enteropathica mutations affect transport activity, localization and zinc-responsive trafficking of the mouse ZIP4 zinc transporter. *Human Molecular Genetics,* 563–571.

Wang, F., Kim, B. E., Petris, M. J., & Eide, D. J. (2004). The mammalian Zip5 protein is a zinc transporter that localizes to the basolateral surface of polarized cells. *The Journal of Biological Chemistry, 49,* 51433–51441.

Wang, K., Pugh, E. W., Griffen, S., Doheny, K. F., Mostafa, W. Z., al-Aboosi, M. M., et al. (2001). Homozygosity mapping places the acrodermatitis enteropathica gene on chromosomal region 8q24.3. *American Journal of Human Genetics, 4,* 1055–1060.

Wang, K., Zhou, B., Kuo, Y. M., Zemansky, J., & Gitschier, J. (2002). A novel member of a zinc transporter family is defective in acrodermatitis enteropathica. *American Journal of Human Genetics, 1,* 66–73.

Wang, L., McDonnell, S. K., Hebbring, S. J., Cunningham, J. M., St Sauver, J., Cerhan, J. R., et al. (2008). Polymorphisms in mitochondrial genes and prostate cancer risk. *Cancer Epidemiology, Biomarkers and Prevention: A Publication of the American Association for Cancer Research, Cosponsored by the American Society of Preventive Oncology, 12,* 3558–3566.

Weaver, B. P., Dufner-Beattie, J., Kambe, T., & Andrews, G. K. (2007). Novel zinc-responsive post-transcriptional mechanisms reciprocally regulate expression of the mouse Slc39a4 and Slc39a5 zinc transporters (Zip4 and Zip5). *Biological Chemistry, 12,* 1301–1312.

Zhao, H., & Eide, D. (1996a). The yeast ZRT1 gene encodes the zinc transporter protein of a high-affinity uptake system induced by zinc limitation. *Proceedings of the National Academy of Sciences of the United States of America, 6,* 2454–2458.

Zhao, H., & Eide, D. (1996b). The ZRT2 gene encodes the low affinity zinc transporter in Saccharomyces cerevisiae. *The Journal of Biological Chemistry, 38,* 23203–23210.

Zhao, N., Gao, J., Enns, C. A., & Knutson, M. D. (2010). ZRT/IRT-like protein 14 (ZIP14) promotes the cellular assimilation of iron from transferrin. *The Journal of Biological Chemistry, 42,* 32141–32150.

SECTION FOUR

Metal Transporters with Broad Specificity

CHAPTER TEN

Nramp: from Sequence to Structure and Mechanism of Divalent Metal Import

Mathieu F. M. Cellier
Institut National de la Research Scientifique—Institut Armand-Frappier, Laval, QC, Canada
Corresponding author: E-mail: mathieu.cellier@iaf.inrs.ca

Contents

Abstract

Mn and Fe are important for energy metabolism and oxidative stress resistance and cells maintain adequate stores for survival and prevention of toxicity. Membrane permeases of the natural resistance-associated macrophage protein (Nramp) family importing protons and divalent metals are conserved from bacteria to man. Nramp hydrophobic core relates structurally to a superfamily of cation-driven carriers with inverted symmetry. Molecular phylogeny and sequence features support Nramp pseudo-symmetric three-dimensional (3D) model, and remote ancestry to the LeuT superfamily. Genetic analyses suggest conservation of Nramp sequence marks the transition from a phylogenetic out-group and may relate to divalent metal selectivity. Three phylogroups of bacterial proton-dependent manganese transporters (MntH) demonstrate specific patterns of sequence conservation suggesting functional constraints linked to ecological or taxonomical distributions, which may contribute to bacterial virulence. Nramp 3D model is supported experimentally by transmembrane topology and structure–function studies of *Escherichia*

Current Topics in Membranes, Volume 69
ISSN 1063-5823,
http://dx.doi.org/10.1016/B978-0-12-394390-3.00010-0

coli and mouse homologs as well as peptide structure analyses. Eukaryotic Nramps are required for Mn and Fe homeostasis, contributing in multicellular organisms to subcellular and systemic metal traffic and intercellular signaling. Nramps are subjected to elaborate regulation including developmental control of gene expression, protein subcellular targeting, dynamic metallo-dependent control of messenger RNA and protein stability and trafficking. Several human pathologies may result from defects in Nramp-dependent Fe^{2+} or Mn^{2+} transport, including iron overload, neurodegenerative diseases and innate susceptibility to infectious diseases.

1. INTRODUCTION

Fe and Mn are metals indispensable to produce cell energy, for metabolic activity and DNA replication and as such represent vital nutrients. These metals have different properties as redox cofactors, iron being more versatile in terms of useful biological redox potentials while Mn is mandatory for oxygenic photosynthesis and some antioxidant defenses. As metabolic cofactors Mn and Fe exert complementary and/or redundant roles and few cases have been reported of bacteria that relinquished iron and rely on Mn-based biochemistry (Aguirre & Culotta, 2012; Granger et al., 2011). Mn is less reactive with oxygen and oxygen radicals so that, in response to oxidative stress, Mn may be used as surrogate metal in enzymes that normally use the more labile cofactor Fe in appropriate conditions (Anjem & Imlay, 2012; Cornelis, Wei, Andrews, & Vinckx, 2011; Kim et al., 2012). Due to their biochemical reactivity these metals must be metabolized adequately within cells to avoid excess accumulation that would be cytotoxic because it induces oxidative stress or poisons enzyme activities (Aguirre & Culotta, 2012; Cornelis et al., 2011). Once taken up Fe and Mn may be incorporated in proteins, stored in special compartments or exported toward soluble carriers that circulate systemically (Hentze, Muckenthaler, Galy, & Camaschella, 2010; Madejczyk & Ballatori, 2012). Thus both Fe and Mn homeostasis are tightly regulated at the level of cell uptake and distribution according to metabolic activity, storage and/or efflux.

Macrophages represent a specialized professional phagocyte cell type in metazoans that has the essential function of ingesting effete and dead cells and cellular debris with a dual purpose: eliminate aberrant structures and recycle nutrients, and contain any potential microbial invaders by limiting their growth (Elliott & Ravichandran, 2010; Gordon & Taylor, 2005). Because the professional phagocyte diet reflects the organism's steady state metabolism, macrophages accomplish through antigen presentation another essential function that contributes to immune surveillance. They produce an immunological diagnostic by presenting antigenic motifs at their surfaces, and secreting cytokines which contribute to maintain tolerance or trigger an immune response induced by detection of pathogen- or danger-associated molecular patterns (Fleming & Mosser, 2011; Geissmann et al., 2010). During infection, both host and microbe compete to acquire vital resources such as Fe and Mn. A general host defense strategy named nutritional immunity aims at limiting microbial growth and virulence by restricting access to key nutrients. However, microbial pathogens can subvert host defenses to gain access to these resources, including Fe and Mn, and for this purpose rely on metal ion transport systems of different types (Ganz, 2009; Kehl-Fie & Skaar, 2010; Weinberg, 1977).

2. A FAMILY OF MEMBRANE PERMEASES HIGHLY CONSERVED FROM BACTERIA TO MAN

Natural resistance-associated macrophage protein (Nramp) is encoded by a gene that was cloned based on its chromosomal location (Vidal et al., 1993). Exchanging a single guanine for an adenine in *Nramp1* is sole responsible for similar quantitative traits measured with diverse parasites (*Mycobacterium bovis* Bacillus Calmette-Guérin (BCG), *Salmonella enterica* serovar Typhimurium and *Leishmania donovani*) (Vidal et al., 1995). Following intravenous infection of susceptible strains of laboratory inbred mice these intracellular parasites accumulate in spleen and liver macrophages, proliferating inside the phagocytic vesicle supposed to contain them (phagosome) (Malo et al., 1994). This point mutation conferring innate susceptibility to infection introduces a potentially negatively charged residue (Asp) instead of a glycyl moiety. Nramp1 is an integral membrane protein with 12 predicted transmembrane segments (TMS); the G169D mutation falls in TMS4 (Fig. 1) and the mutant protein is no longer properly expressed in the membrane of phagosomes (Gruenheid et al., 1997). A deficit in pH-dependent Mn^{2+} efflux from phagosomes has been measured within live primary *Nramp1*$^{-/-}$ macrophages compared to *Nramp1*$^{+/}$-resistant cells (Jabado et al., 2000). The impact of Nramp1 absence on the phagolysosomal maturation depends also on the properties of the engulfed pathogen, including impaired phagosomal acidification (Hackam et al., 1998) and reduced mannose-6P receptor-dependent endosomal trafficking to the phagosome (Cuellar-Mata et al., 2002; Gallant et al., 2007). Importantly, hard X-ray microprobing revealed that the lack of Nramp1 and/or pathogen ability to perturb intracellular metal traffic induce specific variations in phagosomal metal content (Wagner et al., 2005).

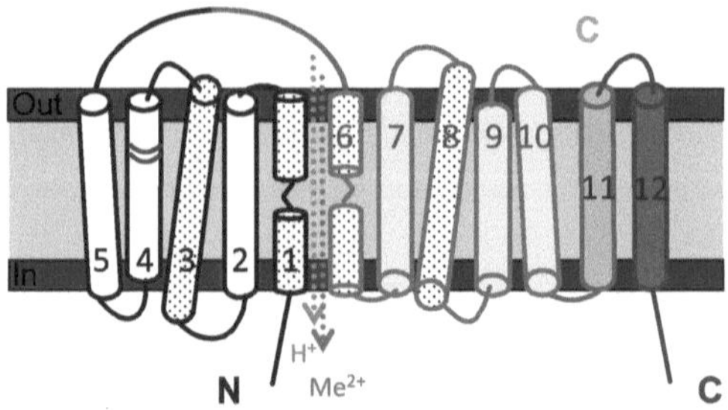

Figure 1 Schematic representation of Slc11 carrier transmembrane topology. A consensus transmembrane topology was deduced from bioinformatic and experimental analyses of mouse Nramp2 and *Escherichia coli* MntH. The conserved hydrophobic core consisting of 10 TMS is followed by one (*E. coli* MntH) or two C-terminal helices (Slc11a2). This conserved core is subdivided in two direct repeats that fold intertwined according to homology threading on structures that share the LeuT fold (Courville et al., 2008; Czachorowski et al., 2009) (see also Fig. 3). The five helix repeats have inverted topologies, and pseudo-symmetric configurations that they exchange during transport. In the LeuT fold, TMS arrangement is dominated by the functional state, i.e. outward- or inward-facing conformation, and the presence or absence of occlusion of the cytoplasmic or extra-cytoplasmic path to the co-substrate-binding sites. Alternating access of the co-substrate-binding sites toward either side of the membrane is mediated by the symmetric interface that acts as a switch for conformational exchange between two structural blocks, a helical bundle (TMS1,2,6,7) and a hash motif (TMS3,4,8,9). Two arms embrace the bundle and gate the transport mechanism (TMS5,10). The central arrangement of TMS1,6 and TMS3,8 contributes to form the co-substrate-binding sites for translocation (Choi et al., 2008; Jeschke, 2012; Khafizov et al., 2010; Morrison et al., 2012). The direction of co-substrate transport is indicated as well as the position of murine Slc11a1 G169D (Vidal et al., 1993) and Slc11a2 G185R mutations (Fleming et al., 1997; Gunshin et al., 1997). See the color plate.

Nramp metal permease defines a small gene family highly conserved in evolution (Cellier et al., 1995) which was renamed Solute carrier 11 (Slc11) family and comprises two genes in humans and mice (Mackenzie & Hediger, 2004). The function of Nramp2, also known as divalent metal transporter 1 (Dmt1, alias Slc11a2) was established by complementary cloning strategies, positional and functional (Fleming et al., 1997; Gunshin et al., 1997). These studies demonstrated the crucial roles of Dmt1 in proton-dependent uptake of Fe^{2+} and Mn^{2+}, both at the level of the apical membrane of intestinal enterocytes (Gunshin et al., 1997) and at the plasma membrane of every cell acquiring iron through the transferrin cycle (Fleming et al., 1997). Nramp1 displayed divalent metal (Me^{2+}) uptake activity comparable to Nramp2, after it reached the plasma membrane following insertion of a hemagglutinin epitope between TMS7 and 8. Both activities depended on temperature, time and acidic pH, implying that Nramp1 and Dmt1 use similar mechanisms to catalyze Me^{2+} transport in the same direction (Forbes & Gros, 2003). In support of this notion, Nramp1 promotes efficient hemoglobin iron recycling through macrophages erythrophagocytosis following hemolytic anemia (Soe-Lin et al., 2009); fish possess only Dmt1 homologs which respond to both infection and iron systemic levels (Elibol-Flemming et al., 2009; Jeon et al., 2011; Neves et al., 2011); and the social amoeba *Dictyostelium discoideum*, which grazes on bacteria, expresses an archetype Nramp in the membrane of phagosomes controlling the replication of pathogens such as *Legionella pneumophila* and *Mycobacterium avium* (Peracino et al., 2006; Peracino, Balest, & Bozzaro, 2010). These pathogens exploit common mechanisms to parasitize amoebas and human macrophages (Schmitz-Esser et al., 2010). Nramp dual role in cell nutrition and intracellular defense has been conserved between amoebozoans and metazoans.

Eukaryotic Nramps are also largely distributed in plants and in protozoan microorganisms (Lin et al., 2011). Plant Nramps contribute to the uptake of various metals including Fe^{2+}, Mn^{2+} and Cd^{2+} (Cailliatte et al., 2010; Colangelo & Guerinot, 2006). *Arabidopsis thaliana* seeds lacking both NRAMP3 and NRAMP4 show deficient plant germination in iron-limited conditions; adult plants display impaired photosynthesis when manganese supply is limited. Both proteins have been localized to the membrane of intracellular vacuoles, implicating these transporters in the mobilization of intracellular Fe^{2+} and Mn^{2+} stores to move them into the cytoplasm (Lanquar et al., 2005, 2010). Similarly, AtNRAMP6 could affect Cd^{2+} intracellular distribution in *A. thaliana* (Cailliatte, Lapeyre, Briat, Mari, & Curie, 2009).

Functional homologs have also been characterized in both Gram-positive and Gram-negative bacteria (Anderson et al., 2009; Hohle & O'Brian, 2009; Kehres et al., 2000; Makui et al., 2000; Que & Helmann, 2000; Runyen-Janecky, Dazenski, Hawkins, & Warner, 2006; Turner et al., 2007). *Escherichia coli* model Nramp, MntH, functions following principles similar to those established in electrophysiological studies using Dmt1 (Mackenzie, Takanaga, Hubert, Rolfs, & Hediger, 2007; Nevo & Nelson, 2006). Slc11 carriers use the energy stored in electrochemical gradients of selected divalent cations and protons to catalyze solute transport by alternating access of co-substrate-binding sites between both sides of the membrane. To allow efficient intracellular Me^{2+} accumulation transport must be directional, implying that once Slc11 carrier has imported Me^{2+} and H^{+} into the cytosol, it cycles back within the membrane without co-substrates. Such transport asymmetry, reflecting differences in the apparent binding affinities of the extra- and intracellular sites, was shown using *E. coli*-oriented vesicle preparations (Courville et al., 2008; Kaback, 1971; Lapointe, Sasseville, & Longpre, 2009). *Escherichia coli* MntH displays temperature dependency expected for a chemiosmotic transporter; proper H^+-dependent Me^{2+} import requires each of the five residues specific of the Nramp family. Nramp conserved role in cell nutrition thus originated in prokaryotic microorganisms.

Sequence conservation between human and *E. coli* Slc11 homologs (35% identity over the 10 TMS hydrophobic core) is unusual for membrane transporters. Though structural conservation is essential to preserve properties such as carrier mechanism of transport, sequence conservation is not mandatory since conserved structural fold may correspond to as little as 10% pairwise amino acid identity or lower, in the case of the Major facilitator superfamily (MFS) family (Radestock & Forrest, 2011). Thus, additional constraints preserved Slc11 sequences, which could relate to the nature of the substrate transported. Other iron transporter families are not particularly conserved sequence, but as Slc11 can also import Cd^{2+} poison, mutations favoring Cd^{2+} uptake may have been counter-selected. Fe^{2+} and Mn^{2+} uptake also have to be tightly regulated thus adding functional constraints on Slc11 sequences. These may be further reinforced by Slc11 H^+-dependent activity, which was suggested to provide an intrinsic feedback control mechanism limiting Me^{2+} entry with a compensatory H^+ influx (Nevo & Nelson, 2006). Two other factors might have restricted Slc11 sequence variation: maintaining intrinsic protein resistance to oxidative environment such as the phagosomal lumen for instance and integrating activity into elaborate regulatory networks within the membrane environment. Hence, the unusual level of Slc11 sequence conservation may result from selective pressure to discriminate Me^{2+} and catalyze efficient transport activity without provoking cytotoxicity, including oxidative stress sensitivity.

3. THE NRAMP FAMILY IS DISTANTLY RELATED TO THE LEUT SUPERFAMILY

The three-dimensional (3D) protein structure necessary to understand catalytic mechanism at the molecular level is in practise difficult to obtain experimentally for membrane transporters. Bioinformatic prediction extracting evolutionary information and experimental validation of 2D topological models are more readily reachable (Daley et al., 2005; Nugent,

Ward, & Jones, 2011). We used evolutionary distributed sequences to predict a consensus transmembrane topology that was tested experimentally using distant homologs (*E. coli* MntH and mouse Nramp2) and both truncated polypeptides and full-length functional proteins containing reporter tags (Courville et al., 2004; Czachorowski, Lam-Yuk-Tseung, Cellier, & Gros, 2009). The combined results produced a consistent transmembrane topology, still approximate in delineating TMS boundaries, with the protein N-terminus intracytoplasmic followed by 11 TMS (*E. coli* MntH) or 12 TMS (mouse Nramp2) (Fig. 1).

Genetic analysis of the amino acid substitutions observed site by site in the multiple sequence alignment revealed positions where amino acid changes correlated with evolutionary groups. Two types of evolutionary rate shifts in amino acid replacement generally suggest possible functional significance: type I, a site conserved within the family varies specifically in a given subgroup, and type II, a site shows distinct residues correlating with different subgroups. It is often assumed that compensatory mutations may explain type I sequence variation, while type II rate shift evokes a radical functional change or divergent protein activity (Gu & Vander Velden, 2002; Knudsen & Miyamoto, 2001). We tested five predicted type II rate shifts distinguishing the Nramp family from its phylogenetic out-group (closest group of sequences showing less than 30% identity). These sites represent Nramp conserved polar residues (Asp, Asn, His) that are membrane embedded; two of them in TMS1 and TMS6 were N-ethylmaleimide (NEM)-accessible (once substituted to Cys in a Cys-less MntH background) and implicated in H^+ influx and H^+-dependent Me^{2+} uptake, respectively (Courville et al., 2008). These data supported a prominent role of TMS1 and TMS6, consistent with their remarkable sequence conservation and coherent with a candidate structural fold proposed in Modbase (Pieper et al., 2011), suggesting a possible evolutionary origin for the Nramp family.

Several evidence support that the Nramp/Slc11 family belongs to the LeuT superfamily of cation-driven transporters with two-fold inverted symmetry. Methods such as sequence clustering (Frickey & Lupas, 2004) and secondary structure similarity search using Hidden Markov Models (e.g. PFAM, PF01566) (Sammut, Finn, & Bateman, 2008) indicated relatively low but seemingly significant sequence relationships (10–15% identity) between Slc11 sequences and families part of the LeuT superfamily (e.g. Slc5, Slc6, Slc7 and Slc38). Members of the LeuT superfamily share a conserved 10 TMS fold, typically superimposable with overall ~3 to 4 Å root mean square deviation (rmsd) and at best ~15% sequence identity (Forrest, Kramer, &

Ziegler, 2011). Hence, both the degree of relationships between Slc11 carriers and the LeuT superfamily and the close superposition of Slc11 3D models with their respective templates are commensurate with similarities in sequence and structure within the LeuT superfamily. Also, 3D fold recognition based on Fold and Function Assignment scores (FFAS) (Jaroszewski, Li, Cai, Weber, & Godzik, 2011) consistently returned the Slc11 profile as one of the closest relative of COGs/KOGs (Tatusov et al., 2003) representing the LeuT fold (Table 1). Lastly, modeling Slc11 carriers on the LeuT fold using diverse templates places the Nramp-specific polar residues within TMS1 and TMS6 in pseudo-symmetric positions, part of small extended peptide chains (non-H bonded) that interrupt transmembrane helices 1 and 6 in their middle. This structural feature reminiscent of other cation transporter families may contribute to bind solute and/or solvent molecules during the transport cycle (Boudker & Verdon, 2010; Courville et al., 2008).

The presence of internal symmetry is a common theme in membrane transporters. It may explain partly why homologous transporters from different life domains, soluble in membrane of different compositions, yet exhibit conserved structures that can be reconstituted using synthetic media in vitro (Sanders & Mittendorf, 2011). Internal symmetry can provide a mechanistic basis for transport such as the oppositely oriented monomers or the tandem inverted repeats in the EmrE and LeuT superfamilies, respectively (Choi, Jeon, Yang, & Kim, 2008; Forrest et al., 2011; Khafizov, Staritzbichler, Stamm, & Forrest, 2010; Morrison et al., 2012). The tandem repeats constituting LeuT carriers comprise five TMS and form intertwined protomers with inverted topologies which coexist in alternative configurations. Superposed structures representing N- and C-protomers from LeuT-type carriers, in conformation "open to out" or "open to in", show two possible protomer configurations: open or closed (Fig. 2). Hence, the transport mechanism relates to co-substrate-driven exchange between each protomer configuration, open or closed, which results in alternating access to opposite sides of the membrane and allows loading/unloading of the co-substrates from their binding sites.

The 3D similarity between N- and C-protomers, however, does not translate into conserved sequences (Forrest et al., 2011). Inverted topologies with respect to the membrane plane and evolving some asymmetry to favor a direction of transport both likely increased sequence divergence between repeats. Slc11 sequence shows little evidence of internal duplication: TMS1 and TMS6 share only one invariant Prolyl part of their helix-interrupting extended peptides while TMS4 and TMS9, folded adjacently as antiparallel

Table 1 Best pairwise FFAS03 scores obtained for clusters of orthologous genes (COG/KOG) representing the Nramp family and the LeuT superfamily demonstrate relatively low-level sequence relationships

		COG1914	KOG1291	COG1953	KOG2466	COG0531	KOG1287	COG4146	KOG2349	COG0733	KOG3660
PF01566	NRAMP	-102	-93.5	-14.4	-15.3			-7.17	-8.89		
COG1914	Mn2+ and Fe2+ transporters of the NRAMP family	-145	-110	-15.4	-13.9			-9.36	-10.4		
KOG1291	Mn2+ and Fe2+ transporters of the NRAMP family	-110	-141	-16.9	-18.2	-9.9	-12.9	-12.7	-12.5	-10.6	
PF02133	Permease for cytosine/purines, uracil, thiamine,allantoin	-14.3	-14.1	-88.4	-79.2	-6.58	-11.3	-14.3	-15.7	-6.71	
COG1953	Cytosine/uracil/thiamine/allantoin permeases	-15.4	-16.9	-147	-115		-13.2	-15.9	-16.5		
KOG2466	Uridine permease/thiamine transporter/allantoin Tp	-13.9	-18.2	-115	-166	-10.8	-15.4	-17.7	-19.4	-13.6	-9.92
2jln_A	MHP1	-13.8	-16.6	-105	-103	-8.26	-13.4	-13.7	-16.2	-8.92	-5.89
COG3949	Uncharacterized membrane protein	-13.6	-13	-27.9	-25.9	-14.3	-18.5	-12.9	-14.7	-12.6	
COG1457	Purine-cytosine permease and related proteins	-13.9	-13.4	-104	-88.1		-11.9	-13.2	-14.9		
PF00324	Amino acid permease	-11.2	-10.2	-11.8	-11.9	-71.9	-65.9	-7.22	-10.5	-10.1	-6.95
COG0531	Amino acid transporters		-9.89		-10.8	-136	-85.4			-10	
KOG1287	Amino acid transporters		-12.9	-13.2	-15.4	-85.4	-142	-9.35	-9.64	-11.3	
3ncy_A	AdiC	-8.75	-10.2	-14.6	-14.3	-79.8	-88.6	-10.4	-9.79	-8.56	
3l1l_A	Arginine/agmatine antiporter	-8.67	-10.4	-14.1	-14.1	-78	-84.9	-10.8	-10.2	-8.83	-6.15
3gi8_C	Uncharacterized protein	-7.52	-11.5	-11	-13.2	-88.2	-91.4	-8.4	-8.06	-8.53	-6.83
PF00474	Sodium:solute symporter family	-13.2	-10.1	-13	-13.2		-8.64	-79.8	-76.5		
COG4146	Predicted symporter	-9.36	-12.7	-15.9	-17.7		-9.36	-145	-115		
KOG2349	Na+:iodide/myo-inositol/multivitamin symporters	-10.4	-12.5	-16.5	-19.4		-9.64	-115	-153		
3dh4_A	Sodium/glucose cotransporter	-10.6	-11.7	-15.8	-16.1		-9.06	-108	-109		
2xq2_A	SODIUM/GLUCOSE COTRANSPORTER	-9.56	-12.7	-13.3	-16.4		-10.4	-120	-125	-5.66	-5.9
COG0591	Na+/proline symporter	-13.1	-13.5	-16.1	-17.4			-96.7	-97.1		
COG4145	Na+/panthothenate symporter	-13.4	-12	-15	-15.5		-9.31	-90.9	-83.7		
KOG2348	Urea transporter	-13.6	-16.8	-18.8	-22.1		-9.72	-92.6	-95.5		
PF00209					-9.16		-8.52			-125	-135
COG0733	Na+-dependent transporters of the SNF family		-10.6		-13.6	-10	-11.3			-150	-110
KOG3659					-12		-9.17			-104	-134
KOG3660	Sodium-neurotransmitter symporter				-9.92					-110	-154
2a65_A	Na(+):NT symporter (Snf family)	-8.86	-10.7	-9.14	-12.6	-10.7	-9.99	-6.17	-8.21	-128	-101
PF02028										-6.28	-6.29
2wsw_A	BCCT FAMILY BETAINE/CARNITINE/CHOLINE Tp						-7.23		-7.87	-6.57	-7.55
2wit_A	GLYCINE BETAINE TRANSPORTER BETP						-7.22	-6.42		-7.85	-8.57
2wsx_A	L-Carnitine/Gamma-Butyrobetaine Antiport						-6.76			-6.34	-7.13

Color code for FFAS03 scores : <-200 <-50 <-18 <-13 <-8

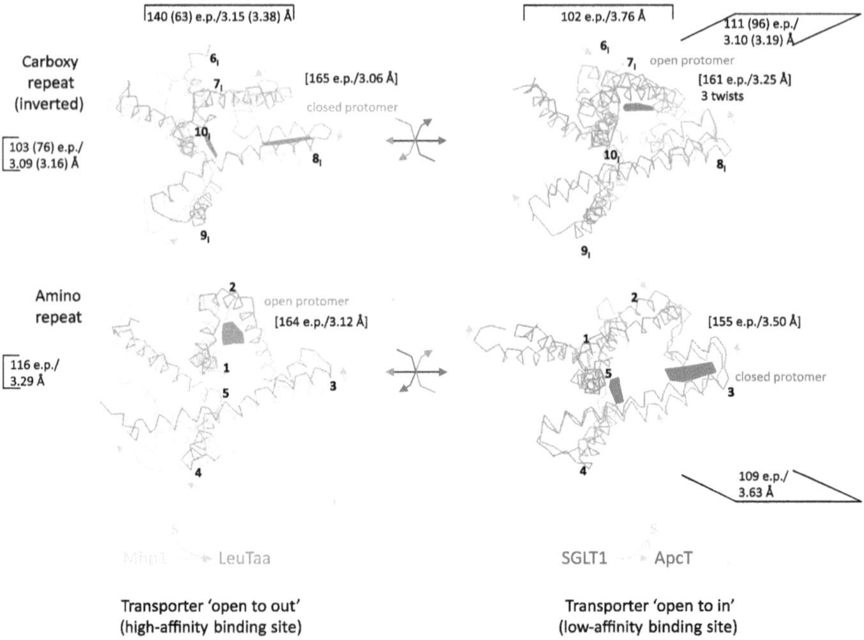

Figure 2 Antiparallel repeats (protomers) alternate between opposite configurations (open/closed) during transport cycling of LeuT-like cation-driven carriers. Coordinates of the crystal structures of cation-driven transporters with inverted symmetry from five different families were aligned using POSA, allowing flexibilities (Ye & Godzik, 2005). Structural superimposition was detailed by visualizing internal amino and carboxy repeats separately for clarity using PyMOL (DeLano, 2002), and by considering the two transport cycle conformations "open to out" and "open to in". TMS numbers are shown (N-end); "i" indicates repeats that were topologically inverted for comparison. Rmsds of pairwise alignments using FATCAT (Ye & Godzik, 2003) are indicated between brackets; e.p., equivalent positions; rmsds of four-repeat POSA alignments are pointed out with three lines showing the pairs of internal repeats that were compared (three superpositions each, vertical, horizontal and diagonal); numbers between parenthesis correspond to the results of superimpositions without flexibility. Cα chain colors: transporters "open to out", LeuT (Yamashita, Singh, Kawate, Jin, & Gouaux, 2005), blue; Mhp1 (Weyand et al., 2008), yellow; transporters "open to in", SGLT1 (Faham et al., 2008), pink; and ApcT (Shaffer, Goehring, Shankaranarayanan, & Gouaux, 2009), green. Gray arrows suggest changes potentially resulting from co-substrate binding/release. See the color plate.

helices on the carrier outer fence, according to the LeuT model (Fig. 3), seem more similar (Fig. 4). Nramp transporters thus plausibly descend from an ancestral family of cation-driven carriers with five TMS inverted repeats, sequence conservation resulting from constraints inherent to Me^{2+} selectivity.

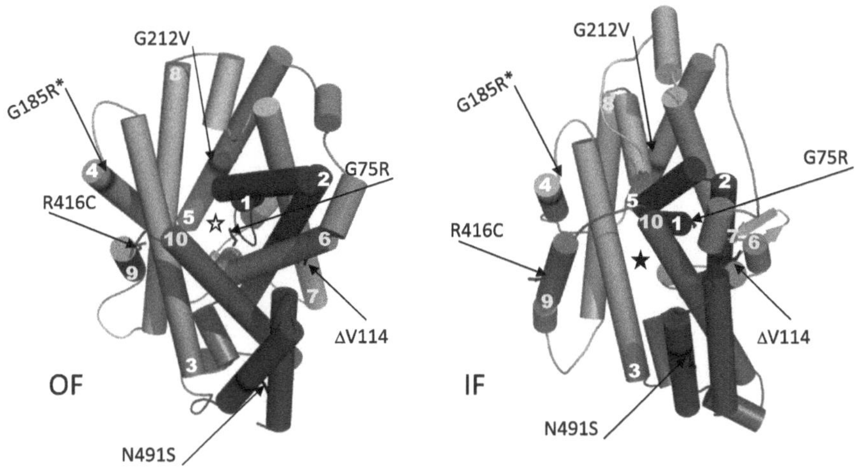

Figure 3 Predicted 3D location of mutations that inactivate NRAMP2/DMT1/SLC11A2. Three-dimensional models were obtained from the Phyre^2 server (Kelley & Sternberg, 2009) by threading the sequence of human NRAMP2 onto crystal structure templates representing two possible states in the carrier cycling mechanism typical of the LeuT superfamily. Outwardly facing (OF; Mhp1, 2JLN) (Weyand et al., 2008) and inwardly facing (IF; vSGLT, 3DH4) (Faham et al., 2008) models are presented as cartoons with NRAMP2 TMS1–12 represented as cylindrical helices. A star indicates the potential transmembrane translocation pathway, white (OF) and black (IF). The predicted location of the mutations associated to the syndrome of microcytic anemia and liver iron overload in humans are indicated (Bardou-Jacquet et al., 2011; Barrios et al., 2012; Beaumont et al., 2006; Blanco et al., 2009; Iolascon et al., 2006; Mims et al., 2005), as well as the mutation linked to microcytic anemia both mice$^{mk/mk}$ and in Belgrade (*b/b*) rats marked with an asterisk (Fleming et al., 1997; Gunshin et al., 1997). See the color plate.

Genetic analysis of site-specific sequence divergence patterns among Nramp phylogroups supports this possibility. To expand previous analyses covering the continuous segment TMS1–3 (Cellier, Bergevin, Boyer, & Richer, 2001) we used Nramp extended hydrophobic core (TMS1–11; Fig. 1). A balanced set of sequences representing the family was initially retrieved by collecting sequences (<90% identity) using Phi-Blast (Jenuth, 2000) strategies distinguishing Nramp phylogroups. These sequences formed clusters linked together when visualizing hits with $-\log P = 45$, five Psi-Blast iterations, the amino acid substitution matrix Blossum 62 and Pfam-a database (Biegert, Mayer, Remmert, Soding, & Lupas, 2006). Multiple sequence alignment and phylogenetic analyses (Tamura et al., 2011) allowed to segregate the most divergent elements from a final set of 665 sequences. The most variable regions were pruned to limit ambiguity; the best evolutionary

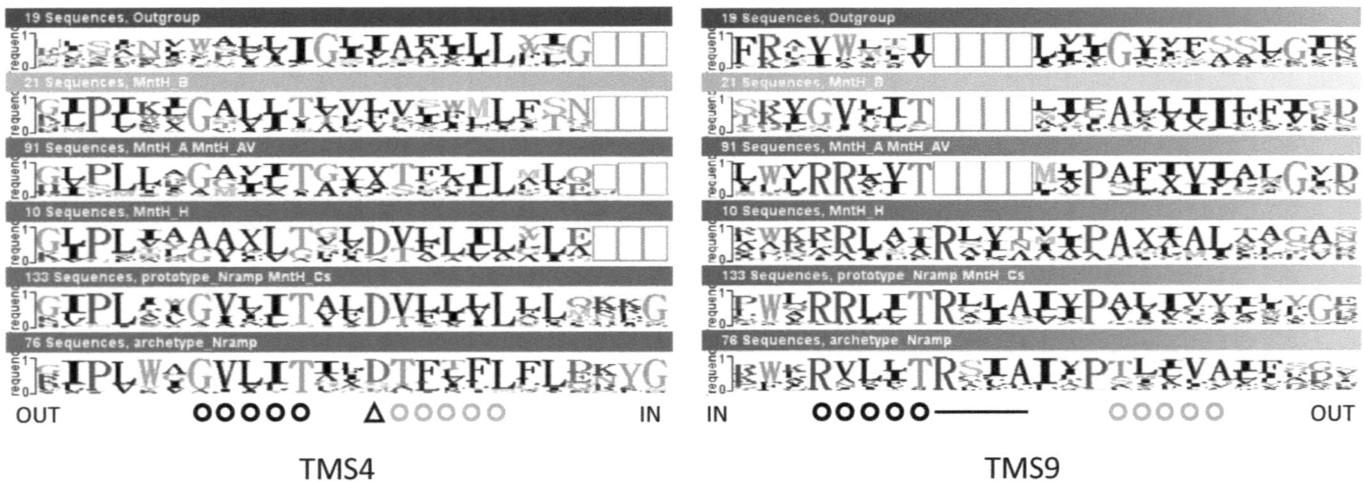

Figure 4 Compared sequence patterns among main Nramp phylogroups between TMS that are predicted pseudo-symmetric based on homology to the LeuT fold. Patterns of conserved sequence among Nramp phylogroups (presented in Fig. 5) were visualized using Phylomlogo (Shih, Lee, Peng, & Wu, 2007). Both the outer and inner halves of the central part of TMS4 demonstrate some conservation, and show similarity with the presumed homologous sections of TMS9 (indicated by black and gray circles, respectively). The introduction of a negative charge in TMS4 (shown by a triangle) correlates with the extension of TMS9 by an extra helix turn, including a positive charge (underlined). These modifications identify the eukaryotic phylogroups, and are shared by the bacterial MntH C cluster, derived from a prototype *Nramp* by horizontal gene transfer. Similar properties also characterize the rather ancestral subgroup MntH H, which stems at the base of the MntH A subtree (cf. Fig. 5) for color version of this figure, the reader is referred to the online version of this book.

model describing sequence variations was tested using a Neighbor-Joining tree, and the final phylogeny was obtained by Minimal Evolution (Fig. 5).

To establish specific variations in amino acid substitution rate along the alignment, either per group or group pairs, we calculated the alpha shape

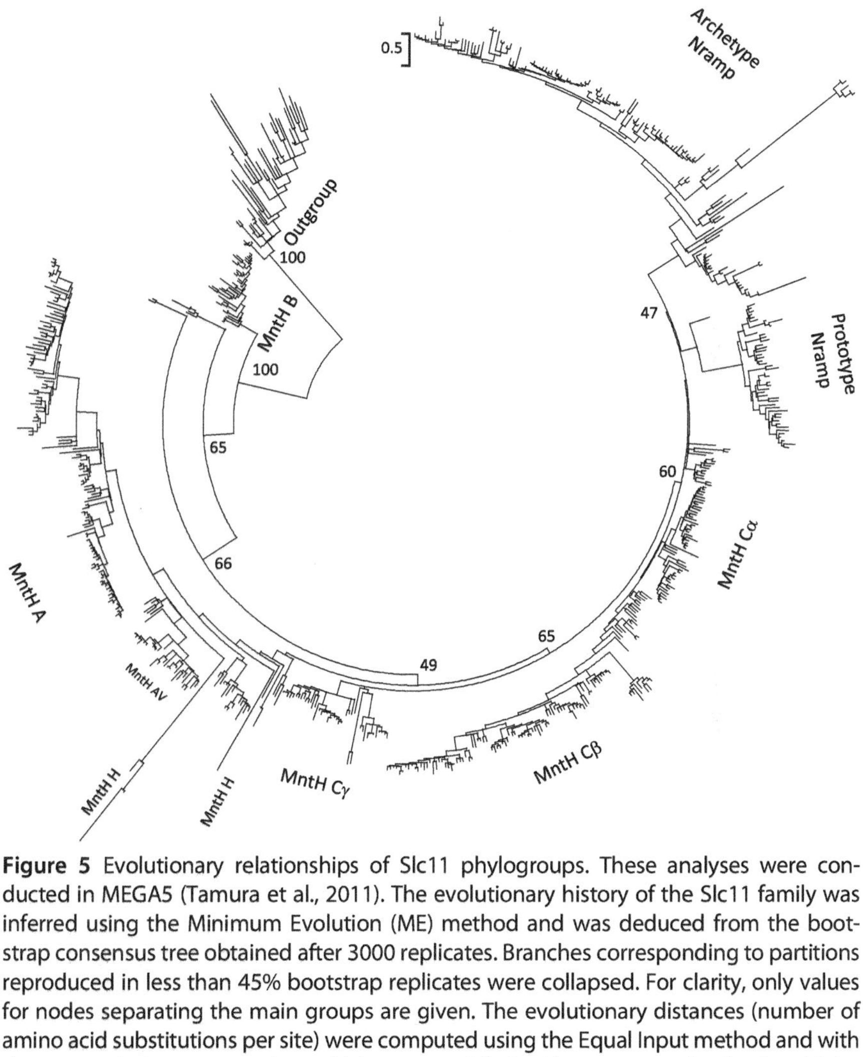

Figure 5 Evolutionary relationships of Slc11 phylogroups. These analyses were conducted in MEGA5 (Tamura et al., 2011). The evolutionary history of the Slc11 family was inferred using the Minimum Evolution (ME) method and was deduced from the bootstrap consensus tree obtained after 3000 replicates. Branches corresponding to partitions reproduced in less than 45% bootstrap replicates were collapsed. For clarity, only values for nodes separating the main groups are given. The evolutionary distances (number of amino acid substitutions per site) were computed using the Equal Input method and with the rate variation among sites which was modeled with a gamma distribution (alpha shape parameter = 0.782). Differences in composition bias among sequences were considered in evolutionary comparisons. The ME tree was searched using the Close-Neighbor-Interchange algorithm. The Neighbor-Joining algorithm was used to generate the initial tree. The analysis involved 665 amino acid sequences. There were a total of 337 positions in the final dataset corresponding to Nramp extended hydrophobic core.

parameter for the gamma distribution best modeling sequence variation. Low alpha values represent sequence heterogeneity along the alignment, suggesting both conserved and nonconserved positions. Observing unchanged or reduced value of alpha upon comparison of two phylogroups indicates evolutionary conservation. To reveal positions where amino acid changes correlate with evolutionary groups, site-specific sequence variations were computed between pairs of groups to calculate coefficients of predicted type I and II functional divergence (Fig. 6, Table 2) (Gu & Vander Velden, 2002; Knudsen & Miyamoto, 2001).

The phylogenetic out-group displayed the highest value for parameter alpha which suggests diverse sequences and reduced functional constraints (Fig. 6). The out-group also demonstrates type I and II divergence from most Nramp phylogroups (Fig. 6, Table 2). MntH B phylogenetic position close to the root of the Nramp family tree (Fig. 5) correlates with its occurrence only in bacterial spp. that are strictly anaerobic (data not shown). MntH B sequences show low proportion of invariant residues (5.5%), little diversity but potential type I divergence (Fig. 6) and low-level predicted type II divergence (Table 2). Phylogenetic reconstruction also shows moderate divergence in MntH B group, indicated by relatively short branch length, compared to the out-group, for instance (Fig. 5).

Genetic analyses indicate that MntH A is diverse, with an alpha parameter value of 1.2 (Fig. 6). MntH A sequences also display less site-specific divergence (both type I and II) compared to MntH B and out-group (Fig. 6, Table 2), consistent with a more central position in the family tree as likely precursor of eukaryotic Nramp (Fig. 5). In contrast, the third bacterial group, MntH C, is found amid eukaryotic sequences and clusters with prototype Nramp, which belong essentially to unicellular eukaryotes (fungi, algae and amoebae). MntH C subdivides in MntH Cα, Cβ and Cγ (Fig. 5). MntH Cα and MntH A have comparable taxonomic distributions (data not shown) but their sequences differ in proportion of invariant sites (17 vs 8.6%), alpha parameter (1.0 vs 1.2) and predicted type I divergence (Fig. 6). These data indicate distinct evolutionary processes shaping MntH A and MntH Cα groups.

Both MntH Cβ and Cγ represent smaller bacterial communities that share specific ecological niches (Suen, Goldman, & Welch, 2007) (Gram-positive lactic bacteria frequently associated with animal gut or plants, and β-proteobacteria associated to soil, plants and opportunistic animal pathogens with large genomes, respectively). Despite narrow taxonomic distribution MntH Cβ and Cγ are diverse, compared to MntH Cα, displaying relatively

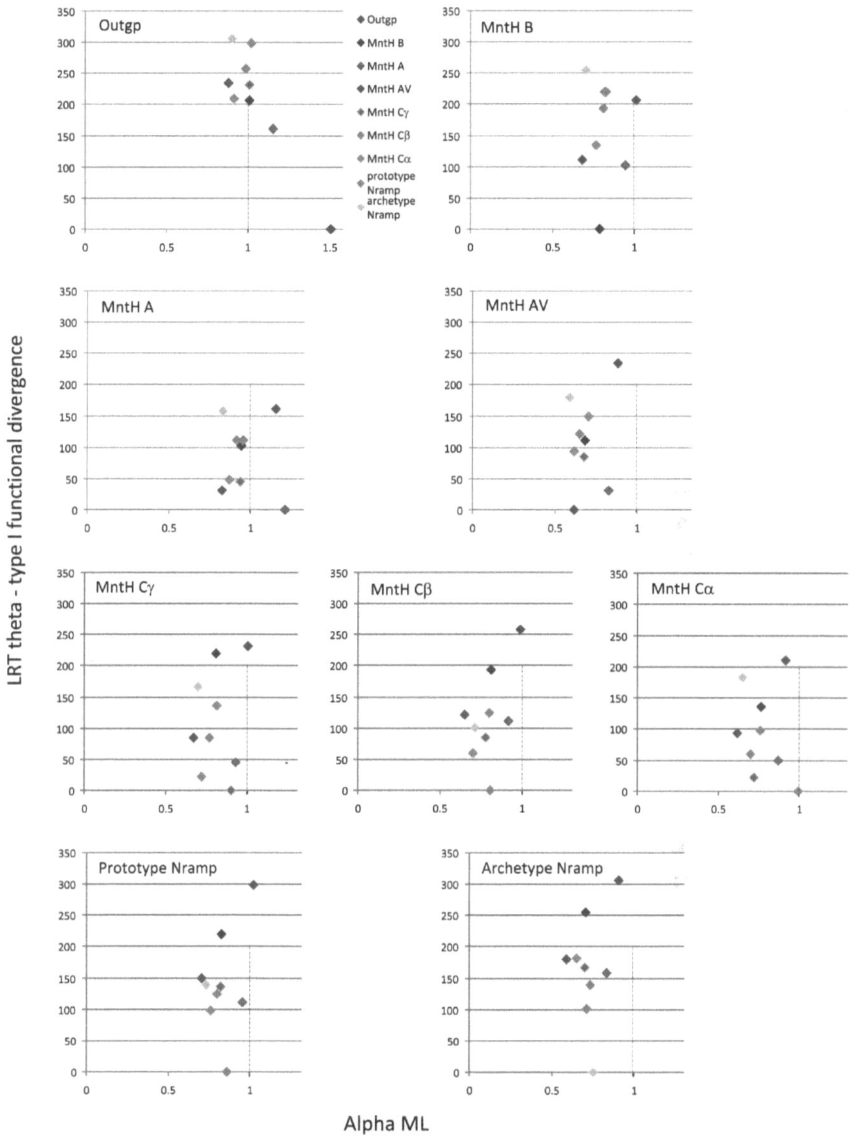

Figure 6 Estimations of amino acid substitution rate variation in eight Nramp family groups and the phylogenetic out-group (outgp) based on maximum likelihood (ML) approaches and using 20 representative sequences per group. Site-to-site substitution rate variation along multiple aligned sequences is described by a gamma distribution of shape parameter alpha, which increases as the rate variation among sites decreases (Gu & Zhang, 1997). The parameter alpha (alpha ML, *x*-axis) was determined for each group (Out-group, 1.507, MntH B, 0.79, MntH A, 1.209, MntH AV, 0.619, MntH Cα, 0.997, MntH Cβ, 0.803 and MntH Cγ, 0.907, prototype Nramp, 0.859 and archetype Nramp, 0.749) and in

homogeneous amino acid substitution rate variation among sites (alpha parameter ~0.8 and 0.9, respectively, Fig. 6), moderate proportion of invariant sites (11.5 and 8%), and higher potential for both type I and II functional divergence (Fig. 6, Table 2). The data indicate that *mntH* Cβ and Cγ are diversifying (encoding four and two subtypes, respectively, Fig. 5), and potential divergence might relate to selective preference for iron (Lopez, Latorre, Reyes-Jara, Cambiazo, & Gonzalez, 2012; Richer, Courville, Bergevin, & Cellier, 2003; Shalom et al., 2007).

With physical evidence of horizontal transfers of *mntH* Cα (Richer et al., 2003), *mntH* Cβ (Richer, Courville, & Cellier, 2004) and *mntH* AV genes (not shown) the data indicate relatively recent origins for MntH C and AV phylogroups, supporting polyphyletic origins and possible functional diversification (Cellier et al., 2001; Richer et al., 2003). In contrast, genetic analysis of MntH B sequences suggests functional constraints that may distinguish an ancient phylogroup and support remote LeuT ancestry.

4. STRUCTURE–FUNCTION RELATIONSHIPS AMONG SLC11 CARRIERS

Several independent experiments aimed at elucidating structure–function relationship among Nramp transporters have produced results supporting structural homology to the LeuT superfamily. Repeated observations demonstrated impaired Me^{2+} and H^{+} transport functions consequently to mutations at the conserved motifs, DPGN in TMS1 and MPHN in TMS6.

Dmt1 metal uptake requires both Asp and Gly residues in TMS1 predicted extended peptides (Cohen, Nevo, & Nelson, 2003; Lam-Yuk-Tseung

pairwise group combinations, as indicated. Functional (type I) divergence between groups was predicted using the coefficient of divergence theta, deduced from site-specific probability to exhibit significantly different substitution rates. A likelihood ratio test (LRT theta, *y*-axis) expresses the probability of divergence between pairs of groups (Gaucher, Gu, Miyamoto, & Benner, 2002). Scatter plot of pairwise group comparisons are represented with point color key to group as indicated, representing the alpha ML and LRT values as *x* and *y* coordinates, respectively. A vertical dotted line (alpha = 1) up to LRT value of 200 shows that all comparisons involving the out-group fall outside this area, and four points fall outside for MntH B, the most divergent group of the Nramp family. In contrast MntH A, almost as diverse as the out-group (high alpha ML), shows comparatively little divergence, while groups of more recent origin (MntH AV and MntH Cs) display relatively low diversity and distributed divergence. The eukaryotic Nramp sequences show similar patterns and significant divergence among them. See the color plate.

Table 2 Coefficient of predicted type II functional divergence

	out	B	A	AV	Cγ	Cβ	Cα	pN	anaN
out		-0.012	0.113	0.421	0.322	0.273	0.329	0.161	0.297
B			-0.21	0.171	0.076	0.07	0.071	0.088	0.356
A				-0.23	0.274	-0.15	-0.16	-0.19	0.268
AV					0.044	0.112	0.082	0.06	0.341
Cg						0.04	-0.2	-0.09	0.183
Cb							-0.19	-0.14	0.297
Ca								-0.15	0.249
pN									0.206
anaN									

nd, not determined

Color code: >250, >100, >0

et al., 2003), together with less conserved polar residues in TMS1 carboxy half, initially predicted extra-cytoplasmic (Cohen et al., 2003). TMS6 His residue in the predicted extended peptide, and another one located six positions downstream in the carboxy half of the helix, affected both Me^{2+} and H^+ uptake (Lam-Yuk-Tseung et al., 2003; Mackenzie, Ujwal, Chang, Romero, & Hediger, 2006). Considering the protistan Nramp homolog reported in *Perkinsus marinus*, mutation of TMS1 and TMS6 motifs, separately or in combination, induced loss of activity due to either defective expression or lack of Me^{2+} transport (Lin et al., 2011).

In bacteria, mutations at TMS1 and TMS6 motifs in *E. coli* homolog MntH also severely reduced H^+ and Me^{2+} uptake activities (Chaloupka et al., 2005; Courville et al., 2008; Haemig & Brooker, 2004; Haemig et al., 2010; Lan et al., 2012). Notably, the lack of Me^{2+}-induced H^+ uptake that results from mutating TMS1 Asp^{34} was observed after measuring pH variations either intracellularly or extracellularly (Courville et al., 2008; Lan et al., 2012). Also, mutants at eight conserved sites in TMS6 indicated the critical importance of residue side chain volume, supporting the notion that Slc11 TMS6 participates in protein conformational changes (Haemig et al., 2010).

In vitro observations using synthetic peptides corresponding to DMT1 TMS1 and TMS6 provided further indications of peculiar structures compatible with the LeuT fold. A peptide spanning TMS1 formed an "α-helix-extended segment-α-helix" configuration in detergent micelles with the amino half deeply embedded, the motif DPGN forming the central flexible region and side chains of negatively charged residues, including the carboxy half, close to the micelle surface binding Mn^{2+} and Co^{2+} ions

(Song et al., 2009; Wang et al., 2011). A peptide representing DMT1 TMS6 also displayed specific features, both in detergent micelles and in aqueous solution, such as a discontinuous helix structure interrupted by an extended segment and self-assembly properties, both altered when the conserved histidines were mutated (Xiao, Li, et al., 2010; Xiao, Wang, Li, & Li 2011).

Such properties were not observed with peptides representing other transmembrane helices, such as TMS3 and TMS4. TMS3 contains three conserved negative charges whereas TMS4 is mutated in murine models of innate susceptibility to intracellular infections (Nramp1 G169D) and microcytic anemia (Dmt1 G185R, Fig. 1). Nramp1 TMS3 folds as a typical amphipathic helix monomeric in aqueous solution, while it aggregates in detergent micelles (Xiao, Wang, et al., 2010). Nramp1 TMS4 peptide displays high helicity and buries deeply in model bilayers independent of pH (Qi et al., 2009). The G169D mutation changed the local conformation near the mutation site, modifying interactions measured with H^+ and Mn^{2+} (Xue et al., 2008, 2009). The properties measured in vitro for TMS3 and TMS4 also appear compatible with LeuT architecture: TMS3 forming a long, bent and amphipathic helix exposed to the solvent, whereas TMS4 is a short highly helical structure at the periphery of the molecule (Fig. 3). Hence so far, in vitro peptide data fit the LeuT-based model with tandem repeats, inverted topology and two-fold symmetry along the membrane normal.

The TMS6 His residue in Slc11 MPH triplet constitutes an evolutionary rate shift between the Nramp and the out-group sequences; it also underwent mutation in a subgroup that emerged in grasses (Xia et al., 2010; Xia, Yamaji, & Ma, 2011). Nrat1 is a plasma membrane-located Slc11 transporter unable to complement Fe^{2+} uptake in *fet3fet4* yeast strain because it transports Me^{3+} instead (Xia et al., 2010, 2011). Site-specific sequence divergence of Nrat1 Me^{3+} carriers from bona fide Nramp Me^{2+} transporters pointed at TMS6 as the most divergent, with two predicted type II rate shifts matching the functional signature MPH (changed for TPY). Also TMS1, 3, 7, 8 and 5 presented divergent sites possibly contributing to Nrat1 preference for Me^{3+}. More generally, the Nramp/Nrat1 sequence comparison showed candidate rate-sifted sites in pairs of TMS predicted to be symmetric and involved in substrate uptake according to LeuT homology (Lin et al., 2011).

Bioinformatic analyses and transmembrane topology mapping studies using functional Slc11 carriers provided further evidence of Slc11 folding similar to the LeuT architecture. *Escherichia coli* MntH polypeptide traverses 11 times the membrane as C-terminal genetic fusions with antibiotic resistance markers produced chimeric proteins actively transporting Me^{2+}

and H^+ and conferring resistance to ampicillin but not to chloramphenicol, thus proving the periplasmic location of MntH terminus (Fig. 1). The loop separating TMS7 and 8 (7/8) also appeared extra-cytoplasmic. Genetic insertion and fusion data supported similar interpretation for the extra-loop 5/6, though insertion of a topological reporter impaired MntH transport activity (Courville et al., 2004). This seemed in line with extra-loop 5/6 sequence conservation and TMS6 proposed role in conformation changes during carrier cycling (Haemig et al., 2010; Xiao et al., 2011). MntH fusion polypeptides showed cytoplasmic exposure of the loops 2/3, 6/7 and 10/11 and periplasmic location of the loops 1/2 and 9/10 (Courville et al., 2004).

Nramp2/Dmt1 transmembrane topology was analyzed by inserting a short immunoreactive peptide tag at sites predicted extra-membranous and showing low sequence conservation. The orientation of tagged constructs showing Me^{2+} uptake was revealed by immunofluorescence with or without cell permeabilization. The Nramp2 termini together with loops 4/5, 6/7 and 9/10 appeared cytoplasmic while loops 5/6, 7/8 and 11/12 were extra-cytoplasmic (Fig. 1). Combined results of MntH and Nramp2 studies leave the loops 3/4 and 8/9 unmapped experimentally. These loops show symmetric locations in LeuT-based 3D models (Fig. 3). Perhaps their rearrangement during transport cycling might explain why tag insertion in these pseudo-symmetric loops impaired function. Experimental studies of Slc11 topology thus support the inverted pseudo-symmetric structure proposed for Nramp based on homology with LeuT (Czachorowski et al., 2009).

The identification of *DMT1* natural mutations responsible for a novel human syndrome of microcytic anemia and liver iron overload provide additional information regarding potential structure–function relationships among Slc11 carriers. DMT1 structural threading consistently ranked members of the LeuT superfamily as best templates (I-Tasser, Roy et al., 2011; Swiss-model, Bordoli & Schwede, 2012; Modeller, Pieper et al., 2011; Phyre^2, Kelley & Sternberg, 2009). Figure 3 shows examples of Nramp2/Dmt1 models used to localize natural mutations causing this syndrome. Two sites occupied by Gly moieties were replaced by nonconservative residues, Arg (TMS1) and Val (TMS5). Both mutations are suspected to impair Nramp2 expression and/or function since replacing Gly residues could perturb inter-helix interactions important for packing or prevent flexibility required for conformation changes (Barrios et al., 2012; Blanco, Kannengiesser, Grandchamp, Tasso, & Beaumont, 2009). According to Dmt1 3D model, residue 75 locates at the N-end of TMS1, close to the membrane/cytosol interface, in position to interact with TMS7 N-end during catalysis (Fig. 3). Gly^{212} occupies a more central position in TMS5 and might possibly interact with TMS1 amino half. The murine mutation inducing microcytic anemia (Fleming et al., 1998) affects another Gly residue at the amino-end of TMS4, proximal to TMS3. Predicted location based on LeuT homology modeling thus suggests that these three mutations could perturb Nramp2 function. The TMS5 mutation G212V was found twice in patients displaying compound heterozygosity: with the V114 in-frame deletion in TMS2 or with TMS11 mutation N491S (Bardou-Jacquet et al., 2011; Beaumont et al., 2006). The latter is one of Nramp-specific sites studied in *E. coli* MntH: N401G or T mutations both preserved protein expression and increased Me^{2+}-induced H^+ uptake (Chaloupka et al., 2005; Courville et al., 2008). The presence of a 12th TMS in eukaryotic proteins may account for phenotypic differences resulting from similar mutations in prokaryotic and eukaryotic homologs.

Lastly, relevant for future Slc11 structure–function studies, several substituted pyrazoles have been identified as potential direct inhibitors of DMT1. Three distinct series of substituted pyrazole blockers were elaborated from the high-throughput screening initial hit pyrazolone; the lead compound was optimized to derive two analogs efficacious in blocking Fe^{2+} uptake by enterocytes in a rodent model of acute iron hyperabsorption. These data suggest that such compounds may represent a promising basis to obtain stereospecific inhibitors of DMT1 (Cadieux et al., 2012).

5. REGULATION OF BACTERIAL *mntH* EXPRESSION AND VIRULENCE

The *E. coli* Slc11 homolog belongs to MntH phylogroup A, possibly evolved from MntH B consequently to the rise in atmospheric oxygen concentration (Cellier et al., 2001). *Escherichia coli mntH* is repressed under normal conditions. Though it transports Mn^{2+} with higher affinity than Fe^{2+}, it is iron that controls *mntH* expression and triggers Mn^{2+} influx (Makui et al., 2000; Patzer & Hantke, 2001). In contrast *Bacillus. subtilis mntH* A is controlled by Mn^{2+} as corepressor of MntR (Que & Helmann, 2000). In fact, *E. coli* and *S.* Typhimurium *mntH* A are dually regulated: Fur–Fe^{2+} complex controls basal transcription level and MntR mediates Mn-dependent repression, insufficient though to extinguish transcription (Kehres et al., 2000, 2002; Patzer & Hantke, 2001).

Since MntH A carriers from Gram-positive and Gram-negative bacteria catalyze Mn^{2+} uptake (Makui et al., 2000; Que & Helmann, 2000), divergent regulation of expression presumably reflects physiological differences in pathways of Fe and Mn metabolisms and prevention of metal-induced oxidative stress. *mntH* mutation confers hydrogen peroxide (H_2O_2) sensitivity (Kehres et al., 2000; Que & Helmann, 2000). *Bacillus* species produce spores to resist extreme environmental conditions, a capacity tributary of adequate Mn supply (Ghosh et al., 2011; Granger et al., 2011). *Escherichia coli* also resists intracellular peroxide stress by accumulating Mn^{2+} in response via MntH A (Anjem et al., 2009), and by further exchanging Mn^{2+} for non-redox Fe^{2+} in mononuclear enzymes contributes to protect their activities (Anjem & Imlay, 2012). The H_2O_2-dependent transcriptional activator OxyR induces *mntH* in the enterobacteria *S.* Typhimurium and *Shigella flexneri* (Kehres et al., 2002; Runyen-Janecky et al., 2006). In *E. coli*, peroxide stress might disrupt Fur- and Fe-dependent regulation and induce *mntH* expression while OxyR stimulates Fur expression (Varghese et al., 2007). Thus, dominant regulation of *mntH* by Fe and Fur contributes to *E. coli* resistance to peroxide stress.

Intracellular accumulation of Mn^{2+} complexes generally confers very high levels of protection from reactive oxygen species (ROS). Bound to a variety of metabolic ligands, Mn^{2+} scavenges different ROS acting either stoichiometrically or catalytically, and can decompose H_2O_2 catalytically. Thus, the intracellular accumulation of Mn^{2+} with P_i together with certain peptide metabolites generally allows combating oxidative stress (Daly, 2009; Granger et al., 2011). Because $Fe^{2+/3+}$ is prone to oxidative redox cycling, regulation of the intracellular ratio of Fe to Mn is an important parameter of homeostasis (Cornelis et al., 2011; Daly, 2009) which can affect in vivo virulence (Veyrier et al., 2011). *Escherichia coli mntH* is part of MntR mini-regulon, characterized by chromatin immunoprecipitation and whole-genome expression analysis (Waters et al., 2011; Yamamoto et al., 2011), which contributes to protect cells against oxidative stress and comprises *mntR*, *mntH*, *yebN*, *dps*, and *mntS*. Dps, the major stationary phase DNA-binding protein, binds Fe^{2+} and mediates chromosome compaction; it is induced with Mn^{2+} uptake as intracellular Mn^{2+} concentration diminishes. MntR-Mn^{2+} also represses *mntS*; MntS overexpression causes Mn^{2+} sensitivity by stimulating MntH activity. In contrast, the Mn^{2+} exporter YebN (MntX) is activated by MntR-Mn^{2+}. These studies implicate *E. coli* MntH A as key factor integrating Fe and Mn homeostasis.

Both nonpathogenic and pathogenic spp. possess MntH A transporters. The frequency of *mntH* genes in pathogenic species increases when considering the phylogroup MntH C (Stubben et al., 2009), derived by horizontal gene transfer of an eukaryotic prototype *Nramp*, and MntH AV, another recently emerged subgroup prone to horizontal gene transfer (Fig. 5) (Richer et al., 2003, 2004). The roles of Mn metabolism in protection against oxidative stress suggest MntH may contribute to bacterial virulence. Enterobacteria may also express a periplasmic binding protein-dependent ATP-binding Cassette (ABC) transporter system that imports Mn^{2+} and Fe^{2+} (SitABCD), often found in genomic islands linked to symbiosis or pathogenicity and on virulence plasmids. Deletion of both *mntH* A and *sit* genes abrogated Mn^{2+} uptake, resistance to oxidative stress and in vivo virulence in *S.* Typhimurium, *S. flexneri*, avian pathogenic *E. coli* and *Yersinia pestis* (Boyer et al., 2002; Perry et al., 2012; Runyen-Janecky et al., 2006; Sabri et al., 2008; Zaharik et al., 2004). *Yersinia pestis* double mutant lost virulence in a model of bubonic plague, but not in a pneumonic plague model, indicating higher Mn requirements for systemic disease. *Yersinia pseudotuberculosis mntH* A deletion limited survival and growth in macrophages expressing a functional Nramp1; mutant bacteria were able to replicate in *Nramp1*$^{-/-}$ macrophages supporting the notion that MntH and Nramp1 compete for intra-phagosomal Mn^{2+} and Fe^{2+} (Champion et al., 2011).

Several *mntH* C genes were characterized regarding the regulation of their expression and contribution to bacterial virulence. The rhizobacterium *Bradyrhizobium japonicum mntH* Cα encodes the Mn^{2+} transporter essential for homeostasis, which is regulated by Mn^{2+} and an Mn-dependent Fur variant (Mur). In contrast, Fe^{2+} was found neither a substrate for MntH nor a regulator of *mntH* expression in vivo (Hohle & O'Brian, 2009). The obligate animal pathogen α-proteobacterium *Brucella abortus* expresses another *mntH* Cα gene encoding a high-affinity Mn^{2+} transporter regulated only by Mn^{2+} and a Mur-like repressor (Menscher et al., 2012). *Brucella abortus mntH* Cα mutant is susceptible to oxidative killing and severely attenuated both in vivo and in cultured macrophages (Anderson et al., 2009). Another pathogen threat, the multi-chromosome β-proteobacterium *Burkholderia pseudomallei*, can survive within macrophages. Though no *mntH* was identified in a genome-wide gain of function screen for *B. pseudomallei* genes conferring *E. coli* the ability to reduce macrophage viability (Dowling et al., 2010), this might reflect problematic heterologous expression of membrane proteins. Using in vivo expression technology, *B. pseudomallei mntH* Cγ1 appeared activated upon macrophage invasion, implying that

Me^{2+} uptake might contribute to intracellular survival. This *mntH* Cγ gene is part of a mono- or bi-cistronic unit, together with a small iron–sulfur protein-coding gene, suggesting that *B. pseudomallei* gene may contribute to Fe^{2+} uptake (Shalom et al., 2007).

An *mntH* Cβ gene is repressed in response to increased intracellular Fe^{2+} in the Lactic bacterium opportunistic pathogen, *Enterococcus faecalis* (Lopez et al., 2012). This MntH Cβ1 protein possesses 12 TMS and a C-terminus predicted cytoplasmic. The gene forms a transcriptional unit with *uspA*, a gene regulated by a variety of stress and nutritional conditions (Nachin, Nannmark, & Nystrom, 2005). Several *mntH* Cα genes also encode 12 TMS proteins genetically fused to UspA. This observation confirms phylogenetic relationships (Fig. 5) with MntH Cβ subgroup stemming from MntH Cα cluster. Decreased *mntH* Cβ1 transcription levels might result ultimately in reduced iron cellular content. Upon heterologous expression in *E. coli*, *E. faecalis mntH* Cβ1 conferred greater sensitivity to iron than *mntH* Cβ2 (Richer et al., 2003). Inactivation of an *mntH* Cβ gene encoding a 12 TMS protein in *Lactococcus lactis* reduced Fe^{2+} import and conferred resistance to tellurite and ROS-induced damage (Chasteen et al., 2009; Turner et al., 2007). Hence, some bacterial *mntH* seem to encode importers with a preference for Fe^{2+}, implying MntH metal specificity may vary with phyletic origin or functional diversification.

6. NRAMP ROLE IN EUKARYOTIC HOST DEFENSE

Protozoans, plants, fungi and metazoans possess eukaryotic Nramp resembling Nramp1/Dmt1 (archetype Nramp, Fig. 5). Homologous functions were evidenced in professional phagocytes such as the amoeba *D. discoideum* and mammalian macrophages, in which overexpression of archetype Nramp blocks bacterial proliferation (Gallant et al., 2007; Govoni et al., 1999; Peracino et al., 2010). *Dictyostelium discoideum* uses phagocytosis to internalize food; after ingestion bacteria are trapped in a vesicle (phagosome) which maturates along an endo-lysosomal pathway leading to acidification, exposure to ROS and content degradation by hydrolytic enzymes. This integrated process relies on efficient signaling, vesicle trafficking and fusion and functioning cytoskeleton, albeit processes that can be subverted by pathogens. Pathogens that are equally well adapted to resist amoeba and animal macrophages may block or manipulate the phagosome maturation, transforming it into a replication niche, or escape into the cytosol (Schmitz-Esser et al., 2010). Hence, *D. discoideum* archetype Nramp function in nutrition and host defense has been evolutionary preserved in the overlapping roles of Nramp1 and Nramp2/Dmt1.

Plant nutritional iron status can control the outcome of an infection. Iron deficiency confers plant resistance to the enterobacterial and fungal pathogens *Dickeya dadantii* and *Botrytis cinerea*. Both disease symptom severity and *D. dadantii* fitness were reduced in iron-deficient *A. thaliana* revealing that iron status influences plant basal immunity (Kieu et al., 2012). Some *A. thaliana Nramp* favor resistance to infection with the enterobacterium *Erwinia chrysanthemi*. AtNRAMP3 and AtNRAMP4 localize to the vacuolar membrane and mobilize Fe^{2+} and Mn^{2+} in plant seed and leaf, respectively. Bacterial challenges with *Pseudomonas syringae* and *E. chrysanthemi* upregulated the expression of AtNRAMP3 in leaves. The double-mutant *nramp3 nramp4* was more susceptible to infection, showing limited production of ROS and the iron storage protein ferritin. NRAMP involvement in innate immunity hence has been also conserved in plants (Segond et al., 2009).

There are three archetype Nramps (SMFs) in the nematode *Caenorhabditis elegans* differing in expression, spatial location, and how they are regulated by and may influence Mn and Fe metabolisms. Iron deficiency activates *Smf3* and represses iron storage protein ferritin genes, effects mediated by the transcription factor hypoxia-inducible factor 1 (HIF-1α) binding cis elements in the target gene promoters, and *C. elegans smf-3* (ko1035) mutant shows reduced iron content indicating a role in intestinal iron absorption (Romney et al., 2011). Loss-of-function alleles of the *smf* genes alter Mn toxicity and decrease innate immunity of animals fed with *Staphylococcus aureus*, suggesting intestinal lumen depletion of Mn^{2+} and Fe^{2+} limits microbial pathogen colonization (Au et al., 2009; Bandyopadhyay et al., 2009). Event-mode X-ray fluorescence detection and tomographic analysis showed ~450-fold higher Mn^{2+} concentration in some intestinal cells compared to the lumen (McColl et al., 2012). These studies indicate that nematode archetype Nramps also sustain host metal nutrition and antimicrobial defense.

In the *Drosophila* model, the behavioral taste defect in *malvolio* mutants (*mvl*) was suppressed by expressing *NRAMP1* (D'Souza, Cheah, Gros, & Rodrigues, 1999) or by supplementing growth media with $MnCl_2$ or $FeCl_2$ (Orgad, Nelson, Segal, & Nelson, 1998), implying that NRAMP1 facilitated Me^{2+} uptake. Analysis of iron accumulation in *mvl* mutants revealed specific deficiency, involving Mvl in systemic iron trafficking (Bettedi et al., 2011). Regarding the mosquito *Anopheles albimanus*, messenger RNA (mRNA) injection in *Xenopus* oocytes and immunolocalization studies suggested mosquito Nramp functions in midgut and Malpighian tubules to transport Fe^{2+} into the cytosol, where it may be stored in cytosolic ferritin (Martinez-Barnetche et al., 2007). Insect Nramps are thus expressed not only in macrophages and in differentiated neurons but also in the midgut, implying that similar to the amoeba, a single archetype Nramp homolog fulfills different physiological roles conserved in mammals. The structural conservation this implies was strikingly demonstrated by observing a mosquito-borne alphavirus Sindbis, related to several emerging human pathogens, which binds and uses Nramp2 or Mvl as receptors to infect cells (Rose et al., 2011).

Nramp1/Nramp2 duplication and further selection for partitioned nutritional and defensive roles may date back to the emergence of terrestrial vertebrates, maybe in response to pressures exerted by aerial microbes and body development (Cellier, Courville, & Campion, 2007). Similarly, duplication of Nramp2/Dmt1 in fish led to divergent regulation in response to iron deficiency in the intestine or to iron overload and bacterial infection in different tissues (Neves et al., 2011). Obvious benefits of *Nramp* duplication and divergence include the possibility to adapt regulations and accumulate genetic variation favorable for separate functions.

The example of tuberculosis (TB) illustrates the point. *Mycobacterium tuberculosis* is a successful obligate pathogen that can infect humans and stay dormant as long as a normal immune response may contain bacterial spread without achieving sterilizing immunity. This translates into 1.7 million deaths each year and more than 9 million new cases worldwide, mostly due to HIV coinfection (Lawn & Zumla, 2011). *Mycobacterium tuberculosis* and pathogenic mycobacteria can survive within the phagosome of professional phagocytes and persist in pulmonary granulomas for decades. The homeostatic balance between host and pathogen is the result of a fine control of inflammatory events, outcome of an elaborate coevolutionary process (Huynh, Joshi, & Brown, 2011).

Particular human genetic variants in immunity-related genes have been associated with different *M. tuberculosis* complex (MTC) lineages (Brites & Gagneux, 2012) but there is not yet evidence of an MTC strain lineage by *NRAMP1* interaction (Stein & Baker, 2011). On the host side, numerous candidate genes have been implicated in association studies testing effector functions in recognition, opsonization and phagocytosis, microbial killing, apoptosis, signaling and secretion of cytokines and chemokines, and antigen presentation, with polymorphism validation in *IFNG*, *NRAMP1* and *NOS2A* (Moller & Hoal, 2010). The association between TB and *NRAMP1* is robust because it was reported in more than 10 studies and TB diagnosis method had no significant impact. This association, measured in different populations, consistently supports the hypothesis that *NRAMP1* may influence host defense against TB though moderate strength limits predictive value (Li et al., 2011; Singh et al., 2011). *NRAMP1* coevolution in adaptation to survive life-threatening pathogens is consistent with its specialized function, which affects relatively little Fe and Mn homeostases that are under the control of Dmt1.

7. NATURAL EVOLUTION OF NRAMP IN NUTRITIONAL IMMUNITY

Nramp/MntH carriers contribute to cell Me^{2+} nutrition; Nramp evolved in eukaryotes into antimicrobial activity while in prokaryotes MntH are diverse, illustrating complex evolutionary history since ancient origins. The prokaryotic ancestor that yielded eukaryotic Nramp is not established, yet, as the precise scenario of eukaryogenesis (Brochier-Armanet, Forterre, & Gribaldo, 2011). Among MntH A group, sequences from *Crenarchaeota* and few other microorganisms (MntH H, Fig. 5) share sequence characteristics including eukaryotic marks in TMS4 and 9 (Fig. 4), which might suggest a possible archaeal ancestor for eukaryotic Nramp.

Before the separation of animal, plant and amoeban kingdoms a gene duplication yielded the prototype and archetype *Nramp*. The best characterized prototype Nramps are *Saccharomyces cerevisiae* suppressors of the lethal mutation *mif*, causing a defect in mitochondrial import of certain proteins, the Smf1–3 proteins (Reddi, Jensen, & Culotta, 2009). Experimental evidence was also obtained for the green alga prototype Nramp complementing yeast growth dependent on manganese and iron (Rosakis & Koster, 2005). In addition, the prototypical *D. discoideum* Nramp is expressed during development when the archetypical Nramp1 is downregulated. The cellular localization of *D. discoideum* prototype Nramps, as well as single and double mutants, are currently being characterized (S. Bozzaro, personal communication). The yeast model of manganese homeostasis is relevant to humans because environmental exposure to toxic levels of manganese causes a syndrome similar to Parkinson's disease (PD), and a number of manganese transporters, metallochaperones, and regulatory mechanisms are being investigated (Chesi, Kilaru, Fang, Cooper, & Gitler, 2012).

Manganese metabolism in yeast showed adaptation to subcellular compartmentalization with transporters operating at the cell surface or at intracellular sites such as the vacuole (Smf1p and Smf2p) and Golgi (Smf1p), mobilizing Mn^{2+} toward the cytosol. Smf2p provides most of cellular Mn^{2+} available to Mn^{2+} requiring enzymes while Smf1p delivers Mn^{2+} to the cytoplasmic manganese antioxidant. Many manganese regulatory pathways in yeast occur at the posttranslational level through changes in Smfp1,2p localization and turnover (Bleackley & MacGillivray, 2011; Reddi et al., 2009). Cells distinguish, using different components of metal-responsive ubiquitination machinery, Mn^{2+}-replete conditions, when cell demand for Mn^{2+} increases, the presence of toxic excess of Mn^{2+} and yet mount another regulatory response to toxic Cd^{2+} exposure (Reddi et al., 2009).

In contrast, yeast Smf3p contributes with other transporters (*FET5, FRE6, FTH1, COT1*) to mobilize Fe^{2+} that is stored in the vacuole. *SMF3* expression is activated in iron-deficient cells as part of Aft1 regulon (Berthelet et al., 2010), and repressed by Met4p, the transcriptional regulator of the sulfur assimilation pathway, which controls several iron homeostasis genes (*FTR1, SMF3, COT1, MRS4, FET3, FRE1*) (Hickman et al., 2011). Together *S. cerevisiae* prototype Nramps contribute to Fe and Mn homeostasis by facilitating Me^{2+} and Fe^{2+} import into the cytosol from various compartments and can be regulated transcriptionally or posttranslationally.

In mammals, the ubiquitin system also contributes to regulate Nramp2/Dmt1 posttranslationally, supplementing the regulations exerted at the posttranscriptional level, by the iron regulating proteins (IRP) binding to a cis iron-responsive element (IRE), and at the transcriptional level. Different isoforms of Nramp2 are expressed at various sites of the body (e.g. intestine, erythrocytes, and kidney) in different subcellular compartments, enabling to adjust the protein expression levels to the body Fe needs (e.g. absorption, distribution, and recycling) (Hentze et al., 2010). Alternative mRNA splicing produces Nramp2 C-terminal isoforms; Nramp2-I for dietary iron uptake in the duodenum and possible reabsorption in the kidney and Nramp2-II for Tf-dependent iron uptake in most peripheral tissues, which is also upregulated by inflammatory stimuli. Nramp2-I is found at the apical membrane of epithelial cells (duodenum, kidney), whereas Nramp2-II with a different C-terminus populates mainly recycling endosomes (Wang & Pantopoulos, 2011).

Nramp2-I regulation by iron may include binding of IRPs to a 3′-end IRE, which stabilizes the mRNA and upregulates protein expression within hours of reduced iron availability (or oxidative stress). Upon prolonged Fe deficiency HIF-2α upregulates *Nramp2-I* transcription together with the duodenal cytochrome b yielding the alternative N-terminus (A) isoform

(Mastrogiannaki et al., 2009; Nunez, 2010). After iron feeding, Nramp2-I is rapidly sequestered from the apical membrane, decreasing duodenal absorption, whereas systemic elevation of iron levels triggers secretion of hepcidin in the blood, mainly by the liver, to regulate iron homeostasis through proteolysis of iron transporters (Brasse-Lagnel et al., 2011; De Domenico, Ward, & Kaplan, 2007).

The Nramp2-II isoform is recycled through the early endocytic network together with the transferrin receptor 1 and the metallo-reductase six-transmembrane epithelial antigen of the prostate 3; acidification and reducing power release Fe^{2+} for translocation into the cytoplasm via Nramp2-II (Ohgami et al., 2005). Though lacking a 3′-IRE, Nramp2-II is down-regulated posttranslationally in response to elevation of intracellular iron, which activates Fe-dependent ubiquitination and proteosomal degradation (Foot et al., 2008, 2011). Nramp2-II (exon 1B) transcription is also stimulated by HIF-1α in response to hypoxia in hepatoma cells (Qian et al., 2011; Wang et al., 2010). In the erythroid lineage, Nramp2 (−IRE) expression level is controlled by the ubiquitous microRNA (miRNA) Let-7d, targeting Nramp2-II mRNA for degradation in RBC precursors (Andolfo et al., 2010).

The parolog Nramp1 is present in tetrapods and protein expression is confined to intracellular phagolysosomes of professional phagocytes (monocytes, macrophages, dendritic cells (DC) and granulocytes). Through phagocytosis Nramp1 Me^{2+}depletion from the phagosomal lumen serves both defense against infectious microbes and nutrient recycling in the steady state (Cellier, 2012). NRAMP1 protein is detected late in myeloid development. A variety of DNA-binding factors and interacting proteins are recruited at *NRAMP1* promoter, including the transcription factors Sp1 and CCAAT/enhancer-binding protein beta , a pioneer factor critical to activate gene expression during monocyte/macrophage development (Magnani, Eeckhoute, & Lupien, 2011), whose binding site overlaps *NRAMP1* major transcription start site (Richer et al., 2008).

Also, two components of the SWI/SNF (switch/sucrose non-fermentable) chromatin remodeling complex, BRG1 and β-actin, are recruited to the promoter via interaction with the transcription factor Activating transcription factor 3 (ATF-3) bound to a 5′ AP-1 element, and activate transcription (Xu et al., 2011). This SWI/SNF complex facilitates lipopolysaccharide (LPS)-dependent gene expression in macrophages (Ostuni & Natoli, 2011). The Activator protein 1 (AP-1) element marks the 5′ boundary of a GT repeat whose polymorphism alters *NRAMP1* expression (Bayele et al., 2007); authentic allelic associations between these polymorphisms and susceptibility to autoimmune or inflammatory diseases have not been reported yet (O'Brien, Archer, Simpson, Torpy, & Nassif, 2008).

IRF8 (interferon regulatory factor 8) is another factor crucial for host defense against intracellular pathogens and myeloid ontogeny (Marquis et al., 2011; Schulz et al., 2012). Mouse IRF8 contributes together with Miz-1 and PU.1 to control *Nramp1* expression in myeloid cells, which is negatively affected by the myelocytomatosis proto-oncogene (c-Myc) but activated with interferon gamma (IFN-γ) (Alter-Koltunoff et al., 2008).

Posttranscriptional control of *NRAMP1* expression includes the mRNA-binding protein HuR, which stabilizes transcripts in differentiated cells (Xu et al., 2011) and posttranslational control of Nramp1 activity through sorting toward the membrane of lysosomes. The N-terminus Nramp1 motif YGSI identified using Nramp1/Nramp2 chimeric proteins allows intracellular sorting of mouse Nramp1 to lysosomes; it is conserved among orthologs (Cellier et al., 2007; Lam-Yuk-Tseung & Gros, 2006). Inventory of sorting signals in bovine Nramp1 using Green fluorescent protein (GFP)-tagged full-length and truncated polypeptides indicated multiple targeting motifs including regions spanning TMS2 and TMS7, predicted pseudo-symmetric and adjacent in LeuT-based Nramp 3D model (Fig. 3). Internal lysosomal targeting motifs were also identified in TMS11 and 12 as additional signals (Cheng & Wang, 2012). Whereas internalization and recycling of Nramp2-II depend on a single carboxyl-terminal cytoplasmic motif (^{555}YLLNT^{559}) Nramp1 harbors several marks for lysosomal targeting.

8. PATHOLOGICAL IMPLICATIONS OF NRAMP FUNCTION

8.1. Iron Overload

Microcytic hypochromic anemia is a common form of anemia in children that is caused by defects in the synthesis of globin, heme or in the acquisition and distribution of iron. Mutations in *DMT1* have been demonstrated in six patients of Czech, Italian, French and Ecuadorian origins presenting a new syndrome of congenital microcytic hypochromic anemia, with early onset, poor response to oral iron treatment, liver iron overload, high transferrin saturation and normal to moderately elevated serum ferritin levels. Presumably, intestinal heme iron absorption compensates at least partly for deficient Fe^{2+} uptake, but impaired Nramp2 endosomal activity in most cells prevents iron efflux to the cytosol. This leads to iron accumulation without triggering ferritin synthesis, resulting in progressive liver iron overload in the face of microcytic anemia (Beaumont et al., 2006; Iolascon et al., 2006; Priwitzerova et al., 2004).

Mutations reported in the *SLC11A2* gene are: c.1197G>C (p.E339D) in homozygous state, c.310-3_5delCTT and c.1246C>T (p.R416C) in compound heterozygous state, c.340_342delGTG and c.635G>T (p.G212V) in compound heterozygous state, c.223G>A (p.G75R) in homozygous state, twice, and c.1472A>G (p.N491S) and c.635G>T (p.G212V) in compound heterozygous state (Bardou-Jacquet et al., 2011; Barrios et al., 2012; Beaumont et al., 2006; Blanco et al., 2009; Iolascon et al., 2006; Mims et al., 2005). Nramp2 mutations were linked to very weak production of full-length functional protein (E399D) and defective cell surface expression (R416C). The mutation G75R is nonconservative, and TMS1 is crucial for substrate translocation, implying the mutant protein is likely nonfunctional; similarly, deletion of one residue in TMS2 is expected to be detrimental. The mutation N491S induced mainly abnormal protein trafficking while the mutation G212V had no detectable impact, implying a rather functional role for Gly^{212} (Bardou-Jacquet et al., 2011), which seems consistent with DMT1 3D model (Fig. 3).

8.2. Neurodegeneration

PD is the most frequent neurodegenerative disorder worldwide. It is characterized by the degeneration of dopaminergic neurons (DNs) in the substantia nigra pars compacta and the presence of Lewy bodies—proteinaceous cytoplasmic inclusions—in the remaining DNs. Aging, iron accumulation, mitochondrial dysfunction, inflammation and oxidative imbalance contribute to its etiology, suggesting that Fe and Mn metabolisms are involved (Kim et al., 2012). Imbalance in Fe to Mn ratio in dopaminergic and glial cells might generate oxidative stress, protein aggregation, and neuronal death. The dopamine oxidation product aminochrome is an endogenous toxin inhibiting mitochondrial complex I that triggers intracellular oxidative stress, retention of Fe and/or Mn by increasing DMT1 expression and reducing FPN1 levels (Aguirre et al., 2012; Madejczyk & Ballatori, 2012). DMT1 represents thus a promising therapeutic target (Cadieux et al., 2012) for slowing PD progression because DMT1 expression increases with age and contributes to iron-dependent pathophysiology (Salazar et al., 2008).

α-Synuclein (α-syn) is the best studied genetic factor linked to PD, causing early onset autosomal dominant forms of familial PD (Chesi et al., 2012).

Mutations in *ATP13A3/PARK9* have also been associated with familial forms of parkinsonism. The encoded lysosomal cationic P-type ATPase pumps Mn^{2+} into lysosomes; PARK9 and yeast homolog YPK9 expressed at the vacuole membrane modify α-syn toxicity which results from mislocalized membrane vesicles from various steps of the endocytic and exocytic pathways (Chesi et al., 2012; Reddi et al., 2009). As human exposure to manganese induces a neurological syndrome, called manganism and presenting symptoms similar to PD, the balance of Fe-based oxidative stress and Mn^{2+} uptake emerges at the heart of PD pathogenesis. Hence, Nramps and the interplay of Fe/Mn metabolisms constitute crucial elements of PD paradigm (Bleackley & MacGillivray, 2011; Kim et al., 2012; Settivari et al., 2009).

Iron transport plays key roles in transmitting nervous signals. N-methyl-D-aspartate (NMDA) receptors activate the neuronal nitric oxide synthase (nNOS) resulting in S-nitrosylation and activation of Dexras1, a small G protein in the Ras subfamily induced by dexamethasone, and connected to nNOS via the adaptor CAPON (carboxy-terminal PDZ ligand of n NOS). Dexras1 also interacts with the peripheral benzodiazepine receptor-associated protein (PAP7), a protein of unknown function, which in turn binds to Nramp2/Dmt1-IB. This signaling cascade culminates with Fe^{2+} import and appears to mediate NMDA neurotoxicity (Cheah et al., 2006). NMDA receptor-dependent plasticity processes stimulate expression of Dmt1-IB + IRE, presumably playing a role in hippocampal spatial memory formation (Haeger et al., 2010). PAP7 interacts also with DMT1 (+IRE); reducing PAP7 levels in erythroid cells downregulates DMT1 (+IRE) protein expression (Okazaki et al., 2012). Physical coupling of Dmt1 to peripheral receptors via PAP7 thus indicates additional layers for posttranslational regulation of Dmt1 activity and functional integration into neural signaling.

Normal brain function depends critically on Fe and Mn homeostasis because imbalance between these redox metals can contribute to oxidative stress and protein aggregation, leading to neuronal death (Aguirre & Culotta, 2012; Kim et al., 2012). Proper regulation of metal transport and trafficking in neurons is of utmost importance; hence Dmt1 (−IRE) is tightly regulated posttranslationally via ubiquitination as Ndfip1 and Ndfip2 interact with Dmt1-II and the E3 ligase WWP2, a member of the Nedd4 family (Foot et al., 2008). Neurons exposed to Fe^{2+} or Co^{2+} upregulates NDFIP1 binding to DMT1, which triggers its degradation and reduces metal import (Howitt et al., 2009). Duodenal enterocytes of *Ndfip1*$^{-/-}$ mice fed a low-iron diet upregulate Dmt1 expression and activity, correlating with increased serum iron levels and transferrin saturation (Foot et al., 2011). Hepcidin-dependent

Fpn internalization is also mediated by lysine-dependent ubiquitination (Qiao et al., 2012; Ross et al., 2012). Also, Tumor Necrosis Factor alpha (TNF-α) and Transforming growth factor, beta 1 (TGF-β1), two prominent cytokines influencing the balance of inflammatory responses, can affect profoundly Fe and Mn homeostasis in astrocytes and microglia (Rathore, Redensek, & David, 2012). Posttranslational regulation of Dmt1 surface expression is crucial to regulate Me^{2+} import dynamically in response to metal bioavailability and to prevent oxidative damage, which are both frequently linked to inflammatory and neurodegenerative disorders.

8.3. Immunopathology

Epidemiological studies indicate the major role of genetic factors in the development of TB. Many association studies of candidate genes were conducted, the most consistent findings pointing at *HLA* class II alleles and *NRAMP1* variants (Moller & Hoal, 2010; Stein & Baker, 2011). Several *NRAMP1* polymorphisms were consistently associated with TB resistance/susceptibility in many studies, including pediatric TB (Malik et al., 2005; Stein & Baker, 2011). Concerning pulmonary TB, associations were established in different populations (mainly African and Asian rather than Western) (Li et al., 2011) involving distinct lineages of *M. tuberculosis* (Stein & Baker, 2011). *SLC11A1* polymorphisms in Asians associate with both pulmonary and extrapulmonary TBs (Meilang et al., 2012). Polymorphisms at *NRAMP1* 3′-end show stronger associations for Western populations, some showing frequency correlating positively with ancient urbanization (Barnes, Duda, Pybus, & Thomas, 2011; Liu et al., 1995). This suggests that human density favored aerosol transmission and evolution of more virulent *M. tuberculosis* strains; in turn, pathogen pressure increased the frequency of *NRAMP1*-resistant alleles by natural selection (Barnes et al., 2011; Brites & Gagneux, 2012; Donoghue, 2011). The function of *NRAMP1* variants was assayed in primary human macrophages by monitoring acquisition of the mannose-6-phosphate receptor in *Salmonella*-containing vacuoles. Individuals homozygous for the *NRAMP1*-274 high-risk allele (CC genotype) showed lower activity, compared to heterozygous individuals (CT genotype), thus linking a common *NRAMP1* variant with susceptibility to pediatric TB and supporting the proposition that *NRAMP1* polymorphism contributes to host defense against TB (Gallant et al., 2007).

A $(GT)_n$ microsatellite repeat polymorphism with capacity for Z-DNA formation was characterized in the promoter region, including nine alleles and modulation of the expression of *SLC11A1* by the number of $(GT)_n$ repeats. The most common promoter allele 3 can drive high-level expression in vitro, while the less frequent allele 2 drives lower expression levels. This sequence-dependent difference in gene expression is amplified by pro-inflammatory stimuli such as LPS, reducing allele 2-driven expression level, and enhancing allele 3 activity. Accordingly, alleles 2 and 3 have been inversely associated with susceptibility to autoimmune or infectious disease, with allele 3 correlating with autoimmunity and inflammation. However, a meta-analysis designed to analyze the link between *SLC11A1* promoter alleles 2 and 3, autoimmunity and inflammation revealed the lack of significant association of allele 3 with immune diseases, contrary to initial reports (O'Brien et al., 2008). Larger, unbiased studies on the relationships between these polymorphisms and autoimmune or inflammatory diseases will be necessary to investigate further the potential link with *NRAMP1*.

9. CONCLUSIONS

Approaching 20 years past its discovery the Nramp family is now well-known to some extent but still holds many secrets. From an evolutionary perspective these include the transition from prokaryotic MntH to eukaryotic Nramp, compared Me^{2+} transport properties of eukaryotic prototype and archetype Nramp coexisting in a single organism, and functional diversification of plant Nramps. In prokaryotes, it will be particularly interesting to study the role of MntH B in anaerobiosis and to further investigate MntH C diversity and possible divergence in Me^{2+} selectivity. Exploring bacterial *mntH* metallo-regulation, contribution to (oxidative) stress resistance, symbiosis or pathogenesis should be rewarding as well.

A number of areas will focus intense experimental activity to determine Slc11 3D structure and to complete accessibility mapping of the transmembrane translocation pathway. These efforts will be necessary to arrive at understanding transmembrane structure–function relationships in the Slc11 carriers. The knowledge gained will allow examining differences in the mechanism of H^{+}-dependent Me^{2+} transport that may exist between phylogroups, in relation to functional roles in nutrition, host defense and intracellular signaling. Other promising research areas include assessing the impact of *NRAMP* genetic polymorphisms and the molecular design of specific inhibitors that interfere directly with H^{+} and Me^{2+} transport.

Future progress will also depend on technological advances allowing to probe metabolic regulation and the dynamics of Nramp turnover, so as to follow Nramp subcellular membrane trafficking and the regulation of metal transport in relation to cell metabolic activity. It will be useful as well to foster broad integration of the pathophysiological implications of the functions of Nramps in systems that control host metabolism, immunity and cognition and which influence prevalent diseases such as iron overload, susceptibility to infectious diseases and neurodegenerative diseases.

REFERENCES

Aguirre, J. D., & Culotta, V. C. (2012). Battles with iron: manganese in oxidative stress protection. *The Journal of Biological Chemistry, 287*, 13541–13548.

Aguirre, P., Urrutia, P., Tapia, V., Villa, M., Paris, I., Segura-Aguilar, J., et al. (2012). The dopamine metabolite aminochrome inhibits mitochondrial complex I and modifies the expression of iron transporters DMT1 and FPN1. *Biometals.* [Epub ahead of print].

Alter-Koltunoff, M., Goren, S., No sbeck, J., Feng, C. G., Sher, A., Ozato, K., et al. (2008). Innate immunity to intraphagosomal pathogens is mediated by interferon regulatory factor 8 (IRF-8) that stimulates the expression of macrophage-specific Nramp1 through antagonizing repression by c-Myc. *The Journal of Biological Chemistry, 283*, 2724–2733.

Anderson, E. S., Paulley, J. T., Gaines, J. M., Valderas, M. W., Martin, D. W., Menscher, E., et al. (2009). The manganese transporter MntH is a critical virulence determinant for Brucella abortus 2308 in experimentally infected mice. *Infection and Immunity, 77*, 3466–3474.

Andolfo, I., De, F. L., Asci, R., Russo, R., Colucci, S., Gorrese, M., et al. (2010). Regulation of divalent metal transporter 1 (DMT1) non-IRE isoform by the micro-RNA Let-7d in erythroid cells. *Haematologica, 95*, 1244–1252.

Anjem, A., Varghese, S., & Imlay, J. A. (2009). Manganese import is a key element of the OxyR response to hydrogen peroxide in Escherichia coli. *Molecular Microbiology, 72*, 844–858.

Anjem, A., & Imlay, J. A. (2012). Mononuclear iron enzymes are primary targets of hydrogen peroxide stress. *The Journal of Biological Chemistry, 287*, 15544–15556.

Au, C., Benedetto, A., Anderson, J., Labrousse, A., Erikson, K., Ewbank, J. J., et al. (2009). SMF-1, SMF-2 and SMF-3 DMT1 orthologues regulate and are regulated differentially by manganese levels in *C. elegans*. *PLoS One, 4*, e7792.

Bandyopadhyay, J., Song, H. O., Park, B. J., Singaravelu, G., Sun, J. L., Ahnn, J., et al. (2009). Functional assessment of Nramp-like metal transporters and manganese in Caenorhabditis elegans. *Biochemical and Biophysical Research Communications, 390*, 136–141.

Bardou-Jacquet, E., Island, M. L., Jouanolle, A. M., Detivaud, L., Fatih, N., Ropert, M., et al. (2011). A novel N491S mutation in the human SLC11A2 gene impairs protein trafficking and in association with the G212V mutation leads to microcytic anemia and liver iron overload. *Blood Cells Molecules and Diseases, 47*, 243–248.

Barnes, I., Duda, A., Pybus, O. G., & Thomas, M. G. (2011). Ancient urbanization predicts genetic resistance to tuberculosis. *Evolution, 65*, 842–848.

Barrios, M., Moreno-Carralero, M. I., Cuadrado-Grande, N., Baro, M., Vivanco, J. L., & Moran-Jimenez, M. J. (2012). The homozygous mutation G75R in the human SLC11A2 gene leads to microcytic anaemia and iron overload. *British Journal of Haematology, 157*, 514–516.

Bayele, H. K., Peyssonnaux, C., Giatromanolaki, A., Arrais-Silva, W. W., Mohamed, H. S., Collins, H., et al. (2007). HIF-1 regulates heritable variation and allele expression phenotypes of the macrophage immune response gene SLC11A1 from a Z-DNA forming microsatellite. *Blood, 110*, 3039–3048.

Beaumont, C., Delaunay, J., Hetet, G., Grandchamp, B., de Montalembert, M., & Tchernia, G. (2006). Two new human DMT1 gene mutations in a patient with microcytic anemia, low ferritinemia, and liver iron overload. *Blood, 107*, 4168–4170.

Berthelet, S., Usher, J., Shulist, K., Hamza, A., Maltez, N., Johnston, A., et al. (2010). Functional genomics analysis of the *Saccharomyces cerevisiae* iron responsive transcription factor Aft1 reveals iron-independent functions. *Genetics, 185*, 1111–1128.

Bettedi, L., Aslam, M. F., Szular, J., Mandilaras, K., & Missirlis, F. (2011). Iron depletion in the intestines of Malvolio mutant flies does not occur in the absence of a multicopper oxidase. *The Journal of Experimental Biology, 214*, 971–978.

Biegert, A., Mayer, C., Remmert, M., Soding, J., & Lupas, A. N. (2006). The MPI Bioinformatics Toolkit for protein sequence analysis. *Nucleic Acids Research, 34*, W335–W339.

Blanco, E., Kannengiesser, C., Grandchamp, B., Tasso, M., & Beaumont, C. (2009). Not all DMT1 mutations lead to iron overload. *Blood Cells Molecules and Diseases, 43*, 199–201.

Bleackley, M. R., & MacGillivray, R. T. (2011). Transition metal homeostasis: from yeast to human disease. *Biometals, 24*, 785–809.

Bordoli, L., & Schwede, T. (2012). Automated protein structure modeling with SWISS-MODEL Workspace and the Protein Model Portal. *Methods in Molecular Biology, 857*, 107–136.

Boudker, O., & Verdon, G. (2010). Structural perspectives on secondary active transporters. *Trends in Pharmacological Sciences, 31*, 418–426.

Boyer, E., Bergevin, I., Malo, D., Gros, P., & Cellier, M. F. (2002). Acquisition of Mn(II) in addition to Fe(II) is required for full virulence of Salmonella enterica serovar. *Typhimurium Infection and Immunity, 70*, 6032–6042.

Brasse-Lagnel, C., Karim, Z., Letteron, P., Bekri, S., Bado, A., & Beaumont, C. (2011). Intestinal DMT1 cotransporter is down-regulated by hepcidin via proteasome internalization and degradation. *Gastroenterology, 140*, 1261–1271.

Brites, D., & Gagneux, S. (2012). Old and new selective pressures on Mycobacterium tuberculosis. *Infection Genetics and Evolution, 12*, 678–685.

Brochier-Armanet, C., Forterre, P., & Gribaldo, S. (2011). Phylogeny and evolution of the Archaea: one hundred genomes later. *Current Opinion in Microbiology, 14*, 274–281.

Cadieux, J. A., Zhang, Z., Mattice, M., Brownlie-Cutts, A., Fu, J., Ratkay, L. G., et al. (2012). Synthesis and biological evaluation of substituted pyrazoles as blockers of divalent metal transporter 1 (DMT1). *Bioorganic and Medicinal Chemistry Letters, 22*, 90–95.

Cailliatte, R., Lapeyre, B., Briat, J. F., Mari, S., & Curie, C. (2009). The NRAMP6 metal transporter contributes to cadmium toxicity. *The Biochemical Journal, 422*, 217–228.

Cailliatte, R., Schikora, A., Briat, J. F., Mari, S., & Curie, C. (2010). High-affinity manganese uptake by the metal transporter NRAMP1 is essential for Arabidopsis growth in low manganese conditions. *Plant Cell, 22*, 904–917.

Cellier, M., Prive, G., Belouchi, A., Kwan, T., Rodrigues, V., Chia, W., et al. (1995). Nramp defines a family of membrane proteins. *Proceedings of the National Academy of Sciences of the United States of America, 92*, 10089–10093.

Cellier, M. F. (2012). Nutritional immunity: homology modeling of Nramp metal import. *Advances in Experimental Medicine and Biology, 946*, 335–351.

Cellier, M. F., Bergevin, I., Boyer, E., & Richer, E. (2001). Polyphyletic origins of bacterial Nramp transporters. *Trends in Genetics, 17*, 365–370.

Cellier, M. F., Courville, P., & Campion, C. (2007). Nramp1 phagocyte intracellular metal withdrawal defense. *Microbes and Infection, 9*, 1662–1670.

Chaloupka, R., Courville, P., Veyrier, F., Knudsen, B., Tompkins, T. A., & Cellier, M. F. (2005). Identification of functional amino acids in the Nramp family by a combination of evolutionary analysis and biophysical studies of metal and proton cotransport in vivo. *Biochemistry, 44*, 726–733.

Champion, O. L., Karlyshev, A., Cooper, I. A., Ford, D. C., Wren, B. W., Duffield, M., et al. (2011). Yersinia pseudotuberculosis mntH functions in intracellular manganese accumulation, which is essential for virulence and survival in cells expressing functional Nramp1. *Microbiology, 157*, 1115–1122.

Chasteen, T. G., Fuentes, D. E., Tantalean, J. C., & Vasquez, C. C. (2009). Tellurite: history, oxidative stress, and molecular mechanisms of resistance. *FEMS Microbiology Reviews, 33*, 820–832.

Cheah, J. H., Kim, S. F., Hester, L. D., Clancy, K. W., Patterson, S. E., III, Papadopoulos, V., et al. (2006). NMDA receptor-nitric oxide transmission mediates neuronal iron homeostasis via the GTPase Dexras1. *Neuron, 51*, 431–440.

Cheng, X., & Wang, H. (2012). Multiple targeting motifs direct NRAMP1 into lysosomes. *Biochemical and Biophysical Research Communications, 419*, 578–583.

Chesi, A., Kilaru, A., Fang, X., Cooper, A. A., & Gitler, A. D. (2012). The role of the Parkinson's disease gene PARK9 in essential cellular pathways and the manganese homeostasis network in yeast. *PLoS One, 7*, e34178.

Choi, S., Jeon, J., Yang, J. S., & Kim, S. (2008). Common occurrence of internal repeat symmetry in membrane proteins. *Proteins, 71*, 68–80.

Cohen, A., Nevo, Y., & Nelson, N. (2003). The first external loop of the metal ion transporter DCT1 is involved in metal ion binding and specificity. *Proceedings of the National Academy of Sciences of the United States of America, 100*, 10694–10699.

Colangelo, E. P., & Guerinot, M. L. (2006). Put the metal to the petal: metal uptake and transport throughout plants. *Current Opinion in Plant Biology, 9*, 322–330.

Cornelis, P., Wei, Q., Andrews, S. C., & Vinckx, T. (2011). Iron homeostasis and management of oxidative stress response in bacteria. *Metallomics, 3*, 540–549.

Courville, P., Chaloupka, R., Veyrier, F., & Cellier, M. F. (2004). Determination of transmembrane topology of the Escherichia coli natural resistance-associated macrophage protein (Nramp) ortholog. *The Journal of Biological Chemistry, 279*, 3318–3326.

Courville, P., Urbankova, E., Rensing, C., Chaloupka, R., Quick, M., & Cellier, M. F. (2008). Solute carrier 11 cations symport requires distinct residues in transmembrane helices 1 and 6. *The Journal of Biological Chemistry, 283*, 9651–9658.

Cuellar-Mata, P., Jabado, N., Liu, J., Furuya, W., Finlay, B. B., Gros, P., et al. (2002). Nramp1 modifies the fusion of Salmonella typhimurium-containing vacuoles with cellular endomembranes in macrophages. *The Journal of Biological Chemistry, 277*, 2258–2265.

Czachorowski, M., Lam-Yuk-Tseung, S., Cellier, M., & Gros, P. (2009). Transmembrane topology of the mammalian Slc11a2 iron transporter. *Biochemistry, 48*, 8422–8434.

D'Souza, J., Cheah, P. Y., Gros, P., & Rodrigues, V. (1999). Functional complementation of the *malvolio* mutation in the taste pathway of *Drosophila melanogaster* by the human natural resistance-associated macrophage protein 1 (Nramp-1). *The Journal of Experimental Biology, 202*, 1909–1915.

Daley, D. O., Rapp, M., Granseth, E., Melen, K., Drew, D., & von Heijne, G. (2005). Global topology analysis of the Escherichia coli inner membrane proteome. *Science, 308*, 1321–1323.

Daly, M. J. (2009). A new perspective on radiation resistance based on Deinococcus radiodurans. *Nature Reviews Microbiology, 7*, 237–245.

De Domenico, I., Ward, D. M., & Kaplan, J. (2007). Hepcidin regulation: ironing out the details. *Journal of Clinical Investigation, 117*, 1755–1758.

DeLano, W. L. (2002). *The PyMOL molecular graphics system*. DeLano Scientific, Sancarlos: California, USA.
Donoghue, H. D. (2011). Insights gained from palaeomicrobiology into ancient and modern tuberculosis. *Clinical Microbiology and Infection, 17*, 821–829.
Dowling, A. J., Wilkinson, P. A., Holden, M. T., Quail, M. A., Bentley, S. D., Reger, J., et al. (2010). Genome-wide analysis reveals loci encoding anti-macrophage factors in the human pathogen Burkholderia pseudomallei K96243. *PLoS One, 5*. e15693.
Elibol-Flemming, B., Waldbieser, G. C., Wolters, W. R., Boyle, C. R., & Hanson, L. A. (2009). Expression analysis of selected immune-relevant genes in channel catfish during Edwardsiella ictaluri infection. *Journal of Aquatic Animal Health, 21*, 23–35.
Elliott, M. R., & Ravichandran, K. S. (2010). Clearance of apoptotic cells: implications in health and disease. *Journal of Cell Biology, 189*, 1059–1070.
Faham, S., Watanabe, A., Besserer, G. M., Cascio, D., Specht, A., Hirayama, B. A., et al. (2008). The crystal structure of a sodium galactose transporter reveals mechanistic insights into Na+/sugar symport. *Science, 321*, 810–814.
Fleming, B. D., & Mosser, D. M. (2011). Regulatory macrophages: setting the threshold for therapy. *European Journal of Immunology, 41*, 2498–2502.
Fleming, M. D., Trenor, C. C., III, Su, M. A., Foernzler, D., Beier, D. R., Dietrich, W. F., et al. (1997). Microcytic anaemia mice have a mutation in Nramp2, a candidate iron transporter gene. *Nature Genetics, 16*, 383–386.
Fleming, M. D., Romano, M. A., Su, M. A., Garrick, L. M., & Garrick, M. D. (1998). Nramp2 is mutated in the anemic Belgrade (b) rat: evidence of a role for Nramp2 in endosomal iron transport. *Proceedings of the National Academy of Sciences, U. S. A, 95*, 1148–1153.
Foot, N. J., Dalton, H. E., Shearwin-Whyatt, L. M., Dorstyn, L., Tan, S. S., Yang, B., et al. (2008). Regulation of the divalent metal ion transporter DMT1 and iron homeostasis by a ubiquitin-dependent mechanism involving Ndfips and WWP2. *Blood, 112*, 4268–4275.
Foot, N. J., Leong, Y. A., Dorstyn, L. E., Dalton, H. E., Ho, K., Zhao, L., et al. (2011). Ndfip1-deficient mice have impaired DMT1 regulation and iron homeostasis. *Blood, 117*, 638–646.
Forbes, J. R., & Gros, P. (2003). Iron, manganese, and cobalt transport by Nramp1 (Slc11a1) and Nramp2 (Slc11a2) expressed at the plasma membrane. *Blood, 102*, 1884–1892.
Forrest, L. R., Kramer, R., & Ziegler, C. (2011). The structural basis of secondary active transport mechanisms. *Biochimica et Biophysica Acta, 1807*, 167–188.
Frickey, T., & Lupas, A. (2004). CLANS: a Java application for visualizing protein families based on pairwise similarity. *Bioinformatics, 20*, 3702–3704.
Gallant, C. J., Malik, S., Jabado, N., Cellier, M., Simkin, L., Finlay, B. B., et al. (2007). Reduced in vitro functional activity of human NRAMP1 (SLC11A1) allele that predisposes to increased risk of pediatric tuberculosis disease. *Genes Immunology, 8*, 691–698.
Ganz, T. (2009). Iron in innate immunity: starve the invaders. *Current Opinion in Immunology, 21*, 63–67.
Gaucher, E. A., Gu, X., Miyamoto, M. M., & Benner, S. A. (2002). Predicting functional divergence in protein evolution by site-specific rate shifts. *Trends in Biochemical Science, 27*, 315–321.
Geissmann, F., Manz, M. G., Jung, S., Sieweke, M. H., Merad, M., & Ley, K. (2010). Development of monocytes, macrophages, and dendritic cells. *Science, 327*, 656–661.
Ghosh, S., Ramirez-Peralta, A., Gaidamakova, E., Zhang, P., Li, Y. Q., Daly, M. J., et al. (2011). Effects of Mn levels on resistance of Bacillus megaterium spores to heat, radiation and hydrogen peroxide. *The Journal of Applied Microbiology, 111*, 663–670.

Gordon, S., & Taylor, P. R. (2005). Monocyte and macrophage heterogeneity. *Nature Reviews Immunology, 5*, 953–964.
Govoni, G., Canonne-Hergaux, F., Pfeifer, C. G., Marcus, S. L., Mills, S. D., Hackam, D. J., et al. (1999). Functional expression of Nramp1 in vitro in the murine macrophage line RAW264.7. *Infection and Immunity, 67*, 2225–2232.
Granger, A. C., Gaidamakova, E. K., Matrosova, V. Y., Daly, M. J., & Setlow, P. (2011). Effects of Mn and Fe levels on Bacillus subtilis spore resistance and effects of Mn2+, other divalent cations, orthophosphate, and dipicolinic acid on protein resistance to ionizing radiation. *Applied Environmental Microbiology, 77*, 32–40.
Gruenheid, S., Pinner, E., Desjardins, M., & Gros, P. (1997). Natural resistance to infection with intracellular pathogens: the Nramp1 protein is recruited to the membrane of the phagosome. *Journal of Experimental Medicine, 185*, 717–730.
Gunshin, H., Mackenzie, B., Berger, U. V., Gunshin, Y., Romero, M. F., Boron, W. F., et al. (1997). Cloning and characterization of a mammalian proton-coupled metal-ion transporter. *Nature, 388*, 482–488.
Gu, X., & Vander Velden, K. (2002). DIVERGE: phylogeny-based analysis for functional-structural divergence of a protein family. *Bioinformatics, 18*, 500–501.
Gu, X., & Zhang, J. (1997). A simple method for estimating the parameter of substitution rate variation among sites. *Molecular Biology and Evolution, 14*, 1106–1113.
Gunshin, H., Mackenzie, B., Berger, U. V., Gunshin, Y., Romero, M. F., Boron, W. F., Nussberger, S., Gollan, J. L., & Hediger, M. A. (1997). Cloning and characterization of a mammalian proton-coupled metal-ion transporter. *Nature, 388*, 482–488.
Hackam, D. J., Rotstein, O. D., Zhang, W., Gruenheid, S., Gros, P., & Grinstein, S. (1998). Host resistance to intracellular infection: mutation of natural resistance-associated macrophage protein 1 (Nramp1) impairs phagosomal acidification. *Journal of Experimental Medicine, 188*, 351–364.
Haeger, P., Alvarez, A., Leal, N., Adasme, T., Nunez, M. T., & Hidalgo, C. (2010). Increased hippocampal expression of the divalent metal transporter 1 (DMT1) mRNA variants 1B and +IRE and DMT1 protein after NMDA-receptor stimulation or spatial memory training. *Neurotoxicity Research, 17*, 238–247.
Haemig, H. A., Moen, P. J., & Brooker, R. J. (2010). Evidence that highly conserved residues of transmembrane segment 6 of Escherichia coli MntH are important for transport activity. *Biochemistry, 49*, 4662–4671.
Haemig, H. A., & Brooker, R. J. (2004). Importance of conserved acidic residues in mntH, the Nramp homolog of Escherichia coli. *Journal of Membrane Biology, 201*, 97–107.
Hentze, M. W., Muckenthaler, M. U., Galy, B., & Camaschella, C. (2010). Two to tango: regulation of mammalian iron metabolism. *Cell, 142*, 24–38.
Hickman, M. J., Petti, A. A., Ho-Shing, O., Silverman, S. J., McIsaac, R. S., Lee, T. A., et al. (2011). Coordinated regulation of sulfur and phospholipid metabolism reflects the importance of methylation in the growth of yeast. *Molecular Biology of the Cell, 22*, 4192–4204.
Hohle, T. H., & O'Brian, M. R. (2009). The mntH gene encodes the major Mn(2+) transporter in Bradyrhizobium japonicum and is regulated by manganese via the Fur protein. *Molecular Microbiology, 72*, 399–409.
Howitt, J., Putz, U., Lackovic, J., Doan, A., Dorstyn, L., Cheng, H., et al. (2009). Divalent metal transporter 1 (DMT1) regulation by Ndfip1 prevents metal toxicity in human neurons. *Proceedings of the National Academy of Sciences of the United States of America, 106*, 15489–15494.
Huynh, K. K., Joshi, S. A., & Brown, E. J. (2011). A delicate dance: host response to mycobacteria. *Current Opinion in Immunology, 23*, 464–472.

Iolascon, A., d'Apolito, M., Servedio, V., Cimmino, F., Piga, A., & Camaschella, C. (2006). Microcytic anemia and hepatic iron overload in a child with compound heterozygous mutations in DMT1 (SCL11A2). *Blood, 107*, 349–354.

Jabado, N., Jankowski, A., Dougaparsad, S., Picard, V., Grinstein, S., & Gros, P. (2000). Natural resistance to intracellular infections: natural resistance- associated macrophage protein 1 (Nramp1) functions as a pH-dependent manganese transporter at the phagosomal membrane. *Journal of Experimental Medicine, 192*, 1237–1248.

Jaroszewski, L., Li, Z., Cai, X. H., Weber, C., & Godzik, A. (2011). FFAS server: novel features and applications. *Nucleic Acids Research, 39*, W38–W44.

Jenuth, J. P. (2000). The NCBI. Publicly available tools and resources on the Web. *Methods in Molecular Biology, 132*, 301–312.

Jeon, S. J., Gonsalves, L. C., Jacobs, J. M., Rhodes, M., Councilman, J., Baya, A., et al. (2011). Short-term infection of striped bass Morone saxatilis with Mycobacterium marinum. *Diseases of Aquatic Organisms, 94*, 117–124.

Jeschke, G. (2012). A comparative study of structures and structural transitions of secondary transporters with the LeuT fold. *European Biophysical Journal.* [Epub ahead of print].

Kaback, H. R. (1971). The transport of sugars across isolated bacterial membranes. *Current Topics in Membranes and Transport, 1*, 35–99.

Kehl-Fie, T. E., & Skaar, E. P. (2010). Nutritional immunity beyond iron: a role for manganese and zinc. *Current Opinion in Chemical Biology, 14*, 218–224.

Kehres, D. G., Janakiraman, A., Slauch, J. M., & Maguire, M. E. (2002). Regulation of Salmonella enterica serovar Typhimurium mntH transcription by H(2)O(2), Fe(2+), and Mn(2+). *The Journal of Bacteriology, 184*, 3151–3158.

Kehres, D. G., Zaharik, M. L., Finlay, B. B., & Maguire, M. E. (2000). The NRAMP proteins of Salmonella typhimurium and Escherichia coli are selective manganese transporters involved in the response to reactive oxygen. *Molecular Microbiology, 36*, 1085–1100, ss.

Kelley, L. A., & Sternberg, M. J. (2009). Protein structure prediction on the Web: a case study using the Phyre server. *Nature Protocols, 4*, 363–371.

Khafizov, K., Staritzbichler, R., Stamm, M., & Forrest, L. R. (2010). A study of the evolution of inverted-topology repeats from LeuT-fold transporters using AlignMe. *Biochemistry, 49*, 10702–10713.

Kieu, N. P., Aznar, A., Segond, D., Rigault, M., Simond-Cote, E., Kunz, C., et al. (2012). Iron deficiency affects plant defence responses and confers resistance to Dickeya dadantii and Botrytis cinerea. *Molecular Plant Pathology.* [Epub ahead of print].

Kim, J., Li, Y., Buckett, P. D., Bohlke, M., Thompson, K. J., Takahashi, M., et al. (2012). Iron-responsive olfactory uptake of manganese improves motor function deficits associated with iron deficiency. *PLoS One, 7*, e33533.

Knudsen, B., & Miyamoto, M. M. (2001). A likelihood ratio test for evolutionary rate shifts and functional divergence among proteins. *Proceedings of the National Academy of Sciences of the United States of America, 98*, 14512–14517.

Lam-Yuk-Tseung, S., Govoni, G., Forbes, J., & Gros, P. (2003). Iron transport by Nramp2/DMT1: pH regulation of transport by 2 histidines in transmembrane domain 6. *Blood, 101*, 3699–3707.

Lam-Yuk-Tseung, S., & Gros, P. (2006). Distinct targeting and recycling properties of two isoforms of the iron transporter DMT1 (NRAMP2, Slc11A2). *Biochemistry, 45*, 2294–2301.

Lan, W. J., Ren, H. L., Pang, Y., Huang, C. S., Xu, Y. F., Brooker, R. J., et al. (2012). A facile transport assay for H+ coupled membrane transport using fluorescence probes. *Analytical Methods, 4*, 44–46.

Lanquar, V., Lelievre, F., Bolte, S., Hames, C., Alcon, C., Neumann, D., et al. (2005). Mobilization of vacuolar iron by AtNRAMP3 and AtNRAMP4 is essential for seed germination on low iron. *EMBO journal, 24*, 4041–4051.

Lanquar, V., Ramos, M. S., Lelievre, F., Barbier-Brygoo, H., Krieger-Liszkay, A., Kramer, U., et al. (2010). Export of vacuolar manganese by AtNRAMP3 and AtNRAMP4 is required for optimal photosynthesis and growth under manganese deficiency. *Plant Physiology, 152*, 1986–1999.

Lapointe, J. Y., Sasseville, L. J., & Longpre, J. P. (2009). Alternating carrier models and the energy conservation laws. *Biophysical Journal, 97*, 2648–2650.

Lawn, S. D., & Zumla, A. I. (2011). *Tuberculosis. Lancet, 378*, 57–72.

Li, X., Yang, Y., Zhou, F., Zhang, Y., Lu, H., Jin, Q., et al. (2011). SLC11A1 (NRAMP1) polymorphisms and tuberculosis susceptibility: updated systematic review and meta-analysis. *PLoS One, 6*, e15831.

Lin, Z., Fernandez-Robledo, J. A., Cellier, M. F., & Vasta, G. R. (2011). The natural resistance-associated macrophage protein from the protozoan parasite Perkinsus marinus mediates iron uptake. *Biochemistry, 50*, 6340–6355.

Liu, J., Fujiwara, T. M., Buu, N. T., Sanchez, F. O., Cellier, M., Paradis, A. J., et al. (1995). Identification of polymorphisms and sequence variants in the human homologue of the mouse natural resistance-associated macrophage protein gene. *American Journal of Human Genetics, 56*, 845–853.

Lopez, G., Latorre, M., Reyes-Jara, A., Cambiazo, V., & Gonzalez, M. (2012). Transcriptomic response of Enterococcus faecalis to iron excess. *Biometals*. [Epub ahead of print].

Mackenzie, B., & Hediger, M. A. (2004). SLC11 family of H(+)-coupled metal-ion transporters NRAMP1 and DMT1. *Pflugers Archiv, 447*, 571–579.

Mackenzie, B., Takanaga, H., Hubert, N., Rolfs, A., & Hediger, M. A. (2007). Functional properties of multiple isoforms of human divalent metal-ion transporter 1 (DMT1). *The Biochemical Journal, 403*, 59–69.

Mackenzie, B., Ujwal, M. L., Chang, M. H., Romero, M. F., & Hediger, M. A. (2006). Divalent metal-ion transporter DMT1 mediates both H+ -coupled Fe2+ transport and uncoupled fluxes. *Pflugers Archiv, 451*, 544–558.

Madejczyk, M. S., & Ballatori, N. (2012). The iron transporter ferroportin can also function as a manganese exporter. *Biochimica et Biophysica Acta, 1818*, 651–657.

Magnani, L., Eeckhoute, J., & Lupien, M. (2011). Pioneer factors: directing transcriptional regulators within the chromatin environment. *Trends in Genetics, 27*, 465–474.

Makui, H., Roig, E., Cole, S. T., Helmann, J. D., Gros, P., & Cellier, M. F. (2000). Identification of the Escherichia coli K-12 Nramp orthologue (MntH) as a selective divalent metal ion transporter. *Molecular Microbiology, 35*, 1065–1078.

Malik, S., Abel, L., Tooker, H., Poon, A., Simkin, L., Girard, M., et al. (2005). Alleles of the NRAMP1 gene are risk factors for pediatric tuberculosis disease. *Proceedings of the National Academy of Science United States of America, 102*, 12183–12188.

Malo, D., Vogan, K., Vidal, S., Hu, J., Cellier, M., Schurr, E., et al. (1994). Haplotype mapping and sequence analysis of the mouse Nramp gene predict susceptibility to infection with intracellular parasites. *Genomics, 23*, 51–61.

Marquis, J. F., Kapoustina, O., Langlais, D., Ruddy, R., Dufour, C. R., Kim, B. H., et al. (2011). Interferon regulatory factor 8 regulates pathways for antigen presentation in myeloid cells and during tuberculosis. *PLoS Genet*, 7. e1002097.

Martinez-Barnetche, J., Garcia, S. M., Neri, L. A., Tello Lopez, A. T., del Carmen, R. M., Gamba, G., et al. (2007). Cloning and functional characterization of the Anopheles albimanus DMT1/NRAMP homolog: implications in iron metabolism in mosquitoes. *Insect Biochemistry Molecular Biology, 37*, 532–539.

Mastrogiannaki, M., Matak, P., Keith, B., Simon, M. C., Vaulont, S., & Peyssonnaux, C. (2009). HIF-2alpha, but not HIF-1alpha, promotes iron absorption in mice. *The journal of clinical investigation, 119*, 1159–1166.

McColl, G., James, S. A., Mayo, S., Howard, D. L., Ryan, C. G., Kirkham, R., et al. (2012). Caenorhabditis elegans maintains highly compartmentalized cellular distribution of metals and steep concentration gradients of manganese. *PLoS One, 7*, e32685.

Meilang, Q., Zhang, Y., Zhang, J., Zhao, Y., Tian, C., Huang, J., et al. (2012). Polymorphisms in the SLC11A1 gene and tuberculosis risk: a meta-analysis update. *The International Journal of Tuberculosis and Lung Disease*. [Epub ahead of print].

Menscher, E. A., Caswell, C. C., Anderson, E. S., & Roop, R. M. (2012). Mur regulates the gene encoding the manganese transporter MntH in Brucella abortus 2308. *The journal of Bacteriology, 194*, 561–566.

Mims, M. P., Guan, Y., Pospisilova, D., Priwitzerova, M., Indrak, K., Ponka, P., et al. (2005). Identification of a human mutation of DMT1 in a patient with microcytic anemia and iron overload. *Blood, 105*, 1337–1342.

Moller, M., & Hoal, E. G. (2010). Current findings, challenges and novel approaches in human genetic susceptibility to tuberculosis. *Tuberculosis (Edinburgh), 90*, 71–83.

Morrison, E. A., DeKoster, G. T., Dutta, S., Vafabakhsh, R., Clarkson, M. W., Bahl, A., et al. (2012). Antiparallel EmrE exports drugs by exchanging between asymmetric structures. *Nature, 481*, 45–50.

Nachin, L., Nannmark, U., & Nystrom, T. (2005). Differential roles of the universal stress proteins of Escherichia coli in oxidative stress resistance, adhesion, and motility. *Journal of Bacteriology, 187*, 6265–6272.

Neves, J. V., Wilson, J. M., Kuhl, H., Reinhardt, R., Castro, L. F., & Rodrigues, P. N. (2011). Natural history of SLC11 genes in vertebrates: tales from the fish world. BMC. *The Evolution Biology, 11*, 106.

Nevo, Y., & Nelson, N. (2006). The NRAMP family of metal-ion transporters. *Biochimica et Biophysica Acta, 1763*, 609–620.

Nugent, T., Ward, S., & Jones, D. T. (2011). The MEMPACK alpha-helical transmembrane protein structure prediction server. *Bioinformatics, 27*, 1438–1439.

Nunez, M. T. (2010). Regulatory mechanisms of intestinal iron absorption-uncovering of a fast-response mechanism based on DMT1 and ferroportin endocytosis. *Biofactors, 36*, 88–97.

O'Brien, B. A., Archer, N. S., Simpson, A. M., Torpy, F. R., & Nassif, N. T. (2008). Association of SLC11A1 promoter polymorphisms with the incidence of autoimmune and inflammatory diseases: a meta-analysis. *Journal of Autoimmunity, 31*, 42–51.

Ohgami, R. S., Campagna, D. R., Greer, E. L., Antiochos, B., McDonald, A., Chen, J., et al. (2005). Identification of a ferrireductase required for efficient transferrin-dependent iron uptake in erythroid cells. *Nature Genetics, 37*, 1264–1269.

Okazaki, Y., Ma, Y., Yeh, M., Yin, H., Li, Z., Yeh, K. Y., et al. (2012). DMT1 (IRE) expression in intestinal and erythroid cells is regulated by peripheral benzodiazepine receptor associated protein 7 (PAP7). *American Journal of Physiology, Gastrointestinal and Liver Physiology, 302*, G1180–G1190.

Orgad, S., Nelson, H., Segal, D., & Nelson, N. (1998). Metal ions suppress the abnormal taste behavior of the *Drosophila* mutant *malvolio*. *The Journal of Experimental Biology, 201*, 115–120.

Ostuni, R., & Natoli, G. (2011). Transcriptional control of macrophage diversity and specialization. *European Journal of Immunology, 41*, 2486–2490.

Patzer, S. I., & Hantke, K. (2001). Dual repression by Fe(2+)-Fur and Mn(2+)-MntR of the mntH gene, encoding an NRAMP-like Mn(2+) transporter in *Escherichia coli*. *The Journal of Bacteriology, 183*, 4806–4813.

Patzer, S. I., & Hantke, K. (2001). Dual repression by Fe(2+)-Fur and Mn(2+)-MntR of the mntH gene, encoding an NRAMP-like Mn(2+) transporter in *Escherichia coli*. *The Journal of Bacteriology, 183*, 4806–4813.

Peracino, B., Balest, A., & Bozzaro, S. (2010). Phosphoinositides differentially regulate bacterial uptake and Nramp1-induced resistance to Legionella infection in Dictyostelium. *Journal of Cell Science, 123*, 4039–4051.

Peracino, B., Wagner, C., Balest, A., Balbo, A., Pergolizzi, B., Noegel, A. A., et al. (2006). Function and mechanism of action of Dictyostelium Nramp1 (Slc11a1) in bacterial infection. *Traffic, 7*, 22–38.

Perry, R. D., Craig, S. K., Abney, J., Bobrov, A. G., Kirillina, O., Mier, I., Jr., et al. (2012). Manganese transporters Yfe and MntH are Fur-regulated and important for the virulence of Yersinia pestis. *Microbiology, 158*, 804–815.

Pieper, U., Webb, B. M., Barkan, D. T., Schneidman-Duhovny, D., Schlessinger, A., Braberg, H., et al. (2011). ModBase, a database of annotated comparative protein structure models, and associated resources. *Nucleic Acids Research, 39*, D465–D474.

Priwitzerova, M., Pospisilova, D., Prchal, J. T., Indrak, K., Hlobilkova, A., Mihal, V., et al. (2004). Severe hypochromic microcytic anemia caused by a congenital defect of the iron transport pathway in erythroid cells. *Blood, 103*, 3991–3992.

Qi, H., Yang, L., Xue, R., Song, Y., Wang, S., & Li, F. (2009). The third and fourth transmembrane domains of Slc11a1: comparison of their structures and positioning in phospholipid model membranes. *Biopolymers, 92*, 52–64.

Qian, Z. M., Wu, X. M., Fan, M., Yang, L., Du, F., Yung, W. H., et al. (2011). Divalent metal transporter 1 is a hypoxia-inducible gene. *The Journal of Cell Physiology, 226*, 1596–1603.

Qiao, B., Sugianto, P., Fung, E., del-Castillo-Rueda, A., Moran-Jimenez, M.-J., Ganz, T., et al. (2012). Hepcidin-induced endocytosis of ferroportin is dependent on ferroportin ubiquitination. *Cell Metabolism, 15*, 918–924.

Que, Q., & Helmann, J. D. (2000). Manganese homeostasis in Bacillus subtilis is regulated by MntR, a bifunctional regulator related to the diphtheria toxin repressor family of proteins. *Molecular Microbiology, 35*, 1454–1468.

Radestock, S., & Forrest, L. R. (2011). The Alternating-access mechanism of MFS transporters arises from inverted-topology repeats. *Journal of Molecular Biology, 407*, 698–715.

Rathore, K. I., Redensek, A., & David, S. (2012). Iron homeostasis in astrocytes and microglia is differentially regulated by TNF-alpha and TGF-beta 1. *Glia, 60*, 738–750.

Reddi, A. R., Jensen, L. T., & Culotta, V. C. (2009). Manganese homeostasis in *Saccharomyces cerevisiae*. *Chemical Reviews, 109*, 4722–4732.

Richer, E., Campion, C. G., Dabbas, B., White, J. H., & Cellier, M. F. (2008). Transcription factors Sp1 and C/EBP regulate NRAMP1 gene expression. *journal, 275*, 5074–5089.

Richer, E., Courville, P., Bergevin, I., & Cellier, M. F. (2003). Horizontal gene transfer of "prototype" Nramp in bacteria. *Journal of Molecular Evolution, 57*, 363–376.

Richer, E., Courville, P., & Cellier, M. F. M. (2004). Molecular evolutionary analysis of the Nramp family. In M. Cellier, & P. Gros (Eds.), *Molecular biology intelligence unit* (pp. 178–194). Springer.

Romney, S. J., Newman, B. S., Thacker, C., & Leibold, E. A. (2011). HIF-1 regulates iron homeostasis in Caenorhabditis elegans by activation and inhibition of genes involved in iron uptake and storage. *PLoS Genetics, 7*. e1002394.

Rosakis, A., & Koster, W. (2005). Divalent metal transport in the green microalga Chlamydomonas reinhardtii is mediated by a protein similar to prokaryotic Nramp homologues. *Biometals, 18*, 107–120.

Rose, P. P., Hanna, S. L., Spiridigliozzi, A., Wannissorn, N., Beiting, D. P., Ross, S. R., et al. (2011). Natural resistance-associated macrophage protein is a cellular receptor for sindbis virus in both insect and mammalian hosts. *Cell Host and Microbe, 10*, 97–104.

Ross, S. L., Tran, L., Winters, A., Lee, K.-J., Plewa, C., oltz, I., et al. (2012). Molecular mechanism of hepcidin-mediated ferroportin internalization requires ferroportin lysines, not tyrosines or JAK-STAT. *Cell Metabolism, 15*, 905–917.

Roy, A., Xu, D., Poisson, J., & Zhang, Y. (2011). A protocol for computer-based protein structure and function prediction. *Journal of Visualized Experiments*, e3259.

Runyen-Janecky, L., Dazenski, E., Hawkins, S., & Warner, L. (2006). Role and regulation of the Shigella flexneri sit and MntH systems. *Infection and Immunity, 74*, 4666–4672.

Sabri, M., Caza, M., Proulx, J., Lymberopoulos, M. H., Bree, A., Moulin-Schouleur, M., et al. (2008). Contribution of the SitABCD, MntH, and FeoB metal transporters to the virulence of avian pathogenic *Escherichia coli* O78 strain chi7122. *Infection and Immunity, 76*, 601–611.

Salazar, J., Mena, N., Hunot, S., Prigent, A., Alvarez-Fischer, D., Arredondo, M., et al. (2008). Divalent metal transporter 1 (DMT1) contributes to neurodegeneration in animal models of Parkinson's disease. *Proceedings of the National Academy of Science U.S.A, 105*, 18578–18583.

Sammut, S. J., Finn, R. D., & Bateman, A. (2008). Pfam 10 years on: 10, 000 families and still growing. *Briefings in Bioinformatics, 9*, 210–219.

Sanders, C. R., & Mittendorf, K. F. (2011). Tolerance to changes in membrane lipid composition as a selected trait of membrane proteins. *Biochemistry, 50*, 7858–7867.

Schmitz-Esser, S., Tischler, P., Arnold, R., Montanaro, J., Wagner, M., Rattei, T., et al. (2010). The genome of the amoeba symbiont "*Candidatus Amoebophilus asiaticus*" reveals common mechanisms for host cell interaction among amoeba-associated bacteria. *The Journal of Bacteriology, 192*, 1045–1057.

Schulz, C., Gomez, P. E., Chorro, L., Szabo-Rogers, H., Cagnard, N., Kierdorf, K., et al. (2012). A lineage of myeloid cells independent of Myb and hematopoietic stem cells. *Science, 336*, 86–90.

Segond, D., Dellagi, A., Lanquar, V., Rigault, M., Patrit, O., Thomine, S., et al. (2009). NRAMP genes function in Arabidopsis thaliana resistance to Erwinia chrysanthemi infection. *Plant Journal, 58*, 195–207.

Settivari, R., Levora, J., and Nass, R. (2009). The divalent metal transporter homologues SMF-1/2 mediate dopamine neuron sensitivity in *caenorhabalitics elegans* models of manganism and parkison disease. *Journal of Biological Chemistry*, 284, 35758–35768.

Shaffer, P. L., Goehring, A., Shankaranarayanan, A., & Gouaux, E. (2009). Structure and mechanism of a Na+-independent amino acid transporter. *Science, 325*, 1010–1014.

Shalom, G., Shaw, J. G., & Thomas, M. S. (2007). In vivo expression technology identifies a type VI secretion system locus in Burkholderia pseudomallei that is induced upon invasion of macrophages. *Microbiology, 153*, 2689–2699.

Shih, A. C., Lee, D. T., Peng, C. L., & Wu, Y. W. (2007). Phylo-mLogo: an interactive and hierarchical multiple-logo visualization tool for alignment of many sequences. *BMC Bioinformatics, 8*, 63.

Singh, A., Gaughan, J. P., & Kashyap, V. K. (2011). SLC11A1 and VDR gene variants and susceptibility to tuberculosis and disease progression in East India. *The International Journal of Tuberculosis and Lung diseases, 15*, 1468–1474, i.

Soe-Lin, S., Apte, S. S., Andriopoulos, B., Jr., Andrews, M. C., Schranzhofer, M., Kahawita, T., et al. (2009). Nramp1 promotes efficient macrophage recycling of iron following erythrophagocytosis in vivo. *Proceedings of the National Academy of Sciences of the United States of America, 106*, 5960–5965.

Song, Y., Wang, D., Qi, H., Xiao, S., Xue, R., & Li, F. (2009). Metal ion binding of the first external loop of DCT1 in aqueous solution. *Metallomics, 1*, 392–394.

Stein, C. M., & Baker, A. R. (2011). Tuberculosis as a complex trait: impact of genetic epidemiological study design. *Mammalian Genome, 22*, 91–99.

Stubben, C. J., Duffield, M. L., Cooper, I. A., Ford, D. C., Gans, J. D., Karlyshev, A. V., et al. (2009). Steps toward broad-spectrum therapeutics: discovering virulence-associated genes present in diverse human pathogens. *BMC Genomics, 10*, 501.

Suen, G., Goldman, B. S., & Welch, R. D. (2007). Predicting prokaryotic ecological niches using genome sequence analysis. *PLoS One, 2*, e743.

Tamura, K., Peterson, D., Peterson, N., Stecher, G., Nei, M., & Kumar, S. (2011). MEGA5: molecular evolutionary genetics analysis using maximum likelihood, evolutionary distance, and maximum parsimony methods. *Molecular Biology of Evolution, 28*, 2731–2739.

Tatusov, R. L., Fedorova, N. D., Jackson, J. D., Jacobs, A. R., Kiryutin, B., Koonin, E. V., et al. (2003). The COG database: an updated version includes eukaryotes. *Bmc Bioinformatics, 4*, 41.

Turner, M. S., Tan, Y. P., & Giffard, P. M. (2007). Inactivation of an iron transporter in Lactococcus lactis results in resistance to tellurite and oxidative stress. *Applied Environmental Microbiology, 73*, 6144–6149.

Varghese, S., Wu, A., Park, S., Imlay, K. R., & Imlay, J. A. (2007). Submicromolar hydrogen peroxide disrupts the ability of Fur protein to control free-iron levels in Escherichia coli. *Molecular Microbiology, 64*, 822–830.

Veyrier, F. J., Boneca, I. G., Cellier, M. F., & Taha, M. K. (2011). A novel metal transporter mediating manganese export (MntX) regulates the Mn to Fe intracellular ratio and Neisseria meningitidis virulence. *PLoS Pathogens, 7*. e1002261.

Vidal, S., Tremblay, M. L., Govoni, G., Gauthier, S., Sebastiani, G., Malo, D., et al. (1995). The Ity/Lsh/Bcg locus: natural resistance to infection with intracellular parasites is abrogated by disruption of the Nramp1 gene. *Journal Experimental Medicine, 182*, 655–666.

Vidal, S. M., Malo, D., Vogan, K., Skamene, E., & Gros, P. (1993). Natural resistance to infection with intracellular parasites: isolation of a candidate for Bcg. *Cell, 73*, 469–485.

Wagner, D., Maser, J., Moric, I., Boechat, N., Vogt, S., Gicquel, B., et al. (2005). Changes of the phagosomal elemental concentrations by Mycobacterium tuberculosis Mramp. *Microbiology, 151*, 323–332.

Wang, D., Song, Y., Li, J., Wang, C., & Li, F. (2011). Structure and metal ion binding of the first transmembrane domain of DMT1. *Biochimica et Biophysica Actal, 1808*, 1639–1644.

Wang, D., Wang, L. H., Zhao, Y., Lu, Y. P., & Zhu, L. (2010). Hypoxia regulates the ferrous iron uptake and reactive oxygen species level via divalent metal transporter 1 (DMT1) Exon1B by hypoxia-inducible factor-1. IUBMB. *Life, 62*, 629–636.

Wang, J., & Pantopoulos, K. (2011). Regulation of cellular iron metabolism. *The Biochemical Journal, 434*, 365–381.

Waters, L. S., Sandoval, M., & Storz, G. (2011). The *Escherichia coli* MntR miniregulon includes genes encoding a small protein and an efflux pump required for manganese homeostasis. *The Journal of Bacteriology, 193*, 5887–5897.

Weinberg, E. D. (1977). Infection and iron metabolism. *American Journal of Clinical Nutrition, 30*, 1485–1490.

Weyand, S., Shimamura, T., Yajima, S., Suzuki, S., Mirza, O., Krusong, K., et al. (2008). Structure and molecular mechanism of a nucleobase-cation-symport-1 family transporter. *Science, 322*, 709–713.

Xia, J., Yamaji, N., Kasai, T., & Ma, J. F. (2010). Plasma membrane-localized transporter for aluminum in rice. *Proceedings of the National Academy of Science U. S. A, 107*, 18381–18385.

Xia, J., Yamaji, N., & Ma, J. F. (2011). Further characterization of an aluminum influx transporter in rice. *Plant Signaling Behavior, 6*, 160–163.

Xiao, S., Li, J., Wang, Y., Wang, C., Xue, R., Wang, S., et al. (2010). Identification of an "alpha-helix-extended segment-alpha-helix" conformation of the sixth transmembrane domain in DMT1. *Biochimica et Biophysica Acta, 1798*, 1556–1564.

Xiao, S., Wang, C., Li, J., & Li, F. (2011). Folding and assembly of TMD 6-related segments of DMT 1 in trifluoroethanol aqueous solution. *Journal of Peptide Science, 17*, 505–511.

Xiao, S., Wang, Y., Yang, L., Qi, H., Wang, C., & Li, F. (2010). Study on structure and assembly of the third transmembrane domain of Slc11a1. *Journal of Peptide Science, 16*, 249–255.

Xu, Y. Z., Thuraisingam, T., Marino, R., & Radzioch, D. (2011). Recruitment of SWI/SNF complex is required for transcriptional activation of the SLC11A1 gene during

macrophage differentiation of HL-60 cells. *The Journal of Biological Chemistry, 286,* 12839–12849.
Xue, R., Wang, S., Qi, H., Song, Y., Wang, C., & Li, F. (2008). Structure analysis of the fourth transmembrane domain of Nramp1 in model membranes. *Biochimica et Biophysica Acta, 1778,* 1444–1452.
Xue, R., Wang, S., Qi, H., Song, Y., Xiao, S., Wang, C., et al. (2009). Structure and topology of Slc11a1(164-191) with G169D mutation in membrane-mimetic environments. *The Journal of Structural Biology, 165,* 27–36.
Yamamoto, K., Ishihama, A., Busby, S. J., & Grainger, D. C. (2011). The Escherichia coli K-12 MntR miniregulon includes dps, which encodes the major stationary-phase DNA-binding protein. *The Journal of Bacteriology, 193,* 1477–1480.
Yamashita, A., Singh, S. K., Kawate, T., Jin, Y., & Gouaux, E. (2005). Crystal structure of a bacterial homologue of Na+/Cl-dependent neurotransmitter transporters. *Nature, 437,* 215–223.
Ye, Y., & Godzik, A. (2003). Flexible structure alignment by chaining aligned fragment pairs allowing twists. *Bioinformatics, 19*(Suppl. 2), ii246–ii255.
Ye, Y., & Godzik, A. (2005). Multiple flexible structure alignment using partial order graphs. *Bioinformatics, 21,* 2362–2369.
Zaharik, M. L., Cullen, V. L., Fung, A. M., Libby, S. J., Kujat Choy, S. L., Coburn, B., et al. (2004). The Salmonella enterica serovar typhimurium divalent cation transport systems MntH and SitABCD are essential for virulence in an Nramp1G169 murine typhoid model. *Infection and Immunity, 72,* 5522–5525.

CHAPTER ELEVEN

Genetic and Biochemical Approaches for Studying the Yellow Stripe-Like Transporter Family in Plants

Sarah S. Conte and Elsbeth L. Walker[1]
Department of Biology, University of Massachusetts, Amherst, MA, USA
[1]Corresponding author: E-mail: ewalker@bio.umass.edu

Contents

Abstract

Iron (Fe) is essential for plants but can be toxic if over-accumulated. Members of the yellow stripe-like (YSL) family of metal transporters play important roles in plant Fe homeostasis, and a great deal of evidence has been gathered over many years that indicates the importance of YSLs in the long distance transport of metals complexed with nicotianamine (NA). This review examines our current knowledge of YSLs, gleaned from both genetic and biochemical approaches. Many unanswered questions remain regarding the substrate specificities of these transporters, which seem to vary widely depending on the individual transporter. Data are also just beginning to become available regarding YSLs in the most basal clade, which may be responsible for intracellular transport of metal–NA complexes. Future research on YSL transporters should focus on utilizing the proven techniques of yeast complementation and *Xenopus* oocyte electrophysiology to examine the substrate specificity of YSLs in greater detail.

Current Topics in Membranes, Volume 69
ISSN 1063-5823,
http://dx.doi.org/10.1016/B978-0-12-394390-3.00011-2

1. INTRODUCTION

Iron (Fe) is an essential element required for many cellular processes in both animals and plants. Fe redox enzymes are involved in essential metabolic pathways including the electron transport chains of respiration and photosynthesis, DNA, lipid and hormone biosynthesis, assimilation of nitrogen, and detoxification of reactive oxygen species (ROS). Because all these processes are physically isolated from one another inside the cell, it is necessary for Fe to be transported effectively between cellular compartments to reach the areas where it is needed. However, only a few transporters have been cloned and characterized that are involved in the intracellular transport of Fe.

Although Fe is the most abundant micronutrient in soil, it is often present in the insoluble Fe(III) oxidation state, which cannot be readily taken up by plant roots. In well-aerated soils, Fe(II) and Fe(III) are present in concentrations that do not typically exceed 10^{-15} M, although plants require concentrations between 10^{-4} and 10^{-9} M for optimal growth (Marschner, 1995). To combat the problem of Fe insolubility, plants have evolved two strategies for iron uptake from the soil. The grasses typically use Strategy II, which involves the synthesis and secretion of phytosiderophores (PS), chelator compounds that bind iron and increase its solubility in the rhizosphere. Iron bound to PS is then taken up at the root epidermis by the yellow stripe (YS) transporters: ZmYS1 in maize (Curie et al., 2001), HvYS1 in barley (Murata et al., 2006), OsYSL15 in rice (Inoue et al., 2009; Lee et al., 2009) and BdYS1 in *Brachypodium distachyon* (Yordem et al., 2011). Interestingly, although much is known about the functionality of the transporters that take up Fe–PS, very little is known about the transporters that function in the secretion of PS. This will be a key area for future research.

Dicotyledonous and non-grass monocotyledonous plants utilize Strategy I for primary iron uptake, which involves lowering the pH of the rhizosphere via proton extrusion, reducing solubilized Fe(III) via the action of a membrane-bound ferric reductase oxidase (FRO2), and uptake of reduced Fe by the ferrous iron transporter, IRT1 (Eide, Broderius, Fett, & Guerinot, 1996; Vert et al., 2002). The uptake of Fe into the plant body must be very tightly regulated, because although Fe is necessary for many cellular processes (in fact, plants are stunted and develop chlorosis if Fe is lacking), it is also toxic if over-accumulated. Fe(II) catalyzes the formation of ROS, including hydroxyl radicals, via the Fenton reaction (Winterbourn, 1995). These free radicals can damage cellular components and cause oxidative stress; thus, excess Fe must either be sequestered into vacuoles or chelated into non-Fenton-reactive forms.

A major *in planta* iron chelation molecule is citrate, which plays a key role in transport of Fe to the above-ground tissues. It is generally well accepted that Fe forms complexes with citrate in the xylem. Indeed, high-pressure liquid chromatography coupled to electrospray ionization time-of-flight mass spectrometry (HPLC-ESI-TOF-MS) and inductively coupled plasma mass spectrometry (HPLC-ICP-MS) were used to positively identify Fe–citrate in xylem as oxo-bridged tri-Fe(III) tri-citrate (Fe_3Cit_3) (Rellan-Alvarez et al., 2010). The complexation of Fe to citrate helps to maintain the solubility of Fe as well as to maintain Fe in a non-Fenton active form, thus preventing it from causing oxidative damage.

Upon arriving in the leaves, Fe must move from the apoplastic space to the symplast. Although citrate is known to be important for Fe movement throughout the xylem based on the studies of the *frd3* mutant (Durrett, Gassmann, & Rogers, 2007; Green & Rogers, 2004; Roschzttardtz, Seguela-Arnaud, Briat, Vert, & Curie, 2011), the role of citrate in xylem unloading is still unclear. The metal chelator nicotianamine (NA) occurs ubiquitously in higher plants and forms stable complexes in vitro with manganese (Mn), Fe(II), cobalt (Co), zinc (Zn), nickel (Ni) and copper (Cu), in increasing order of affinity (Anderegg & Ripperger, 1989). Studies have shown that NA can also bind Fe(III) in vitro with high affinity, and computer analyses predicted that the stability of metal–NA complexes is maximal at pH values above 6.5, while at pH 5.5, organic acids like citrate would be the main chelation molecules (von Wiren et al., 1999). Interestingly, when NA and deoxymugineic acid (DMA, which is a PS derived from NA) were mixed, Fe was predicted to complex DMA from pH 3.5 to 5.5, whereas above pH 5.5, NA was the major complexor of Fe. This indicates that ligand exchanges are likely to occur when Fe moves from apoplast to symplast. These predictions were later confirmed using electrospray ionization time-of-flight mass spectrometry (ESI-TOF-MS) (Rellan-Alvarez, Abadia, & Alvarez-Fernandez, 2008), indicating that NA is a good candidate for a symplastic chelator of metals.

Members of the yellow stripe-like (YSL) family of metal transporters are important players in plant metal homeostasis. A great deal of evidence has been gathered over many years that indicates the importance of YSLs in the long distance transport of metals complexed with NA. The YSL transporters, along with their founding member ZmYS1, belong to the oligopeptide transporter (OPT) family, members of which are involved in the transport of tri-, tetra-, penta- and hexapeptides and amino acid derivatives (Yen, Tseng, & Saier, 2001), and may play roles in plant growth and development (Lubkowitz, 2006). One OPT member in Arabidopsis, AtOPT6, can mediate the uptake of glutathione derivatives and metal complexes (Cagnac, Bourbouloux, Chakrabarty, Zhang, & Delrot, 2004).

2. FOUNDERS OF THE FAMILY: THE YS TRANSPORTERS

As mentioned previously, the YS transporters are responsible for primary uptake of Fe–PS complexes into plant roots. The most well characterized of these is ZmYS1 from *Zea mays*, named as such due to the mutant phenotype of yellow-striped leaves that results from the disruption of the *ZmYS1* gene. Early experiments showed that *ys1* mutants were defective in the uptake of Fe–PS (von Wiren, Mori, Marschner, & Romheld, 1994) and Zn–DMA (von Wiren, Marchner, & Romheld, 1996). The mutation was later mapped, and the YS1 gene was found to encode an Fe–PS transporter based on its ability to complement the yeast *fet3fet4* mutant, which is defective in both low- and high-affinity iron uptake, in the presence of Fe–DMA (Curie et al., 2001). *ZmYS1* is also upregulated by iron deficiency in roots and leaves (Curie et al., 2001; Roberts, Pierson, Panaviene, & Walker, 2004; Ueno, Yamaji, & Ma, 2009) and localizes to the distal side of epidermal cells of the crown and lateral roots (Ueno et al., 2009). In the leaves, ZmYS1 localizes to mesophyll but not epidermal cells, implicating it in intracellular transport of iron in maize (Ueno et al., 2009).

ZmYS1 has been extensively characterized on a biochemical level. Transport of Fe(III)–DMA is dependent on a proton gradient, and its substrate specificity has been studied using yeast complementation, yeast uptake assays and electrophysiological studies (Roberts et al., 2004; Schaaf et al., 2004). Schaaf et al. determined that the substrate specificity of ZmYS1 is quite broad, and it can transport Fe(II)–NA and Fe(III)–NA based on the complementation of *fet3fet4*, Zn(II)–DMA based on the complementation of the zinc-deficient yeast mutant *zap1*, and Ni(II)–DMA and Ni(II)–NA based on the uptake of ^{63}Ni–DMA into wild-type yeast. Additionally, Roberts et al. demonstrated that ZmYS1 transports Cu–mugineic acid (MA) based on the complementation of the copper uptake defective *ctr1* yeast strain (Roberts et al., 2004). Electrophysiological studies using *Xenopus laevis* oocytes provided information about the affinity of ZmYS1 for these various substrates. Currents induced by Ni(II)–DMA, Zn(II)–DMA and Cu(II)–DMA were smaller than those induced by Fe(III)–DMA, and metal–DMA complexes were found to be more favored than metal–NA complexes as substrates for ZmYS1 (Schaaf et al., 2004). Work by Roberts et al. added support to these findings, showing that 10-fold excesses of Ni–MA, Zn–MA, Cu–MA, Mn–MA and Cd–MA were unable to inhibit Fe–MA uptake in radioactive iron uptake assays (Roberts et al., 2004). Thus, despite its apparent broad substrate specificity, the primary function of ZmYS1 seems to be the transport of Fe–PS complexes.

In rice, there are 18 predicted YSL family members, and the member most closely related to ZmYS1 is OsYSL15 (Inoue et al., 2009). *OsYSL15* is upregulated in roots under iron deficiency (Inoue et al., 2009; Lee et al., 2009) and transports Fe(III)–DMA based on the *fet3fet4* yeast complementation (Inoue et al., 2009; Lee et al., 2009) and electrophysiology with *Xenopus* oocytes (Inoue et al., 2009). In the latter experiment, no currents were induced with Fe(II)–NA, Fe(III)–NA or Mn(II)–NA; however, Lee et al. found that OsYSL15 could transport Fe(II)–NA based on the *fet3fet4* complementation (Lee et al., 2009). Both groups showed that OsYSL15 localizes to the plasma membrane based on the bombardment of onion epidermal cells with OsYSL15–GFP (green fluorescent protein) constructs (Inoue et al., 2009; Lee et al., 2009). Similar to *ZmYS1*, *OsYSL15* is also upregulated in shoots during iron deficiency, implicating it in movement of metals throughout the plant body. Two *ysl15* mutants exhibited chlorosis under Fe deficiency and had decreased concentrations of Fe in shoots, roots and seeds. Additionally, overexpression of *OsYSL15* increased the Fe concentration in leaves and seeds (Lee et al., 2009).

The ZmYS1 homolog in barley, HvYS1, exhibits a more narrow substrate specificity than either ZmYS1 or OsYSL15. In electrophysiological experiments, current was not induced in *Xenopus* oocytes injected with *HvYS1* cRNA in the presence of Cu, Zn, Ni, Mn or Co complexed with MA. Similarly, current was not induced in the presence of Fe(II)–NA; however, a strong voltage-dependent current was observed in the presence of Fe(III)–MA (Murata et al., 2006). These authors also confirmed previous data indicating that ZmYS1 has a broad substrate specificity and can transport Cu(II)–MA, Zn(II)–MA and Fe(II)–NA, with smaller currents induced by Ni(II)–MA, Mn(II)–MA and Co(II)–MA. Yeast complementation data using *fet3fet4* confirmed the ability of HvYS1 to transport Fe(III)–MA.

HvYS1 is more strongly expressed in roots, where it is upregulated by Fe deficiency (Murata et al., 2006; Ueno et al., 2009) but not by deficiency of

Mn, Zn or Cu (Ueno et al., 2009). Expression in roots seems to peak prior to the onset of PS secretion, a phenomenon not observed for ZmYS1 (Ueno et al., 2009). Because PS is quickly degraded by soil microorganisms, a peak in expression prior to secretion should increase the efficiency of Fe–PS uptake. As expected, the HvYS1 protein localizes to the plasma membrane (Murata et al., 2006).

The model grass *B. distachyon* (brachypodium) contains 19 proteins that show strong similarity to ZmYS1, and the two that were most likely to be YS1 orthologs were designated as BdYS1A and BdYS1B (Yordem et al., 2011). However, *BdYS1B* contains a retained intron and a 400-bp duplication, is expressed at significantly lower levels compared to *BdYS1A*, and exhibits only slightly induced expression (2.7-fold) under Fe deficiency. Additionally, while BdYS1A was able to complement yeast *fet3fet4*, BdYS1B was not. Taken together, these data indicate that BdYS1B is likely to be a pseudogene. Further studies on BdYS1A must be undertaken in order to understand whether its substrate specificity is broad, as in the case of ZmYS1, or narrow, as in the case of HvYS1.

3. GENETIC STUDIES OF YSL GENES

3.1. Expression Analysis

Consistent with their function in the uptake of Fe(III)–PS under Fe-deficient conditions, it is not surprising that *ZmYS1*, *HvYS1*, *OsYSL15* and *BdYS1A* are strongly upregulated in response to low Fe (Curie et al., 2001; Koike et al., 2004; Murata et al., 2006; Ueno et al., 2009; Yordem et al., 2011). However, many of the YSLs are either downregulated by Fe deficiency or not regulated by Fe status (Table 1). Data for the rice YSLs vary in different studies, and there are consistent data available for only three genes: *OsYSL14*, which several groups have agreed is not regulated by Fe status (Inoue et al., 2009; Koike et al., 2004), *OsYSL15*, which is upregulated by low Fe (Inoue et al., 2009; Koike et al., 2004; Lee et al., 2009), and *OsYSL18*, which is not regulated by low Fe, Mn or Zn (Aoyama et al., 2009; Inoue et al., 2009). The tissue expression pattern of *OsYSL15*, as revealed by promoter–GUS (beta-glucuronidase) analysis, indicates that it may be involved not only in Fe uptake from the rhizosphere (as discussed previously) but also in phloem transport of Fe (Inoue et al., 2009). Tissue expression patterns of *OsYSL18* implicate it in translocation of iron in reproductive organs and phloem in leaf joints (Aoyama et al., 2009). Although the regulation of *OsYSL2* in the roots

Table 1 An overview of the YS and YSL transporters characterized thus far in various plant species

YSL	Gene regulation pattern	Subcellular localization	Evidence of transport function: Fe(II)–NA	Fe(III)–MA	Fe(III)–DMA	Other metal	References
Arabidopsis thaliana							
AtYSL1	UR in S by +Fe						Le Jean et al. (2005)
	DR in S by −Fe						Waters et al. (2006)
AtYSL2	DR in S and R by −Fe and −Zn	PM	YC				Chu et al. (2010)
							Schaaf et al. (2005)
	DR in R by −Fe	LPM	YC			YC (Cu–NA)	DiDonato et al. (2004)
	DR in S by −Fe or +Cu						
AtYSL3	DR in S by −Fe						Waters et al. (2006)
		LPM	YC		YC		Chu et al. (2010)
AtYSL4		Predicted TP					Jaquinod et al. (2007)
AtYSL6		Predicted TP					Jaquinod et al. (2007)
Brachypodium distachyon							
BdYS1A	UR in R by −Fe				YC		Yordem et al. (2011)
BdYS1B	UR in R by −Fe						Yordem et al. (2011)
BdYSL1	DR in S by −Fe						Yordem et al. (2011)
BdYSL2	DR in S by −Fe						Yordem et al. (2011)
BdYSL3	DR in S by −Fe						Yordem et al. (2011)
BdYSL6A	UR in R by −Fe						Yordem et al. (2011)
BdYSL6B	UR in R by −Fe						Yordem et al. (2011)

(Continued)

Table 1 An overview of the YS and YSL transporters characterized thus far in various plant species—cont'd

YSL	Gene regulation pattern	Subcellular localization	Evidence of transport function: Fe(II)–NA	Fe(III)–MA	Fe(III)–DMA	Other metal	References
BdYSL9	UR in R by −Fe						Yordem et al. (2011)
BdYSL13	DR in R by −Fe						Yordem et al. (2011)
Brassica juncea							
BjYSL2.1	NR by Fe or Zn status						Das et al. (2011)
BjYSL5.8	UR in R by −Zn or +Pb DR in R by +Cd						Das et al. (2011)
BjYSL6.1	UR in S by −Fe or +Cd DR in R by +Cd						Das et al. (2011)
Hordeum vulgare							
HvYS1	UR in R by −Fe UR in R by −Fe	PM		YC, XO			Murata et al. (2006) Ueno et al. (2009)
HvYSL2	UR in R by −Fe and −Zn				XO	XO (Zn–DMA) XO (Cu–DMA) XO (Ni–DMA) XO (Mn–DMA) XO (Co–DMA)	Araki et al. (2011)
HvYSL5	UR in R by −Fe	Ves or TP					Zheng et al. (2011)

			Oryza sativa			
OsYSL2	UR in L by −FE	PM	XO		XO (Mn−NA)	Koike et al. (2004)
	UR in S and R by −Fe					Lee et al. (2009)
	NR by Fe status in R					Inoue et al. (2009)
OsYSL5	NR by Fe status					Lee et al. (2009)
OsYSL6	DR in R and L by −Fe					Koike et al. (2004)
	NR by Fe status in R					Inoue et al. (2009)
	NR by Fe, Mn, Zn or Cu				RTA (Mn−NA)	Sasaki et al. (2011)
	NR by Fe status					Lee et al. (2009)
OsYSL8	DR in R by −Fe					Inoue et al. (2009)
OsYSL9	UR in S and R by −Fe					Lee et al. (2009)
	NR by Fe status in R					Inoue et al. (2009)
OsYSL12	NR by Fe status					Lee et al. (2009)
OsYSL13	DR in L by −Fe					Koike et al. (2004)
	NR by Fe status in R					Inoue et al. (2009)
	NR by Fe status					Lee et al. (2009)
OsYSL14	NR by Fe status					Koike et al. (2004)
	NR by Fe status in R					Inoue et al. (2009)
OsYSL15	UR in R by −Fe					Koike et al. (2004)
	UR in R by −Fe	PM		YC, XO		Inoue et al. (2009)
	UR in S and R by −Fe	PM	YC	YC		Lee et al. (2009)

(*Continued*)

Table 1 An overview of the YS and YSL transporters characterized thus far in various plant species—cont'd

YSL	Gene regulation pattern	Subcellular localization	Evidence of transport function: Fe(II)–NA	Fe(III)–MA	Fe(III)–DMA	Other metal	References
OsYSL16	UR in R by −Fe						Koike et al. (2004)
	UR in R stele by −Fe						Inoue et al. (2009)
	NR by Fe status						Lee et al. (2009)
	NR by metals	PM					Lee et al. (2012)
OsYSL18	NR by −Fe, −Mn or −Zn	PM			XO		Aoyama et al. (2009)
	NR by Fe status in R						Inoue et al. (2009)
Thlaspi caerulescens							
TcYSL3	NR by metals		YC, RTA			RTA (Ni–NA)	Gendre et al. (2007)
TcYSL5							Gendre et al. (2007)
TcYSL7							Gendre et al. (2007)
Zea mays							
ZmYS1	UR in R by −Fe	Distal side of epidermal root cells; leaf mesophyll cells	YC, XO	YC, XO	YC, XO	YC (Zn–DMA)	Curie et al. (2001)

UR in S by −Fe		XO (Zn−MA; Zn−DMA) YC (Cu−MA) XO (Cu−MA; Cu−DMA) RTA (Ni−NA; Ni−DMA) XO (Ni−MA; Ni−DMA) XO (Cd−DMA) XO (Mn−MA; Mn−DMA) XO (Co−MA) Co inhibits Fe−MA RTA	Roberts et al. (2004) Schaaf et al. (2004) Murata et al. (2006) Ueno et al. (2009)

Data regarding gene regulation pattern, subcellular localization, and evidence of Fe and/or other metal transport function are provided for each transporter. Proteins designated in boldface type have been experimentally determined to be YS1 orthologs responsible for primary iron uptake from the soil. DR, downregulated; UR, upregulated; NR, not regulated; R, roots; S, shoots; L, leaves; −Fe, iron-deficient conditions; −Mn, manganese-deficient conditions; −Zn, zinc-deficient conditions; +Fe, excess iron; +Cd, excess cadmium; +Pb, excess lead; YC, yeast complementation; RTA, radiolabeled transport assay; XO, currents induced in *Xenopus* oocytes; PM, plasma membrane; LPM, lateral plasma membrane; Ves, vesicles; TP, tonoplast.

is contested, with one group stating that *OsYSL2* is not regulated by Fe status (Inoue et al., 2009) and another group stating that it is upregulated in roots by low Fe (Lee et al., 2009), a third group was unable to detect OsYSL2 expression in roots via Northern blot but observed a very large upregulation in leaves under Fe-deficient conditions (Koike et al., 2004). The latter group used promoter–GUS analysis to show phloem-specific expression in Fe-sufficient roots, which was increased under Fe-deficient conditions. In leaves, GUS staining was observed in phloem cells of the vascular bundle in Fe-sufficient plants; this staining spread to all tissues under Fe-deficient conditions, although particularly strong staining was observed in companion cells. Recently, *OsYSL6* was found to be constitutively expressed in both shoots and roots, and expression was not regulated by deficiency of Zn, Fe, Mn or Cu. Based on promoter–GFP assays, OsYSL6 is expressed in all cells of shoots and roots (Sasaki, Yamaji, Xia, & Ma, 2011). This is quite different from the expression pattern of other YSLs, and could indicate that OsYSL6 has a distinct function.

Of the eight YSLs in Arabidopsis, only three have been well characterized. AtYSL2 is the most similar protein to ZmYS1, and the expression of *AtYSL2* is downregulated in roots under –Fe (DiDonato, Roberts, Sanderson, Eisley, & Walker, 2004; Schaaf et al., 2005) and –Zn (Schaaf et al., 2005) and downregulated in shoots under –Fe or excess Cu (DiDonato et al., 2004). *YSL2p:GUS* was expressed in xylem-associated cells and in leaf mesophyll, with broad expression observed throughout the root (DiDonato et al., 2004). *AtYSL1* was expressed in the xylem parenchyma of leaves, pollen and young siliques, and it was induced by Fe excess in shoots (Le Jean, Schikora, Mari, Briat, & Curie, 2005). Both *AtYSL1* and *AtYSL3* were downregulated in shoots under –Fe (Waters et al., 2006) and their expression patterns were similar in leaves, with GUS staining observed primarily in the veins and the cells surrounding the veins. Older leaves stained more intensely than younger leaves, and flower-specific expression was observed for both genes (Waters et al., 2006).

Metal hyperaccumulator plant species like *Arabidopsis halleri* and *Thlaspi caerulescens* (renamed *Noccaea caerulescens*) have proven to be useful tools for the study of metal tolerance and hyperaccumulation mechanisms, and recent work has highlighted the role of NA in these species. Both *A. halleri* and *N. caerulescens* exhibit elevated nicotianamine synthase (*NAS*) expression relative to nonaccumulators (Deinlein et al., 2012; Mari et al., 2006), and suppression of *AhNAS2* expression via RNA interference (RNAi) led to a decrease in the NA accumulation in roots, a decrease in root-to-shoot translocation of Zn and a decrease in leaf Cd accumulation (Deinlein et al., 2012). Earlier studies indicated that heterologous expression of *AhNAS2* in a zinc-sensitive *Schizosaccaromyces pombe* strain caused an increase in zinc tolerance (Weber, Harada, Vess, Roepenack-Lahaye, & Clemens, 2004), and overexpression of *TcNAS* in *Arabidopsis thaliana* conferred resistance to Ni (Pianelli, Mari, Marquès, Lebrun, & Czernic, 2005). Thus, NA is likely to play a role in metal tolerance and hyperaccumulation, and members of the YSL family may function in the transport of metal–NA chelates in hyperaccumulator species.

Three YSL genes have been isolated thus far from the metal hyperaccumulator *N. caerulescens*. *TcYSL7* is highly expressed in flowers, *TcYSL5* is highly expressed in shoots, and *TcYSL3* has broad expression throughout the plant body. The expression of *TcYSL7* and *TcYSL5* is associated with the vasculature in the shoots and the central cylinder in the roots, although exposure to Zn, Cd or Ni does not affect the expression of any of the *TcYSLs* (Gendre et al., 2007). In *Brassica juncea*, partial clones of 27 YSLs were obtained that were similar to AtYSLs and TcYSLs (Das et al., 2011). *BjYSL6.1* and *BjYSL5.8* showed elevated expression in cadmium-treated shoots and lead-treated roots, respectively. These data provide evidence that YSLs play roles in metal transport in hyperaccumulator species, and these YSLs may have a slightly broader substrate specificity than YSLs in non-hyperaccumulators like Arabidopsis.

Characterization of YSLs has also recently begun in brachypodium and maize. Yordem et al. (2011) have identified 18 YSLs in maize and 19 in brachypodium. All the brachypodium YSLs were found to be regulated by Fe status to varying extents, with *BdYSL1*, *2* and *3* downregulated in shoots under low Fe (similar to what was observed for Arabidopsis YSL1, 2 and 3), while *BdYSL13* was downregulated in roots under low Fe. As expected, the YS1 orthologs *BdYS1A* and *BdYS1B* were upregulated in roots under Fe deficiency, and *BdYSL6A*, *6B* and *9* were regulated similarly. Gene expression data remain to be obtained for the maize YSLs, and the brachypodium YSLs await further biochemical characterization.

3.2. Mutant Analysis and Complementation Studies

As mentioned previously, the *Yellow Stripe1* gene in maize was named as such due to the *ys1* mutant phenotype of interveinal chlorosis, manifesting as obvious "yellow stripes" on the leaves. Unfortunately, phenotypes of YSL knockouts are not always that obvious. A knockdown in the barley *HvYSL5* gene did not show any detectable phenotypes (Zheng et al., 2011). A similar phenomenon was observed for a T-DNA insertion line in *AtYSL2* (DiDonato et al., 2004). However, two independent Arabidopsis *ysl1* single mutants contained increased levels of NA in their shoots, and seeds of both mutants contained less iron and NA compared to wild-type seeds. Additionally, *ysl1* seeds germinated slowly on iron-deficient media (Le Jean et al., 2005). However, it was only when the double-mutant *ysl1ysl3* was created that the characteristic interveinal chlorosis phenotype emerged after 2 weeks of growth on MS agar plates or soil (Waters et al., 2006).

Many other interesting phenotypes were observed in *ysl1ysl3* double mutants. Leaf Fe concentration was decreased, while the concentrations of Mn, Zn and Cu were elevated. In seeds, the concentrations of Fe, Zn and Cu were low. These results, taken together with the increased expression of *AtYSL1* and *AtYSL3* during senescence in leaves, indicate a role for AtYSL1 and AtYSL3 in mobilization of metals from leaves during senescence. Another interesting phenotype was the reduced fertility of the double mutant due to defective anther and embryo development. When *ysl1ysl3* primary inflorescence stems were grafted onto wild-type rosettes, anther stunting was alleviated and pollen was released normally from the anthers of the grafted flowers (Chu et al., 2010). Additionally, seed set, seed weight and seed germination were restored to the levels observed in wild-type self-grafts. In contrast, the levels of Fe, Zn and Cu in seeds from grafted plants remained low. Thus, AtYSL1 and AtYSL3 are required in the rosettes to establish normal pollen and seed development, but are likely to be required in the inflorescence stems, cauline leaves, flowers and/or developing seeds in order for the seeds to contain normal metal levels. Interestingly, although *AtYSL1*, *2* and *3* are closely related and have similar patterns of expression, neither *AtYSL1* nor *AtYSL2* is able to completely functionally complement the reproductive defects of *ysl1ysl3* double-mutant plants when driven by the AtYSL3 promoter. This indicates that these genes have distinct functions *in planta*.

In rice, only a few YSL mutants have been studied thus far. A Tos-17 insertion line (NC6189; *osysl6*) that produces no *OsYSL6* transcript was identified that exhibited decreased growth in both roots and shoots under high Mn conditions. Although there was no difference in the total Mn concentration between wild-type rice and the knockout line, a higher Mn concentration was measured in the leaf apoplastic solution and a lower Mn concentration was measured in the symplastic solution of *osysl6* plants compared to wild type. This provides evidence that OsYSL6 is required for a net movement of Mn from the apoplast to the symplast. Variable phenotypes have been observed for *OsYSL15* mutant and knockdown plants. Silenced lines generated via RNAi (OsYSL15i plants) exhibited severe growth defects when grown on MS medium; these defects were corrected when OsYSL15i plants were grown on medium supplemented with 0.5 mM Fe(II)–citrate (Inoue et al., 2009). Growth was inversely correlated with expression of *OsYSL15*, with the least suppressed line showing the best growth and the most suppressed line showing the least growth. Different results were obtained with two independent T-DNA knockout alleles, *osysl15-1* and *osysl15-2*. These plants did not differ phenotypically from controls when grown on standard MS medium, but did show reductions in height, fresh weight and chlorophyll concentration when grown on Fe-deficient medium (Lee et al., 2009). Nitric oxide, which improves the internal availability of Fe, alleviated the chlorosis phenotype. No differences were observed when plants were grown on Zn-deficient medium. Fe concentrations were also decreased in the shoots and roots of knockout plants at the seedling stage, but Zn, Mn and Cu remained unchanged. Overexpression lines were also generated and found to have increased Fe concentrations in leaves and seeds. These lines flowered 10 days later than wild-type plants, had reduced plant height and fewer total spikelets. However, grain yields were not significantly different from wild-type plants, indicating that overexpression of *OsYSL15* could be useful in biofortification strategies.

3.3. Subcellular Localization Studies

Most of the YSL transporters that have been studied to date have been experimentally shown to localize to the plasma membrane of cells. This is true for the rice OsYSL2 (Koike et al., 2004), OsYSL16 (Lee, Ryoo, Jeon, Guerinot, & An, 2012) and OsYSL18 (Aoyama et al., 2009) as well as the YS1 homologs OsYSL15 (Inoue et al., 2009; Lee et al., 2009) and HvYS1 (Murata et al., 2006). Interestingly, while also plasma membrane localized, ZmYS1 exhibited a specific pattern of polar localization to the distal side of epidermal cells in crown and lateral roots (Ueno et al., 2009). In Arabidopsis, AtYSL1 and AtYSL3 localize to the plasma membrane based on the expression of 35S-YSL1–GFP in onion skin cells and expression of genomic YSL3 (including its native promoter) fused to GFP in Arabidopsis roots (Chu et al., 2010). Interestingly, both AtYSL3 and AtYSL2 were shown to localize almost exclusively to lateral plasma membranes and were rarely observed at the apical or basal end of cells. This points to a role for these proteins in the lateral movement of metals within veins, which is consistent with their gene expression patterns (Chu et al., 2010; DiDonato et al., 2004; Waters et al., 2006).

The most basal clade of the YSL family tree contains YSLs from the moss *Physcomitrella patens* and the lycophyte *Selaginella moellendorffii*, and also contains OsYSL5 and 6 from rice, AtYSL4 and 6 from Arabidopsis, ZmYSL6 from maize, and HvYSL5 from barley (Yordem et al., 2011; Zheng et al., 2011). Contrary to what was observed for other YSLs, a recent study indicated that HvYSL5 localizes either to vesicles or the tonoplast based on the bombardment of onion skin cells with the open reading frame (ORF) of *HvYSL5* fused to *smGFP*. However, OsYSL6 localization was inconclusive; bombardment experiments indicated that regardless of whether GFP was fused to the N- or C-terminus, the GFP signal appeared cytoplasmic (Sasaki et al., 2011). Unpublished data from our laboratory indicate that when AtYSL4 and AtYSL6 were tagged with GFP at the C-terminus, the GFP signal was again cytoplasmic, indicating that proteins in this group may be difficult to localize using conventional methodology. Interestingly, Jaquinod et al. (2007) identified AtYSL4 and AtYSL6 as members of the tonoplast proteome. Taken together, these localization data suggest that YSLs in the most basal clade may play roles in the intracellular transport of metal chelates.

4. BIOCHEMICAL STUDIES OF YSL GENES

4.1. Yeast Complementation

Saccharomyces cerevisiae mutants that are defective in uptake of various metals have been very useful tools for studying the transport functionality of YSLs. The *S. cerevisiae* mutant strain *fet3fet4* (DEY1453), which is defective in both high- and low-affinity iron uptake, was initially used to identify the major root iron transporter IRT1 (Eide et al., 1996). Typical *fet3fet4* complementation experiments involve omitting Fe from the yeast growth media and resupplying it as Fe(II)–NA, Fe(III)–MA or Fe(III)–DMA. If transporters that are capable of moving these substrates are located on the yeast plasma membrane, complementation will occur. Other strains that have been used similarly are *ctr1*, which is defective in copper uptake (Dancis, Haile, Yuan, & Klausner, 1994; Dancis, Yuan, et al., 1994), and *zap1*, which is defective in zinc uptake (Zhao & Eide, 1997).

Yeast complementation has been frequently used to test the functionality of Arabidopsis YSLs. AtYSL1 and AtYSL3 were able to complement *fet3fet4* in the presence of Fe(II)–NA (Chu et al., 2010), and AtYSL2 was able to complement *fet3fet4* in the presence of Fe(II)–NA and *ctr1* in the presence of Cu(II)–NA (DiDonato et al., 2004). Interestingly, AtYSL3 could also complement *fet3fet4* in the presence of Fe(III)–DMA (Chu et al., 2010). This was an unusual finding given that PS (including DMA) are not produced in Arabidopsis; however, it may indicate that AtYSL3 can transport Fe(III)–NA complexes, which are expected to occur in all plants (von Wiren et al., 1999). Further experiments must be carried out in order to truly understand the substrate specificity of AtYSL3.

As mentioned previously, three YSLs have been cloned in the hyper-accumulator *T. caerulescens* (TcYSL3, 5 and 7). Of these, only TcYSL3 was able to complement *fet3fet4* in the presence of Fe(II)–NA (Gendre et al., 2007). TcYSL3 was unable to complement Zn and Cu uptake mutants in the presence of Zn(II)–NA and Cu(II)–NA, but uptake experiments using reconstituted metal–NA complexes containing ^{63}Ni or ^{55}Fe indicated that TcYSL3 catalyzed the uptake of both Ni–NA and Fe–NA into yeast cells.

4.2. *Xenopus* Oocyte Electrophysiology

Xenopus laevis oocytes represent another heterologous system in which YSL transporter functionality has been studied. Using this system, information can be obtained about ion coupling, stoichiometry, substrate specificity, substrate affinity and sensitivity to inhibitors. All these attributes make the *Xenopus* oocyte an attractive system in which to study YSLs, but electro-physiological experiments require specialized tools, equipment and expertise in order to be performed correctly. Once a method is successfully developed, a great deal of data can be obtained for the transporter in question by bathing oocytes in different metal chelate solutions and/or adding various inhibitors.

As mentioned earlier, ZmYS1 has been successfully studied using electrophysiological methods (Schaaf et al., 2004). HvYS1 and OsYSL15 have also been shown to transport Fe(III)–MA (Murata et al., 2006) and Fe(III)–DMA (Inoue et al., 2009), respectively, using electrophysiology. Recently, Araki, Murata, and Murata (2011) used electrophysiological methods to study the substrate specificity of HvYSL2. In addition to Fe(III)–DMA, HvYSL2 can transport Zn(II)–DMA, Ni(II)–DMA, Cu(II)–DMA, Mn(II)–DMA and Co(II)–DMA at fairly similar levels, which makes it functionally more similar to ZmYS1 than HvYS1, the latter of which has a narrow substrate specificity (Harada, Sugase, Namba, Iwashita, & Murata, 2007; Murata et al., 2006).

In rice, electrophysiological experiments were used to examine OsYSL2 and OsYSL18. It was observed that OsYSL2 can transport Fe(II)–NA and Mn(II)–NA, but not Fe(III)–DMA or Mn(II)–DMA, among various other metals tested (Koike et al., 2004). Thus, OsYSL2 maintains a fairly narrow substrate specificity and is likely to function in phloem transport of Fe and Mn. On the other hand, OsYSL18 can transport Fe(III)–DMA but not Fe(II)–NA, Zn(II)–DMA or Zn(II)–NA (Aoyama et al., 2009). Although PS have historically been known to be responsible for Fe acquisition from the rhizosphere, PS have been detected in the shoots of barley and rice (Higuchi et al., 2001), DMA was detected in rice phloem sap (Mori et al., 1991), and the genes involved in DMA biosynthesis in rice were co-expressed in the phloem companion cells of roots, leaves and reproductive organs (Bashir et al., 2006; Inoue et al., 2003, 2008). Because OsYSL18 is expressed in phloem parenchyma, companion cells, flowers and developing seeds, it may be involved in long distance transport of Fe(III)–DMA via phloem, and thus Fe(III)–DMA may also contribute to seed Fe loading in rice.

4.3. Relating YSL Protein Structure to Function

Despite their high degree of homology, members of the YSL family have rather variable substrate specificities. As mentioned previously, ZmYS1 has a broad substrate specificity and is capable of transporting Fe(III), Fe(II), Ni(II), Zn(II), Cu(II), Mn(II) and Cd(II) complexed with PS, in addition to Fe(II)–NA and Ni(II)–NA (Roberts et al., 2004; Schaaf et al., 2004). However, HvYS1, despite having 72.7% identity and 95.0% similarity to ZmYS1, specifically transports Fe(III)–PS complexes (Murata et al., 2006). Although the substrate specificity of the YSL family has not yet been fully explored, biochemical experiments have shown that several members of clade 1 are capable of transporting other metals in addition to Fe. These include AtYSL2, which transports Cu(II)–NA, HvYSL2, which transports Zn, Cu, Ni, Mn and Co complexed with DMA, OsYSL2, which transports Mn(II)–NA, and TcYSL3, which transports Ni(II)–NA (Table 1). OsYSL6, which is a member of clade 2, transports Mn(II)–NA (Table 1). Expression data indicate that YSLs from *B. juncea* are regulated by Cd and lead (Pb) in addition to Fe (Table 1), although biochemical data from yeast complementation, yeast uptake and/or *Xenopus* oocyte experiments will be required to confirm that these metals are indeed being specifically transported by these YSLs.

Structural modeling of the YSL family has not generated a consistent consensus topology for the family. SOSUI (Hirokawa, Boon-Chieng, & Mitaku, 1998), which uses average hydrophobicity and average amphiphilicity of helices to arrive at a predicted topology, predicts 12 transmembrane (TM) domains for ZmYS1 and HvYS1 (Harada et al., 2007). However, for other members of the YSL family, SOSUI predicts anywhere between 11 and 16 TM domains, calling into question the validity of some of the models. The TOPCONS software package (Hennerdal & Elofsson, 2011) predicts a topology for ZmYS1 and HvYS1 with 15 TM domains; this represents a consensus topology from models generated using Hidden Markov Models (S-THMM and HMMTOP; Tusnády & Simon, 2001; Viklund & Elofsson, 2004), model recognition approaches (Jones, Taylor, & Thornton, 1994) and prediction from first principles (Bernsel et al., 2008). A comparison of SOSUI and TOPCONS topology predictions for ZmYS1 is shown in Fig. 1. As with SOSUI, TOPCONS predictions for other family members range from 12 to 16 TM domains. Examination of the predicted topologies for all YSL family members from Arabidopsis, brachypodium, rice, and maize shows that the amino termini of the proteins are predicted to be intracellular. The two exceptions are the OsYSL17 and ZmYSL6 proteins, which are predicted to have extracellular N-termini. Thus, the true topology for the YSL family remains unknown, and awaits experimental investigation.

Two regions in YSL proteins appear to be hypervariable, and could account for the substrate specificities of particular YSL family members. The first is the outer membrane loop predicted to lie between the sixth and seventh TM segment according to the SOSUI model for HvYS1 and ZmYS1. To identify the region responsible for Fe(III)–PS specificity in HvYS1 relative to ZmYS1, Harada et al. (2007) constructed HvYS1–ZmYS1 chimeric proteins by exchanging this region in the two proteins. Expression of chimeras in *Xenopus* oocytes revealed that residues 314–385 of HvYS1 were responsible for the selectivity of HvYS1 for Fe(III)–DMA vs. Fe(II)–NA. Further experiments using circular dichroism spectra indicated that the region forms an alpha helix in HvYS1, while the corresponding loop in ZmYS1 is flexible (Harada et al., 2007). Alpha helix percentage predictions were made for the corresponding regions of AtYSL2, OsYSL2 and TcYSL3. HvYS1 was found to have a much higher alpha helix percentage (32.1) compared to the other transporters (1.9–5.1), but further experimental studies are warranted in order to confirm the correlation between helicity of this loop and substrate specificity. It is also interesting to note that the loops between TM domains contain a higher number of charged residues compared to other regions. Analysis of additional YSL family members from rice, maize, brachypodium and Arabidopsis indicates that, within each clade, some sequence conservation exists in this region (Fig. 2). This conservation between multiple species is likely to imply some conservation of function beyond the simple capacity to form ordered or unordered domains. Thus, another possibility is that charged residues in this region may participate in substrate recognition of trivalent and divalent metal–chelator complexes, or in interactions with other proteins.

The second hypervariable region occurs at the N-termini of the YSL proteins. An abundance of glutamic acid residues was noted in ZmYS1 and the Arabidopsis YSLs when ZmYS1 was first identified (Curie et al., 2001). Also, the sequence REKELELELER was noted as being

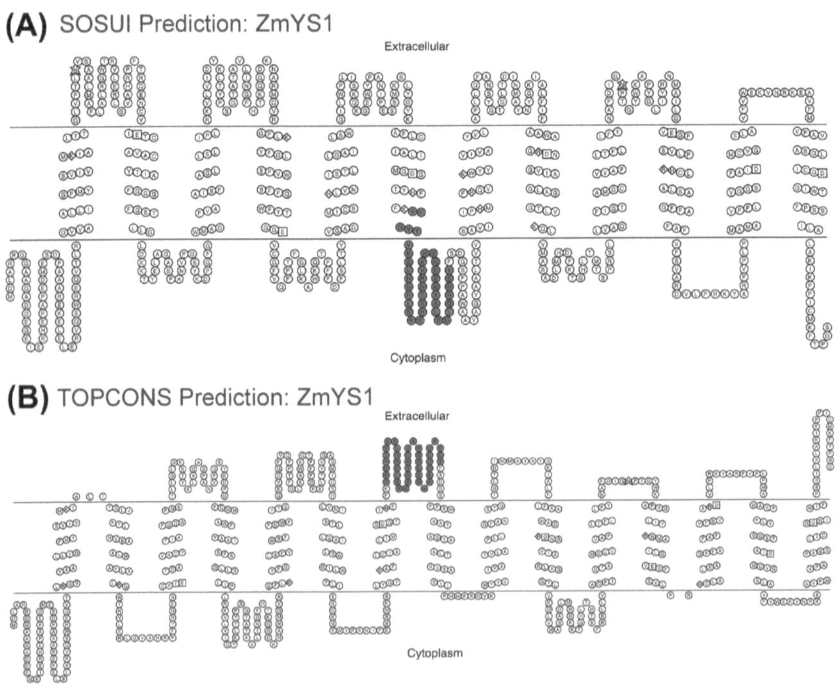

Figure 1 Proposed topologies for ZmYS1. (A) SOSUI prediction of ZmYS1 topology. (B) TOPCONS prediction of ZmYS1 topology. The variable loop domain is shaded. For color version of this figure, the reader is referred to the online version of this book.

(A) YS1 orthologs

```
HvYS1      KIVGITAMSMYRQFSHKQVNNK-AKNADDTVSLEELHRQEIF
BdYS1A     KIIGITSKSMYRQFNHKRVENRGVRNVDNTIALDELQRVEVF
ZmYS1      KVFGVTVKSLHQRLSRKRATNRVANGGDEMAALDDLQRDEIF
OsYSL15    KVTGVTAKSLHNRFNRKSVSNTASEEGD-MVSLDDLQRDEVF
           *: *:*  *::.::.:* . *   .  *   :*::*:* *:*
```

(B) Clade 1

```
BdYSL1     KIIAFTVKSLLGRSKLKNTK--------KEED-IPVLDDLHRN-EVF
ZmYSL1     KVVAFTVKSLLGRSRLKNAK--------KEED-IPVLDEIQRN-EIF
OsYSL9     KIVALTIKNLFDSSKLKNAK--------KGED-MPVLDELHRN-EVF
BdYSL9     KIIVSTTMNLLDKSKLKNTK--------KEED-ILPLDELHRN-EVF
OsYSL16    KVIVVTLKSVRERSRRRGLNN-----RVADAD-TMAIDDMQRN-EVF
BdYSL3     KVIIITLKSMHERSKRGRLN------RVADAD-SVSIDDRQRD-EVF
ZmYSL3     KVIAITVKNIRERSRRKNLN------RVADAD-TMALDDMQRD-EVF
OsYSL2     KVTGITAKSLHERSNRRHAK------KATDED-TFVIADMQRD-EFF
BdYSL2     KVTGITAKSLHGRSNRKHVK------RAANEG-TVAIDDMQRD-EVF
ZmYSL2     KVLGVTAKSLHERSKLRRSNNRGSSSRVADEDNAAAIDSLRRD-EVF
AT2        KILFFTGRSFHSRLSKTNSIS---TLVEVPEDSTKESDNLKRENEVF
AT3        KILFKTGINMYVKLNNRNSGK---SNSEKDKQSIAD---LKRD-EIF
AT1        KILFVTIANVNARLKNKPND-------LDDVGHKKQRKDLKED-ENF
           *:   *  ..                              :.: * *
```

(C) Clade 2 (basal clade)

```
Pp111567   KIIFKSARQLYVHYKTNNQLPVTAPGRH---HLHNQSSSTNHARRTEIF
Pp122023   VILYKSGKELHAHYASHNQLPVSNAGAHPTLPCKFTLISSDHARRSEIF
Sm179245   KVVIVSVWFMYKHYKTD-LLPITAGADS-----RDGVLTTDEKKRNETF
OsYSL5     KVFLIIAKEICNARSKEHDLPVQALLQDDD---SSRQLLDE-KRQTEIF
BdYSL4     RIFFVIAREIWYVQPKRSNHPVQALQ-DNK---GSKQLMDE-KLQTEIF
BdYSL6A    KIIYATIKEIVNARSKQGRLPLVQ-VQDDD---EGSKLSAEEKFLNEVF
BdYSL6B    KIIYATIKEIMNARSKQGKLPLVC-VLYDD---EGSKVSAEEKLLNEVF
OsYSL6     KIIYATIKEVMNARSKQGRLPLVR-VHDDD---EGSKLSAEEKLRNDTF
ZmYSL6     KIIYATIKEIMNARAKQGRLPLAQ-VQDGD---EGSNLSAEEKHLNETF
AT4        KIIVVTVKEICNKSSRQHNLPVFTDILDKS---KTSVLMREKKKRDIIF
AT6        KIIAVTVKELCSSRSRRLNLPIVTDGVDDS---EASEILLVKKKRDEVF
Sm269976   KIAVISLRAAMIHKAGGDDATGVP------------LLDLDRDKLTETF
            :                  .                            *
```

Figure 2 Variable loop domain. This domain was defined as the sequence bordered by a highly conserved lysine and a perfectly conserved phenylalanine. Each YSL subgroup was analyzed separately using CLUSTALW2 in order to highlight regions of the variable domain that are conserved among proteins in each subgroup. The analysis included all known YSLs from Arabidopsis (AT), maize (Zm), *Brachypodium distachyon* (Bd), rice (Os), the moss *Physcomitrella patens* (Pp) and the lycophyte *Selligenella mullendorfii*. An "*" (asterisk) indicates positions which have a single, fully conserved residue; a ":" (colon) indicates conservation between groups of strongly similar properties (scoring >0.5 in the Gonnet PAM 250 matrix); a "." (period) indicates conservation between groups of weakly similar properties (scoring ≤ 0.5 in the Gonnet PAM 250 matrix). (A) Fe(III)–PS transporters experimentally determined to be orthologous to ZmYS1. (B) Clade 1 YSL proteins. This subgroup is found in both grasses and non-grasses and includes the YS1 orthologs shown separately in (A). (C) Clade 2. This subgroup is the most basal of the family. (D) Clade 3. This subgroup is found in grasses and non-grasses (Yordem et al., 2011). (E) Clade 4. This clade is specific to the grasses, and does not contain non-grass members *(Yordem et al., 2011).*

(D) Clade 3

```
OsYSL10     KVMTRTTTALVMQVR---AMMSEPTLPVSGGGGQ-------TPEETFDDKRRTELF
ZmYSL5      KVMTRTVTALVVQVR---RMMSEPTLPISGGGGEGGGIPLPAPEETFDDRRRTELF
BdYSL5      KVMTKTMIALAAQVR---FMLSGPSLPISNADDQD-----QTPAETFDDQRRKEVF
AT5         KVLSRTLSGLFVQLRGPTTSISRTSFTLEEDP----HASPLSPKQSYDDQRRTRFF
AT8         KVLIRTFSGLISQIRGKAG--SRSSLAHKEDPPAS-PASPLTPRISYDDQRRTRFF
AT7         KVLGRTVFGLYKQFK------NKDVLPINDHT----STAPVT--ISYDDKRRTELF
OsYSL11     KVFGYTIKGFIVMYK---NKNSN-TLPISDNGTPA----NATEEESFDDKRRNELF
BdYSL11     KVFGKTIIALIEMHK---KKNSK-ALPVSDDGTPG----ATVEAESFDDKRRTELF
ZmYSL11     KVFGHTIGALISMYW---KKHAN-RLPVSDNMTPP----ATVETESFDDKRRIELF
ZmYSL11A    KVLYSTIEAFVSRYR---NRKAN-TLPVSNDGTPAD---ATTETESFDDKRRLELF
BdYSL10     KVTLRMVRAFMAASR---KR----SALVSGDGGLA-------LVRSFDEKRRAEFF
OsYSL12     KVLIRKTAGFVVMMK---KNS---TLPVSNNGSPM----VATEAISFDDERRTELF
ZmYSL12     KVLIRTTAGFISMMK---KNS---TLPVSNNGSP------ITEGVSFDDERRRDLF
BdYSL12     KVLVRTTAAFATMVK---KNN---MIPVSDEGGAM----VTGEAVSYDDQRRTELF
OsYSL14     KVLIRTIAGFISMVQ---NNSKG-MLPVSDNGMSM----STAEEVSFDDERRTEIF
BdYSL14     KVLIRTIAGFISMVQ---KNSKA-MLPVSDNGR------PTSEAVSFDDERRTELF
ZmYSL14     KVLIRTIAGFISMVQ---QNSKG-MLPVSDNGSSM----SATEALSFDDERRTEIF
ZmYSL14A    KVLIRTIAGFISMVQ---QNSKS-MLPVSDNGSSM----SSMEAVSFDDERRIENF
OsYSL13     KMIILTAFSLRSQIK---KKNAS-TLPVSDDGMV-----TTTAAVSYDEERRNELF
ZmYSL13     KIFILTVVSLRSQVR---KANAS-TLPISDDGTV-----TSTAPTSYDEERRNEHF
BdYSL13     KMLFLTVRSLRSQLK---KSNAS-TLPVSDDETS-----NSNEAISYDEERRNELF
ZmYSL10     TVLLRTVASIAASTE---KRKFFGALPVNGDGSNGAA--APAPPPSFDDARRTEFF
OsYSL17     SALVRTACAMRKRRR----GASTVTAAVPPFQCLS---ATERTMQSFDDRRRAQVF
            .        .:                                  ::*: **   *
```

(E) Grass specific clade

```
OsYSL1     KVPMGVSMVLGDCLFQLGAITVKAVQHYRKGRQEQK-----------LAVDGAADDGGGGCVPDDDDENKWHATYDERRRNQVF
BdYSL15    KVPMGIAMVLGDCLFQLGAITIRAANHFHKNRQEQK-----------LSGASMPDDR--------NSDEQTSLSYDDCRRSKIF
OsYSL18    KVFCAIAMILGDGIFQLVAISLRTIHTVRHHQV--------------------AAETLRSFS-DVDAMPRPVMSFDDRRRTQVF
ZmYSL18    KVFSAIGMILGDGILQLVVISLKTVHTMRHHQL--------------------AAETIRSFS-NVDSMPRPVLSFDDRRRTQVF
BdYSL18    KVFGAIGMILGDGIFQLVVISVKTMNTMRQHQLE-------------------AAESMRSFSANIDAPPRPVLSFDDRRRTQVF
OsYSL7     KVFIGLSMIMADCLFVFLCIMVRTTCAMIKRR-RQ---AMQGGGG-----------NAQPFQGIDIAD-QPVKSFDDRRRAQVF
ZmYSL17    KVFIGVSMILADGLFNFLSIVAYTLHTMRKRRSRQQPIIIQGGGCGGGSEADDGDTQPLPFHCLNAAAEQPKKSFDDRRRAQVF
OsYSL8     QVFISIAMIVGDGLFNFFSIVLRTAYDLYLKRRGG-------------ASKQPQETPFAGA----TGTERQVLSFDDRRRTQVF
BdYSL17    QVFISIAMIVGDGLFNFFSILLRTSYDMYLKRTGR-------------SKASP--MPFAGAGVAINATERQALSFDDRRRTQIF
OsYSL3     K-------------------------------------------------------------------KHPSLNYDDRKRIEVF
OsYSL4     KVFISVTLIVTDGLINFLILVTSAAINFYHIRQQQQ----------------------QTSGLASYISKNPSMNYDERKRIEMF
BdYSL16    KIFITVALIVTDGLYTLIMLLITAYNDYRLKRQE------------------------TDSVVANYLKRHPSLNYDDRKRIEVF
           :                                                                    .   .:*: :* ::*
```

Figure 2 (*continued*).

reminiscent of the REGLE sequence involved in Fe(III) transport (Ramanan & Wang, 2000) and iron binding by the ferritin light chain (Chasteen & Harrison, 1999).

Further examination of the YSL family in maize, rice, brachypodium and Arabidopsis shows that while members from clades I and III have acidic N-termini (Fig. 3B and D, boxes), members from clade II, which is the most basal clade for the family, generally lack this feature (Fig. 3C). Notably, the YS1 orthologous group shares moderate sequence conservation at the amino terminus, but the REKELELELER sequence is absent in the rice, barley and brachypodium proteins. YSLs in the grass-specific clade noted by (Yordem et al. 2011) have moderately acidic N-termini that are generally small compared to the N-terminal domains of other YSL groups. However, this clade shares some sequence conservation at the N-termini (Fig. 3E). The significance of this is not currently understood.

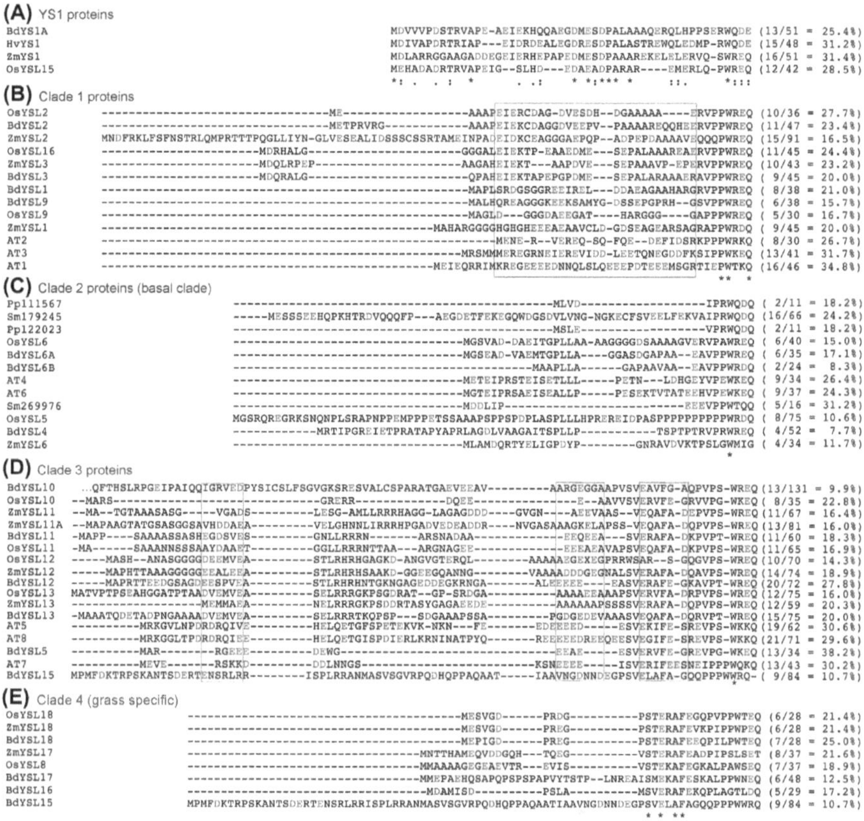

Figure 3 Amino termini of YSL proteins. YSL family members were considered as subgroups according to the phylogenetic tree generated by Yordem et al. (2011). Alignments were made using CLUSTALW2 (Larkin et al., 2007). The analysis included all known YSLs from Arabidopsis (AT), maize (Zm), *Brachypodium distachyon* (Bd), rice (Os), the moss *Physcomitrella patens* (Pp) and the lycophyte *Selligenella mullendorfii*. An "*" (asterisk) indicates positions that have a single, fully conserved residue; a ":" (colon) indicates conservation between groups of strongly similar properties (scoring >0.5 in the Gonnet PAM 250 matrix); a "." (period) indicates conservation between groups of weakly similar properties (scoring ≤0.5 in the Gonnet PAM 250 matrix). Boxes are drawn around regions with conserved acidic amino acid composition. (A) Fe(III)–PS transporters experimentally determined to be orthologous to ZmYS1. (B) Clade 1 YSL proteins. This subgroup is found in both grasses and non-grasses and includes the YS1 orthologs shown separately in (A). (C) Clade 2. This subgroup is the most basal of the family. (D) Clade 3. This subgroup is found in grasses and non-grasses (Yordem et al., 2011). (E) Clade 4. This clade is specific to the grasses, and does not contain non-grass members (Yordem et al., 2011). For color version of this figure, the reader is referred to the online version of this book.

5. CONCLUSIONS AND FUTURE PERSPECTIVES

A synergy of genetic and biochemical approaches must be used in order to fully understand the function of YS1 homologs and members of the YSL family in various plant species. Although quite a bit of data are available regarding YSL gene regulation in various plant species, data on rice YSL gene expression conflict between different research groups. These conflicting data could be due to the differential methodologies employed, differences in the ages of the plants, or other factors. Localization data are available for many YSLs, but localization data for YSLs in the most basal clade are either unavailable or inconclusive. The fusion of GFP to a membrane protein that localizes to an intracellular compartment may place GFP in an environment in which the fluorescence is not easily observable due to alterations in pH. This is just one potential explanation for why OsYSL6, AtYSL4 and AtYSL6 have been difficult to localize and why HvYSL5 localization is somewhat unclear. Major unanswered questions remain regarding the substrate specificity of YSL transporters. Because this specificity seems to vary widely, it is important that transporters be studied on a case-by-case basis. Yeast complementation and *Xenopus* oocyte electrophysiology are proven techniques that can provide a great deal of information regarding substrate preferences of YSLs. These techniques will be critical for our further understanding of YSLs, especially for YSLs in brachypodium and *B. juncea*, for which little data are currently available.

REFERENCES

Anderegg, G., & Ripperger, H. (1989). Correlation between metal complex formation and biological activity of nicotianamine analogues. *Journal of the Chemical Society. Chemical Communications, 10*, 647–650.

Aoyama, T., Kobayashi, T., Takahashi, M., Nagasaka, S., Usuda, K., Kakei, Y., et al. (2009). OsYSL18 is a rice iron(III)-deoxymugineic acid transporter specifically expressed in reproductive organs and phloem of lamina joints. *Plant Molecular Biology, 70*(6), 681–692.

Araki, R., Murata, J., & Murata, Y. (2011). A novel barley yellow stripe 1-like transporter (HvYSL2) localized to the root endodermis transports metal-phytosiderophore complexes. *Plant and Cell Physiology, 52*(11), 1931–1940.

Bashir, K., Inoue, H., Nagasaka, S., Takahashi, M., Nakanishi, H., Mori, S., et al. (2006). Cloning and characterization of deoxymugineic acid synthase genes from graminaceous plants. *The Journal of Biological Chemistry, 281*(43), 32395–32402.

Bernsel, A., Viklund, H., Falk, J., Lindahl, E., von Heijne, G., & Elofsson, A. (2008). Prediction of membrane-protein topology from first principles. *Proceedings of the National Academy of the Sciences of United States of America, 105*(20), 7177–7181.

Cagnac, O., Bourbouloux, A., Chakrabarty, D., Zhang, M. Y., & Delrot, S. (2004). AtOPT6 transports glutathione derivatives and is induced by primisulfuron. *Plant Physiology, 135*(3), 1378–1387.

Chasteen, N. D., & Harrison, P. M. (1999). Mineralization in ferritin: an efficient means of iron storage. *Journal of Structural Biology, 126*(3), 182–194.

Chu, H. H., Chiecko, J., Punshon, T., Lanzirotti, A., Lahner, B., Salt, D. E., et al. (2010). Successful reproduction requires the function of Arabidopsis YELLOW STRIPE-LIKE1 and YELLOW STRIPE-LIKE3 metal-nicotianamine transporters in both vegetative and reproductive structures. *Plant Physiology, 154*(1), 197–210.

Curie, C., Panaviene, Z., Loulergue, C., Dellaporta, S. L., Briat, J. F., & Walker, E. L. (2001). Maize *yellow stripe1* encodes a membrane protein directly involved in Fe(III) uptake. *Nature, 409*, 346–349.

Dancis, A., Haile, D., Yuan, D. S., & Klausner, R. D. (1994). The Saccharomyces cerevisiae copper transport protein (Ctr1p). Biochemical characterization, regulation by copper, and physiologic role in copper uptake. *The Journal of Biological Chemistry, 269*(41), 25660–25667.

Dancis, A., Yuan, D. S., Haile, D., Askwith, C., Eide, D., Moehle, C., et al. (1994). Molecular characterization of a copper transport protein in S. cerevisiae: an unexpected role for copper in iron transport. *Cell, 76*(2), 393–402.

Das, S., Sen, M., Saha, C., Chakraborty, D., Das, A., Banerjee, M., et al. (2011). Isolation and expression analysis of partial sequences of heavy metal transporters from Brassica juncea by coupling high throughput cloning with a molecular fingerprinting technique. *Planta, 234*, 139–156.

Deinlein, U., Weber, M., Schmidt, H., Rensch, S., Trampczynska, A., Hansen, T. H., et al. (2012). Elevated nicotianamine levels in Arabidopsis halleri roots play a key role in zinc hyperaccumulation. *The Plant Cell, 24*(2), 708–723.

DiDonato, R. J., Jr., Roberts, L. A., Sanderson, T., Eisley, R. B., & Walker, E. L. (2004). Arabidopsis yellow stripe-like2 (YSL2): a metal-regulated gene encoding a plasma membrane transporter of nicotianamine-metal complexes. *The Plant Journal, 39*(3), 403–414.

Durrett, T. P., Gassmann, W., & Rogers, E. E. (2007). The FRD3-mediated efflux of citrate into the root vasculature is necessary for efficient iron translocation. *Plant Physiology, 144*(1), 197–205.

Eide, D., Broderius, M., Fett, J., & Guerinot, M. L. (1996). A novel iron-regulated metal transporter from plants identified by functional expression in yeast. *Proceedings of the National Academy of the Sciences of United States of America, 93*, 5624–5628.

Gendre, D., Czernic, P., Conejero, G., Pianelli, K., Briat, J. F., Lebrun, M., et al. (2007). TcYSL3, a member of the YSL gene family from the hyper-accumulator Thlaspi caerulescens, encodes a nicotianamine-Ni/Fe transporter. *The Plant Journal, 49*(1), 1–15.

Green, L. S., & Rogers, E. E. (2004). FRD3 controls iron localization in Arabidopsis. *Plant Physiology, 136*(1), 2523–2531.

Harada, E., Sugase, K., Namba, K., Iwashita, T., & Murata, Y. (2007). Structural element responsible for the Fe(III)-phytosiderophore specific transport by HvYS1 transporter in barley. *FEBS Letters, 581*(22), 4298–4302.

Hennerdal, A., & Elofsson, A. (2011). Rapid membrane protein topology prediction. *Bioinformatics, 27*(9), 1322–1323.

Higuchi, K., Watanabe, S., Takahashi, M., Kawasaki, S., Nakanishi, H., Nishizawa, N. K., et al. (2001). Nicotianamine synthase gene expression differs in barley and rice under Fe-deficient conditions. *The Plant Journal, 25*(2), 159–167.

Hirokawa, T., Boon-Chieng, S., & Mitaku, S. (1998). SOSUI: classification and secondary structure prediction system for membrane proteins. *Bioinformatics, 14*, 378–379.

Inoue, H., Higuchi, K., Takahashi, M., Nakanishi, H., Mori, S., & Nishizawa, N. K. (2003). Three rice nicotianamine synthase genes, OsNAS1, OsNAS2, and OsNAS3 are expressed in cells involved in long-distance transport of iron and differentially regulated by iron. *The Plant Journal, 36*(3), 366–381.

Inoue, H., Kobayashi, T., Nozoye, T., Takahashi, M., Kakei, Y., Suzuki, K., et al. (2009). Rice OsYSL15 is an iron-regulated iron(III)-deoxymugineic acid transporter expressed in the roots and is essential for iron uptake in early growth of the seedlings. *The Journal of Biological Chemistry, 284*(6), 3470–3479.

Inoue, H., Takahashi, M., Kobayashi, T., Suzuki, M., Nakanishi, H., Mori, S., et al. (2008). Identification and localisation of the rice nicotianamine aminotransferase gene OsNAAT1 expression suggests the site of phytosiderophore synthesis in rice. *Plant Molecular Biology, 66*(1–2), 193–203.

Jaquinod, M., Villiers, F., Kieffer-Jaquinod, S., Hugouvieux, V., Bruley, C., Garin, J., et al. (2007). A proteomics dissection of Arabidopsis thaliana vacuoles isolated from cell culture. *Molecular and Cellular Proteomics, 6*(3), 394–412.

Jones, D. T., Taylor, W. R., & Thornton, J. M. (1994). A model recognition approach to the prediction of all-helical membrane protein structure and topology. *Biochemistry, 33*(10), 3038–3049.

Koike, S., Inoue, H., Mizuno, D., Takahashi, M., Nakanishi, H., Mori, S., et al. (2004). OsYSL2 is a rice metal-nicotianamine transporter that is regulated by iron and expressed in the phloem. *The Plant Journal, 39*(3), 415–424.

Larkin, M. A., Blackshields, G., Brown, N. P., Chenna, R., McGettigan, P. A., McWilliam, H., et al. (2007). Clustal W and Clustal X version 2.0. *Bioinformatics, 23*(21), 2947–2948.

Le Jean, M., Schikora, A., Mari, S., Briat, J. F., & Curie, C. (2005). A loss-of-function mutation in AtYSL1 reveals its role in iron and nicotianamine seed loading. *The Plant Journal, 44*(5), 769–782.

Lee, S., Chiecko, J. C., Kim, S. A., Walker, E. L., Lee, Y., Guerinot, M. L., et al. (2009). Disruption of OsYSL15 leads to iron inefficiency in rice plants. *Plant Physiology, 150*(2), 786–800.

Lee, S., Ryoo, N., Jeon, J. S., Guerinot, M. L., & An, G. (2012). Activation of rice yellow stripe1-like 16 (OsYSL16) enhances iron efficiency. *Molecular Cell, 33*(2), 117–126.

Lubkowitz, M. (2006). The OPT family functions in long-distance peptide and metal transport in plants. *Genetic Engineering (N Y), 27*, 35–55.

Mari, S., Gendre, D., Pianelli, K., Ouerdane, L., Lobinski, R., Briat, J. F., et al. (2006). Root-to-shoot long-distance circulation of nicotianamine and nicotianamine-nickel chelates in the metal hyperaccumulator Thlaspi caerulescens. *Journal of Experimental Botany, 57*(15), 4111–4122.

Marschner, H. (1995). *Mineral nutrition of higher plants* (2nd ed.). San Diego, CA: Academic Press, Inc.

Mori, S., Nishizawa, N., Hayashi, H., Chino, E., Yoshimura, E., & Ishihara, J. (1991). Why are young rice plants highly susceptible to iron deficiency? *Plant and Soil, 130*, 143–156.

Murata, Y., Ma, J. F., Yamaji, N., Ueno, D., Nomoto, K., & Iwashita, T. (2006). A specific transporter for iron(III)-phytosiderophore in barley roots. *The Plant Journal, 46*(4), 563–572.

Pianelli, K., Mari, S., Marquès, L., Lebrun, M., & Czernic, P. (2005). Nicotianamine over-accumulation confers resistance to nickel in Arabidopsis thaliana. *Transgenic Research, 14*, 739–748.

Ramanan, N., & Wang, Y. (2000). A high-affinity iron permease essential for Candida albicans virulence. *Science, 288*(5468), 1062–1064.

Rellan-Alvarez, R., Abadia, J., & Alvarez-Fernandez, A. (2008). Formation of metal-nicotianamine complexes as affected by pH, ligand exchange with citrate and metal

exchange. A study by electrospray ionization time-of-flight mass spectrometry. *Rapid Communications in Mass Spectrometry, 22*(10), 1553–1562.
Rellan-Alvarez, R., Giner-Martinez-Sierra, J., Orduna, J., Orera, I., Rodriguez-Castrillon, J. A., Garcia-Alonso, J. I., et al. (2010). Identification of a tri-iron(III), tri-citrate complex in the xylem sap of iron-deficient tomato resupplied with iron: new insights into plant iron long-distance transport. *Plant and Cell Physiology, 51*(1), 91–102.
Roberts, L. A., Pierson, A. J., Panaviene, Z., & Walker, E. L. (2004). Yellow stripe1. Expanded roles for the maize iron-phytosiderophore transporter. *Plant Physiology, 135*(1), 112–120.
Roschzttardtz, H., Seguela-Arnaud, M., Briat, J. F., Vert, G., & Curie, C. (2011). The FRD3 citrate effluxer promotes iron nutrition between symplastically disconnected tissues throughout Arabidopsis development. *The Plant Cell, 23*(7), 2725–2737.
Sasaki, A., Yamaji, N., Xia, J., & Ma, J. F. (2011). OsYSL6 is involved in the detoxification of excess manganese in rice. *Plant Physiology, 157*(4), 1832–1840.
Schaaf, G., Ludewig, U., Erenoglu, B. E., Mori, S., Kitahara, T., & von Wirén, N. (2004). ZmYS1 functions as a proton-coupled symporter for phytosiderophore- and nicotianamine-chelated metals. *The Journal of Biological Chemistry, 279*, 9091–9096.
Schaaf, G., Schikora, A., Haberle, J., Vert, G., Ludewig, U., Briat, J. F., et al. (2005). A putative function for the Arabidopsis Fe-phytosiderophore transporter homolog AtYSL2 in Fe and Zn homeostasis. *Plant and Cell Physiology, 46*(5), 762–774.
Tusnády, G. E., & Simon, I. (2001). The HMMTOP transmembrane topology prediction server. *Bioinformatics, 17*(9), 849–850.
Ueno, D., Yamaji, N., & Ma, J. F. (2009). Further characterization of ferric-phytosiderophore transporters ZmYS1 and HvYS1 in maize and barley. *Journal of Experimental Botany, 60*(12), 3513–3520.
Vert, G., Grotz, N., Dedaldechamp, F., Gaymard, F., Guerinot, M. L., Briat, J. F., et al. (2002). IRT1, an Arabidopsis transporter essential for iron uptake from the soil and for plant growth. *The Plant Cell, 14*(6), 1223–1233.
Viklund, H., & Elofsson, A. (2004). Best alpha-helical transmembrane protein topology predictions are achieved using hidden Markov models and evolutionary information. *Protein Science, 13*(7), 1908–1917.
von Wiren, N., Klair, S., Bansal, S., Briat, J.-F., Khodr, H., Shioiri, T., et al. (1999). Nicotianamine chelates both Fe[III] and Fe[II]. Implications for metal transport in plants. *Plant Physiology, 119*, 1107–1114.
von Wiren, N., Marchner, H., & Romheld, V. (1996). Roots of iron-efficient maize also absorb phytosiderophore-chelated zinc. *Plant Physiology, 111*, 1119–1125.
von Wiren, N., Mori, S., Marschner, H., & Romheld, V. (1994). Iron inefficiency in maize mutant *ys1* (Zea mays L. cv yellow-stripe) is caused by a defect in uptake of iron phytosiderophores. *Plant Physiology, 106*, 71–77.
Waters, B. M., Chu, H. H., Didonato, R. J., Roberts, L. A., Eisley, R. B., Lahner, B., et al. (2006). Mutations in Arabidopsis *yellow stripe-like1* and *yellow stripe-like3* reveal their roles in metal ion homeostasis and loading of metal ions in seeds. *Plant Physiology, 141*(4), 1446–1458.
Weber, M., Harada, E., Vess, C., Roepenack-Lahaye, E., & Clemens, S. (2004). Comparative microarray analysis of Arabidopsis thaliana and Arabidopsis halleri roots identifies nicotianamine synthase, a ZIP transporter and other genes as potential metal hyperaccumulation factors. *The Plant Journal, 37*, 269–281.
Winterbourn, C. C. (1995). Toxicity of iron and hydrogen peroxide: the Fenton reaction. *Toxicology Letters, 82–83*, 969–974.
Yen, M.-R., Tseng, Y.-H., & Saier, M. H., Jr. (2001). Maize *Yellow Stripe1*, and iron-phytosiderophore uptake transporter, is a member of the oligopeptide transporter (OPT) family. *Microbiology, 147*, 2881–2883.
Yordem, B. K., Conte, S. S., Ma, J. F., Yokosho, K., Vasques, K. A., Gopalsamy, S. N., et al. (2011). Brachypodium distachyon as a new model system for understanding iron

homeostasis in grasses: phylogenetic and expression analysis of yellow stripe-like (YSL) transporters. *Annals of Botany, 108*(5), 821–833.
Zhao, H., & Eide, D. J. (1997). Zap1p, a metalloregulatory protein involved in zinc-responsive transcriptional regulation in Saccharomyces cerevisiae. *Molecular Cell Biology, 17*(9), 5044–5052.
Zheng, L., Fujii, M., Yamaji, N., Sasaki, A., Yamane, M., Sakurai, I., et al. (2011). Isolation and characterization of a barley yellow stripe-like gene, HvYSL5. *Plant and Cell Physiology, 52*(5), 765–774.

SECTION FIVE

Unique Transport Systems with Essential Roles

CHAPTER TWELVE

Pathways of Arsenic Uptake and Efflux

Hung-Chi Yang*, Hsueh-Liang Fu†, Yung-Feng Lin‡ and Barry P. Rosen$,1

*Department of Medical Biotechnology and Laboratory Sciences, Chang-Gung University, Tao-Yuan, Taiwan
†Department of Pathology, Wayne State University School of Medicine, Detroit, MI, USA
‡Department of Medical Laboratory Science and Biotechnology, Taipei Medical University, Taipei, Taiwan
$Department of Cellular Biology and Pharmacology, Florida International University College of Medicine, Miami, FL, USA
[1]Corresponding author: E-mail: brosen@fiu.edu

Contents

Abstract

Arsenic is the most prevalent environmental toxic substance and ranks first on the U.S. Environmental Protection Agency's Superfund List. Arsenic is a carcinogen and a causative agent of numerous human diseases. Paradoxically arsenic is used as

Current Topics in Membranes, Volume 69
ISSN 1063-5823,
http://dx.doi.org/10.1016/B978-0-12-394390-3.00012-4

a chemotherapeutic agent for treatment of acute promyelocytic leukemia. Inorganic arsenic has two biological important oxidation states: As(V) (arsenate) and As(III) (arsenite). Arsenic uptake is adventitious because the arsenate and arsenite are chemically similar to required nutrients. Arsenate resembles phosphate and is a competitive inhibitor of many phosphate-utilizing enzymes. Arsenate is taken up by phosphate transport systems. In contrast, at physiological pH, the form of arsenite is $As(OH)_3$, which resembles organic molecules such as glycerol. Consequently, arsenite is taken into cells by aquaglyceroporin channels. Arsenic efflux systems are found in nearly every organism and evolved to rid cells of this toxic metalloid. These efflux systems include members of the multidrug resistance protein family and the bacterial exchangers Acr3 and ArsB. ArsB can also be a subunit of the ArsAB As(III)-translocating ATPase, an ATP-driven efflux pump. The ArsD metallochaperone binds cytosolic As(III) and transfers it to the ArsA subunit of the efflux pump. Knowledge of the pathways and transporters for arsenic uptake and efflux is essential for understanding its toxicity and carcinogenicity and for rational design of cancer chemotherapeutic drugs.

1. INTRODUCTION

Environmental arsenic comes primarily from natural geological sources such as volcanoes and hot springs (Inskeep & McDermott, 2005). An arsenic biogeocycle results from biological transformations of arsenic (Bhattacharjee & Rosen, 2007; Rensing & Rosen, 2009; Ye, Rensing, Rosen, & Zhu, in press). The Environmental Protection Agency (EPA) asserts that it pervades our drinking water (Council, 2001) and imperils the safety of our food supply (Stone, 2008). The contamination of groundwater by arsenic in Bangladesh has been called the largest poisoning of a population in history (Smith, Lingas, & Rahman, 2000). As a result of persistent human exposure and health hazard, arsenic ranks first on the Superfund List (http://www.atsdr.cdc.gov/spl/). Arsenic is a carcinogen and a causative agent of cardiovascular and peripheral vascular disease, neurological disorders, diabetes mellitus and various forms of cancer (Abernathy, Thomas, & Calderon, 2003; Tchounwou, Centeno, & Patlolla, 2004; Tseng et al., 2002). Arsenic exposure during pregnancy contributes to low birth weight and fetal loss, as well as delayed infant development (Tofail et al., 2009).

Arsenic is introduced anthropogenically as herbicides and pesticides, wood preservatives, animal feeds and semiconductors. For example, copper–chromium–arsenic (CCA) treatment was used for decades to preserve wood from fungi and insects. Although CCA is no longer used in the United States, playgrounds, household decks, boardwalks, telephone poles and other places where CCA-treated wood was used means that arsenic will contaminate our food and water supplies for the foreseeable future. Inorganic and organic arsenicals have been and are still used for agriculture and animal husbandry. Arsenic acid in the form of herbicides such as Desiccant L-10 by Atochem/Elf Aquitaine was used to defoliate cotton fields so that the next crops could be planted. This arsenic continues to contaminate those former cotton fields in the southern states in the US. Many of those fields are now planted with rice, which is the largest non-seafood source of arsenic in the American diet (Williams et al., 2005; Williams et al., 2007). Arsenic-contaminated brown rice syrup is used in baby food, a serious exposure problem for the most vulnerable (Jackson et al., 2012). Lead arsenate has been one of the most extensively used of arsenical insecticides. It was first used in 1892 against the gypsy moth (*Lymantria dispar*) (http://soils.tfrec.wsu.edu/leadhistory.htm). According to the EPA, lead arsenate was routinely used as a growth regulator on 17% of the U.S. grapefruit crop in the 1980s. Ten thousand pounds of lead arsenate were also used annually for control of cockroaches, silverfish and crickets (http://pmep.cce.cornell.edu/profiles/insect-mite/fenitrothion-methylpara/lead-arsenate/insect-prof-leadars.html). Even today the arsenic is finding its way into apple juice (http://www.doctoroz.com/videos/arsenic-apple-juice). In addition to inorganic arsenic, organic arsenicals are used in agriculture and animal husbandry. These include salts of methylarsenate [MAs(V)] and dimethylarsenate [DMAs(V)] (for example, one form of Ortho Weed-B-Gone Crabgrass Killer). Although they are no longer being produced, DMAs(V) and MAs(V) are still used as herbicides and fungicides for golf courses in Florida, and runoff from these golf courses introduce these arsenicals into drinking water (http://www.naplesnews.com/news/2006/oct/22/arsenic_levels_get_herbicide_pulled/?local_news). The organic arsenical roxarsone (4-hydroxy-3-nitrophenylarsonic acid) is fed to chickens as feed supplement. In addition to arsenic contamination of the chicken, arsenic-contaminated chicken litter is used as a fertilizer.

Paradoxically, arsenic and the related metalloid antimony have been used for centuries as an herbal remedy for a number of illnesses and more recently as chemotherapeutic agents (Kwong & Todd, 1997). Arsenic- and antimony-containing drugs are currently used for treating acute promyelocytic leukemia (APL) (the arsenical Trisenox) (Liu, Zhou, Chen, & Chen, 2012) and diseases caused by protozoan parasites (the antimonial Pentostam) (Carter & Fairlamb, 1993), and resistance to metalloid drugs is a serious clinical problem.

Inorganic arsenic has two biological important oxidation states: As(V) (arsenate) and As(III) (arsenite). At neutral pH, the arsenate oxyanion resembles phosphate and is a competitive inhibitor of many phosphate-utilizing enzymes. In contrast, As(III) is uncharged $As(OH)_3$ at neutral pH (pK_a 9.2). Consequently, As(V) uptake systems are arsenate anion transporters, while the substrate of As(III) transport systems is $As(OH)_3$. This is significant, as discussed below, that most metalloids are taken up by aquaglyceroporin (AQP) channels as neutral species.

Arsenic uptake is by-and-large adventitious—cells have no reason to take up this poisonous metalloid (Bhattacharjee, Rosen, & Mukhopadhyay, 2009). In contrast, arsenic efflux systems are found in nearly every organism and evolved to rid cells of toxic metalloids (Rensing & Rosen, 2009). In bacteria and archaea, the genes for arsenic detoxification are usually encoded by arsenic resistance (*ars*) operons. Multidrug resistance protein (MRP) orthologues remove arsenite from the cytosol of fungi, plants and animals, including humans, as the reduced glutathione (GSH) conjugate $As(GS)_3$. Knowledge of the pathways and transporters for metalloid uptake and efflux is essential for understanding toxicity and for rational design of metalloid drugs and for treating drug-resistant microorganisms and tumor cells.

2. UPTAKE SYSTEMS FOR ARSENIC

2.1. Arsenate Uptake

The arsenate oxyanion is chemically similar to phosphate, and its toxicity is based primarily on the competitive inhibition of proteins that use phosphate, as do many of the enzymes in intermediary metabolism and oxidative phosphorylation. However, the intracellular concentration of phosphate is usually quite high, so most cells are relatively insensitive to arsenate unless phosphate starved. In most organisms, arsenate is taken up adventitiously by phosphate transporters. For example, in *Escherichia coli* arsenate is taken up by two phosphate uptake systems (Willsky, Bennett, & Malamy, 1973; Willsky & Malamy, 1980), Pst and Pit. Pst is a high-affinity, low-capacity system induced by phosphate starvation, while Pit is a low-affinity, high-capacity constitutive system that is also the major uptake system for arsenate (Rosenberg, Gerdes, & Chegwidden, 1977).

2.2. Bidirectional Transport of Arsenite by Bacterial AQPs

Trivalent arsenite is considerably more toxic than pentavalent arsenate due to its tendency to react as a soft metal with thiols. Arsenite forms relatively weak bonds with monothiols, and high intracellular concentrations of arsenite can deplete cells of glutathione. It forms strong bonds with dithiols in small molecules such as the lipoic acid cofactor and with vicinal thiols in proteins, leading to inactivation of various enzymes and receptors. Trivalent arsenic is often referred to in the literature as the anion arsenite. But, in contrast to arsenate, which is a negatively charged oxyanion in solution, trivalent inorganic arsenic is, in fact, the neutral undissociated acid $As(OH)_3$ at physiological pH (Ramírez-Solis, Mukopadhyay, Rosen, & Stemmler, 2004). For convenience the terms $As(OH)_3$, As(III) and arsenite will be used interchangeably in this chapter.

Unlike arsenate, which was long known to be taken up by phosphate transporters, how the more toxic arsenite enters the cell was unknown until the ground-breaking study by Sanders, Rensing, Kuroda, Mitra, and Rosen (1997). Using transposon mutagenesis to identify genes associated with accumulation of metalloids in *E. coli*, they isolated an Sb(III)-tolerant strain in which the *glpF* gene was disrupted by insertion of the transposon. Antimony is a related metalloid that shares chemical properties with arsenic. Subsequently, Meng et al. (2004) showed that the *glpF* mutant also had a significantly reduced rate of arsenic uptake. GlpF is a bacterial glycerol facilitator (Heller, Lin, & Wilson, 1980) that is the first identified member of the AQP or Major intrinsic membrane protein (MIP) superfamily, for which Peter Agre was awarded the Nobel Prize in Chemistry in 2003 (Agre, 2004). There are two main branches in the superfamily: true aquaporins, which are water channels, and AQPs, channels with larger pores that conduct water, glycerol and other small neutral solutes (Agre et al., 2002). These findings suggest that both arsenite and antimonite can be recognized by GlpF as polyol forms related to glycerol. With a pK_a of 9.2, arsenite is not an anion but was shown to be uncharged $As(OH)_3$ under physiological conditions (Ramírez-Solis et al., 2004). Although GlpF is responsible for adventitious uptake of As(III) in *E. coli*, the legume symbiont *Sinorhizobium meliloti* uses a clever strategy for resistance to arsenate (but not arsenite) using a GlpF orthologue, AqpS, for efflux of As(III) (Fig. 1) (Yang, Cheng, Finan, Rosen, & Bhattacharjee, 2005). When *S. meliloti* is exposed to environmental arsenate, it is taken up by phosphate transporters and reduced inside the cell to trivalent arsenite by an arsenate reductase, ArsC. The internally generated trivalent arsenite then flows downhill out of the cell through AqpS. Together, AqpS and ArsC constitute a novel arsenate detoxification pathway in *S. meliloti*. The presence of *aqpS* and *arsC* in the *meliloti* arsenic resistance operon is not an isolated observation, and the combination of AQP and arsenate reductase

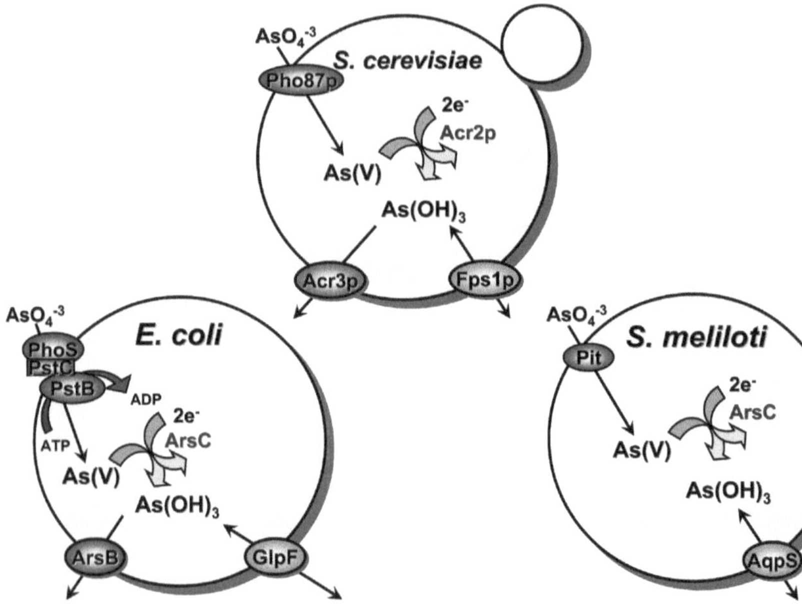

Figure 1 ***Pathways of arsenic uptake and efflux.*** Arsenate [As(V)] is taken up by phosphate transporters, and As(III) is conducted up by AQPs (GlpF in *Escherichia coli*, Fps1p in yeast, and AQP7 and AQP9 in mammals). In both *E. coli* and *Saccharomyces cerevisiae*, arsenate is reduced to arsenite by arsenate reductases. In *E. coli* and many other bacteria, arsenite is extruded from the cells by ArsB alone or by the ArsAB ATPase. In many other bacteria, Acr3 replaces ArsB as the plasma membrane arsenite efflux protein. (Some bacteria have both ArsB and Acr3!) Fungi also use Acr3 for arsenite efflux, and yeast Ycf1p, which is a member of the MRP group of the ABC superfamily of drug resistance ATPases, pumps $As(GS)_3$ into the vacuole. The legume symbiotic bacterium *Sinorhizobium meliloti* is unique in that it uses an AQP for arsenite efflux rather than uptake. In contrast to most bacteria, *S. meliloti* is very sensitive to arsenite because it has neither ArsB nor Acr3. It is resistant to arsenate, which, as in other organisms, is brought into cells by the phosphate transporters. The first step of detoxification involves reduction of arsenate to arsenite by ArsC. The AqpS channel facilitates downhill transport of internally generated As(III). Thus, this detoxification mechanism functions to confer arsenate but not arsenite resistance in *S. meliloti*. See the color plate.

may confer arsenate tolerance in other bacterial species. The presence of orthologous glycerol facilitators in *E. coli* and *S. meliloti* that produce either metalloid sensitivity or resistance demonstrates that bidirectional AQPs can participate in either arsenite uptake or efflux, depending on the direction of the concentration gradient of arsenite.

2.3. AQPs Are As(III) Transporters in Eukaryotes

Just as prokaryotes use AQPs for As(III) uptake, eukaryotes have AQPs that facilitate the uptake of arsenic and other metalloids (Bhattacharjee, Mukhopadhyay, Thiyagarajan, & Rosen, 2008). For example, the yeast Fps1p is an orthologue of GlpF that is responsible for uptake and efflux of glycerol in *Saccharomyces cerevisiae*. Deletion of *fps1* enhances the tolerance to arsenite and potassium antimonyl tartrate (Liu et al., 2002; Wysocki et al., 2001), suggesting that Fps1p mediates the influx of trivalent metalloid in yeast, so that its absence confers tolerance. Under elevated osmolarity conditions, when the Fps1p channel is closed, wild-type cells show a similar degree of metalloid tolerance as does an *fps1* deletion mutant. In contrast, yeast cells expressing a constitutively open form of the Fps1p are extremely sensitive to both arsenite and antimonite. Direct transport assays also indicate that arsenite uptake is mediated by Fps1p. A mitogen-activated protein kinase Hog1p phosphorylates Fps1p on Thr 231 *in vivo* and regulates influx of arsenic in response to arsenite (Thorsen et al., 2006). The arsenic sensing and signaling events mediated by activation of Hog1p and Fps1p contribute to the tolerance of arsenic acquisition. Wysocki, R., Bobrowicz, P., & Ulaszewski, S. (1997) first showed that Fps1p facilitates efflux of As(III) out of cells and is essential to maintain As(V) resistance (Maciaszczyk-Dziubinska, E., Migocka, M., & Wysocki, R. (2012)). As(V) enters the yeast cell through phosphate transporters (Bun-ya et al., 1996). The arsenate reductase Acr2p converts As(V) to As(III) followed by efflux As(III) out of the cells by both Fps1p and arsenite permease Acr3p (Fig. 1) (Ghosh, Shen, & Rosen, 1999; Liu et al., 2002). This scenario is reminiscent to the As(V) tolerance conferred by the concerted action of AqpS and ArsC in *S. meliloti* (Yang et al., 2005). It is tempting to speculate that bidirectional As(III) movement by AQPs is involved in both arsenic uptake and efflux in other organisms.

Plants have many AQPs, and a number of these have been shown to facilitate uptake of metalloids, including toxic arsenic and antimony, as well as boron and silicon, which are required in plants, and probably germanium (Bhattacharjee et al., 2008). This has led to the postulate that AQPs may have evolved in the first cells to allow uptake of required metalloids rather than organic polyols, which were not likely to be present in high concentrations in primordial oceans. Vertebrates have fewer; of more than 13 members of AQPs identified in mammals, four (AQP3, AQP7, AQP9 and AQP10) are classical AQPs (Ishibashi et al., 1997, 1998; Ishibashi, Imai, & Sasaki, 2000; Ishibashi, Morinaga, Kuwahara, Sasaki, & Imai, 2002). A yeast

Fps1p deletion mutant expressing rat AQP9 was used as a model for studying the function of mammalian AQPs (Liu et al., 2002). The *fps1* mutant showed enhanced tolerance to arsenite and antimonite, and had diminished uptake of both metalloids. Complementation of rat AQP9 significantly reversed tolerance and transport of metalloids. Notably, higher rates of metalloid uptake and greater metalloid susceptibility were observed. These reports indicate that AQP9 is a better metalloid channel than Fps1p.

Xenopus laevis oocytes microinjected with either rat AQP7 or AQP9 complementary RNA (cRNA) demonstrated increased uptake of arsenite by AQP7 and a lesser degree by AQP9 (Liu et al., 2002). Likewise, the functions of four known human AQPs hAQP3, hAQP7, fAQP9 and hAQP10 in arsenite transport were determined in *X. laevis* oocytes (Liu, Carbrey, Agre, & Rosen, 2004). The human AQP9 is a more effective arsenite transporter than that of hAQP7, while little or no uptake was detected in hAQP3 and hAQP10. AQP9 residues Phe64 and Arg219 were predicted to be part of the selectivity filter based on the comparison with the crystal structure of the bacterial homologue GlpF and bovine erythrocyte water channel bAQP1. To investigate the requirement of these residues in the substrate selectivity and conductivity between glycerol and $As(OH)_3$, site-directed mutagenesis was employed. The results from uptake assay showed that a positive-charged lysine residue could substitute for Ala219, which is required at the entry to the AQP9 channel. On the other hand, a hydrophobic phenylalanine residue predicted to position glycerol near the conserved arginine residue is actually dispensable for uptake of substrate (Liu et al., 2004). Taken together, these results suggest that arsenite shares the same translocation pathway with glycerol in AQP9.

The conversion of inorganic arsenic to methylated arsenic is a common metabolic reaction in mammals (Thomas, Waters, & Styblo, 2004). For example, arsenic trioxide is methylated by then enzyme arsenite-*S*-adenosylmethionine methyltransferase (AS3MT) in mammalian liver. One of the initial products of arsenic methylation is methylarsonous acid, MAs(III), a more toxic and carcinogenic form than inorganic arsenite. Subsequently, these methylated arsenicals are released into circulation and excreted as oxidized species in urine and feces. However, questions on how methylated arsenicals pass through liver cells and translocate to other tissues and the underlying mechanism remain unanswered. Since AQP9 is the major liver isoform, the uptake of methylated arsenicals by GlpF, Fps1p and rat AQP9 was investigated (Liu, Styblo, & Rosen, 2006). Rat AQP9 facilitates uptake of MAs(III) at a higher rate than As(III). In addition, rat AQP9 is a better channel for MAs(III) than that of Fps1p. Mechanistically, the Phe64 is not required, whereas Arg219 is indispensable for MAs(III) translocation. These findings indicate that inorganic and methylated As(III) use the identical translocation pathway in AQP9. Since AQPs have been shown to be bidirectional in bacteria and yeast, $As(OH)_3$ has been proposed to flow down its concentration into hepatocytes, where it is methylated to $MAs(OH)_2$ by AS3MT, and the methylated product then flows out of the cell again by AQP9, where it is excreted in urine either before or after nonenzymatic and spontaneous oxidation to the urinary species MAs(V) (Fig. 2) (Liu et al., 2006).

Since AQP9 is expressed in liver, brain, testes and leukocytes, these tissues may be highly sensitive to arsenic toxicity and carcinogenesis. The susceptibility of AQP9 expressed tissues may be further complicated by a variety of chemical modifications of arsenic, including methylation, oxidation, reduction, and glutathionylation. Carbrey et al. (2003) constructed AQP9-null mice as animal model for the evaluation of arsenic toxicity *in vivo*. Upon injection of sodium arsenite ($NaAsO_2$), AQP9-null mice display reduced survival. The enhanced accumulation of arsenic is detected in several organs, with the highest in heart and up to 20 times higher than wild-type control. In

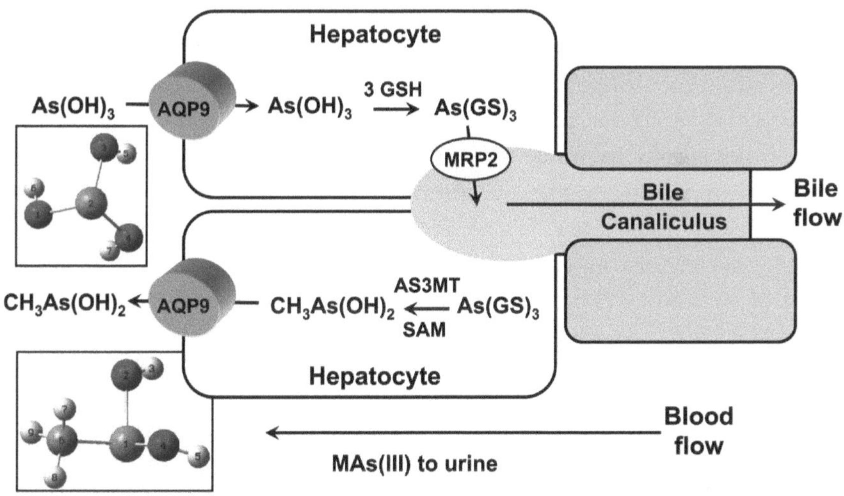

Figure 2 ***Uptake and efflux of trivalent arsenic species in liver.*** Inorganic $As(OH)_3$ flows down its concentration gradient from blood into hepatocytes through AQP9, the liver AQP isoform. In the cytosol of the hepatocyte, As(III) can be either glutathionylated or methylated to MAs(V), which is reduced to MAs(III). $As(GS)_3$ is pumped into bile by MRP2. Alternatively, As(III) is methylated by the enzyme As(III) SAM methyltransferase (AS3MT) to $CH_3As(OH)_2$, which then flows down its concentration gradient via AQP9 into blood, to the kidney and eliminated in the urine. See the color plate.

addition, the AQP9-null mice sustain profound bradycardia within hours after sodium arsenite injection. Despite the arsenic level is detected in multiple organs, the excretion of arsenic in urine and feces is significantly reduced in AQP9-null mice with only one-tenth arsenic level of wild-type mice. The reduced clearance of different arsenicals by AQP9-null mice indicates that AQP9 is involved in the export of multiple arsenic species. Taken together, these results clearly demonstrate that AQP9 facilitates *in vivo* excretion of arsenicals by liver and suggest a role of AQP9 in partial protection against arsenic toxicity.

The role of AQPs in arsenic transport is an uncharted territory in emerging animal models. Zebra fish (*Danio rerio*) may develop into a valuable aquatic vertebrate model for investigating the environmental effects of arsenic. Among seven AQPs annotated in the zebra fish genome project, Hamdi et al. (2009) cloned five genes (*aqp3*, *aqp3l*, *aqp9a*, *aqp9b* and *aqp10*), which encode homologues of human AQP3, AQP9 and AQP10. Using Reverse transcription polymerase chain reaction (RT-PCR), these AQP genes were detected in multiple tissues and exhibited differential tissue expression. As in humans, liver is the primary organ of arsenic accumulation and detoxification in zebra fish. These zebra fish AQPs show substantial glycerol transport at equivalent rates. In addition, these water channels facilitate the uptake of As(III), MAs(III) and Sb(III). These findings suggest that AQPs are the major pathways for the accumulation of inorganic and methylated arsenicals in zebra fish tissues. However, it is not clear whether AQPs facilitate translocation of pentavalent form of methylated arsenicals such as MAs(V) and DMAs(V), which are the primary forms of organoarsenicals excreted in the urine of mammals. More recently, McDermott, Jiang, Beene, Rosen, and Liu (2010) demonstrated that *X. laevis* oocytes microinjected with human AQP9 cRNA showed enhanced uptake of both MAs(V) and DMAs(V) at acidic pH (pH 5.5) compared with neutral pH. Hg(II), an aquaporin inhibitor, prevented conduction of arsenicals through AQP9. In contrast, phloretin, an inhibitor of water and glycerol permeation via AQP9, inhibited translocation of pentavalent MAs(V) and DMAs(V) but not trivalent As(III) and MAs(III). These results imply that the translocation pathway for the pentavalent arsenical species is different from that of the trivalent.

2.4. Arsenic and AQPs in Cancer Chemotherapy

In a series of clinical studies evaluating the efficacy of arsenic in treatment of various cancers, intravenous injection of arsenic trioxide (As_2O_3) was shown to induce a complete remission in patients of newly diagnosed or relapsed

APL (Emadi & Gore, 2010). APL is featured with the t(15;17) (q22;q21) chromosomal translocation which fuses the promyelocytic leukemia gene (*PML*) with the retinoic acid receptor α gene (*RAR* α) (Au et al., 2011). The *PML–RAR* fusion gene encodes a chimeric protein that leads to arrest of maturation at the promyelocyte stage of myeloid cell development (Soignet, 2001). Low concentrations of arsenic (0.1–0.5 μM) have been shown to induce differentiation of malignant promyelocyte by inactivation of *PML–RAR* fusion protein, whereas higher concentration of arsenic (0.5–2.0 μM) causes apoptosis through different signaling pathways (Dilda & Hogg, 2007). These signaling pathways include modulation of redox signaling by arsenic-generated reactive oxygen species (Dai, Weinberg, Waxman, & Jing, 1999) and inactivation of Nuclear factor-kappaB (NF-kB) pathway through binding Inhibitor of nuclear factor-kappaB (IkB) kinase with arsenic (Kapahi et al., 2000). Additionally, arsenic also induces proliferation arrest in several kinds of cancer cells, such as multiple myeloma and myelodysplastic syndromes (Dilda & Hogg, 2007; Emadi & Gore, 2010). To understand the mode of action of the arsenic-based drug, it is essential to delineate the absorption, distribution, metabolism, excretion and toxicity of arsenic drugs in cancer treatment.

Transfection of AQP9 into human leukemia cells renders them hypersensitive to Trisenox (arsenic trioxide) (Bhattacharjee, Carbrey, Rosen, & Mukhopadhyay, 2004). This drug sensitivity is attributed to increased rates of arsenic uptake and proportional to the expression level of AQP9 in different lineages of leukemia cells (Leung, Pang, Yuen, Kwong, & Tse, 2007). HL60 cells treating with vitamin D displayed enhanced sensitivity to arsenic and elevated expression of AQP9. Sensitivity was a consequence of higher rates of arsenic uptake due to elevated levels of AQP9. Similarly, upregulation of AQP9 correlated with increased arsenic uptake in HL60 cells treated with all-trans retinoic acid (ATRA) (Leung et al., 2007). The elevated arsenic uptake led to arsenic-induced cytotoxicity. This may explain the synergism between ATRA and arsenic, where APL patients show improved response upon concomitant treatment with both (Hu et al., 2009). These observations indicate the potential for application of pharmacological agents to enhance AQP9 expression for improving treatment of leukemia.

As chronic exposure of arsenic in drinking water leads to a variety of clinical symptoms, it becomes an emerging epidemic in many regions of the world. Even worse, the clinical symptoms of arsenic poisoning are exacerbated by poor nutritional status. The consequence of arsenic poisoning in populations exposed to high levels of arsenic in the ecosystem is mainly determined by the interaction of multiple factors, including gene expression, environment and nutrition. These factors may dictate the arsenic level in different tissues through modulating the expression level of AQPs. For example, starvation and diabetes mellitus upregulate AQP9 expression up to 20-fold in liver. In addition, AQP9 levels correlate with the nutritional status of the subject and circulating insulin levels (Carbrey et al., 2003). As a result, people with malnutrition are more susceptible to arsenic-induced hepatic toxicity upon drinking arsenic-contaminated water (Agre & Kozono, 2003). Moreover, AQP9 levels are also regulated by steroid hormones (Pastor-Soler et al., 2010) and human chorionic gonadotrophin (Marino, Castro-Parodi, Dietrich, & Damiano, 2010). Further studies on the relationship among AQP expression, arsenic transport and subsequent pharmacological response will elucidate the roles of AQPs in modulating the effects of arsenic in human health and diseases.

3. EFFLUX SYSTEMS FOR ARSENIC

3.1. A Multiplicity of As(III) Efflux Systems

As a consequence of the ubiquity of arsenic, nearly every organism has intrinsic or acquired mechanisms for arsenic detoxification. This includes biotransformations such as oxidation and reduction (Bhattacharjee & Rosen, 2007; Mukhopadhyay & Rosen, 2002), methylation (Ye et al., in press) and demethylation (Yoshinaga, Cai, & Rosen, 2011). But, by far, the most common detoxification mechanism is removal of As(III) from the cytosol, by either efflux out of the cell or sequestration in an organelle (Bhattacharjee & Rosen, 2007). To date, all identified arsenic efflux systems are for trivalent As(III); no As(V) efflux systems have been found. Considering that As(III) is more toxic and carcinogenic than As(V), it is counterintuitive that extrusion systems for As(V) do not exist. We have postulated that one of the earliest challenges of the first cells was to grow in the presence of heavy metals and metalloids that were likely present at high concentrations in primordial oceans (Rosen, 1999). The atmosphere was neutral, and arsenite would have predominated over arsenate, favoring the evolution of transporters that extrude arsenite out of the cell. As the atmosphere became oxidizing, most As(III) would have become oxidized to As(V), so proteins evolved that reduce arsenate to arsenite, the substrate of the existing transport systems.

In many eukaryotes As(IIII) resistance is conferred by members of the MRP group of the Adenosine triphosphate binding cassette (ABC) superfamily of transport ATPases that are physiologically involved in export of GS conjugates such as leukotriene C4 (Haimeur, Conseil, Deeley, & Cole, 2004). MRP1-catalyzed export of glutathione from cells was increased by arsenite, suggesting that MRP1 functions as an ATP-driven $As(GS)_3$ pump (Zaman et al., 1995). In liver, MRP2 extrudes arsenic–glutathione complexes into bile and may be a major route of arsenic detoxification in humans (Fig. 2) (Cui, Kobayashi, Hayakawa, & Hirano, 2004). MRP homologues have been shown to confer arsenic resistance in eukaryotic microbes. Metalloid-containing drugs are still the first-line therapy for trypanosomiasis and leishmaniasis, and clinical resistance is a serious problem in treatment. In arsenite-resistant strains selected in vitro, there is increased expression of *pgpA*, which encodes an MRP orthologue that pumps arsenite as $As(GS)_3$ (Dey, S., Papadopoulou, B., Haimeur, A., Roy, G., Grondin, K., Dou, D., et al. (1994); Legare et al., 2001). In *S. cerevisiae*, an MRP orthologue, Ycf1p, confers Cd(II) resistance by pumping $Cd(GS)_2$ into the vacuole (Li et al., 1997). Ycf1p also transports $As(GS)_3$ into the vacuole and confers arsenite resistance in yeast (Ghosh et al., 1999). In addition to Ycf1p, *S. cerevisiae* has a gene cluster of three ACR genes that also confers arsenic resistance (Bobrowicz, Wysocki, Owsianik, Goffeau, & Ulaszewski, 1997). Acr1p is a transcription factor, and Acr2p is an arsenate reductase (Mukhopadhyay & Rosen, 1998), and Acr3p is an arsenite permease (Ghosh et al., 1999; Wysocki, Bobrowicz, & Ulaszewski, 1997).

3.2. The Acr3 Permease

In bacteria and archaea, arsenic resistance genes are usually organized in *ars* operons that are controlled by an ArsR As(III)-responsive transcriptional repressor (San Francisco, Hope, Owolabi, Tisa, & Rosen, 1990). Nearly every *ars* operon has a gene encoding one of two unrelated As(III) efflux proteins, Acr3 or ArsB. The Acr3 permease is a member of the bile/arsenite/riboflavin transporter (BART) superfamily (Mansour, Sawhney, Tamang, Vogl, & Saier, 2007). The Acr3 subfamily includes members found in bacteria, archaea and fungi and is at least as widespread and perhaps more so than members of the ArsB family (Achour, Bauda, & Billard, 2007; Mansour et al., 2007). Unfortunately, the literature is confused by the fact that many members of the Acr3 family are annotated as ArsB even though they exhibit no significant sequence similarity to ArsB. The first identified member of this family is encoded by the *ars* operon of the skin (sigK intervening) element in the chromosome of *Bacillus subtilis* (Sato & Kobayashi, 1998).

There are no protein structural data available for Acr3 or related BART proteins, but the transmembrane topology of Acr3 has been explored. The membrane topology of the *B. subtilis* Acr3 was investigated using translational fusions, but the results could not distinguish between 8 and 10 transmembrane segments (Aaltonen & Silow, 2008). Cysteine scanning mutagenesis and accessibility to thiol-reactive probes were used to determine the transmembrane topology of Acr3-1 from iron-reducing alkaliphile, *Alkaliphilus metalliredigens* (Fu et al., 2009). The results clearly demonstrate that AmAcr3-1 has 10 transmembrane spanning segments, with the N- and C-termini localized in the cytosol. This method is more likely to yield an accurate topology since it utilizes active membrane proteins. A highly conserved cysteine residue in the P/R-C-T/I-AMV motif of the third transmembrane segment (TM3) of *Corynebacterium glutamicum* Acr3-1 (CgAcr3-1) and a glutamate in the TM9 may be involved in metalloid translocation. Replacing either Cys129 or Glu305 led to loss of transport activity in CgAcr3-1. The requirement for a thiol suggests that the transport mechanism of Acr3 may be different from that of ArsB, in which no cysteine residues are required for As(III) transport (Chen, Dey, & Rosen, 1996). The role of these residues in mediating transport is still unknown, but it is reasonable to speculate that both residues may serve as selectivity filter for As (III).

The properties of a more distant homologue from *Shewanella oneidensis* were examined (Xia et al., 2008). The *S. oneidensis* homologue confers resistance to arsenate but not arsenite. The purified protein binds arsenate, not arsenite, indicating that it is not a true Acr3 orthologue. There is, however, debate over the substrate specificity for Acr3. Fungal members of this family include the ScAcr3p metalloid efflux protein from *S. cerevisiae*, which was proposed to be selective for As(III) over Sb(III) (Maciaszczyk, Wysocki, Golik, Lazowska, & Ulaszewski, 2004). Recently, ScAcr3p has been suggested to transport Sb(III) in vivo as well (Maciaszczyk-Dziubinska et al., 2012), and yeast plasma membrane vesicles accumulate both As(III) and Sb(III) by exchange with protons (Maciaszczyk-Dziubinska et al., 2012). However, other data indicating that Sb(III) is not transported in yeast plasma membrane vesicles are not in agreement with that conclusion (Ghosh et al., 1999), and bacterial Acr3 orthologues appear to be specific for As(III) and do not transport Sb(III) (Fu et al., 2009; Villadangos et al., 2012).

Cells of *E. coli* expressing AmAcr3-1 or CgAcr3-1 confer As(III) resistance (Fu et al., 2009). Everted membrane vesicles prepared from *C. glutamicum* expressing Acr3-1 were shown to transport As(III) but not Sb(III) using Nicotinamide adenine dinucleotide (NADH) as an energy source. However, neither dissipating the positive interior $\Delta\psi$ alone nor the acid interior ΔpH alone had a significant effect on NADH-driven accumulation of As(III) in everted vesicles, indicating that either individual component of the electrochemical proton gradient, $\Delta\psi$ or ΔpH, is capable of energizing As(III) transport via CgAcr3-1. These results demonstrate that Acr3 is an electrophoretic metalloid–proton exchanger with $As(OH)_3/H^+$ antiporter activity (Villadangos et al., 2012).

Some plants also have proteins related to Acr3 that confer arsenic tolerance. For example, an Acr3 orthologue was recently identified in the vacuole of the brake fern *Pteris vittata* and is responsible for arsenic tolerance in the plant (Indriolo, Na, Ellis, Salt, & Banks, 2010). In contrast, rice (*Oryza sativa*) does not have Acr3, yet arsenic in the rice grain is a serious health hazard (Zhao, McGrath, & Meharg, 2010). To attempt to mitigate this problem, yeast ScAcr3 was expressed in rice, with the result that the arsenic content in the grain was reduced (Duan, Kamiya, Ishikawa, Arao, & Fujiwara, 2012).

3.3. The ArsB Permease and the ArsAB As(III)-Translocating ATPase

Many *ars* operons have three genes, *arsRBC*. There are at least three related ArsR transcriptional repressors with As(III)-binding sites in different locations in the protein (Ordóñez et al., 2008; Qin et al., 2007; Shi, Dong, Scott, Ksenzenko, & Rosen, 1996). ArsB, like Acr3, is a $As(OH)_3/H^+$ antiporter that extrudes As(III), conferring resistance (Meng et al., 2004). Although it has not been identified in animals, an ArsB homologue exists in plants (Ma et al., 2007). ArsC is an arsenate reductase that converts As(V) to As(III), the substrate of ArsB, extending the range of resistance to include As(V) (Gladysheva, Oden, & Rosen, 1994; Ji & Silver, 1992). Some have two more genes, *arsD* and *arsA*, such as the *arsRDABC* operon in *E. coli* plasmid R773 (Chen, Misra, Silver, & Rosen, 1986; San Francisco et al., 1990). ArsA and ArsB form the ArsAB complex, a pump utilizing the energy of ATP hydrolysis for As(III) or Sb(III) extrusion (Hsu & Rosen, 1989; Mobley & Rosen, 1982; Rosen, Weigel, Karkaria, & Gangola, 1988). ArsD is a metallochaperone that transfers As(III) or Sb(III) to ArsA (Lin, Walmsley, & Rosen, 2006). Cells expressing the five genes *arsRDABC* are more resistant to As(V) and As(III) than those expressing only the *arsRBC* genes (Dey & Rosen, 1995).

3.3.1. The ArsB Permease

Nearly every prokaryote has either an *arsB* or *acr3* gene, which are found in roughly equal frequencies. Often there are multiple copies of either, and sometimes both within a single organism. For example, *E. coli* has a chromosomal *arsB* gene as well as plasmids that have additional copy or copies. ArsB from plasmid R773 is a 429-residue membrane protein with 12 transmembrane segments (Tisa & Rosen, 1990; Wu, Tisa, & Rosen, 1992). ArsB is an antiporter that extrudes As(III) or Sb(III) from cells by H^+/$As(OH)_3$ exchange coupled to the electrochemical proton gradient (Meng et al., 2004). Many proteins with As(III)-binding sites have groups or pairs of cysteine residues in the binding site. In contrast, ArsB has only a single cysteine residue that could be changed to serine or alanine residues without effect. Thus, unlike most other As(III) detoxification proteins, ArsB does not use soft metal chemistry in As(III) transport.

ArsB is unique in having a dual mode of energy coupling depending on its association with the ArsA ATPase. *In vivo* studies on the energetics of arsenite extrusion showed that expression of both *arsA* and *arsB* genes resulted in ATP-coupled arsenite extrusion independent of the electrochemical proton gradient (Dey & Rosen, 1995). In contrast, in cells expressing only the *arsB* gene, arsenite extrusion was coupled to electrochemical energy and independent of ATP. Everted membrane vesicles from *E. coli* cells expressing the ArsA–ArsB complex catalyze ATP-coupled uptake of As(III) (Dey, Dou, Tisa, & Rosen, 1994). On the other hand, everted membrane vesicles prepared from cells expressing only *arsB* exhibited As(III) uptake coupled to electrochemical energy (Kuroda, Dey, Sanders, & Rosen, 1997). These results suggest that ArsB can be either a secondary antiporter coupled to the proton motive force or an obligatory ATP-coupled pump depending on the subunit composition of the transport complex.

3.3.2. The ArsA As(III)-Activated ATPase

The *arsRDABC* operon of *E. coli* plasmid R773 encodes the ArsAB ATPase, a metalloid pump that confers resistance by actively extruding As(III) or Sb(III) from cells (Rosen, 2002). The 583-residue ArsA is a metalloid-activated ATPase that comprises the catalytic subunit of the pump (Hsu & Rosen, 1989). It is normally bound to ArsB (Dey, S., Dou, D., Tisa, L. S., & Rosen, B. P. (1994)), but, in the absence of ArsB, ArsA is found in the cytosol and can be purified as a soluble protein (Hsu & Rosen, 1989). ArsA has two homologous halves, the N-terminal A1 domain (residues 1–288) and the C-terminal A2 domain (residues 314–583), which are connected by a flexible 25-residue linker peptide (residues 289–313). Each half has a consensus nucleotide-binding domain (NBD). At the interface of the two domains is a binding site for As(III) or Sb(III), the metalloid-binding domain

(MBD) involved in activation of ATPase activity. Connecting the MBD to the two NBDs are signal transduction domains (STDs) in each half of ArsA. Although the NBDs show considerable sequence similarity, they are not identical. Both are required for resistance, transport and ATPase activity (Karkaria, Chen, & Rosen, 1990; Kaur & Rosen, 1992), and they interact with each other (Li, Liu, & Rosen, 1996; Zhou, Radaev, Rosen, & Gatti, 2000). Mg^{2+} is required for ArsA ATPase activity (Hsu & Rosen, 1989). Conformational changes at the nucleotide-binding site, as evidenced by an increase in intrinsic tryptophan fluorescence, were observed only on the addition of MgATP, suggesting that Mg^{2+} binds to ArsA as a complex with ATP (Zhou, Liu, & Rosen, 1995). An aspartic residue in ATP- and GTP-binding proteins serves as Mg^{2+} ligand. Sequence alignment of ArsA with other ATPase such as nitrogenase iron protein (NifH) (Georgiadis, Komiya, Woo, Kornuc, & Rees, 1992), RecA (Story & Steitz, 1992) and Ras p21 (Pai et al., 1990) suggested that Asp45 might be an Mg^{2+} ligand in NBD1. Cells expressing Asp45 mutants lost arsenite resistance. Purified D45A and D45N ArsAs were inactive, whereas the D45E enzyme exhibited 5% of wild-type activity, with a five-fold decrease in affinity for Mg^{2+}, demonstrating that Asp45 is part of the Mg^{2+}-binding site.

3.3.2.1. The Nucleotide-Binding Domains

Both NBD1 and NBD2 hydrolyze ATP, with steady state hydrolysis dominated by the activity of NBD1. The overall rate of ATP hydrolysis is slow in the absence of metalloid and is accelerated by metalloid binding. NBD1 and NBD2 appear to play different roles in catalysis. ATP binding and hydrolysis at the individual NBDs were examined by photolabeling with the ATP analogue 8-azido-5′-[α-(32)P]-ATP at 4 °C (Jiang et al., 2005). Metalloid activation correlated with a >10-fold increase in affinity for nucleotide. To investigate the relative contributions of the two NBDs in catalysis, a thrombin site was introduced in the linker between ArsA1 and ArsA2. This allowed discrimination between incorporation of labeled nucleotides into the two halves of ArsA. The results indicate that both NBD1 and NBD2 bind and hydrolyze ATP even in the absence of metalloid. Sb(III) increases the affinity of NBD1 to a greater extent than NBD2. 8-Azido-5′-[γ-(32)P]-ATP was used to measure ATP hydrolysis at 37 °C. Under these catalytic conditions, both nucleotide-binding domains hydrolyze ATP, but hydrolysis in A1 is stimulated to a greater degree by Sb(III) than A2. These results are strong support for the hypothesis that the two homologous halves of the ArsA are functionally nonequivalent.

Other lines of evidence for supporting different roles for the two NBDs came from biochemical studies of the two deviant lysines in the Walker A motifs of NBD1 and NBD2 (Fu, Ajees, Rosen, & Bhattacharjee, 2010). Many ATPases have a Walker A ATP-binding motif (also called the P-loop or phosphate-binding loop), G/AxxxxGKT/S (where *x* is any amino acid), that makes direct contact with the phosphates of the bound ATP. This family of ATPases has been subdivided into those containing a majority consensus Walker A motif and a subgroup with the "deviant" Walker A motif subgroup, xKGGxxK(T/S) (Koonin, 1993). ArsA is a member of the deviant Walker A motif family, whose members have diverse functions including arsenite transport (ArsA), nitrogen fixation (NifH), DNA segregation (ParA) and spatial regulation of cell division (MinD). An emerging theme for this subgroup is that the role of ATP is to modulate the interaction of the protein with itself and its partner (Lutkenhaus & Sundaramoorthy, 2003; Schindelin, Kisker, Schlessman, Howard, & Rees, 1997). For NifH and ArsA, the dimer is in a more open conformation in the presence of ADP and contains two nucleotide-binding sites at the interface of the two monomers. The binding sites are composed primarily of amino acid residues coming from one of the monomers (Schindelin et al., 1997). Importantly, NifH has also been captured in the transition state, which reveals a more closed conformation, with important new intermonomer contacts. The most important of these is made by the signature "deviant" lysine, which contacts a phosphate of the nucleotide bound to the opposite monomer. To investigate the role of deviant lysines in the Walker A motif of ArsA, Lys16 in NBD1 and Lys335 in NBD2 were mutated for functional analysis (Fu et al., 2010). Phenotypic and biochemical data suggested that Lys335 is necessary for completing NBD1, which is required for ArsA to go through the catalytic cycle by hydrolysis of the ATP in the NBD1. By extrapolation, the role of Lys16 would be to complete NBD2, but the function of NBD2 is not clear.

3.3.2.2. The Metalloid-Binding Domain

In the closed ArsA structure, the two NBDs are located at the interface between A1 and A2, in close proximity to each other (Zhou et al., 2000). Over 20 Å distant from the NBDs is a metalloid-binding domain (MBD), where three Sb(III) or As(III) are bound at the A1–A2 interface. Antimonite or arsenite activates ArsA ATPase activity. In the absence of the metalloids, ArsA has a low level of ATPase activity. Arsenite stimulates ATP hydrolysis by 3- to 5-fold, whereas a 10- to 20-fold stimulation is observed with antimonite as the activator. All other oxyanions tested had no effect on

ATPase activity (Hsu & Rosen, 1989). In the crystal structure, one Sb(III) is connected to Cys113 from A1 and Cys422 from the A2 (Site 1), a second to Cys172 from A1 and His-453 from A2 (Site 2), and the third to His-148 from A1 and Ser-420 from A2 (Site 3). Thus, the three metalloid atoms act as molecular glue to bring the A1 and A2 halves of ArsA together, an event that is linked to activation of ATP hydrolysis (Bhattacharjee, Li, Ksenzenko, & Rosen, 1995; Bhattacharjee & Rosen, 1996). ArsA binds a single Sb(III) with high affinity only in the presence of Mg^{2+}-ATPγS. Mutation of the codons for Cys113 and Cys422 eliminated Sb(III) binding to purified ArsA (Ruan, Bhattacharjee, & Rosen, 2006). C113A/C422A ArsA has basal ATPase activity similar to that of the wild type but lacks metalloid-stimulated activity. Cells expressing the $arsA_{C113A/C422A}B$ genes had an intermediate level of metalloid resistance and accumulation between those expressing only *arsB* alone and those expressing wild-type *arsAB* genes, indicating that whereas metalloid stimulation of ArsA activity enhances the ability of the pump to reduce the intracellular concentration of metalloid, high-affinity binding of metalloid by ArsA is not obligatory for transport or resistance. However, cells bearing wild-type *arsAB* replaced cells with mutant $arsA_{C113A/C422A}B$ in mixed populations of cells in a sublethal concentration of arsenite, suggesting that the metalloid-binding site confers an evolutionary advantage. What is the role of the other two metalloids observed in the crystal structure of ArsA, which shows one liganded to Cys172 and His453 and the other liganded to His148 and Ser420? The contribution of those putative metalloid sites was examined (Ruan, Bhattacharjee, & Rosen, 2008). There was little effect of mutagenesis of residues His148 and Ser420 on Sb(III) binding. A C172A ArsA mutant and C172A/H453A double mutant exhibited decreased affinity for Sb(III). Thus, while there appears only a single high-affinity metalloid-binding site in ArsA, the results suggest that Cys172 controls the affinity of this site for metalloid and hence the efficiency of As(III) to activate the ArsAB efflux pump.

3.3.2.3. The Signal Transduction Domain

Connecting the MBD to the two NBDs are the STDs, one in each half of the protein (Zhou & Rosen, 1997). The STDs each have a 12-residue signature sequence $D_{142}TAPTGH_{148}TIRLL$ (STD1) and $D_{447}TAPTGH_{453}TLLLL$ (STD2), which correspond to the Switch II region of other nucleotide-binding proteins and have been proposed to be involved in transmission of the energy of ATP hydrolysis to metalloid transport. ArsA homologues have been found in every sequenced genome of eubacteria, archaea, fungi, plants and animals. This sequence is highly conserved in ArsA homologues from every kingdom, indicating that this common motif has a conserved function. In the structure, the two DTAPTGHT sequences clearly span the spaces between the MgADP-filled NBDs and the MBD (Zhou et al., 2000). During ATP hydrolysis, the carboxy-terminal end of the A1 sequence becomes exposed to a less polar environment, whereas the amino-terminal end becomes exposed to a more hydrophilic environment as the product, ADP, is formed, indicating that this sequence has considerable conformational mobility during the catalytic cycle. From these results the DTAPTGHT sequences have been proposed to be STDs involved in cross talk between the NBDs and the MBDs (Bhattacharjee & Rosen, 2000; Zhou & Rosen, 1997).

3.3.2.4. The Linker Regions

Connecting A1 and A2 is a linker region of 25 residues. In the absence of ATP and metalloid, ArsA is hypersensitive to trypsin (Hsu & Rosen, 1989), with the initial attack within the linker at residue Arg290 (Ramaswamy & Kaur, 1998). Binding of both nucleotide and metalloid provides substantial synergistic protection from protease. The requirement for this linker sequence was examined by the mutagenic insertion of five glycine residues or by the deletion of 5, 10, 15 or 23 residues (Li & Rosen, 2000). Cells expressing *arsA* with the five-residue insertion had wild-type arsenite resistance. Resistance of cells expressing modified *arsA* genes with deletions was dependent on the length of the linker but independent of the actual sequence. Cells with 5 or 10 residues deleted exhibited slightly reduced resistance. Further deletions decreased resistance significantly. The purified mutant with the five-residue insertion had the same affinity for ATP and Sb(III) as the wild-type enzyme. Mutants with 10-, 15- or 23-residue deletions exhibited decreased affinity for both Sb(III) and ATP. The enzyme with a 23-residue deletion exhibited only basal ATPase activity and was unable to be activated by Sb(III). These results suggest that the linker brings the two halves of the protein into proper contact with each other, facilitating catalysis.

4. THE ARSD AS(III) METALLOCHAPERONE AND ITS INTERACTION WITH THE ARSA ATPASE

4.1. Properties of Metallochaperones

All cells regulate the intracellular concentration of metals to prevent toxicity. This is necessary because even essential metals become toxic to the cell in excess due to their ability to catalyze cytotoxic reactions. It was estimated that total cytoplasmic concentration of free copper, an essential metal, to be less than 10^{-18} M, which is several orders of magnitude less than one copper atom per cell (Rae, Schmidt, Pufahl, Culotta, & O'Halloran, 1999). In other words, less than 0.01% of the total cellular copper is ever free in the cytoplasm at any one time. Yet, most copper enzymes bind copper ions with K_d values that are much higher than 10^{-18} M. For example, the copper- and zinc-dependent enzyme superoxide dismutase (SOD1) binds copper ions with a K_d around 10^{-15} M in vitro. These considerations suggest that copper enzymes (and perhaps other metalloproteins) rely upon accessory factors for acquiring metals and delivering them to partner proteins without the necessity for free (and toxic) cytosolic metal ions (O'Halloran & Culotta, 2000).

Most identified metallochaperones are for copper, the arsenic chaperone, ArsD, is the first and, to date, only characterized chaperone for the toxic metalloid arsenic (Lin, Walmsley and Rosen, 2006). ArsD sequesters cytosolic As(III) and delivers it to the ArsAB pump for efflux and detoxification. Most *ars* operons have three genes, *arsRBC*, but others have two additional genes, *arsD* and *arsA*, such as the *arsRDABC* of plasmid R773. Interestingly, these two genes are nearly always adjacent to each other in the operons and gene clusters, indicating that they were coevolved as a unit. It also suggests that ArsD and ArsA have associated functions in arsenic resistance.

4.2. Properties of the ArsD As(III) Metallochaperone

Cells expressing *arsAB* are resistant to arsenite even in the absence of an *arsD* gene ($IC_{50} = 9$ mM), and expression of *arsD* along with *arsAB* produced only a slight increase in resistance ($IC_{50} = 13$ mM) (Lin et al., 2007). So what is the primary physiological role of ArsD? Typical resistance experiments are performed under laboratory conditions, e.g. high concentrations (millimolar) of arsenite for short periods of time (hours). In the environment, bacteria are exposed to low levels (micromolar) of arsenite for long periods of time (days, months, years). In a molecular competition experiment designed to more closely resemble environmental conditions, a mixed culture of cells of *E. coli* expressing *arsDAB* and *arsAB* were grown at 10 μM arsenite, a sub-toxic concentration but one that is found in arsenic-contaminated regions of the world. After a week, the only cells remaining had all three genes, *arsDAB*. Thus, having an *arsD* gene gives bacteria a competitive advantage over cells having only *arsAB* when growing in subtoxic concentrations of arsenic. Furthermore, cells co-expressing *arsD* and *arsAB* genes accumulated less As(III) than those expressing only *arsAB* genes (Lin et al., 2006).

These results indicate that ArsD acts to increase the efficiency of arsenic extrusion by the ArsAB pump and imply that the two proteins physically interact. Association of ArsD and ArsA was demonstrated by yeast two-hybrid analysis (Lin et al., 2006). Among the four soluble proteins encoded by the R773 *arsRDABC* operon, ArsD was shown to interact only with ArsA and with itself (because ArsD is a homodimer), but not with the ArsR repressor or the ArsC arsenate [As(V)] reductase. Likewise ArsA only interacts with ArsD. Notably no As(III) or Sb(III) was added to the yeast culture media, indicating that ArsD and ArsA interact to some extent in the absence of metalloid.

4.2.1. Metalloid Binding by ArsD

ArsD is a 120-residue protein that can dimerize, although it is not clear whether the monomer or dimer (or both) interact(s) with ArsA. It has three cysteine pairs, Cys12–Cys13, Cys112–Cys113 and Cys119–120, but only the Cys12–Cys13 pair and a nearby residue, Cys18, are required for ArsD chaperone activity, so most experiments were performed with an ArsD derivate truncated at residue 109 (Lin et al., 2007). The structure of As(III)-binding site was examined by extended X-ray absorption fine structure spectroscopy (EXAFS), which showed that Cys12, Cys13 and Cys18 form three-coordinate binding site in which each cysteine thiolate is 2.24 Å from the centrally bound arsenic atom in what can be predicted to be a pyramidal structure (Yang, Rawat, Stemmler, & Rosen, 2010) (Fig. 3). ArsD has an affinity for Sb(III) of approximately 0.3 μM (Lin et al., 2007). In contrast, its partner, ArsA, has a K_d for Sb(III) approximately 9 μM (Ruan et al., 2006). Thus, ArsA has about 30-fold lower affinity for metalloid than that of ArsD, which raises the question of how metalloid can be transferred quantitatively to ArsA. One possibility is that the process is kinetically, not thermodynamically, driven. The ArsAB ATPase pumps metalloid out of the cells, so, by mass action, metalloid could be pulled from ArsD to the ArsAB pump.

Binding of As(III) by ArsD was characterized using tryptophan fluorescence spectroscopy (Yang et al., 2010). ArsD has only two tryptophan residues located distant from the As(III)-binding site. By site-directed mutagenesis, those were changed to tyrosine residues, and single

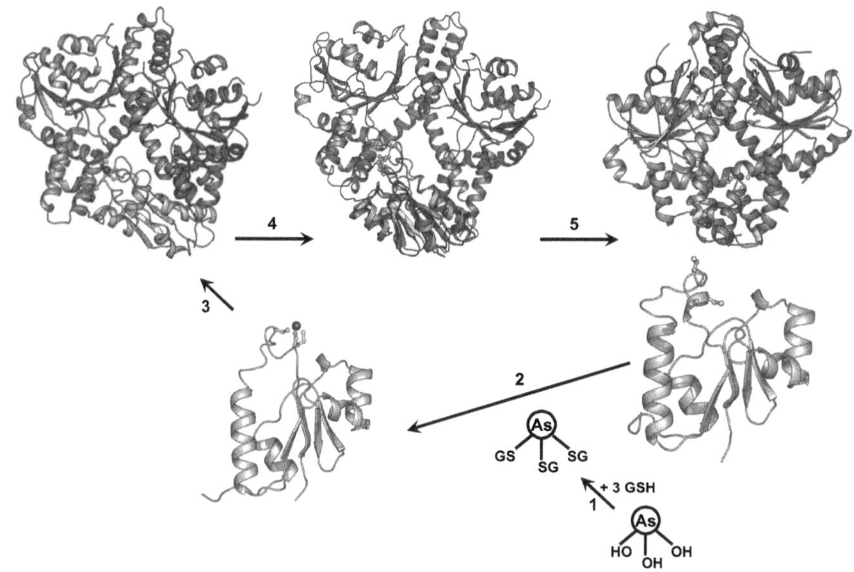

Figure 3 ***Transfer of As(III) from ArsD to ArsA.*** The model shows a hypothetical five-step reaction scheme for transfer of As(III) from ArsD to ArsA for extrusion from the cell. Step 1: Intracellular $As(OH)_3$ complexes with GSH to form $As(GS)_3$. Step 2: ArsD binds As(III) by exchange of the three thiols of $As(GS)_3$ for the three thiols of residues Cys12, Cys13 and Cys18. Step 3: The ArsD–As(III) fits into the cavity of the open form of ArsA, with the three thiols of ArsD juxtaposed to the three thiols of the As(III)-binding site of ArsA. Step 4: In a three-step thiol exchange reaction, As(III) is transferred from ArsD to ArsA. Step 5: As(III) binding to ArsA induces a conformational change that increases the rate of ATP hydrolysis and, consequently, the rate of As(III) extrusion by the ArsAB pump. See the color plate.

tryptophan derivatives within the As(III)-binding site were constructed by changing either Thr15 or Val17 to tryptophan residues. Either T15W or V17W exhibited quenching of intrinsic protein fluorescence upon binding of As(III) or Sb(III). From the As(III) dependence of quenching, a K_d of approximately 2 μM for arsenite was determined. Mutants lacking any one of the cysteines had 10-fold decreased affinity for As(III). Notably inorganic As(III) quenched fluorescence very slowly. When the glutathione conjugate, $As(GS)_3$, was added, the rate was at least 1000-fold faster. Considering that intracellular GSH concentrations are usually in the millimolar range, nearly all As(III) would be in the form of $As(GS1')_3$ in vivo, with essentially no free As(III). These results indicate that physiologically glutathione-conjugated arsenic is the species recognized by ArsD (Fig. 3, step 1).

4.2.2. Interactions between ArsD and the ArsA ATPase

A combination of recent biochemical, molecular genetic and structural results have provided insights into the details of the ArsD–ArsA interaction. Interaction of ArsD and ArsA was examined in vitro using purified proteins. This was first demonstrated by reaction with the bifunctional cysteine cross-linking reagent dibromobimane (Lin et al., 2006). Dibromobimane cross-linking was increased by MgATP, suggesting that the two proteins interact during ATP hydrolysis by ArsA. To examine the mechanism of metalloid transfer, an active form of ArsD fused to the maltose-binding protein was constructed. The chimeric protein was loaded with either As(III) or Sb(III), and the metalloid protein bound to amylose column. ArsA was then flowed through the column with a variety of ligands, and, finally, ArsD was eluted with maltose (Lin et al., 2006; Yang et al., 2010). The concentrations of eluted proteins and metalloids were determined, and transfer of metalloid from ArsD to ArsA was compared under catalytic or non-catalytic conditions. The results showed that ATP hydrolysis facilitated metalloid transfer from ArsD to ArsA, suggesting that ArsA is in a transient conformation during the catalytic cycle when transfer takes place.

One important question is whether transfer of As(III) from ArsD to ArsA is mediated by direct interaction of the two proteins that allows for channeling between the thiols of ArsD and ArsA or by a thermodynamic dissociation/reassociation process. The As(III) chelator dimercaptosuccinate was used to differentiate the two mechanisms. The results demonstrated that the As(III) transfer from ArsD to ArsA was insensitive to dimercaptosuccinate, consistent with channeling of metalloid from one to the other, and suggest that ArsD and ArsA form an interface at their metal-binding sites (Yang et al., 2010). At the biochemical level, it appears that transfer of As(III) from ArsD enhances ArsA ATPase activity by increasing its affinity for metalloids (Lin et al., 2006). In the presence of ArsD, the apparent affinity of ArsA for As(III) is increased nearly two orders of magnitude, but the V_{max} for ATP hydrolysis remains unchanged. ArsD effectively increases the affinity of ArsA for As(III), making the ArsAB pump more effective at concentrations of As(III) commonly found in the environment.

4.2.3. Structural Basis of ArsD Function

The ArsD structure was recently solved by nuclear magnetic resonance (NMR) and by X-ray crystallization at 1.6 Å (Ye et al., 2010; Ye et al., 2010). The ArsD monomer has a core of four β-strands flanked by four α-helices (Fig. 3). Although the region of ArsD cysteines that form the arsenic-binding site was not visible in that structure, a structure of an oxidized form of an ArsD homologue from *Bacteroides vulgatus*, in which the cysteines are in disulfide bonds, has been deposited (PDB code: 3KTB; Kim, Y., Tesar, C., Feldmann, B., Joachimiak, A., Midwest Center for Structural Genomics, unpublished data). From superimposition of R773 ArsD and the *B. vulgatus* orthologue, a model of the metallated form of ArsD109 was constructed (Ye et al., 2010). In the *B. vulgatus* structure, the Cys12 sulfur is oriented away from the Cys13 and Cys18, which are in a disulfide bond with each other. However, EXAFS data with R773 ArsD indicate that all three sulfur atoms are equidistant to the arsenic atom at 2.24 Å. Consequently, when the unliganded form of ArsD is presented with As(III), it is hypothesized to form an intermediate in which As(III) is weakly bound to the sulfur atoms of Cys13 and Cys18 (Ye et al., 2010). Cys12 could then reorient to become the third ligand, completing the pyramidal high-affinity binding site.

To understand how ArsD functions in solution, an NMR characterization was also performed (Ye et al., 2010). Overall the NMR and crystallographic information were in reasonable agreement. The NMR assignments indicate that the protein is a well-folded homodimer with conserved residues Cys12, Cys13 and Cys18 in an unstructured environment. The unfolded metalloid-binding site is between the structured β-strand 1 and α-helix 1, suggesting that this region of apoArsD is normally flexible and folds around metalloid, allowing the three cysteine residues to form the binding site only when As(III) is present.

ArsA undergoes a number of conformational changes during the catalytic cycle, and, in particular, cycles between an open (unliganded and basal activity) and closed (ligand bound and activated) forms. The crystal structure of a closed form of the R773 ArsA has been solved at 2.1 Å (Zhou et al., 2000). Recently structures of a yeast homologue of ArsA termed Arr4p (Shen, Hsu, Kang, Rosen, & Bhattacharjee, 2003) or Get3 (Auld et al., 2006) that is involved in membrane insertion of tailed-anchored proteins were solved in both closed and open forms (Bozkurt et al., 2009; Mateja et al., 2009). The open form of ArsA was modeled on the open structure of Get3 (Yang, Salam, & Rosen, 2011). The open model of ArsA displays a cavity into which fits ArsD well, and predicts that the ArsD–ArsA interface involves 15 residues from ArsD and 13 from ArsA, with their metalloid-binding sites in close proximity (Yang et al., 2011) (Fig. 3).

A genetic approach was applied to examine the interaction of ArsD and ArsA. In addition to the three conserved N-terminal cysteine residues, 15 ArsD residues were identified as being involved in ArsD function by genetic mapping (Yang et al., 2011). Applying a repressed transactivator yeast two-hybrid selection, nine ArsD mutants, S14R, T20I, D28T, D28V, T31A, Q34R, Q38R and V61A, were identified as increasing interaction with ArsA. Five mutants, V17A, V22A, V27D, Q51H and F55L, were found to decrease ArsA interaction by reverse yeast two-hybrid assays. Two additional mutations in Lys37 and Lys62 were identified by a combination of site-directed mutagenesis and chemical modification. Alanine substitutions in Lys37 and Lys62 were shown to lose interaction with ArsA, and reduce the activity on ArsA ATPase stimulation; however, substitution with arginine at either position was tolerated, suggesting that a positive charge is involved. All 15 residues were mapped on the surface of ArsD, in which four are close to the metal-binding loop and seven are on the outer surface of α-helix 1. Their locations are consistent with the structural model generated by *in silico* docking. These results strongly point to an interface involves one surface of helix 1 and the metalloid-binding site.

4.3. Do Higher Organisms Have an As(III) Metallochaperone?

To date, the ArsD As(III) metallochaperone has only been identified in prokaryotes. Are there As(III) metallochaperones in eurkaryotes? The ArsD monomer has a thioredoxin fold (Ye et al., 2010). Trx plays an important role in maintaining redox balances in prokaryotic and eukaryotic cells. It includes a group of small proteins of ~12 kDa that have a highly conserved sequence (Cys-Gly-Pro-Cys) that forms a redox-active site of dithiol or disulfide to protect cytosolic proteins from aggregation or inactivation (Arner & Holmgren, 2000). Trx gene expression is induced by As(III) (Hu, Jin, & Snow, 2002). Arsenic–Trx complexes have been detected by liquid chromatography with inductively coupled plasma mass spectrometry (Wang, Zhang, Li, & Le, 2007). Binding stoichiometry for different arsenic species, including inorganic and methylated As(III), were consistent with the available cysteine residues in the protein. It is possible that Trx or another protein with a thioredoxin fold binds and transfers As(III) to other proteins in an analogous fashion as the ArsD As(III) metallochaperone.

ACKNOWLEDGMENTS

The research described in this study was supported by U.S. Public Health Service Grant GM55425 to B.P.R. and Taiwan National Science Council NSC100-2320-B-038-018 to Y.F.L.

REFERENCES

Aaltonen, E. K., & Silow, M. (2008). Transmembrane topology of the Acr3 family arsenite transporter from *Bacillus subtilis*. *Biochimica et Biophysica Acta, 1778*(4), 963–973.

Abernathy, C. O., Thomas, D. J., & Calderon, R. L. (2003). Health effects and risk assessment of arsenic. *Journal of Nutrition, 133*(5 Suppl. 1), 1536S–1538S.

Achour, A. R., Bauda, P., & Billard, P. (2007). Diversity of arsenite transporter genes from arsenic-resistant soil bacteria. *Research in Microbiology, 158*(2), 128–137.

Agre, P. (2004). Nobel lecture. Aquaporin water channels. *Bioscience Reports, 24*(3), 127–163.

Agre, P., King, L. S., Yasui, M., Guggino, W. B., Ottersen, O. P., Fujiyoshi, Y., et al. (2002). Aquaporin water channels—from atomic structure to clinical medicine. *Journal of Physiology, 542*(Pt 1), 3–16.

Agre, P., & Kozono, D. (2003). Aquaporin water channels: molecular mechanisms for human diseases. *FEBS Letters, 555*(1), 72–78.

Arner, E. S., & Holmgren, A. (2000). Physiological functions of thioredoxin and thioredoxin reductase. *European Journal of Biochemistry, 267*(20), 6102–6109.

Au, W. Y., Kumana, C. R., Lee, H. K., Lin, S. Y., Liu, H., Yeung, D. Y., et al. (2011). Oral arsenic trioxide-based maintenance regimens for first complete remission of acute promyelocytic leukemia: a 10-year follow-up study. *Blood, 118*(25), 6535–6543.

Auld, K. L., Hitchcock, A. L., Doherty, H. K., Frietze, S., Huang, L. S., & Silver, P. A. (2006). The conserved ATPase Get3/Arr4 modulates the activity of membrane-associated proteins in *Saccharomyces cerevisiae*. *Genetics, 174*(1), 215–227.

Bhattacharjee, H., Carbrey, J., Rosen, B. P., & Mukhopadhyay, R. (2004). Drug uptake and pharmacological modulation of drug sensitivity in leukemia by AQP9. *Biochemical and Biophysical Research Communications, 322*(3), 836–841.

Bhattacharjee, H., Li, J., Ksenzenko, M. Y., & Rosen, B. P. (1995). Role of cysteinyl residues in metalloactivation of the oxyanion-translocating ArsA ATPase. *The Journal of Biological Chemistry, 270*(19), 11245–11250.

Bhattacharjee, H., Mukhopadhyay, R., Thiyagarajan, S., & Rosen, B. P. (2008). Aquaglyceroporins: ancient channels for metalloids. *Journal of Biology,* 7(9), 33.

Bhattacharjee, H., & Rosen, B. P. (1996). Spatial proximity of Cys113, Cys172, and Cys422 in the metalloactivation domain of the ArsA ATPase. *The Journal of Biological Chemistry, 271*(40), 24465–24470.

Bhattacharjee, H., & Rosen, B. P. (2000). Role of conserved histidine residues in metalloactivation of the ArsA ATPase. *Biometals, 13*(4), 281–288.

Bhattacharjee, H., & Rosen, B. P. (2007). Arsenic metabolism in prokaryotic and eukaryotic microbes. InNies, D H. S., & Simon. (Eds.). (2007). *Molecular microbiology of heavy metals,* Vol. 6 (pp. 371–406). Heidelberg/New York: Springer-Verlag.

Bhattacharjee, H., Rosen, B. P., & Mukhopadhyay, R. (2009). Aquaglyceroporins and metalloid transport: implications in human diseases. *Handbook of Experimental Pharmacology, 190,* 309–325.

Bobrowicz, P., Wysocki, R., Owsianik, G., Goffeau, A., & Ulaszewski, S. (1997). Isolation of three contiguous genes, ACR1, ACR2 and ACR3, involved in resistance to arsenic compounds in the yeast *Saccharomyces cerevisiae*. *Yeast, 13*(9), 819–828.

Bozkurt, G., Stjepanovic, G., Vilardi, F., Amlacher, S., Wild, K., Bange, G., et al. (2009). Structural insights into tail-anchored protein binding and membrane insertion by Get3. *Proceedings of the National Academy of Sciences of the United States of America, 106*(50), 21131–21136.

Bun-ya, M., Shikata, K., Nakade, S., Yompakdee, C., Harashima, S., & Oshima, Y. (1996). Two new genes, PHO86 and PHO87, involved in inorganic phosphate uptake in *Saccharomyces cerevisiae*. *Current Genetics, 29*(4), 344–351.

Carbrey, J. M., Gorelick-Feldman, D. A., Kozono, D., Praetorius, J., Nielsen, S., & Agre, P. (2003). Aquaglyceroporin AQP9: solute permeation and metabolic control of expression in liver. *Proceedings of the National Academy of Sciences of the United States of America, 100*(5), 2945–2950.

Carter, N. S., & Fairlamb, A. H. (1993). Arsenical-resistant trypanosomes lack an unusual adenosine transporter. *Nature, 361*(6408), 173–176.

Chen, C. M., Misra, T. K., Silver, S., & Rosen, B. P. (1986). Nucleotide sequence of the structural genes for an anion pump. The plasmid-encoded arsenical resistance operon. *The Journal of Biological Chemistry, 261*(32), 15030–15038.

Chen, Y., Dey, S., & Rosen, B. P. (1996). Soft metal thiol chemistry is not involved in the transport of arsenite by the Ars pump. *Journal of Bacteriology, 178*(3), 911–913.

Council, N. R. (2001). *Arsenic in drinking water: 2001 update.* http://www.nap.edu/books/0309076293/html/.

Cui, X., Kobayashi, Y., Hayakawa, T., & Hirano, S. (2004). Arsenic speciation in bile and urine following oral and intravenous exposure to inorganic and organic arsenics in rats. *Toxicological Sciences, 82*(2), 478–487.

Dai, J., Weinberg, R. S., Waxman, S., & Jing, Y. (1999). Malignant cells can be sensitized to undergo growth inhibition and apoptosis by arsenic trioxide through modulation of the glutathione redox system. *Blood, 93*(1), 268–277.

Dey, S., Dou, D., & Rosen, B. P. (1994). ATP-dependent arsenite transport in everted membrane vesicles of *Escherichia coli. The Journal of Biological Chemistry, 269*(41), 25442–25446.

Dey, S., Dou, D., Tisa, L. S., & Rosen, B. P. (1994). Interaction of the catalytic and the membrane subunits of an oxyanion-translocating ATPase. *Archives of Biochemistry and Biophysics, 311*(2), 418–424.

Dey, S., Papadopoulou, B., Haimeur, A., Roy, G., Grondin, K., Dou, D., et al. (1994). High level arsenite resistance in *Leishmania tarentolae* is mediated by an active extrusion system. *Molecular and Biochemical Parasitology, 67*(1), 49–57.

Dey, S., & Rosen, B. P. (1995). Dual mode of energy coupling by the oxyanion-translocating ArsB protein. *Journal of Bacteriology, 177*(2), 385–389.

Dilda, P. J., & Hogg, P. J. (2007). Arsenical-based cancer drugs. *Cancer Treatment Reviews, 33*(6), 542–564.

Duan, G., Kamiya, T., Ishikawa, S., Arao, T., & Fujiwara, T. (2012). Expressing ScACR3 in rice enhanced arsenite efflux and reduced arsenic accumulation in rice grains. *Plant and Cell Physiology, 53*(1), 154–163.

Emadi, A., & Gore, S. D. (2010). Arsenic trioxide—an old drug rediscovered. *Blood Review, 24*(4–5), 191–199.

Fu, H. L., Ajees, A. A., Rosen, B. P., & Bhattacharjee, H. (2010). Role of signature lysines in the deviant walker a motifs of the ArsA ATPase. *Biochemistry, 49*(2), 356–364.

Fu, H. L., Meng, Y., Ordonez, E., Villadangos, A. F., Bhattacharjee, H., Gill, J. A., et al. (2009). Properties of arsenite efflux permeases (Acr3) from *Alkaliphilus metalliredigens* and *Corynebacterium glutamicum. The Journal of Biological Chemistry, 284*(30), 19887–19895.

Georgiadis, M. M., Komiya, H. P. C., Woo, D., Kornuc, J. J., & Rees, D. C. (1992). Crystallographic structure of the nitrogenase iron protein from *Azotobacter vinelandii. Science, 257*(5077), 1653–1659.

Ghosh, M., Shen, J., & Rosen, B. P. (1999). Pathways of As(III) detoxification in *Saccharomyces cerevisiae. Proceedings of the National Academy of Sciences of the United States of America, 96,* 5001–5006.

Gladysheva, T. B., Oden, K. L., & Rosen, B. P. (1994). Properties of the arsenate reductase of plasmid R773. *Biochemistry, 33*(23), 7288–7293.

Haimeur, A., Conseil, G., Deeley, R. G., & Cole, S. P. (2004). The MRP-related and BCRP/ABCG2 multidrug resistance proteins: biology, substrate specificity and regulation. *Current Drug Metabolism, 5*(1), 21–53.

Hamdi, M., Sanchez, M. A., Beene, L. C., Liu, Q., Landfear, S. M., Rosen, B. P., et al. (2009). Arsenic transport by zebrafish aquaglyceroporins. *BMC Molecular Biology, 10*, 104.

Heller, K. B., Lin, E. C., & Wilson, T. H. (1980). Substrate specificity and transport properties of the glycerol facilitator of *Escherichia coli*. *Journal of Bacteriology, 144*(1), 274–278.

Hsu, C. M., & Rosen, B. P. (1989). Characterization of the catalytic subunit of an anion pump. *The Journal of Biological Chemistry, 264*(29), 17349–17354.

Hu, J., Liu, Y. F., Wu, C. F., Xu, F., Shen, Z. X., Zhu, Y. M., et al. (2009). Long-term efficacy and safety of all-trans retinoic acid/arsenic trioxide-based therapy in newly diagnosed acute promyelocytic leukemia. *Proceedings of the National Academy of Sciences of the United States of America, 106*(9), 3342–3347.

Hu, Y., Jin, X., & Snow, E. T. (2002). Effect of arsenic on transcription factor AP-1 and NF-kappaB DNA binding activity and related gene expression. *Toxicological Letters, 133*(1), 33–45.

Indriolo, E., Na, G., Ellis, D., Salt, D. E., & Banks, J. A. (2010). A vacuolar arsenite transporter necessary for arsenic tolerance in the arsenic hyperaccumulating fern Pteris vittata is missing in flowering plants. *The Plant Cell, 22*(6), 2045–2057.

Inskeep, W. P., & McDermott, T. R. (2003). Thermal biology, I. In *Geothermal biology and geochemistry in Yellowstone National Park: Proceeding of the Thermal Biology Institute workshop, Yellowstone National Park, WY, October 2003, Bozeman, MT, 2005, 2003* (p. 352). Bozeman, MT: Montana State University Publications.

Ishibashi, K., Imai, M., & Sasaki, S. (2000). Cellular localization of aquaporin 7 in the rat kidney. *Experimental Nephrology, 8*(4–5), 252–257.

Ishibashi, K., Kuwahara, M., Gu, Y., Kageyama, Y., Tohsaka, A., Suzuki, F., et al. (1997). Cloning and functional expression of a new water channel abundantly expressed in the testis permeable to water, glycerol, and urea. *The Journal of Biological Chemistry, 272*(33), 20782–20786.

Ishibashi, K., Kuwahara, M., Gu, Y., Tanaka, Y., Marumo, F., & Sasaki, S. (1998). Cloning and functional expression of a new aquaporin (AQP9) abundantly expressed in the peripheral leukocytes permeable to water and urea, but not to glycerol. *Biochemical and Biophysical Research Communications, 244*(1), 268–274.

Ishibashi, K., Morinaga, T., Kuwahara, M., Sasaki, S., & Imai, M. (2002). Cloning and identification of a new member of water channel (AQP10) as an aquaglyceroporin. *Biochimica et Biophysica Acta, 1576*(3), 335–340.

Jackson, B. P., Taylor, V. F., Karagas, M. R., Punshon, T., & Cottingham, K. L. (2012). Arsenic, organic foods, and brown rice syrup. *Environmental Health Perspectives*.

Ji, G., & Silver, S. (1992). Reduction of arsenate to arsenite by the ArsC protein of the arsenic resistance operon of *Staphylococcus aureus* plasmid pI258. *Proceedings of the National Academy of Sciences of the United States of America, 89*(20), 9474–9478.

Jiang, Y., Bhattacharjee, H., Zhou, T., Rosen, B. P., Ambudkar, S. V., & Sauna, Z. E. (2005). Nonequivalence of the nucleotide binding domains of the ArsA ATPase. *The Journal of Biological Chemistry, 280*(11), 9921–9926.

Kapahi, P., Takahashi, T., Natoli, G., Adams, S. R., Chen, Y., Tsien, R. Y., et al. (2000). Inhibition of NF-kappa B activation by arsenite through reaction with a critical cysteine in the activation loop of Ikappa B kinase. *The Journal of Biological Chemistry, 275*(46), 36062–36066.

Karkaria, C. E., Chen, C. M., & Rosen, B. P. (1990). Mutagenesis of a nucleotide-binding site of an anion-translocating ATPase. *The Journal of Biological Chemistry, 265*(14), 7832–7836.

Kaur, P., & Rosen, B. P. (1992). Mutagenesis of the C-terminal nucleotide-binding site of an anion-translocating ATPase. *The Journal of Biological Chemistry, 267*(27), 19272–19277.

Koonin, E. V. (1993). A superfamily of ATPases with diverse functions containing either classical or deviant ATP-binding motif. *Journal of Molecular Biology, 229*(4), 1165–1174.

Kuroda, M., Dey, S., Sanders, O. I., & Rosen, B. P. (1997). Alternate energy coupling of ArsB, the membrane subunit of the Ars anion-translocating ATPase. *The Journal of Biological Chemistry, 272*(1), 326–331.

Kwong, Y. L., & Todd, D. (1997). Delicious poison: arsenic trioxide for the treatment of leukemia. *Blood, 89*(9), 3487–3488.

Legare, D., Richard, D., Mukhopadhyay, R., Stierhof, Y. D., Rosen, B. P., Haimeur, A., et al. (2001). The *Leishmania* ATP-binding cassette protein *PGPA* is an intracellular metal-thiol transporter ATPase. *The Journal of Biological Chemistry, 276*(28), 26301–26307.

Leung, J., Pang, A., Yuen, W. H., Kwong, Y. L., & Tse, E. W. (2007). Relationship of expression of aquaglyceroporin 9 with arsenic uptake and sensitivity in leukemia cells. *Blood, 109*(2), 740–746.

Li, J., Liu, S., & Rosen, B. P. (1996). Interaction of ATP binding sites in the ArsA ATPase, the catalytic subunit of the Ars pump. *The Journal of Biological Chemistry, 271*(41), 25247–25252.

Li, J., & Rosen, B. P. (2000). The linker peptide of the ArsA ATPase. *Molecular Microbiology, 35*(2), 361–367.

Li, Z. S., Lu, Y. P., Zhen, R. G., Szczypka, M., Thiele, D. J., & Rea, P. A. (1997). A new pathway for vacuolar cadmium sequestration in *Saccharomyces cerevisiae*: YCF1-catalyzed transport of bis(glutathionato)cadmium. *Proceedings of the National Academy of Sciences of the United States of America, 94*(1), 42–47.

Lin, Y. F., Walmsley, A. R., & Rosen, B. P. (2006). An arsenic metallochaperone for an arsenic detoxification pump. *Proceedings of the National Academy of Sciences of the United States of America, 103*(42), 15617–15622.

Lin, Y. F., Yang, J., & Rosen, B. P. (2007). ArsD residues Cys12, Cys13, and Cys18 form an As(III)-binding site required for arsenic metallochaperone activity. *The Journal of Biological Chemistry, 282*(23), 16783–16791.

Liu, J. X., Zhou, G. B., Chen, S. J., & Chen, Z. (2012). Arsenic compounds: revived ancient remedies in the fight against human malignancies. *Current Opinion in Chemical Biology.*

Liu, Y., Promeneur, D., Rojek, A., Kumar, N., Frokiaer, J., Nielsen, S., et al. (2007). Aquaporin 9 is the major pathway for glycerol uptake by mouse erythrocytes, with implications for malarial virulence. *Proceedings of the National Academy of Sciences of the United States of America, 104*(30), 12560–12564.

Liu, Z., Carbrey, J. M., Agre, P., & Rosen, B. P. (2004). Arsenic trioxide uptake by human and rat aquaglyceroporins. *Biochemical and Biophysical Research Communications, 316*(4), 1178–1185.

Liu, Z., Shen, J., Carbrey, J. M., Mukhopadhyay, R., Agre, P., & Rosen, B. P. (2002). Arsenite transport by mammalian aquaglyceroporins AQP7 and AQP9. *Proceedings of the National Academy of Sciences of the United States of America, 99*(9), 6053–6058.

Liu, Z., Styblo, M., & Rosen, B. P. (2006). Methylarsonous acid transport by aquaglyceroporins. *Environmental Health Perspectives, 114*(4), 527–531.

Lutkenhaus, J., & Sundaramoorthy, M. (2003). MinD and role of the deviant Walker A motif, dimerization and membrane binding in oscillation. *Molecular Microbiology, 48*(2), 295–303.

Ma, J. F., Yamaji, N., Mitani, N., Tamai, K., Konishi, S., Fujiwara, T., et al. (2007). An efflux transporter of silicon in rice. *Nature, 448*(7150), 209–212.

Maciaszczyk, E., Wysocki, R., Golik, P., Lazowska, J., & Ulaszewski, S. (2004). Arsenical resistance genes in Saccharomyces douglasii and other yeast species undergo rapid evolution involving genomic rearrangements and duplications. *FEMS Yeast Research, 4*(8), 821–832.

Maciaszczyk-Dziubinska, E., Migocka, M., & Wysocki, R. (2012). Acr3p is a plasma membrane antiporter that catalyzes As(III)/H^+ and Sb(III)/H^+ exchange in *Saccharomyces cerevisiae*. *Biochimica et Biophysica Acta, 1808*(7), 1855–1859.

Maciaszczyk-Dziubinska, E., Wawrzycka, D., Sloma, E., Migocka, M., & Wysocki, R. (2012). The yeast permease Acr3p is a dual arsenite and antimonite plasma membrane transporter. *Biochimica et Biophysica Acta, 1798*(11), 2170–2175.

Mansour, N. M., Sawhney, M., Tamang, D. G., Vogl, C., & Saier, M. H., Jr. (2007). The bile/arsenite/riboflavin transporter (BART) superfamily. *FEBS Journal, 274*(3), 612–629.

Marino, G. I., Castro-Parodi, M., Dietrich, V., & Damiano, A. E. (2010). High levels of human chorionic gonadotropin (hCG) correlate with increased aquaporin-9 (AQP9) expression in explants from human preeclamptic placenta. *Reproductive Science, 17*(5), 444–453.

Mateja, A., Szlachcic, A., Downing, M. E., Dobosz, M., Mariappan, M., Hegde, R. S., et al. (2009). The structural basis of tail-anchored membrane protein recognition by Get3. *Nature, 461*(7262), 361–366.

McDermott, J. R., Jiang, X., Beene, L. C., Rosen, B. P., & Liu, Z. (2010). Pentavalent methylated arsenicals are substrates of human AQP9. *Biometals, 23*(1), 119–127.

Meng, Y. L., Liu, Z., & Rosen, B. P. (2004). As(III) and Sb(III) uptake by GlpF and efflux by ArsB in *Escherichia coli*. *The Journal of Biological Chemistry, 279*, 18334–18341.

Mobley, H. L., & Rosen, B. P. (1982). Energetics of plasmid-mediated arsenate resistance in *Escherichia coli*. *Proceedings of the National Academy of Sciences of the United States of America, 79*(20), 6119–6122.

Mukhopadhyay, R., & Rosen, B. P. (2002). Arsenate reductases in prokaryotes and eukaryotes. *Environmental Health Perspectives, 110*(Suppl. 5), 745–748.

Mukhopadhyay, R., & Rosen, B. P. (1998). The *Saccharomyces cerevisiae ACR2* gene encodes an arsenate reductase. *FEMS Microbiology Letters, 168*, 127–136.

O'Halloran, T. V., & Culotta, V. C. (2000). Metallochaperones: an intracellular shuttle service for metal ions. *The Journal of Biological Chemistry, 275*(33), 25057–25060.

Ordóñez, E., Thiyagarajan, S., Cook, J. D., Stemmler, T. L., Gil, J. A., Mateos, L. M., et al. (2008). Evolution of metal(loid) binding sites in transcriptional regulators. *The Journal of Biological Chemistry, 283*(37), 25706–25714.

Pai, E. F., Krengel, U., Petsko, G. A., Goody, R. S., Kabsch, W., & Wittinghofer, A. (1990). Refined crystal structure of the triphosphate conformation of H-ras p21 at 1.35 A resolution: implications for the mechanism of GTP hydrolysis. *EMBO Journal, 9*(8), 2351–2359.

Pastor-Soler, N. M., Fisher, J. S., Sharpe, R., Hill, E., Van Hoek, A., Brown, D., et al. (2010). Aquaporin 9 expression in the developing rat epididymis is modulated by steroid hormones. *Reproduction, 139*(3), 613–621.

Qin, J., Fu, H. L., Ye, J., Bencze, K. Z., Stemmler, T. L., Rawlings, D. E., et al. (2007). Convergent evolution of a new arsenic binding site in the ArsR/SmtB family of metalloregulators. *The Journal of Biological Chemistry, 282*(47), 34346–34355.

Rae, T. D., Schmidt, P. J., Pufahl, R. A., Culotta, V. C., & O'Halloran, T. V. (1999). Undetectable intracellular free copper: the requirement of a copper chaperone for superoxide dismutase. *Science, 284*(5415), 805–808.

Ramaswamy, S., & Kaur, P. (1998). Nucleotide binding to the C-terminal nucleotide binding domain of ArsA: studies with an ATP analogue, 5' -p-fluorosulfonylbenzoyladenosine. *The Journal of Biological Chemistry, 273*(15), 9243–9248.

Ramírez-Solis, A., Mukopadhyay, R., Rosen, B. P., & Stemmler, T. L. (2004). Experimental and theoretical characterization of arsenite in water: insights into the coordination environment of As-O. *Inorganic Chemistry, 43*(9), 2954–2959.

Rensing, C., & Rosen, B. P. (2009). Heavy metals cycles (arsenic, mercury, selenium, others). In M. Schaechter (Ed.), *Encyclopedia of microbiology* (pp. 205–219). Oxford, U.K.: Elsevier.

Rosen, B. P. (1999). Families of arsenic transporters. *Trends in Microbiology, 7*, 207–212.
Rosen, B. P. (2002). Biochemistry of arsenic detoxification. *FEBS Letters, 529*(1), 86–92.
Rosen, B. P., Weigel, U., Karkaria, C., & Gangola, P. (1988). Molecular characterization of an anion pump. The *arsA* gene product is an arsenite(antimonate)-stimulated ATPase. *The Journal of Biological Chemistry, 263*(7), 3067–3070.
Rosenberg, H., Gerdes, R. G., & Chegwidden, K. (1977). Two systems for the uptake of phosphate in *Escherichia coli*. *Journal of Bacteriology, 131*(2), 505–511.
Ruan, X., Bhattacharjee, H., & Rosen, B. P. (2008). Characterization of the metalloactivation domain of an arsenite/antimonite resistance pump. *Molecular Microbiology, 67*(2), 392–402.
Ruan, X., Bhattacharjee, H., & Rosen, B. P. (2006). Cys-113 and Cys-422 form a high affinity metalloid binding site in the ArsA ATPase. *The Journal of Biological Chemistry, 281*(15), 9925–9934.
San Francisco, M. J., Hope, C. L., Owolabi, J. B., Tisa, L. S., & Rosen, B. P. (1990). Identification of the metalloregulatory element of the plasmid-encoded arsenical resistance operon. *Nucleic Acids Research, 18*(3), 619–624.
Sanders, O. I., Rensing, C., Kuroda, M., Mitra, B., & Rosen, B. P. (1997). Antimonite is accumulated by the glycerol facilitator GlpF in *Escherichia coli*. *Journal of Bacteriology, 179*(10), 3365–3367.
Sato, T., & Kobayashi, Y. (1998). The *ars* operon in the skin element of *Bacillus subtilis* confers resistance to arsenate and arsenite. *Journal of Bacteriology, 180*(7), 1655–1661.
Schindelin, H., Kisker, C., Schlessman, J. L., Howard, J. B., & Rees, D. C. (1997). Structure of ADP x AIF4(-)-stabilized nitrogenase complex and its implications for signal transduction. *Nature, 387*(6631), 370–376.
Shen, J., Hsu, C. M., Kang, B. K., Rosen, B. P., & Bhattacharjee, H. (2003). The *Saccharomyces cerevisiae* Arr4p is involved in metal and heat tolerance. *Biometals, 16*, 369–378.
Shi, W., Dong, J., Scott, R. A., Ksenzenko, M. Y., & Rosen, B. P. (1996). The role of arsenic-thiol interactions in metalloregulation of the *ars* operon. *The Journal of Biological Chemistry, 271*(16), 9291–9297.
Smith, A. H., Lingas, E. O., & Rahman, M. (2000). Contamination of drinking-water by arsenic in Bangladesh: a public health emergency. *Bulletin of the World Health Organization, 78*(9), 1093–1103.
Soignet, S. L. (2001). Clinical experience of arsenic trioxide in relapsed acute promyelocytic leukemia. *Oncologist, 6*(Suppl. 2), 11–16.
Stone, R. (2008). Food safety. Arsenic and paddy rice: a neglected cancer risk? *Science, 321*(5886), 184–185.
Story, R. M., & Steitz, T. A. (1992). Structure of the recA protein-ADP complex. *Nature, 355*(6358), 374–376.
Tchounwou, P. B., Centeno, J. A., & Patlolla, A. K. (2004). Arsenic toxicity, mutagenesis, and carcinogenesis—a health risk assessment and management approach. *Molecular and Cellular Biochemistry, 255*(1–2), 47–55.
Thomas, D. J., Waters, S. B., & Styblo, M. (2004). Elucidating the pathway for arsenic methylation. *Toxicology and Applied Pharmacology, 198*(3), 319–326.
Thorsen, M., Di, Y., Tangemo, C., Morillas, M., Ahmadpour, D., Van der Does, C., et al. (2006). The MAPK Hog1p modulates Fps1p-dependent arsenite uptake and tolerance in yeast. *Molecular Biology of the Cell, 17*(10), 4400–4410.
Tisa, L. S., & Rosen, B. P. (1990). Molecular characterization of an anion pump. The ArsB protein is the membrane anchor for the ArsA protein. *The Journal of Biological Chemistry, 265*(1), 190–194.
Tofail, F., Vahter, M., Hamadani, J. D., Nermell, B., Huda, S. N., Yunus, M., et al. (2009). Effect of arsenic exposure during pregnancy on infant development at 7 months in rural Matlab, Bangladesh. *Environmental Health Perspectives, 117*(2), 288–293.

Tseng, C. H., Tseng, C. P., Chiou, H. Y., Hsueh, Y. M., Chong, C. K., & Chen, C. J. (2002). Epidemiologic evidence of diabetogenic effect of arsenic. *Toxicological Letters, 133*(1), 69–76.
Villadangos, A. F., Fu, H. L., Gil, J. A., Messens, J., Rosen, B. P., & Mateos, L. M. (2012). Efflux permease CgAcr3-1 of Corynebacterium glutamicum is an arsenite-specific antiporter. *The Journal of Biological Chemistry, 287*(1), 723–735.
Wang, Z., Zhang, H., Li, X. F., & Le, X. C. (2007). Study of interactions between arsenicals and thioredoxins (human and *E. coli*) using mass spectrometry. *Rapid Communications in Mass Spectrometry, 21*(22), 3658–3666.
Williams, P. N., Price, A. H., Raab, A., Hossain, S. A., Feldmann, J., & Meharg, A. A. (2005). Variation in arsenic speciation and concentration in paddy rice related to dietary exposure. *Environmental Science and Technology, 39*(15), 5531–5540.
Williams, P. N., Raab, A., Feldmann, J., & Meharg, A. A. (2007). Market basket survey shows elevated levels of As in South Central U.S. processed rice compared to California: consequences for human dietary exposure. *Environmental Science and Technology, 41*(7), 2178–2183.
Willsky, G. R., Bennett, R. L., & Malamy, M. H. (1973). Inorganic phosphate transport in *Escherichia coli*: involvement of two genes which play a role in alkaline phosphatase regulation. *Journal of Bacteriology, 113*(2), 529–539.
Willsky, G. R., & Malamy, M. H. (1980). Effect of arsenate on inorganic phosphate transport in *Escherichia coli*. *Journal of Bacteriology, 144*(1), 366–374.
Wu, J., Tisa, L. S., & Rosen, B. P. (1992). Membrane topology of the ArsB protein, the membrane subunit of an anion-translocating ATPase. *The Journal of Biological Chemistry, 267*(18), 12570–12576.
Wysocki, R., Bobrowicz, P., & Ulaszewski, S. (1997). The *Saccharomyces cerevisiae ACR3* gene encodes a putative membrane protein involved in arsenite transport. *The Journal of Biological Chemistry, 272*(48), 30061–30066.
Wysocki, R., Chery, C. C., Wawrzycka, D., Van Hulle, M., Cornelis, R., Thevelein, J. M., et al. (2001). The glycerol channel Fps1p mediates the uptake of arsenite and antimonite in *Saccharomyces cerevisiae*. *Molecular Microbiology, 40*(6), 1391–1401.
Xia, X., Postis, V. L., Rahman, M., Wright, G. S., Roach, P. C., Deacon, S. E., et al. (2008). Investigation of the structure and function of a Shewanella oneidensis arsenical-resistance family transporter. *Molecular Membrane Biology, 25*(8), 691–705.
Yang, H. C., Cheng, J., Finan, T. M., Rosen, B. P., & Bhattacharjee, H. (2005). Novel pathway for arsenic detoxification in the legume symbiont Sinorhizobium meliloti. *Journal of Bacteriology, 187*(20), 6991–6997.
Yang, J., Rawat, S., Stemmler, T. L., & Rosen, B. P. (2010). Arsenic binding and transfer by the ArsD As(III) metallochaperone. *Biochemistry, 49*(17), 3658–3666.
Yang, J., Salam, A. A., & Rosen, B. P. (2011). Genetic mapping of the interface between the ArsD metallochaperone and the ArsA ATPase. *Molecular Microbiology, 79*(4), 872–881.
Ye, J., Ajees, A. A., Yang, J., & Rosen, B. P. (2010). The 1.4 Å crystal structure of the ArsD arsenic metallochaperone provides insights into its interaction with the ArsA ATPase. *Biochemistry, 49*(25), 5206–5212.
Ye, J., He, Y., Skalicky, J., Rosen, B. P., & Stemmler, T. L. (2010). Resonance assignments and secondary structure prediction of the As(III) metallochaperone ArsD in solution. *Biomolecular NMR Assignments, 2*(3), 211–219.
Ye, J., Rensing, C., Rosen, B. P., & Zhu, Y. G. (2012). Arsenic biomethylation by photosynthetic organisms. *Trends in Plant Science, 17*(3), 155–162.
Yoshinaga, M., Cai, Y., & Rosen, B. P. (2011). Demethylation of methylarsonic acid by a microbial community. *Environmental Microbiology, 13*(5), 1205–1215.

Zaman, G. J., Lankelma, J., van Tellingen, O., Beijnen, J., Dekker, H., Paulusma, C., et al. (1995). Role of glutathione in the export of compounds from cells by the multidrug-resistance-associated protein. *Proceedings of the National Academy of Sciences of the United States of America, 92*(17), 7690–7694.

Zhao, F. J., McGrath, S. P., & Meharg, A. A. (2010). Arsenic as a food chain contaminant: mechanisms of plant uptake and metabolism and mitigation strategies. *Annual Review of Plant Biology, 61*, 535–559.

Zhou, T., Liu, S., & Rosen, B. P. (1995). Interaction of substrate and effector binding sites in the ArsA ATPase. *Biochemistry, 34*(41), 13622–13626.

Zhou, T., Radaev, S., Rosen, B. P., & Gatti, D. L. (2000). Structure of the ArsA ATPase: the catalytic subunit of a heavy metal resistance pump. *EMBO Journal, 19*(17), 1–8.

Zhou, T., & Rosen, B. P. (1997). Tryptophan fluorescence reports nucleotide-induced conformational changes in a domain of the ArsA ATPase. *The Journal of Biological Chemistry, 272*(32), 19731–19737.

CHAPTER THIRTEEN

Extracellular Heme Uptake and the Challenges of Bacterial Cell Membranes

Aaron D. Smith and Angela Wilks[1]
Department of Pharmaceutical Sciences, School of Pharmacy, University of Maryland, Baltimore, USA
[1]Corresponding author: E-mail: awilks@rx.umaryland.edu

Contents

Abstract

In bacteria, the fine balance of maintaining adequate iron levels while preventing the deleterious effects of excess iron has led to the evolution of sophisticated cellular mechanisms to obtain, store, and regulate iron. Iron uptake provides a significant challenge given its limited bioavailability and need to be transported across the bacterial cell wall and membranes. Pathogenic bacteria have circumvented the iron-availability issue by utilizing the hosts' heme-containing proteins as a source of iron. Once internalized, iron is liberated from the porphyrin enzymatically for cellular processes within the bacterial cell. Heme, a lipophilic and toxic molecule, poses a significant challenge in terms of transport given its chemical reactivity. As such, pathogenic bacteria have evolved sophisticated membrane transporters to coordinate, sequester, and transport heme. Recent advances in the biochemical and structural characterization of the membrane-bound heme transport proteins are discussed in the context of ligand coordination, protein–protein interaction, and heme transfer.

Current Topics in Membranes, Volume 69
ISSN 1063-5823,
http://dx.doi.org/10.1016/B978-0-12-394390-3.00013-6

1. INTRODUCTION

Aerobic respiration has led to natural selection of elements essential for the diversity of chemical reactions required for life in an oxygen environment. Iron due to its versatility in ligand coordination and tunable redox state is an essential element for the survival of most organisms. It facilitates many cellular processes ranging from dioxygen transport, energy transducing pathways, and nitrogen and hydrogen fixation (Aisen, Enns, & Wessling-Resnick, 2001; Papanikolaou & Pantopoulos, 2005). Iron despite being the fourth most abundant metal in the earth's crust is not readily available at physiological pH as it forms insoluble iron (II/III) oxides, hematite (Fe_2O_3) and magnetite (Fe_3O_4). Alternatively, when completely reduced, Fe(II) is soluble and promotes the generation of toxic-free radicals through Fenton chemistry. These reactive oxygen species (ROS) can have deleterious effects on the cell damaging lipids, proteins, and nucleic acids. For bacterial pathogens, iron is most often the limiting factor for growth and a key virulence determinant (Anzaldi & Skaar, 2010; Braun, 2005; Brown & Holden, 2002; Bullen, Rogers, Spalding, & Ward, 2006; Cornelissen & Sparling, 1994; Payne, 1993). Bacterial pathogens face even greater hardship obtaining iron due to the host innate immune response (Doherty, 2007; Ganz, 2009; Nairz, Schroll, Sonnweber, & Weiss, 2010). This response is orchestrated by the combined action of the key regulators of iron homeostasis, hepcidin and ferroportin (De Domenico, Ward, & Kaplan, 2011; Ganz, 2011). Indeed, hepcidin was first isolated as an antimicrobial peptide produced in the liver and macrophages that directly sequestered iron (Verga Falzacappa & Muckenthaler, 2005). Thus, hepcidin has a dual function as an iron chelator and regulator of iron homeostasis by inducing hypoferremia or more specifically a low-serum iron-binding capacity and increased levels of ferritin within the tissues. In addition, there are other frontline defense proteins that reduce the levels of iron including lipocalin-2, which can directly bind bacterial siderophores such as enterobactin, and lactoferrin (Lf) that sequester any free iron (Bachman, Miller, & Weiser, 2009; Borregaard & Cowland, 2006; Clifton, Corrent, & Strong, 2009). Lipocalin-2 [also known as siderocalin or neutrophil gelatinase-associated protein (NGAL)] is secreted by neutrophils, macrophages, and other cell types and is upregulated as a component of the acute-phase response to infection. Lf, a glycoprotein similar to transferrin (Tf), is found in high concentrations in neutrophils and mucous secretions. Lf possesses a wide range of antimicrobial and immunomodulaory properties in addition to the ability to bind iron (Gonzalez-Chavez, Arevalo-Gallegos, & Rascon-Cruz, 2009; Jenssen & Hancock, 2009).

As a means of circumventing the innate immune response, bacterial pathogens have evolved several redundant mechanisms to obtain iron including the utilization of heme. With 75% of total body iron being complexed in heme, this presents a prime reservoir of iron for invading pathogens (Otto, Sparrius, Wors, de Graaf, & MacLaren, 1994). Recently, it has been shown in *Staphylococcus aureus* and *Bordetella pertussis* that heme is the preferred source of iron during specific stages of infection (Brickman, Vanderpool, & Armstrong, 2006; Skaar, Humayun, Bae, DeBord, & Schneewind, 2004). As would be expected for a biologically active iron chelate, heme, if not adequately sequestered, can generate ROS, and due to its hydrophobic properties, heme can readily intercalate into membrane lipids disrupting organelle and cell structure (Balla et al., 1993; Belcher, Beckman, Balla, Balla, & Vercellotti, 2010; Halliwell & Gutteridge, 1984). As such heme biosynthesis and degradation are tightly regulated and coupled to the cellular heme protein levels. Therefore, capture of heme from cellular heme proteins, most notably hemoglobin (Hb), presents a major challenge to bacterial pathogens. Additionally, heme must then traverse the thick peptidoglycan cell wall of gram-positive organisms and outer membrane (OM) and cytoplasmic membrane (CM) of gram-negative bacteria. The heme uptake systems comprise a concerted network of highly evolved proteins that (a) sequester heme with high affinity in a redox inert state and (b) utilize both the free energy of protein–protein interaction and the energy generating systems of the cell membrane to translocate heme into the cytoplasm.

2. HEME CAPTURE AND TRANSPORT ACROSS THE CELL WALL AND MEMBRANES

2.1. Active Transport Across the Gram-Negative OM

Gram-negative pathogens must actively transport heme through the OM and shuttle it across the periplasmic space (Fig. 1). The OM an asymmetric lipid bilayer protects the organism from the external environment including from the effects of extracellular toxins. The inner layer of the OM is

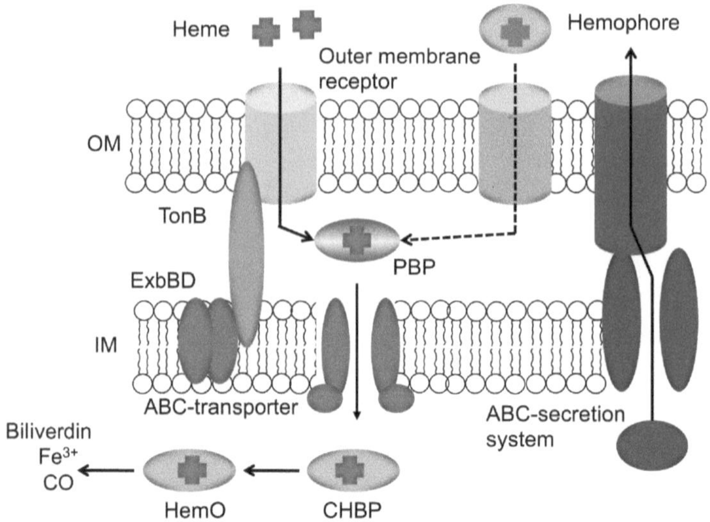

Figure 1 Heme transport across the gram-negative OM. Heme is transported across the OM by a TonB-dependent OM receptor, shuttled to an ABC transporter by the PBP, and actively transported across the inner membrane (IM). In the cytoplasm, heme is sequestered by the cytoplasmic heme-binding protein (CHBP) for transfer to HemO. In some bacteria, such as *Serratia marcescens* and *Pseudomonas aeruginosa*, a secreted hemophore sequesters heme for transport via a hemophore-dependent OM receptor.

composed of phospholipids and the outer layer of lipopolysaccharide that provides the bacteria with a negatively charged polar surface preventing the influx of lipophilic compounds. Therefore, within the membrane, there are numerous transporters for the active influx and efflux of molecules such as heme. All heme OM transporters have a similar closed β-barrel architecture consisting of 22 antiparallel β-strands and a globular domain most often referred to as the plug, cork, or hatch that occludes the pore of the β-barrel. The β-strands are connected by short turns in the periplasm and by long, hydrophilic loops at the cell surface that play a critical role in protein–protein interaction and heme binding (Fig. 2A). The TonB-dependent OM receptors bear significant sequence and structural homology to the more well-characterized iron–siderophore receptors (Braun & Endriss, 2007; Wandersman & Delepelaire, 2004; Wilks & Burkhard, 2007). In addition to high structural similarities, multiple sequence alignment of all the known heme OM receptors identified two absolutely conserved histidine residues located in a conserved FRAP/NPNL domains that are essential for receptor function (Bracken, Baer, Abdur-Rashid, Helms, & Stojiljkovic, 1999). The

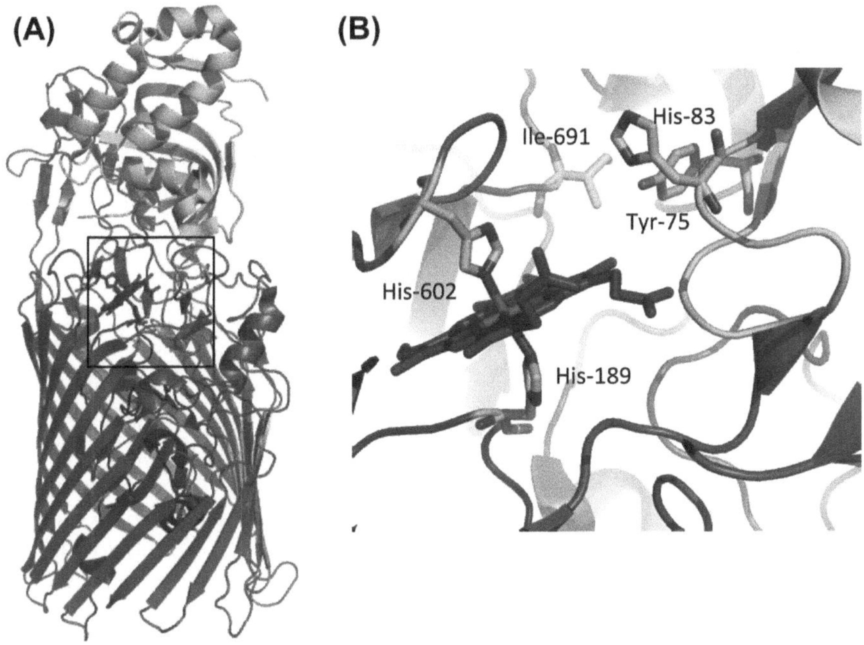

Figure 2 Structure of the HasA–HasR complex of *Serratia marcescens*. (A) HasA docked to the HasR through the extracellular loops. (B) Enlarged view of the heme-binding site. Heme is coordinated in HasR by His-602 from extracellular loop 7 and His-189 on the N-terminal plug. *Adapted from PDB file 3CSL.* See the color plate.

OM heme receptors of gram-negative organisms fall into two classes, those that scavenge heme through the secretion of a hemophore (Cescau et al., 2007; Wandersman & Delepelaire, 2004; Wandersman & Stojiljkovic, 2000) and those that recognize specific heme protein complexes such as Hb or Hb–haptoglobin (Burkhard & Wilks, 2007; Leduc et al., 2008; Rohde & Dyer, 2004; Simpson, Olczak, & Genco, 2000). As is the case with iron–siderophore uptake, there is a level of redundancy whereby bacteria encode several distinct heme uptake systems. In *Pseudomonas aeruginosa*, two heme uptake operons have been characterized, the *has* (*heme acquisition system*) which encodes a hemophore (HasA) and the *phu* (*Pseudomonas* heme utilization) system that utilizes the hosts' heme proteins directly (Ochsner, Johnson, & Vasil, 2000).

The analogous hemophore system in *Serratia marcescens* has been extensively studied and much of our mechanistic understanding of these heme transporters derives from this system. The OM receptor, HasR, of *S. marcescens* can acquire free heme at micromolar concentrations (Letoffe,

Ghigo, & Wandersman, 1994). However, the efficiency of heme uptake is greatly enhanced in the presence of the hemophore, HasA, with heme or Hb (Ghigo, Letoffe, & Wandersman, 1997). HasA binds heme with a very high affinity ($K_a = 5.3 \times 10^{10}$/M) (Letoffe et al., 1994). However, there is no apparent affinity for HasA with Hb nor does it seem to increase the rate of heme dissociation from Hb, and it is proposed that heme is captured by HasA as it spontaneously dissociates from Hb (Yukl et al., 2010). HasA coordinates heme axially through His-32 and Tyr-75, with His-83 forming a hydrogen bond to Tyr-75, strengthening the Tyr–iron bond contributing to this high affinity for heme (Deniau et al., 2003; Letoffe et al., 2001). The H83A mutation resulted in a 256-fold decrease in affinity for heme, indicating its importance for stabilizing heme in the binding pocket (Letoffe et al., 1994). The function of holo-HasA is to deliver heme to the OM receptor HasR and it has been shown to form an extremely tight complex with the HasR receptor (Izadi-Pruneyre et al., 2006; Letoffe, Nato, Goldberg, & Wandersman, 1999). HasR, however, has a much lower affinity for heme ($K_a = 5 \times 10^6$/M) (Izadi-Pruneyre et al., 2006). Surprisingly, heme transfer from HasA to HasR occurs spontaneously and independently of TonB; this otherwise thermodynamically disfavorable transfer of heme from a very high-affinity site to a lower affinity site is driven by the highly exothermic and exergonic protein–protein interaction (Izadi-Pruneyre et al., 2006). Once heme is transferred to HasR, heme is coordinated by two conserved histidine residues, His-189 and His-603. Molecular modeling of the HasR receptor identified His-603 as being located on the extracellular loop L7 and His-189 resides on the extracellular face of the N-terminal "plug" (Barjon, Wecker, Izadi-Pruneyre, & Delepelaire, 2007). These observations have been corroborated by UV-vis and resonance Raman (rR) spectroscopic studies identifying a ferric 6-coordinate low spin (6CLS) (Izadi-Pruneyre et al., 2006).

The mechanism of heme transfer from the high-affinity hemophore to the OM receptor was further revealed on crystallization of the HasA/holo–HasR complex. Sequence alignment of HasR with the closely related siderophore TonB-dependent OM receptors identified exceptionally large extracellular loops L2, L6, and L9 of HasR. Mutagenesis studies revealed that two of these loops, L6 and L9, along with L8 were critical for binding HasA (Barjon et al., 2007). The recently solved crystal structure of the HasA:HasR:heme complex further shows that loops L6, L9, and L8 all form contacts with HasA and bury a large surface area of the protein (Krieg et al.,

2009). In the structure, heme is located in the channel of HasR, coordinated by His-189 and His-603, which is a 9.2 Å translation from the binding pocket of HasA, along with rotation of the porphyrin ring and exposing the proprionates to solvent (Fig. 2B). Additionally, the loop bearing the coordinating histidine in HasA (His-32) is displaced from the binding pocket by loops L7 and L8 of HasR, thereby weakening the affinity of heme for the hemophore. Interestingly, superposition of heme in the binding pocket of HasA in the HasA:HasR complex reveals that Ile-671 on extracellular loop L8 of HasR would involve a steric clash with the heme (Krieg et al., 2009). This observation led to a hypothesis for heme release from the high-affinity HasA site to the lower affinity HasR site through protein–protein-induced conformational change. The protein–protein-induced conformational change weakens the His-32 heme ligand of HasA followed by steric displacement of heme by Ile-671 triggering heme release to HasR. Furthermore, His-83 in HasA which hydrogen bonds to the axial Tyr-75 ligand is rotated away form the active site in the HasA/HasR structure which would impede heme sliding back into the HasA binding pocket (Krieg et al., 2009). The mechanisms were further supported on mutation of HasRIle-671 to Gly where the ability to extract heme from HasA was abolished while retaining its ability to bind free heme. Furthermore, the crystal structures of holo-HasA/HasR-I671G revealed that in contrast to the wild-type heme remained bound to HasA. Upon release of heme from the hemophore and capture by the receptor, the energy provided through TonB then supposedly facilitates translocation of heme into the periplasm and returning the receptor to the ground state.

The second class of OM receptors do not rely on recognition of heme-loaded hemophores but instead interact with and directly capture heme from host heme proteins. OM receptors of this type have been identified in many gram-negative pathogens such as *Yersinia* sp. (Rossi et al., 2001; Stojiljkovic and Hantke, 1992, 1994), *Neisseria* sp. (Stojiljkovic, Larson, Hwa, Anic, & So, 1996; Stojiljkovic et al., 1995), *Vibrio* sp. (Henderson & Payne, 1994), and *P. aeruginosa* (Ochsner et al., 2000). These receptors have high sequence identity (25–65%) and structural similarity to the hemophore-dependent receptors, and their related iron–siderophore OM receptors (17–23%), suggestive of a common mechanism of heme transport (Braun & Endriss, 2007; Wandersman & Delepelaire, 2004; Wandersman & Stojiljkovic, 2000; Wilks & Burkhard, 2007).

The non-hemophore receptors are typified by ShuA from *Shigella dysenteriae* which interacts and strips heme directly from heme proteins such as Hb. Similar to the bis–His coordination in HasR, ShuA has a conserved His-86 on the N-terminal plug and His-420 on extracellular loop L7. The recent crystal structure of apo-ShuA confirms His-86 and His-420 location on the N-terminal plug and extracellular loop, respectively. However, His-86 and His-420 are 9.86 Å apart, suggesting that a significant conformational change must take place upon heme binding (Cobessi, Meksem, & Brillet, 2010). A mechanism based on the accumulated biochemical data and the HasA/HasR structure has His-420 coordinating to the heme in the first ligation step with hydrophobic residues on the L7 loop interacting with the porphyrin and displacing the heme which is coordinated to the N-terminal plug His which then forms a bis–histidine ligation through the loop His sequestering heme within ShuA and preventing back transfer of the heme. In this way, the heme is coordinated in a redox-inactive state excluded from the extracellular environment by closure of the loops over the barrel. Kinetic analysis of heme transfer from Hb revealed that transfer from oxidized form of the Hb dimer ($metHb_2$) or tetramer ($metHb_4$) is unidirectional and on the order of 10^4 faster than the rate of free heme association and 10^3 faster than heme transfer from oxyhemoglobin, indicating that the oxidized form of Hb is most likely the physiological substrate for ShuA (Burkhard & Wilks, 2007). Biochemical characterization of the wild type, His-86 Ala and His-420 Ala mutants, revealed that both residues were required for heme acquisition from methemoglobin (Burkhard & Wilks, 2007). Kinetic studies indicated that heme could be released from Hb to the ShuA H86A or ShuA H420A receptor but in the absence of the second His residue the heme is transferred back to methemoglobin. In contrast, the H86A/H420A ShuA mutant had no residual heme bound to the receptor and was also unable to extract heme from methemoglobin. Thus, both His residues are essential for the unidirectional transfer of heme from methemoglobin to the OM receptor, ShuA.

The mechanism of translocation through the OM β-barrel receptors has been the subject of intense study with conflicting mechanisms having been proposed. As no energy sources are accessible in the periplasm (e.g. ATP, protein gradients), the OM heme receptors must harness energy generated by the proton-motive force of the CM to transport ligands into the periplasm. Heme transport is driven by the CM electrochemical gradient that is then transduced across the periplasm to the receptor through the TonB–ExbB–ExbD complex. The TonB protein interacts with the OM receptor through a consensus sequence on the receptor known as the "TonB box" (Braun & Endriss, 2007). The first step in heme transport across the OM most likely involves a conformational change on heme binding that exposes the TonB box and promotes an interaction between the receptor and TonB (Khursigara, De Crescenzo, Pawelek, & Coulton, 2004, 2005). Crystal structures of the closely related ferrichrome (FhuA) and cobalamin (BtuB) OM receptors show that, upon binding of TonB, an interprotein β-sheet is formed at the TonB box (Pawelek et al., 2006; Shultis, Purdy, Banchs, & Wiener, 2006). The formation of this β-sheet positions the TonB α-helix 1a in proximity to the plug where extensive electrostatic interactions drive conformational changes within the plug (Pawelek et al., 2006; Shultis et al., 2006). Structural and computational experiments to date suggest a shearing or mechanical pulling mechanism, where hydration of the central β-sheet of the cork domain disrupts interactions with the inner face of the β-barrel. The perpendicular force applied to the β-sheet on interaction with TonB results in local unfolding of the cork creating a channel for passage of the ligand (Chimento, Kadner, & Wiener, 2005; Faraldo-Gomez, Smith, & Sansom, 2003; Gumbart, Wiener, & Tajkhorshid, 2007; Pawelek et al., 2006; Shultis et al., 2006). Heme translocated through the OM receptor is captured by the periplasmic heme-binding protein that then acts as a soluble receptor for a heme-dependent ABC (ATP-binding cassette) transporter in the CM (Koster, 2005; Krewulak & Vogel, 2008). The mechanism of heme release from the OM receptors to the periplasmic binding proteins (PBPs) is not known. However, studies have shown that the TonB protein acts as a scaffold for recruitment of the PBP placing it in the vicinity to readily access the ligand and transport it to the ABC transporter (Carter et al., 2006). Presumably, the heme affinity for the OM receptor must decrease upon interaction with the PBP with the energetic cost of heme transfer being met by the free energy from complexation.

2.2. ATP-Dependent Transport Across the Cytoplasmic Membrane

The soluble periplasmic and lipid-anchored substrate-binding proteins deliver heme to a specific ABC transporter in gram-negative and gram-positive pathogens, respectively. The gram-negative bacterial PBPs are critical in shuttling substrates across the periplasm from the OM to the CM. PBPs are divided into three classes based on the number of interdomain connections (Borths, Locher, Lee, & Rees, 2002; Dwyer & Hellinga, 2004). All three classes of PBPs are comprised of N- and C-terminal globular domains that on closure form ligand-binding cleft between the domains. The Class I and II PBPs have two or three interdomain connections, respectively. Class I proteins, including the maltose-binding protein, and Class II proteins, such as ribonucleotide reductase, undergo large structural changes which are required for substrate binding and release (Quiocho & Ledvina, 1996). Class III proteins have one interdomain connection between the globular domains and undergo relatively small conformational changes (Borths et al., 2002; Clarke, Braun, Winkelmann, Tari, & Vogel, 2002; Karpowich, Huang, Smith, & Hunt, 2003). Proteins included in Class III are BtuF (Borths et al., 2002), a vitamin B_{12}-binding protein, and FhuD (Clarke et al., 2002), a protein that binds hydroxamic acid siderophores. The structurally determined heme-binding proteins ShuT and PhuT have very similar architecture to BtuF where the N- and C-domains consist of a five-stranded β-sheet flanked by helices (Fig. 3A). Heme is ligated through a Tyr located on the N-terminal domain at the interface of the heme-binding cleft (Fig. 3B). The Tyr ligand points away from the binding pocket in the apoprotein and on heme coordination reorients toward the pocket but there is little overall change in the structure. A similar reorganization of the proximal ligand is observed in the ferric–siderophore FhuD protein and has been proposed to play a role in guiding the ligand into the pocket (Zhu, Alexeev, Hunter, Campopiano, & Sadler, 2003). In the *S. dysenteriae* system, ShuT accepts heme from ShuA and forms a highly stable complex with Tyr-94 (Eakanunkul et al., 2005; Ho et al., 2007). Binding affinities for the PBP ShuT have not been measured directly due to residually bound heme

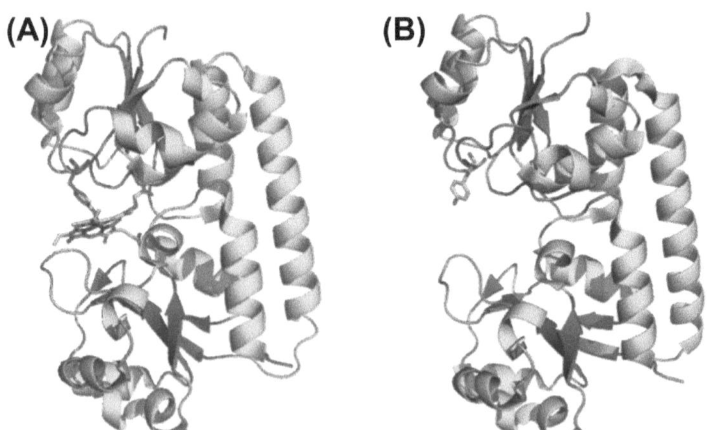

Figure 3 Overall structural fold of the holo- and apo-ShuT. (A) Holo-ShuT with the proximal Tyr-94 highlighted. (B) Apo-ShuT with the proximal Tyr-94 pointing outward. *Adapted from PDB file 2R7A.* See the color plate.

present following purification of the protein; however, rate constants of heme extraction by apo-hemoglobin ($K_d = 10^{14}$) indicate a high affinity of ShuT for heme (Eakanunkul et al., 2005). The apparent rate constant of heme exchange from holo-ShuT to apo-myoglobin is 4×10^{-5} M (Eakanunkul et al., 2005). This rate decreases 10^4-fold when Tyr-94 is mutated to Ala (Eakanunkul et al., 2005). Although mechanistic details of ShuT release from ShuA are not yet known, it is expected that conformational changes resulting from heme coordination to ShuT play a role in release of the PBP to the periplasm where it can interact with the ABC transporter to release heme for transport across the CM. While the proximal face of the heme-binding cleft of ShuT and PhuT are similar, the distal side has significant differences. The largest of these differences is a hairpin loop after the β-6 strand that is pointing up in PhuT and down in ShuT. Due to the positioning of the loop in the ShuT cleft, it has the potential to make van der Waals contacts with the heme. A second hairpin loop in PhuT protrudes into the heme cleft providing an Arg that is able to interact with one of the heme propionates. It has been proposed, based on computational studies, that such hydrogen bonding networks play an important role in heme binding and release, suggesting that sequence diversity in the active site can be tolerated as long as a suitable hydrogen bond network is retained (Liu, Su, et al., 2008). The lipoprotein relay network of gram-positive organisms will be discussed in the following section.

ABC transporters constitute one of the largest protein superfamilies and are primarily responsible for the ATP-dependent transport of substrate molecules across cellular membranes (Higgins, 1992, 2001). The ABC transporters are composed of two transmembrane domains (TMDs), which form the translocation channel, and two cytoplasmic nucleotide-binding domains (NBDs) that bind and hydrolyze ATP. The TMD vary in sequence and chain length allowing for specificity of substrate, while the ATP-binding domains are highly conserved with the Walker A and Walker B ATP-binding motifs, a glutamine residue on the Q-loop, a highly conserved histidine residue in the Switch region, and the LSGGQ "signature motif" which is unique to NBD proteins (Holland & Blight, 1999; Hopfner et al., 2000; Schneider & Hunke, 1998; Walker, Saraste, & Gay, 1982).

Transport of heme across the CM by the ABC transporters is proposed to be similar to that of the previously characterized Vitamin B_{12} transport system BtuCD. The ABC transporter comprises two transmembrane BtuC subunits and two corresponding ATP-binding BtuD subunits (Hvorup et al., 2007). A general mechanism for ligand translocation in the type II

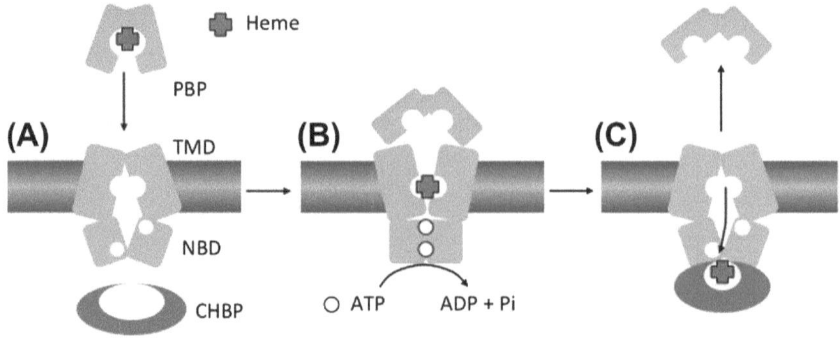

Figure 4 Proposed ATP-switching model for the heme ABC transporters. (A) PBP interacts with the TMD transporter causing a conformational change. (B) Conformational changes in TMD drive ATP binding and release of the ligand to the transporter channel. (C) ATP hydrolysis then couples translocation heme to the cytoplasmic heme-binding protein (CHBP). *Adapted from Burkhard and Wilks (2008).*

ABC transporters is proposed to be via an ATP-switching model (Fig. 4) (Higgins & Linton, 2004). In such a model, the ligand-bound PBP binds to the TMD of the ABC transporter inducing a conformational change to the outside facing state. This lowers the activation energy for ATP binding to the NBD. ATP binding induces the closed dimer formation of the NBD inducing the inward facing conformation of the TMD for active transport to the cytoplasm. ATP hydrolysis drives release of the ligand to the cytoplasm, following which the NBD dimer is destabilized resetting the transporter. Recent molecular dynamic simulations based on the various structures of the BtuCD complex support the transition from the outward to inward facing state being dependent on conformational changes in the NBDs (Weng, Fan, & Wang, 2012).

In the case of the heme–ShuUV transporter, release of heme from ShuT to ShuU presumably correlates with an opening of the lobes of ShuT and ShuU upon binding. This hypothesis is supported by crystallographic data of the vitamin B_{12} transporter, BtuCD, complexed with BtuF (Hvorup et al., 2007). Based on current findings with the ShuUV system, this step in ligand release occurs in the absence of ATP and is consistent with a model in which conformational changes reduce the heme affinity of ShuT, releasing heme to the lower affinity site within the transporter (Fig. 4). A sequence alignment of all predicted heme ABC transporters reveals two conserved His residues in ShuU His-252 and His-262 that are also present in the vitamin B_{12} transporter, BtuC. Crystal structures of BtuCD reveal the His residues to be located in the translocation channel (Hvorup et al., 2007; Locher, Lee, &

Rees, 2002). Heme translocation studies show that both His residues are required for transfer of heme through the ABC transporter, ShuUV; however, the detailed molecular mechanism of transport has not yet been determined (Burkhard & Wilks, 2008).

Unlike the well-characterized vitamin B_{12} ABC transporters release of heme to the cytoplasm requires interaction of the ShuUV transporter with the cytoplasmic heme-binding protein, ShuS (Burkhard & Wilks, 2008). The ShuUV transporter when reconstituted in liposomes requires the cytoplasmic heme-binding protein ShuS to be present in the external media for active translocation of heme from the lumen (Burkhard & Wilks, 2008). In the absence of the cytoplasmic heme-binding protein, no ATP hydrolysis or heme translocation occurs. This was the first direct evidence that the physiological function of the cytoplasmic heme-binding proteins is to capture exogenously acquired heme following translocation through the CM. Earlier studies suggested a homolog of the *S. dysenteriae* ShuS protein from *Y. enterocolitica* (HemS) encoded a hemeoxygenase (HO) required for iron release (Stojiljkovic & Hantke, 1994). However, this annotation was based on the observation that deletion of the gene caused heme toxicity. The ChuS protein of *Escherichia coli* 0157:H7 has been reported to be an HO based on in vitro assays reporting degradation of heme in the presence of a reducing agent (Suits, Jaffer, & Jia, 2006; Suits et al., 2005). However, it has not been determined if the catalytic turnover is greater than one or that the product of the reaction is indeed biliverdin. The enzymatic activity remains questionable given the nonenzymatic "coupled oxidation" of heme described almost half a century ago (O'Carra & Colleran, 1969, 1970). Furthermore, it has recently been shown that the role of the cytoplasmic heme-binding proteins is to sequester and traffic heme within the cell (Lansky et al., 2006). In vivo and in vitro studies suggest that once the cytoplasmic heme-binding protein has obtained heme from the ABC transporter, it shuttles the heme to an iron-regulated hemeoxygenase (HemO; previously annotated as both *pa-HO* and PigA) for degradation (Bhakta & Wilks, 2006; Block et al., 2007; Kaur, Lansky, & Wilks, 2009; Lansky et al., 2006).

2.3. Transport Across the Gram-Positive Cell Wall

The extensive peptidoglycan cell wall in gram-positive organisms is required for structural rigidity and protection from osmotic lysis; however, these properties pose a significant hurdle to ligand transport. Moreover, the redox reactivity of heme requires sequential delivery via specific protein–protein interactions to prevent oxidation of membrane lipids. The transport of heme through the peptidoglycan cell wall and across the membrane involves several proteins from surface-exposed lipoproteins that are anchored in the cell wall to the ABC transporters in the cell membrane (Grigg, Ukpabi, Gaudin, & Murphy, 2010; Maresso & Schneewind, 2006) (Fig. 4). The most well-characterized gram-positive heme transport system is that of *S. aureus* (Maresso & Schneewind, 2006; Mazmanian et al., 2003) although others have recently been identified in *Streptococcus pyogenes* (Bates et al., 2003; Ekins & Niven, 2003; Hanks, Liu, McClure, & Lei, 2005; Liu & Lei, 2005; Sook et al., 2008; Zhu, Liu, & Lei, 2008), *Listeria monocytogenes* (Jin et al., 2006), and *Corynebacterium diphtheria* (Drazek, Hammack, & Schmitt, 2000). A similar system has been identified in *Mycobacterium tuberculosis* and encodes a secreted protein and two membrane proteins required for heme transport across the cell wall (Tullius et al., 2011).

The *S. aureus* iron-regulated heme uptake gene cluster (*isd*) encodes a sortase (*srtB*) that is required for sorting and anchoring the surface-exposed heme transport proteins to the cell wall (Mazmanian, Ton-That, Su, & Schneewind, 2002) (Fig. 4). Heme acquisition by *S. aureus* involves the cell wall-anchored IsdA, IsdB, IsdC, and IsdH (HarA) and IsdD, IsdE, and IsdF that comprise the transmembrane ABC transporter (Mazmanian et al., 2003; Skaar & Schneewind, 2004). In addition to the transport proteins, the *isd* gene cluster encodes the heme-degrading IsdG and IsdI proteins. The surface proteins IsdH and IsdB directly extract heme from Hb and Hb–haptoglobin complexes (Dryla, Gelbmann, von Gabain, & Nagy, 2003; Mazmanian et al., 2003; Pilpa et al., 2009; Torres, Pishchany, Humayun, Schneewind, & Skaar, 2006). Several studies have shown sequential transfer of heme from the surface proteins IsdH and IsdB through IsdA to IsdC which then delivers heme to the membrane proximal IsdE (Fig. 4) (Liu, Tanaka, et al., 2008; Muryoi et al., 2008). Holo-IsdA cannot deliver heme to apo-IsdE and requires facilitation by apo-IsdC (Zhu, Xie, et al., 2008). An alternative study suggests that IsdA and IsdB physically interact and colocalize to a distinct site within the cell wall coordinating the transfer of heme from Hb at the cell surface to the cell wall proteins (Pishchany, Dickey, & Skaar, 2009). In the final uptake step, IsdE, the cell membrane lipoprotein component of the permease transporter (IsdDEF) transfers heme to the membrane-bound transporter IsdD/F.

All cell wall-anchored Isd proteins contain one to three copies of the near iron transport (NEAT) domains (Fig. 5A) (Andrade, Ciccarelli, Perez-Iratxeta, & Bork, 2002; Grigg et al., 2010). The domains acquired their

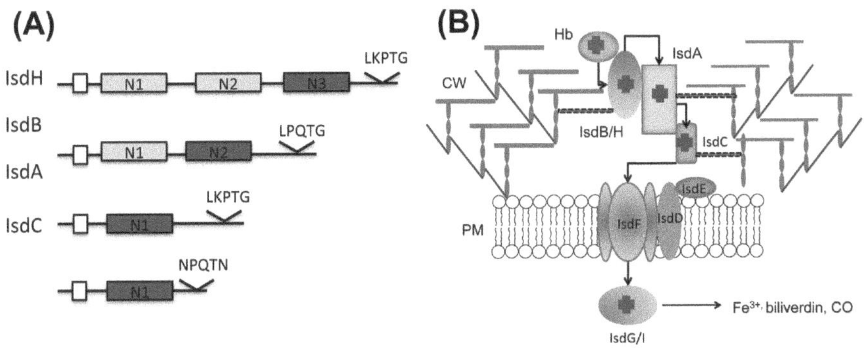

Figure 5 Heme uptake system of gram-positive organisms. (A) Schematic representation of the Isd NEAT domain proteins with the secretion signal represented by a white box. The NEAT domains are numbered from the N-terminus and the heme-binding NEAT domains are shown by the dark boxes. The sortase anchoring sequence for the respective proteins are indicated at the C-terminus. (B) Heme transfer across the cell wall (CW) and periplasmic membrane (PM). IsdB/H, IsdA, and IsdB are anchored to the cell wall and IsdD, IsdE and IsdF are the heme-binding protein and permease components. Heme transported into the cytoplasm is degraded by the IsdG/I system. *Adapted from Grigg et al. (2010).*

names as a result of their vicinity in the genome to putative iron chelate ABC transporters. The Isd surface protein has a secretion signal peptide for targeting, a sortase anchoring signal for attachment to the cell wall, and at least one NEAT domain. The multiple NEAT domain proteins are separated by sequences (15–20 residues) that are often highly charged and predicted to be disordered (Grigg et al., 2010). In proteins with multiple NEAT domains, such as IsdH, not all domains bind heme but are required to increase the binding affinity to the target protein through increased protein–protein interaction. A number of NEAT domains have been structurally characterized including IsdA N1 (holo and apo), IsdC N1 (holo and apo), IsdH N1 (apo), IsdH N3 (holo and apo) (Grigg, Vermeiren, Heinrichs, & Murphy, 2007; Pilpa et al., 2006; Sharp, Schneider, Cockayne, & Paoli, 2007; Villareal, Pilpa, Robson, Fadeev, & Clubb, 2008; Watanabe et al., 2008), and IsdB N2 (Gaudin, Grigg, Arrieta, & Murphy, 2011). All the NEAT domains have the same overall eight-stranded β-sandwich fold architecture with an α-helix connecting the β1- and β2-strands that in the heme-coordinating NEAT domains is stabilized on ligand binding (Fig. 6B) (Pilpa et al., 2006; Villareal et al., 2008). Structural and spectroscopic studies of the holo-NEAT domains reveal an absolutely conserved Tyr ligand located in a hydrophobic pocket on the β8-strand (Fig. 6A) (Grigg et al., 2007; Mack,

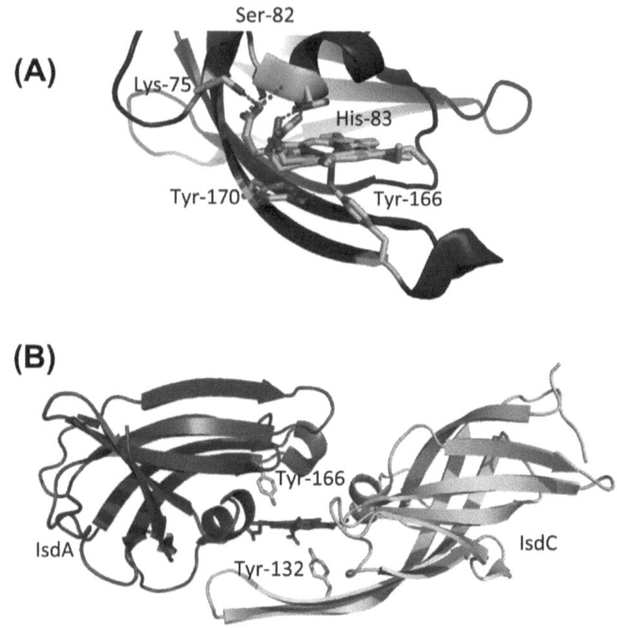

Figure 6 The NEAT domains of IsdA and IsdC. (A) Heme-binding site of IsdA highlighting hydrogen bond interactions of Ser-82, Lys-75, and Tyr-170 with the heme propionates and proximal Tyr-166, respectively. (B) Heme transfer from IsdA to IsdC via the "handclasp" model. Heme slides 18 Å from one binding pocket to the other. *Adapted from PDB files 3QZO and 2O6P.* See the color plate.

Vermeiren, Heinrichs, & Stillman, 2004; Muryoi et al., 2008; Sharp et al., 2007; Tiedemann, Muryoi, Heinrichs, & Stillman, 2008; Vermeiren, Pluym, Mack, Heinrichs, & Stillman, 2006; Watanabe et al., 2008). A second Tyr, located four residues downstream, hydrogen bonds with the coordinating Tyr and forms a π-bond interaction with the heme pyrrole ring. A conserved Ser and several hydrophilic residues on the distal side of the pocket form H-bonds with at least one heme propionate providing a high-affinity site.

Despite a similar overall architecture, minor differences among the Isd proteins are observed in the heme-binding pocket with the major variation in the composition of loops and chain termini. These variations play a role in the unique function of each protein. For example, the nonheme-binding IsdHN1 (Sharp et al., 2007) does not contain Tyr-52, Tyr-132, His-134, and Tyr-136 residues that are involved in heme binding (Pilpa et al., 2006; Sharp et al., 2007). Additionally, the N1 domain of IsdH has a series of disordered aromatic residues (Tyr-125, Tyr-126, His-127, Phe-128,

Phe-129) in loop 2 between the β1b- and β2-strands (Pilpa et al., 2006). These residues are conserved in all Hb-binding NEAT domains. It had been proposed that the function of IsdH is to tether Hb to the surface of the cell wall to enable heme transfer through IsdA to IsdC (Sharp et al., 2007). Recent spectroscopic and structural studies of the IsdH N1 domain in complex with Hb revealed a highly specific protein–protein interaction involving significant contributions from the conserved aromatic loops 2 and 4 with the hemoglobin α-subunit (αHb) (Krishna Kumar et al., 2011). Conservation of loops 2 and 4 in IsdH N2 and IsdB N1 would suggest that they also have similar affinity for binding to αHb. Furthermore, heme dissociation from Hb on interaction with the IsdH/B N1 domains alone is three to four times slower than for the intact receptors, suggesting that interaction of the Hb-binding domains enhances interaction with the heme acceptor NEAT domains N3 of IsdH and N2 of IsdB domains promoting release of heme from Hb. The structure of the IsdB N2 heme-binding domain was recently solved, revealing a unique heme ligand set contributed by Tyr-440 and Met-362 (Gaudin et al., 2011). Interestingly, the crystal structure revealed a mixture of pentacoordinate and hexacoordinate states with Met-362 partially occupying the sixth ligand position. Based on spectroscopic studies of the Met-362 Leu variant, the increased heme off rates, combined with the decrease in transfer rate, suggest that Met-362 is crucial for heme transfer to IsdA.

The relative heme affinities of the Isd proteins have been estimated from the apparent rate of heme transfer from the proteins to Mb, where IsdC > IsdE > IsdB ≈ IsdA (Zhu, Xie, et al., 2008). Kinetic studies of the Isd system in combination with circular dichroism and mass spectral data provided *in vitro* evidence that supports a model (Mazmanian et al., 2003) in which heme transport across the envelope of gram-positive bacteria is based on localization of the proteins in the envelope (Fig. 5B) (Liu, Tanaka, et al., 2008; Muryoi et al., 2008; Zhu, Xie, et al., 2008). Heme is transferred from metHb to IsdB at a rate of 0.31 per second (Zhu, Xie, et al., 2008). This is 80 and 2200 times greater than the rate for dissociation from the β and α subunits of human metHb, respectively. This transfer is a single exponential process unlike that of simple heme dissociation from metHb and transfer between metHb and IsdA, IsdC, and IsdE, indicating that metHb transfers heme to IsdB specifically (Zhu, Xie, et al., 2008). An in-depth analysis of the kinetics of heme transfer from intact IsdA to IsdC (which have a single NEAT domain) revealed a two-step process, namely complex formation followed by heme transfer and dissociation (Liu, Tanaka, et al.,

2008). In contrast, a one-step kinetic model is observed for heme transfer between the isolated IsdA N1 and IsdC N1 domains, indicating that protein–protein interaction is rate limiting for heme transfer between holo-IsdA N1 and apo-IsdC N1. IsdA accepts heme from IsdH/B and variants of the proximal Tyr-166 significantly reduce the rate of heme transfer to and from IsdA, despite the fact that the off rates increased significantly on loss of the proximal Tyr (Grigg, Mao, & Murphy, 2011). The data confirmed the Tyr ligand was essential for directional transfer and a model whereby heme is directly shuffled from the Tyr ligand of the donor protein to the Tyr on the acceptor protein. Whether heme is passed off via a hexacoordinate di-Tyr intermediate between the donor and acceptor protein or rapid release from the donor followed by coordination of the Tyr in the acceptor protein has not been determined. A recent paramagnetic relaxation enhancement NMR study has proposed a weak transient "handclasp" model for heme transfer from IsdA to IsdC N1 where the binding affinity between the proteins is extremely weak and estimated to be in the low millimolar range, thus facilitating rapid transfer (Fig. 6B) (Villareal, Spirig, Robson, Liu, Lei, & Clubb, 2011).

Heme is transferred from the cell wall-anchored Isd proteins to the Isd membrane ABC transporter for transport into the cytoplasm. The ABC transporter has three components: (1) a heme-binding protein, (2) a membrane permease, and (3) an ATP hydrolase (Jones, O'Mara, & George, 2009). The structure of the heme-binding protein IsdE of *S. aureus* like all NEAT domains has an immunoglobin-like eight-stranded β-sandwich fold with two domains linked by an α-helix (Grigg et al., 2007). Heme binds in a cleft between the two domains where the heme is six coordinate with an axial ligand coming from each domain, Met-78 and His-229. Very little is known about the remaining components of the ABC transporter in *S. aureus*; however, IsdF shares 22% identity with the structurally characterized Vitamin B_{12} transporter, BtuC. However, the *isd* locus of *S. aureus* does not encode an ATPase (Mazmanian et al., 2003) similar to the siderophore transporters Hts and Sir (Dale, Sebulsky, & Heinrichs, 2004; Speziali, Dale, Henderson, Vines, & Heinrichs, 2006). Genetic studies have provided some evidence that these systems utilize FhuC as the ATPase portion of the ABC transporter. The same is likely the case for the Isd system. *S. pyogenes* (Bates et al., 2003), *L. monocytogenes* (Jin et al., 2006), and *C. diphtheriae* (Drazek et al., 2000) all encode putative ATPases; however, no structural data are available for any of these ABC transporter systems. Despite the lack of structural data on the transmembrane transporters, they are likely to be similar to the ABC transporter systems of gram-negative bacteria described in the previous sections.

Although the mechanism by which gram-negative and gram-positive pathogens acquire heme has fundamental differences due to the nature of the cells architecture, it is clear that common motifs in heme binding are emerging. All NEAT domains of gram-positive bacteria have an immunoglobin-like β-sandwich fold comprised of two antiparallel β-sheets bridged by an α-helix. This architecture is very similar to gram-negative Class III PBPs ShuT, PhuT, BtuF, and FhuD that are comprised of two globular β-domains linked by a single α-helix. In both cases, heme is coordinated in a cleft between the two β-domains. Both the cell wall-anchored Isd proteins and the PBPs coordinate the heme through a Tyr axial ligand. Furthermore, the unusual His–Tyr ligation observed in the HasA hemophores of *S. marcescens* and *P. aeruginosa* has significance in the mechanism of heme capture and transfer to the OM receptor. In contrast to the gram-positive cell surface-anchored proteins, the translocation of heme through the membrane is driven primarily via bis–Histidine coordination in both the gram-negative OM receptors and ABC transporters, a theme most likely conserved in gram-positive ABC transporters.

3. TRANSMEMBRANE HEME SENSORS AND ATP-DEPENDENT EFFLUX PUMPS

3.1. Transmembrane Two-Component Heme Sensor Kinase Systems

Bacteria respond to chemical and physical environmental signals through two-component systems comprising a membrane sensor kinase and a soluble response regulator (Krell et al., 2010; Szurmant, White, & Hoch, 2007). The ability of the invading pathogen to respond to the extracellular environment is crucial to the acquisition of heme as well as the prevention of heme toxicity. Heme import is required to meet the iron needs of the pathogen; however, as heme accumulates, the propensity for toxic side effects increases. Pathogens respond to intracellular heme accumulation through such a two-component sensor kinase system that reduces intracellular heme by signaling expression of heme-degrading enzymes (Bibb, King, Kunkle, & Schmitt, 2005; Bibb, Kunkle, & Schmitt, 2007) or heme efflux transporters (Stauff, Torres, & Skaar, 2007; Stauff et al., 2008). In *C. diphtheriae*, two heme-dependent sensor kinases have been described the

Corynebacterium heme response system (ChrAS) and heme response regulator (HrrAS) where both systems are required for the optimal expression of HmuO under low iron conditions. Furthermore, the ChrAS system is responsible for 80% of the activation of *hmuO* under low iron conditions and in high iron in the presence of heme appears to partially reverse the effects of iron. Recently, the ChrAS system was also shown to regulate the ATP-dependent heme-regulated transporter HrtAB responsible for the efflux of heme from the cell (Bibb & Schmitt, 2010). Similarly, the *S. aureus* heme sensor system (HssRS) responds to elevated heme levels by increasing expression of a similar heme efflux system HrtAB (Stauff et al., 2008). The sensor kinase presumably responds through direct binding of heme to the extracytoplasmic domain of HssS, although an indirect mechanism cannot be ruled out (Stauff et al., 2007). Ligand binding to the extracytoplasmic domain of the histidine kinase HssS triggers phosphorylation of an aspartate on the cognate receptor, HssR. Phosphorylated HssR then binds to a consensus sequence within the *hrtAB* promoter upregulating transcription of the heme efflux pump. The catalytic domains of the sensor kinases, namely the ATP-binding domain and HisKA phosphorylation domain, are highly conserved with some variation in the cytoplasmic signaling and extracytoplasmic sensor domains (Szurmant et al., 2007). Based on sequence identity, HssS shares significant homology with the homodimer sensor kinases belonging to the HAMP (for histidine kinases, adenylyl cyclases, methyl-accepting proteins, and other prokaryotic signaling proteins) domain family. The HAMP domain is thought to be involved in transducing the signal from the sensor domain through rotational changes in conformation applied through a four-helix bundle composed of two helices from each monomer (Ferris et al., 2011; Hulko et al., 2006). The structure of the heme-binding domain and detailed mechanism of activation remain to be determined.

3.2. ATP-Dependent Efflux Pumps

Heme efflux transporters have been identified in *S. aureus* (Stauff et al., 2008; Stauff & Skaar, 2009), *C. diphtheriae* (Bibb & Schmitt, 2010), and *Lactococcus lactis* (Lechardeur et al., 2012). In contrast to the gram-positive *C. diphtheriae* and *S. aureus* HrtAB efflux pumps which are regulated by membrane-associated two-component sensor kinase systems in gram-negative organisms such as *L. lactis*, the efflux system responds to intracellular heme through a transcriptional regulator, HrtR. Despite advances in the regulation of the efflux pumps, there is currently no structural details available although sequence alignment suggests that the heme efflux systems are similar to the macrolide-specific ABC transporters (MacAB) (Routh et al., 2011; Xu et al., 2009, 2010). The recent identification and characterization of heme efflux systems and their role in virulence beg further structural and mechanistic analysis as potential future antimicrobial targets. To date, heme utilization as an antimicrobial target has solely been focused on uptake and utilization (see Section 5).

4. HEME UPTAKE AND UTILIZATION IN ANTIMICROBIAL DRUG DEVELOPMENT

4.1. OM Heme Receptors as Vaccine Targets

As a result of the wealth of knowledge on the surface-exposed TonB-dependent receptors of gram-negative pathogens, they have been extensively investigated as potential vaccine targets (Rohde & Dyer, 2003; Schryvers & Stojiljkovic, 1999). The surface accessibility and role in virulence of the Hb and transferrin receptors made them early candidates for vaccine development (Afonina et al., 2006; Cornelissen, 2003). The transferrin and Hb receptors of *Haemophilus influenza*, *Neisseria meningitidis*, and *Neisseria gonorrhoeae* were the early focus of such studies (Holland et al., 1996; Lewis et al., 1999; Richardson & Stojiljkovic, 1999; Stevenson, Williams, & Griffiths, 1992; Webb & Cripps, 1999). However, a potential drawback has been the high level of redundancy and phase variation in the transferrin and heme uptake systems (Ekins, Bahrami, Sijercic, Maret, & Niven, 2004; Richardson & Stojiljkovic, 1999, 2001). For example, *N. meningitidis* encodes two Hb receptors HpuAB and HmbR that undergo phase variation controlled by a translational frameshift (Lewis et al., 1999; Richardson & Stojiljkovic, 1999) and similar phase variation has also been observed in *N. gonorrhoeae* Hb receptors (Chen, Elkins, & Sparling, 1998). In addition, phase variation allows the bacteria to adjust rapidly to the host environment allowing specific receptors to play a greater or lesser role during infection. Despite this drawback, both *N. meningitidis* and *N. gonorrhoeae* have been the focus of vaccine development (Cornelissen, 2008). In contrast, the *N. gonorrhoeae* transferrin-binding proteins TbpA and TbpB were shown to be expressed in all clinical isolates, while showing little strain to strain variation, and do not undergo high-frequency phase variation (Cornelissen, Anderson, Boulton, & Sparling, 2000; Cornelissen, Anderson, & Sparling, 1997).

These proteins have shown partial success as potential vaccine candidates in animal models of infection (Cornelissen, 2008; Price, 2007; Price et al., 2007). *Hemophilus ducreyi* has three TonB-dependent receptors of which only the Hb receptor HgbA is required for acquisition of iron during infection (Leduc et al., 2008). Initial immunization studies with HgbA in a swine model for chancroid provided some protection from infection (Afonina et al., 2006). Subsequent studies have focused on identification of the core domain essential in Hb recognition in order to develop a subunit vaccine (Nepluev et al., 2009).

In the gram-positive arena, much of the focus on vaccine development has been directed toward *S. aureus* (Clarke et al., 2006). Culturing *S. aureus* in a mouse tissue cage model versus iron-rich media identified over 54 genes, 36 of which were upregulated during infection and in iron-restricted media (Allard et al., 2006). From this initial study, six proteins were selected for further evaluation including the Hb receptor, IsdB, its coreceptor IsdH, and the sortase, SrtB, which is required in anchoring IsdC to the cell wall. From this study, IsdH appeared to be the best vaccine candidate when screened in a mouse mastitis model (Ster et al., 2010). In a separate study, the IsdA and IsdB receptors were investigated as potential vaccine targets in the prevention of abscess development in mice challenged by intravenous *S. aureus* (Kim et al., 2010). The conclusion from this study indicated that antibodies against IsdB and IsdA in part protected against staphylococcal infection by interfering with heme uptake and utilization.

The significant advances in the genomics of surface-exposed receptors have identified potential vaccine candidates from *in vivo* transcriptional profiles of the surface-exposed proteins during infection (Kurz et al., 2003). While, to date, there have been no clinical trials of human vaccine candidates as a result of such studies, it is inevitable that genomics and proteomics will provide a more sophisticated and detailed picture of the role of the OM heme receptors in colonization, infection, and disease (Clarke et al., 2006). Such studies will provide a framework for the future development of vaccine development.

4.2. Metalloporphyrin-Based Inhibitors of Heme Utilization

It has been known for decades that pathogenic bacteria are susceptible to porphyrin-based compounds (Begum, Kim, Kumar, Stojiljkovic, & Wataya, 2003; Bozja, Yi, Shafer, & Stojiljkovic, 2004; Stojiljkovic, Evavold, & Kumar, 2001). The antimicrobial activity of porphyrins can arise from their photoactive properties or by specifically targeting heme uptake systems. While there are many examples of photodynamic antimicrobial porphyrin and chlorin derivatives (Embleton, Nair, Cookson, & Wilson, 2002; Farrell, Robbins, Rhys-Williams, & Love, 2010; Huang et al., 2010; Nitzan, Balzam-Sudakevitz, & Ashkenazi, 1998), we restrict the current review to metalloporphyrins that specifically target heme uptake across the OM (Bozja et al., 2004; Stojiljkovic et al., 2001). Early studies showed that noniron metalloporphyrins have broad spectrum antibacterial activity against both gram-positive and gram-negative organisms, including *S. aureus*, *Yersinia enterocolitica*, *N. meningitidis*, and *N. gonorrhoeae* (Stojiljkovic, Kumar, & Srinivasan, 1999). A series of noniron metalloporphyrins (PPIX) were tested against a panel of bacterial pathogens with several having bacteriostatic activity (Mn-PPIX, Zn-PPIX, Ga-PPIX, and Ru-PPIX). In *Y. enterocolitica*, Ga-PPIX and Ru-PPIX at higher concentrations were shown to be bactericidal and mutants devoid of the OM receptor HemR were resistant to Ga-PPIX. In contrast, the cytoplasmic heme-dependent ABC transporter system was not required for sensitivity. In these initial studies, it was not determined if Ga-PPIX was actively transported into the periplasm via the OM heme receptor or by passive diffusion through the membrane. Furthermore, it was not definitively shown that Ga-PPIX was present in the periplasm, largely due to inconsistencies in the recovery of extractable Ga-PPIX from the cells. However, the fact that metal-free porphyrins have no antimicrobial activity and the inhibitory effect could be alleviated with increasing heme concentrations suggests that the noniron PPIXs are competing for binding to the OM receptor. In a recent study, the oral pathogen *P. gingivalis* when treated with In-PPIX showed retarded growth, which on the basis of transcriptional analysis appeared to be due to an inhibition of nucleotide metabolism and protein processing following internalization of In-PPIX through the Iht receptor system (Yukitake et al., 2011).

A model for the antimicrobial activity of noniron porphyrins has been proposed whereby exposure to such inhibitors induces production of ROS and oxidative stress (Stojiljkovic et al., 1999, 2001). This model is based on the fact that superoxide dismutase and catalase mutants have increased sensitivity to the noniron porphyrins and that active respiration is required for the antimicrobial effects. The authors propose that this generation of ROS may be due to insertion of noniron porphyrins directly into respiratory proteins by the cytochrome assembly machinery. Although it has been shown that *E. coli* overexpressing the Hb receptor of *P. gingivalis* binds both

Cu- and Zn-PPIX with similar affinity as Fe-PPIX, their remains no direct evidence for active transport of noniron porphyrins into the periplasm (Olczak, Dixon, & Genco, 2001). Despite promising early studies on the OM heme receptors as Trojan horses in delivering noniron metalloporphyrin antibacterials, most of the current focus has been directed at utilizing the photodynamic properties of light-activated Sn-chlorins as antibacterials (Embleton et al., 2002, 2005; Suci, Kang, Gmur, Douglas, & Young, 2010).

5. FUTURE PERSPECTIVES

The challenges of transporting a highly reactive and toxic cofactor such as heme across the physical barrier of the cell wall and lipid membrane require specialized systems. Our current knowledge of the mechanisms by which heme is transported has been aided greatly by the wealth of structural data on the cell wall lipoproteins and OM receptors of gram-positive and gram-negative bacteria, respectively. However, as we define the common structural motifs of heme transporters, there are significant challenges remaining, such as understanding the complex nature of TonB-coupled conformational changes in driving heme transport across the gram-negative OM and the nature of the protein–protein interactions of the cell wall-anchored lipoproteins in heme transfer within the peptidoglycan cell wall. Future studies must encompass the challenges of understanding the nature of protein–protein-driven allosteric changes that trigger ligand coordination and spin-state changes required for heme translocation. Structural determination of protein complexes combined with advanced spectroscopic studies will greatly enhance our understanding of heme uptake in bacterial pathogens. The significance of heme uptake and utilization in bacterial pathogenesis and virulence combined with a structural and mechanistic understanding of such processes may provide a platform for future antimicrobial therapeutic intervention.

ACKNOWLEDGMENTS

The authors would like to thank past and present members of the Wilks' Laboratory for all their contributions to the understanding of heme uptake in bacterial pathogens. We would also like to thank our collaborators Tom Poulos, Kent Rodgers, Mario Rivera, John Dawson, Mike Schmitt, Amanda Oglesby-Sherrouse, Shelley Payne, and the late Igor Stojiljkovic.

REFERENCES

Afonina, G., Leduc, I., Nepluev, I., Jeter, C., Routh, P., Almond, G., et al. (2006). Immunization with the *Haemophilus ducreyi* hemoglobin receptor HgbA protects against infection in the swine model of chancroid. *Infection and Immunity, 74*, 2224–2232.

Aisen, P., Enns, C., & Wessling-Resnick, M. (2001). Chemistry and biology of eukaryotic iron metabolism. *The International Journal of Biochemistry and Cell Biology, 33*, 940–959.

Allard, M., Moisan, H., Brouillette, E., Gervais, A. L., Jacques, M., Lacasse, P., et al. (2006). Transcriptional modulation of some *Staphylococcus aureus* iron-regulated genes during growth in vitro and in a tissue cage model *in vivo*. *Microbes and Infection, 8*, 1679–1690.

Andrade, M. A., Ciccarelli, F. D., Perez-Iratxeta, C., & Bork, P. (2002). NEAT: a domain duplicated in genes near the components of a putative Fe^{3+} siderophore transporter from Gram-positive pathogenic bacteria. *Genome Biology, 3*, 0047.

Anzaldi, L. L., & Skaar, E. P. (2010). Overcoming the heme paradox: heme toxicity and tolerance in bacterial pathogens. *Infection and Immunity, 78*, 4977–4989.

Bachman, M. A., Miller, V. L., & Weiser, J. N. (2009). Mucosal lipocalin 2 has pro-inflammatory and iron-sequestering effects in response to bacterial enterobactin. *PLoS Pathogens, 5*, e1000622.

Balla, J., Jacob, H. S., Balla, G., Nath, K., Eaton, J. W., & Vercellotti, G. M. (1993). Endothelial-cell heme uptake from heme proteins: induction of sensitization and desensitization to oxidant damage. *Proceedings of the National Academy of Sciences of the United States of America, 90*, 9285–9289.

Barjon, C., Wecker, K., Izadi-Pruneyre, N., & Delepelaire, P. (2007). Mutagenesis and molecular modeling reveal three key extracellular loops of the membrane receptor HasR that are involved in hemophore HasA binding. *Journal of Bacteriology, 189*, 5379–5382.

Bates, C. S., Montanez, G. E., Woods, C. R., Vincent, R. M., & Eichenbaum, Z. (2003). Identification and characterization of a *Streptococcus pyogenes* operon involved in binding of hemoproteins and acquisition of iron. *Infection and Immunity, 71*, 1042–1055.

Begum, K., Kim, H. S., Kumar, V., Stojiljkovic, I., & Wataya, Y. (2003). In vitro antimalarial activity of metalloporphyrins against *Plasmodium falciparum*. *Parasitology Research, 90*, 221–224.

Belcher, J. D., Beckman, J. D., Balla, G., Balla, J., & Vercellotti, G. (2010). Heme degradation and vascular injury. *Antioxidants and Redox Signaling, 12*, 233–248.

Bhakta, M. N., & Wilks, A. (2006). The mechanism of heme transfer from the cytoplasmic heme binding protein PhuS to the delta-regioselective heme oxygenase of *Pseudomonas aeruginosa*. *Biochemistry, 45*, 11642–11649.

Bibb, L. A., King, N. D., Kunkle, C. A., & Schmitt, M. P. (2005). Analysis of a heme-dependent signal transduction system in *Corynebacterium diphtheriae*: deletion of the chrAS genes results in heme sensitivity and diminished heme-dependent activation of the hmuO promoter. *Infection and Immunity, 73*, 7406–7412.

Bibb, L. A., Kunkle, C. A., & Schmitt, M. P. (2007). The ChrA-ChrS and HrrA-HrrS signal transduction systems are required for activation of the hmuO promoter and repression of the hemA promoter in *Corynebacterium diphtheriae*. *Infection and Immunity, 75*, 2421–2431.

Bibb, L. A., & Schmitt, M. P. (2010). The ABC transporter HrtAB confers resistance to hemin toxicity and is regulated in a hemin-dependent manner by the ChrAS two-component system in *Corynebacterium diphtheriae*. *Journal of Bacteriology, 192*, 4606–4617.

Block, D. R., Lukat-Rodgers, G. S., Rodgers, K. R., Wilks, A., Bhakta, M. N., & Lansky, I. B. (2007). Identification of two heme-binding sites in the cytoplasmic heme-trafficking protein PhuS from *Pseudomonas aeruginosa* and their relevance to function. *Biochemistry, 46*, 14391–14402.

Borregaard, N., & Cowland, J. B. (2006). Neutrophil gelatinase-associated lipocalin, a siderophore-binding eukaryotic protein. *Biometals, 19*, 211–215.
Borths, E. L., Locher, K. P., Lee, A. T., & Rees, D. C. (2002). The structure of Escherichia coli BtuF and binding to its cognate ATP binding cassette transporter. *Proceedings of the National Academy of Sciences of the United States of America, 99*, 16642–16647.
Bozja, J., Yi, K., Shafer, W. M., & Stojiljkovic, I. (2004). Porphyrin-based compounds exert antibacterial action against the sexually transmitted pathogens *Neisseria gonorrhoeae* and *Haemophilus ducreyi*. *International Journal of Antimicrobial Agents, 24*, 578–584.
Bracken, C. S., Baer, M. T., Abdur-Rashid, A., Helms, W., & Stojiljkovic, I. (1999). Use of heme-protein complexes by the *Yersinia enterocolitica* HemR receptor: histidine residues are essential for receptor function. *Journal of Bacteriology, 181*, 6063–6072.
Braun, V. (2005). Bacterial iron transport related to virulence. *Contributions to Microbiology, 12*, 210–233.
Braun, V., & Endriss, F. (2007). Energy-coupled outer membrane transport proteins and regulatory proteins. *Biometals, 20*, 219–231.
Brickman, T. J., Vanderpool, C. K., & Armstrong, S. K. (2006). Heme transport contributes to in vivo fitness of *Bordetella pertussis* during primary infection in mice. *Infection and Immunity, 74*, 1741–1744.
Brown, J. S., & Holden, D. W. (2002). Iron acquisition by Gram-positive bacterial pathogens. *Microbes and Infection, 4*, 1149–1156.
Bullen, J. J., Rogers, H. J., Spalding, P. B., & Ward, C. G. (2006). Natural resistance, iron and infection: a challenge for clinical medicine. *Journal of Medical Microbiology, 55*, 251–258.
Burkhard, K. A., & Wilks, A. (2007). Characterization of the outer membrane receptor ShuA from the heme uptake system of *Shigella dysenteriae*. Substrate specificity and identification of the heme protein ligands. *The Journal of Biological Chemistry, 282*, 15126–15136.
Burkhard, K. A., & Wilks, A. (2008). Functional characterization of the *Shigella dysenteriae* heme ABC transporter. *Biochemistry, 47*, 7977–7979.
Carter, D. M., Miousse, I. R., Gagnon, J. N., Martinez, E., Clements, A., Lee, J., et al. (2006). Interactions between TonB from *Escherichia coli* and the periplasmic protein FhuD. *The Journal of Biological Chemistry, 281*, 35413–35424.
Cescau, S., Cwerman, H., Letoffe, S., Delepelaire, P., Wandersman, C., & Biville, F. (2007). Heme acquisition by hemophores. *Biometals, 20*, 603–613.
Chen, C. J., Elkins, C., & Sparling, P. F. (1998). Phase variation of hemoglobin utilization in *Neisseria gonorrhoeae*. *Infection and Immunity, 66*, 987–993.
Chimento, D. P., Kadner, R. J., & Wiener, M. C. (2005). Comparative structural analysis of TonB-dependent outer membrane transporters: implications for the transport cycle. *Proteins, 59*, 240–251.
Clarke, S. R., Brummell, K. J., Horsburgh, M. J., McDowell, P. W., Mohamad, S. A., Stapleton, M. R., et al. (2006). Identification of in vivo-expressed antigens of *Staphylococcus aureus* and their use in vaccinations for protection against nasal carriage. *Journal of Infectious Diseases, 193*, 1098–1108.
Clarke, T. E., Braun, V., Winkelmann, G., Tari, L. W., & Vogel, H. J. (2002). X-ray crystallographic structures of the *Escherichia coli* periplasmic protein FhuD bound to hydroxamate-type siderophores and the antibiotic albomycin. *The Journal of Biological Chemistry, 277*, 13966–13972.
Clifton, M. C., Corrent, C., & Strong, R. K. (2009). Siderocalins: siderophore-binding proteins of the innate immune system. *Biometals, 22*, 557–564.
Cobessi, D., Meksem, A., & Brillet, K. (2010). Structure of the heme/hemoglobin outer membrane receptor ShuA from *Shigella dysenteriae*: heme binding by an induced fit mechanism. *Proteins, 78*, 286–294.

Cornelissen, C. N. (2003). Transferrin-iron uptake by gram-negative bacteria. *Frontiers in Bioscience, 8*, d836–d847.

Cornelissen, C. N. (2008). Identification and characterization of gonococcal iron transport systems as potential vaccine antigens. *Future Microbiology, 3*, 287–298.

Cornelissen, C. N., Anderson, J. E., Boulton, I. C., & Sparling, P. F. (2000). Antigenic and sequence diversity in gonococcal transferrin-binding protein A. *Infection and Immunity, 68*, 4725–4735.

Cornelissen, C. N., Anderson, J. E., & Sparling, P. F. (1997). Characterization of the diversity and the transferrin-binding domain of gonococcal transferrin-binding protein 2. *Infection and Immunity, 65*, 822–828.

Cornelissen, C. N., & Sparling, P. F. (1994). Iron piracy: acquisition of transferrin-bound iron by bacterial pathogens. *Molecular Microbiology, 14*, 843–850.

Dale, S. E., Sebulsky, M. T., & Heinrichs, D. E. (2004). Involvement of SirABC in iron-siderophore import in *Staphylococcus aureus*. *Journal of Bacteriology, 186*, 8356–8362.

De Domenico, I., Ward, D. M., & Kaplan, J. (2011). Hepcidin and ferroportin: the new players in iron metabolism. *Seminars in Liver Disease, 31*, 272–279.

Deniau, C., Gilli, R., Izadi-Pruneyre, N., Letoffe, S., Delepierre, M., Wandersman, C., et al. (2003). Thermodynamics of heme binding to the HasA (SM) hemophore: effect of mutations at three key residues for heme uptake. *Biochemistry, 42*, 10627–10633.

Doherty, C. P. (2007). Host-pathogen interactions: the role of iron. *Journal of Nutrition, 137*, 1341–1344.

Drazek, E. S., Hammack, C. A., & Schmitt, M. P. (2000). Corynebacterium diphtheriae genes required for acquisition of iron from haemin and haemoglobin are homologous to ABC haemin transporters. *Molecular Microbiology, 36*, 68–84.

Dryla, A., Gelbmann, D., von Gabain, A., & Nagy, E. (2003). Identification of a novel iron regulated staphylococcal surface protein with haptoglobin-haemoglobin binding activity. *Molecular Microbiology, 49*, 37–53.

Dwyer, M. A., & Hellinga, H. W. (2004). Periplasmic binding proteins: a versatile superfamily for protein engineering. *Current Opinion in Structural Biology, 14*, 495–504.

Eakanunkul, S., Lukat-Rodgers, G. S., Sumithran, S., Ghosh, A., Rodgers, K. R., Dawson, J. H., et al. (2005). Characterization of the periplasmic heme-binding protein ShuT from the heme uptake system of *Shigella dysenteriae*. *Biochemistry, 44*, 13179–13191.

Ekins, A., Bahrami, F., Sijercic, A., Maret, D., & Niven, D. F. (2004). Haemophilus somnus possesses two systems for acquisition of transferrin-bound iron. *Journal of Bacteriology, 186*, 4407–4411.

Ekins, A., & Niven, D. F. (2003). Transferrin-dependent expression of TbpA by Histophilus ovis involves a poly G tract within tbpA. *FEMS Microbiology Letters, 220*, 95–98.

Embleton, M. L., Nair, S. P., Cookson, B. D., & Wilson, M. (2002). Selective lethal photosensitization of methicillin-resistant *Staphylococcus aureus* using an IgG-tin (IV) chlorin e6 conjugate. *Journal of Antimicrobial Chemotherapy, 50*, 857–864.

Embleton, M. L., Nair, S. P., Heywood, W., Menon, D. C., Cookson, B. D., & Wilson, M. (2005). Development of a novel targeting system for lethal photosensitization of antibiotic-resistant strains of *Staphylococcus aureus*. *Antimicrobial Agents and Chemotherapy, 49*, 3690–3696.

Faraldo-Gomez, J. D., Smith, G. R., & Sansom, M. S. (2003). Molecular dynamics simulations of the bacterial outer membrane protein FhuA: a comparative study of the ferrichrome-free and bound states. *Biophysical Journal, 85*, 1406–1420.

Farrell, D. J., Robbins, M., Rhys-Williams, W., & Love, W. G. (2010). In vitro activity of XF-73, a novel antibacterial agent, against antibiotic-sensitive and -resistant Gram-positive and Gram-negative bacterial species. *International Journal of Antimicrobial Agents, 35*, 531–536.

Ferris, H. U., Dunin-Horkawicz, S., Mondejar, L. G., Hulko, M., Hantke, K., Martin, J., et al. (2011). The mechanisms of HAMP-mediated signaling in transmembrane receptors. *Structure, 19*, 378–385.

Ganz, T. (2009). Iron in innate immunity: starve the invaders. *Current Opinion in Microbiology, 21*, 63–67.

Ganz, T. (2011). Hepcidin and iron regulation, 10 years later. *Blood, 117*, 4425–4433.

Gaudin, C. F., Grigg, J. C., Arrieta, A. L., & Murphy, M. E. (2011). Unique heme-iron coordination by the hemoglobin receptor IsdB of *Staphylococcus aureus*. *Biochemistry, 50*, 5443–5452.

Ghigo, J. M., Letoffe, S., & Wandersman, C. (1997). A new type of hemophore-dependent heme acquisition system of Serratia marcescens reconstituted in *Escherichia coli*. *Journal of Bacteriology, 179*, 3572–3579.

Gonzalez-Chavez, S. A., Arevalo-Gallegos, S., & Rascon-Cruz, Q. (2009). Lactoferrin: structure, function and applications. *International Journal of Antimicrobial Agents, 33*, 301.e1–301.e8.

Grigg, J. C., Mao, C. X., & Murphy, M. E. (2011). Iron-coordinating tyrosine is a key determinant of NEAT domain heme transfer. *Journal of Molecular Biology, 413*, 684–698.

Grigg, J. C., Ukpabi, G., Gaudin, C. F., & Murphy, M. E. (2010). Structural biology of heme binding in the Staphylococcus aureus Isd system. *Journal of Inorganic Biochemistry, 104*, 341–348.

Grigg, J. C., Vermeiren, C. L., Heinrichs, D. E., & Murphy, M. E. (2007). Heme coordination by Staphylococcus aureus IsdE. *The Journal of Biological Chemistry, 282*, 28815–28822.

Gumbart, J., Wiener, M. C., & Tajkhorshid, E. (2007). Mechanics of force propagation in TonB-dependent outer membrane transport. *Biophysical Journal, 93*, 496–504.

Halliwell, B., & Gutteridge, J. M. (1984). Oxygen toxicity, oxygen radicals, transition metals and disease. *Biochemical Journal, 219*, 1–14.

Hanks, T. S., Liu, M., McClure, M. J., & Lei, B. (2005). ABC transporter FtsABCD of Streptococcus pyogenes mediates uptake of ferric ferrichrome. *BMC Microbiology, 5*, 62.

Henderson, D. P., & Payne, S. M. (1994). Vibrio cholerae iron transport systems: roles of heme and siderophore iron transport in virulence and identification of a gene associated with multiple iron transport systems. *Infection and Immunity, 62*, 5120–5125.

Higgins, C. F. (1992). ABC transporters: from microorganisms to man. *Annual Review of Cell Biology, 8*, 67–113.

Higgins, C. F. (2001). ABC transporters: physiology, structure and mechanism—an overview. *Research in Microbiology, 152*, 205–210.

Higgins, C. F., & Linton, K. J. (2004). The ATP switch model for ABC transporters. *Nature Structural and Molecular Biology, 11*, 918–926.

Ho, W. W., Li, H., Eakanunkul, S., Tong, Y., Wilks, A., Guo, M., et al. (2007). Holo- and apo-bound structures of bacterial periplasmic heme-binding proteins. *The Journal of Biological Chemistry, 282*, 35796–35802.

Holland, I. B., & Blight, M. A. (1999). ABC-ATPases, adaptable energy generators fuelling transmembrane movement of a variety of molecules in organisms from bacteria to humans. *Journal of Molecular Biology, 293*, 381–399.

Holland, J., Parsons, T. R., Hasan, A. A., Cook, S. M., Stevenson, P., Griffiths, E., et al. (1996). Conservation and antigenic cross-reactivity of the transferrin-binding proteins of *Haemophilus influenzae, Actinobacillus pleuropneumoniae* and *Neisseria meningitidis*. *Microbiology, 142*(Pt 12), 3505–3513.

Hopfner, K. P., Karcher, A., Shin, D. S., Craig, L., Arthur, L. M., Carney, J. P., et al. (2000). Structural biology of Rad50 ATPase: ATP-driven conformational control in DNA double-strand break repair and the ABC-ATPase superfamily. *Cell, 101*, 789–800.

Huang, L., Huang, Y. Y., Mroz, P., Tegos, G. P., Zhiyentayev, T., Sharma, S. K., et al. (2010). Stable synthetic cationic bacteriochlorins as selective antimicrobial photosensitizers. *Antimicrobial Agents and Chemotherapy, 54*, 3834–3841.

Hulko, M., Berndt, F., Gruber, M., Linder, J. U., Truffault, V., Schultz, A., et al. (2006). The HAMP domain structure implies helix rotation in transmembrane signaling. *Cell, 126*, 929–940.

Hvorup, R. N., Goetz, B. A., Niederer, M., Hollenstein, K., Perozo, E., & Locher, K. P. (2007). Asymmetry in the structure of the ABC transporter-binding protein complex BtuCD-BtuF. *Science, 317*, 1387–1390.

Izadi-Pruneyre, N., Huche, F., Lukat-Rodgers, G. S., Lecroisey, A., Gilli, R., Rodgers, K. R., et al. (2006). The heme transfer from the soluble HasA hemophore to its membrane-bound receptor HasR is driven by protein–protein interaction from a high to a lower affinity binding site. *The Journal of Biological Chemistry, 281*, 25541–25550.

Jenssen, H., & Hancock, R. E. (2009). Antimicrobial properties of lactoferrin. *Biochimie, 91*, 19–29.

Jin, B., Newton, S. M., Shao, Y., Jiang, X., Charbit, A., & Klebba, P. E. (2006). Iron acquisition systems for ferric hydroxamates, haemin and haemoglobin in *Listeria monocytogenes. Molecular Microbiology, 59*, 1185–1198.

Jones, P. M., O'Mara, M. L., & George, A. M. (2009). ABC transporters: a riddle wrapped in a mystery inside an enigma. *Trends in Biochemical Sciences, 34*, 520–531.

Karpowich, N. K., Huang, H. H., Smith, P. C., & Hunt, J. F. (2003). Crystal structures of the BtuF periplasmic-binding protein for vitamin B_{12} suggest a functionally important reduction in protein mobility upon ligand binding. *The Journal of Biological Chemistry, 278*, 8429–8434.

Kaur, A. P., Lansky, I. B., & Wilks, A. (2009). The role of the cytoplasmic heme-binding protein (PhuS) of Pseudomonas aeruginosa in intracellular heme trafficking and iron homeostasis. *The Journal of Biological Chemistry, 284*, 56–66.

Khursigara, C. M., De Crescenzo, G., Pawelek, P. D., & Coulton, J. W. (2004). Enhanced binding of TonB to a ligand-loaded outer membrane receptor: role of the oligomeric state of TonB in formation of a functional FhuA.TonB complex. *The Journal of Biological Chemistry, 279*, 7405–7412.

Khursigara, C. M., De Crescenzo, G., Pawelek, P. D., & Coulton, J. W. (2005). Kinetic analyses reveal multiple steps in forming TonB-FhuA complexes from *Escherichia coli. Biochemistry, 44*, 3441–3453.

Kim, H. K., DeDent, A., Cheng, A. G., McAdow, M., Bagnoli, F., Missiakas, D. M., et al. (2010). IsdA and IsdB antibodies protect mice against *Staphylococcus aureus* abscess formation and lethal challenge. *Vaccine, 28*, 6382–6392.

Koster, W. (2005). Cytoplasmic membrane iron permease systems in the bacterial cell envelope. *Frontiers in Bioscience, 10*, 462–477.

Krell, T., Lacal, J., Busch, A., Silva-Jimenez, H., Guazzaroni, M. E., & Ramos, J. L. (2010). Bacterial sensor kinases: diversity in the recognition of environmental signals. *Annual Review of Microbiology, 64*, 539–559.

Krewulak, K. D., & Vogel, H. J. (2008). Structural biology of bacterial iron uptake. *Biochim et Biophysica Acta, 1778*, 1781–1804.

Krieg, S., Huche, F., Diederichs, K., Izadi-Pruneyre, N., Lecroisey, A., Wandersman, C., et al. (2009). Heme uptake across the outer membrane as revealed by crystal structures of the receptor-hemophore complex. *Proceedings of the National Academy of Sciences of the United States of America, 106*, 1045–1050.

Krishna Kumar, K., Jacques, D. A., Pishchany, G., Caradoc-Davies, T., Spirig, T., Malmirchegini, G. R., et al. (2011). Structural basis for hemoglobin capture by *Staphylococcus aureus* cell-surface protein, IsdH. *The Journal of Biological Chemistry, 286*, 38439–38447.

Kurz, S., Hubner, C., Aepinus, C., Theiss, S., Guckenberger, M., Panzner, U., et al. (2003). Transcriptome-based antigen identification for *Neisseria meningitidis*. *Vaccine, 21*, 768–775.

Lansky, I. B., Lukat-Rodgers, G. S., Block, D., Rodgers, K. R., Ratliff, M., & Wilks, A. (2006). The cytoplasmic heme-binding protein (PhuS) from the heme uptake system of Pseudomonas aeruginosa is an intracellular heme-trafficking protein to the delta-regioselective heme oxygenase. *The Journal of Biological Chemistry, 281*, 13652–13662.

Lechardeur, D., Cesselin, B., Liebl, U., Vos, M. H., Fernandez, A., Brun, C., et al. (2012). Discovery of intracellular heme-binding protein HrtR, which controls heme efflux by the conserved HrtB-HrtA transporter in Lactococcus lactis. *The Journal of Biological Chemistry, 287*, 4752–4758.

Leduc, I., Banks, K. E., Fortney, K. R., Patterson, K. B., Billings, S. D., Katz, B. P., et al. (2008). Evaluation of the repertoire of the TonB-dependent receptors of *Haemophilus ducreyi* for their role in virulence in humans. *Journal of Infectious Diseases, 197*, 1103–1109.

Letoffe, S., Deniau, C., Wolff, N., Dassa, E., Delepelaire, P., Lecroisey, A., et al. (2001). Haemophore-mediated bacterial haem transport: evidence for a common or overlapping site for haem-free and haem-loaded haemophore on its specific outer membrane receptor. *Molecular Microbiology, 41*, 439–450.

Letoffe, S., Ghigo, J. M., & Wandersman, C. (1994). Iron acquisition from heme and hemoglobin by a *Serratia marcescens* extracellular protein. *Proceedings of the National Academy of Sciences of the United States of America, 91*, 9876–9880.

Letoffe, S., Nato, F., Goldberg, M. E., & Wandersman, C. (1999). Interactions of HasA, a bacterial haemophore, with haemoglobin and with its outer membrane receptor HasR. *Molecular Microbiology, 33*, 546–555.

Lewis, L. A., Gipson, M., Hartman, K., Ownbey, T., Vaughn, J., & Dyer, D. W. (1999). Phase variation of HpuAB and HmbR, two distinct haemoglobin receptors of Neisseria meningitidis DNM2. *Molecular Microbiology, 32*, 977–989.

Liu, M., & Lei, B. (2005). Heme transfer from streptococcal cell surface protein Shp to HtsA of transporter HtsABC. *Infection and Immunity, 73*, 5086–5092.

Liu, M., Su, J. G., Kong, R., Sun, T. G., Tan, J. J., Chen, W. Z., et al. (2008). Molecular dynamics simulations of the bacterial periplasmic heme binding proteins ShuT and PhuT. *Biophysical Chemistry, 138*, 42–49.

Liu, M., Tanaka, W. N., Zhu, H., Xie, G., Dooley, D. M., & Lei, B. (2008). Direct hemin transfer from IsdA to IsdC in the iron-regulated surface determinant (Isd) heme acquisition system of *Staphylococcus aureus*. *The Journal of Biological Chemistry, 283*, 6668–6676.

Locher, K. P., Lee, A. T., & Rees, D. C. (2002). The *E. coli* BtuCD structure: a framework for ABC transporter architecture and mechanism. *Science, 296*, 1091–1098.

Mack, J., Vermeiren, C., Heinrichs, D. E., & Stillman, M. J. (2004). In vivo heme scavenging by Staphylococcus aureus IsdC and IsdE proteins. *Biochemical and Biophysical Research Communications, 320*, 781–788.

Maresso, A. W., & Schneewind, O. (2006). Iron acquisition and transport in Staphylococcus aureus. *Biometals, 19*, 193–203.

Mazmanian, S. K., Skaar, E. P., Gaspar, A. H., Humayun, M., Gornicki, P., Jelenska, J., et al. (2003). Passage of heme-iron across the envelope of *Staphylococcus aureus*. *Science, 299*, 906–909.

Mazmanian, S. K., Ton-That, H., Su, K., & Schneewind, O. (2002). An iron-regulated sortase anchors a class of surface protein during *Staphylococcus aureus* pathogenesis. *Proceedings of the National Academy of Sciences of the United States of America, 99*, 2293–2298.

Muryoi, N., Tiedemann, M. T., Pluym, M., Cheung, J., Heinrichs, D. E., & Stillman, M. J. (2008). Demonstration of the iron-regulated surface determinant (Isd) heme transfer pathway in *Staphylococcus aureus*. *The Journal of Biological Chemistry, 283*, 28125–28136.

Nairz, M., Schroll, A., Sonnweber, T., & Weiss, G. (2010). The struggle for iron—a metal at the host-pathogen interface. *Cellular Microbiology, 12*, 1691–1702.

Nepluev, I., Afonina, G., Fusco, W. G., Leduc, I., Olsen, B., Temple, B., et al. (2009). An immunogenic, surface-exposed domain of *Haemophilus ducreyi* outer membrane protein HgbA is involved in hemoglobin binding. *Infection and Immunity, 77*, 3065–3074.

Nitzan, Y., Balzam-Sudakevitz, A., & Ashkenazi, H. (1998). Eradication of *Acinetobacter baumannii* by photosensitized agents in vitro. *Journal of Photochemistry and Photobiology B, 42*, 211–218.

O'Carra, P., & Colleran, E. (1969). Coupled oxidation of myoglobin with ascorbate as a model of haem breakdown in vivo. *Biochemical Journal, 115*, 13P–14P.

O'Carra, P., & Colleran, E. (1970). Methine-bridge specificity of the coupled oxidation of myoglobin and haemoglobin with ascorbate. *Biochemical Journal, 119*, 42P–43P.

Ochsner, U. A., Johnson, Z., & Vasil, M. L. (2000). Genetics and regulation of two distinct haem-uptake systems, *phu* and *has*, in *Pseudomonas aeruginosa*. *Microbiology, 146*(Pt 1), 185–198.

Olczak, T., Dixon, D. W., & Genco, C. A. (2001). Binding specificity of the Porphyromonas gingivalis heme and hemoglobin receptor HmuR, gingipain K, and gingipain R1 for heme, porphyrins, and metalloporphyrins. *Journal of Bacteriology, 183*, 5599–5608.

Otto, B. R., Sparrius, M., Wors, D. J., de Graaf, F. K., & MacLaren, D. M. (1994). Utilization of haem from the haptoglobin-haemoglobin complex by *Bacteroides fragilis*. *Microbial Pathogenesis, 17*, 137–147.

Papanikolaou, G., & Pantopoulos, K. (2005). Iron metabolism and toxicity. *Toxicology and Applied Pharmacology, 202*, 199–211.

Pawelek, P. D., Croteau, N., Ng-Thow-Hing, C., Khursigara, C. M., Moiseeva, N., Allaire, M., et al. (2006). Structure of TonB in complex with FhuA, *E. coli* outer membrane receptor. *Science, 312*, 1399–1402.

Payne, S. M. (1993). Iron acquisition in microbial pathogenesis. *Trends in Microbiology, 1*, 66–69.

Pilpa, R. M., Fadeev, E. A., Villareal, V. A., Wong, M. L., Phillips, M., & Clubb, R. T. (2006). Solution structure of the NEAT (NEAr transporter) domain from IsdH/HarA: the human hemoglobin receptor in *Staphylococcus aureus*. *Journal of Molecular Biology, 360*, 435–447.

Pilpa, R. M., Robson, S. A., Villareal, V. A., Wong, M. L., Phillips, M., & Clubb, R. T. (2009). Functionally distinct NEAT (NEAr transporter) domains within the *Staphylococcus aureus* IsdH/HarA protein extract heme from methemoglobin. *The Journal of Biological Chemistry, 284*, 1166–1176.

Pishchany, G., Dickey, S. E., & Skaar, E. P. (2009). Subcellular localization of the *Staphylococcus aureus* heme iron transport components IsdA and IsdB. *Infection and Immunity, 77*, 2624–2634.

Price, A. A. (2007). Meningococcal vaccines. *Current Pharmaceutical Design, 13*, 2009–2014.

Price, G. A., Masri, H. P., Hollander, A. M., Russell, M. W., & Cornelissen, C. N. (2007). Gonococcal transferrin binding protein chimeras induce bactericidal and growth inhibitory antibodies in mice. *Vaccine, 25*, 7247–7260.

Quiocho, F. A., & Ledvina, P. S. (1996). Atomic structure and specificity of bacterial periplasmic receptors for active transport and chemotaxis: variation of common themes. *Molecular Microbiology, 20*, 17–25.

Richardson, A. R., & Stojiljkovic, I. (1999). HmbR, a hemoglobin-binding outer membrane protein of *Neisseria meningitidis*, undergoes phase variation. *Journal of Bacteriology, 181*, 2067–2074.

Richardson, A. R., & Stojiljkovic, I. (2001). Mismatch repair and the regulation of phase variation in *Neisseria meningitidis*. *Molecular Microbiology, 40*, 645–655.

Rohde, K. H., & Dyer, D. W. (2003). Mechanisms of iron acquisition by the human pathogens *Neisseria meningitidis* and *Neisseria gonorrhoeae*. *Frontiers in Bioscience, 8*, d1186–d1218.

Rohde, K. H., & Dyer, D. W. (2004). Analysis of haptoglobin and hemoglobin-haptoglobin interactions with the *Neisseria meningitidis* TonB-dependent receptor HpuAB by flow cytometry. *Infection and Immunity, 72*, 2494–2506.

Rossi, M. S., Fetherston, J. D., Letoffe, S., Carniel, E., Perry, R. D., & Ghigo, J. M. (2001). Identification and characterization of the hemophore-dependent heme acquisition system of *Yersinia pestis*. *Infection and Immunity, 69*, 6707–6717.

Routh, M. D., Zalucki, Y., Su, C. C., Zhang, Q., Shafer, W. M., & Yu, E. W. (2011). Efflux pumps of the resistance-nodulation-division family: a perspective of their structure, function, and regulation in gram-negative bacteria. *Advances in Enzymology and Related Areas of Molecular Biology, 77*, 109–146.

Schneider, E., & Hunke, S. (1998). ATP-binding-cassette (ABC) transport systems: functional and structural aspects of the ATP-hydrolyzing subunits/domains. *FEMS Microbiology Review, 22*, 1–20.

Schryvers, A. B., & Stojiljkovic, I. (1999). Iron acquisition systems in the pathogenic *Neisseria*. *Molecular Microbiology, 32*, 1117–1123.

Sharp, K. H., Schneider, S., Cockayne, A., & Paoli, M. (2007). Crystal structure of the heme-IsdC complex, the central conduit of the Isd iron/heme uptake system in *Staphylococcus aureus*. *The Journal of Biological Chemistry, 282*, 10625–10631.

Shultis, D. D., Purdy, M. D., Banchs, C. N., & Wiener, M. C. (2006). Outer membrane active transport: structure of the BtuB:TonB complex. *Science, 312*, 1396–1399.

Simpson, W., Olczak, T., & Genco, C. A. (2000). Characterization and expression of HmuR, a TonB-dependent hemoglobin receptor of *Porphyromonas gingivalis*. *Journal of Bacteriology, 182*, 5737–5748.

Skaar, E. P., Humayun, M., Bae, T., DeBord, K. L., & Schneewind, O. (2004). Iron-source preference of *Staphylococcus aureus* infections. *Science, 305*, 1626–1628.

Skaar, E. P., & Schneewind, O. (2004). Iron-regulated surface determinants (Isd) of *Staphylococcus aureus*: stealing iron from heme. *Microbes and Infection, 6*, 390–397.

Sook, B. R., Block, D. R., Sumithran, S., Montanez, G. E., Rodgers, K. R., Dawson, J. H., et al. (2008). Characterization of SiaA, a streptococcal heme-binding protein associated with a heme ABC transport system. *Biochemistry, 47*, 2678–2688.

Speziali, C. D., Dale, S. E., Henderson, J. A., Vines, E. D., & Heinrichs, D. E. (2006). Requirement of *Staphylococcus aureus* ATP-binding cassette-ATPase FhuC for iron-restricted growth and evidence that it functions with more than one iron transporter. *Journal of Bacteriology, 188*, 2048–2055.

Stauff, D. L., Bagaley, D., Torres, V. J., Joyce, R., Anderson, K. L., Kuechenmeister, L., et al. (2008). *Staphylococcus aureus* HrtA is an ATPase required for protection against heme toxicity and prevention of a transcriptional heme stress response. *Journal of Bacteriology, 190*, 3588–3596.

Stauff, D. L., & Skaar, E. P. (2009). The heme sensor system of *Staphylococcus aureus*. *Contributions to Microbiology, 16*, 120–135.

Stauff, D. L., Torres, V. J., & Skaar, E. P. (2007). Signaling and DNA-binding activities of the *Staphylococcus aureus* HssR-HssS two-component system required for heme sensing. *The Journal of Biological Chemistry, 282*, 26111–26121.

Ster, C., Beaudoin, F., Diarra, M. S., Jacques, M., Malouin, F., & Lacasse, P. (2010). Evaluation of some *Staphylococcus aureus* iron-regulated proteins as vaccine targets. *Veterinary Immunology and Immunopathology, 136*, 311–318.

Stevenson, P., Williams, P., & Griffiths, E. (1992). Common antigenic domains in transferrin-binding protein 2 of *Neisseria meningitidis, Neisseria gonorrhoeae*, and *Haemophilus influenzae* type b. *Infection and Immunity, 60*, 2391–2396.

Stojiljkovic, I., Evavold, B. D., & Kumar, V. (2001). Antimicrobial properties of porphyrins. *Expert Opinion on Investigational Drugs, 10*, 309–320.

Stojiljkovic, I., & Hantke, K. (1992). Hemin uptake system of *Yersinia enterocolitica*: similarities with other TonB-dependent systems in Gram-negative bacteria. *The EMBO Journal, 11*, 4359–4367.

Stojiljkovic, I., & Hantke, K. (1994). Transport of haemin across the cytoplasmic membrane through a haemin-specific periplasmic binding-protein-dependent transport system in *Yersinia enterocolitica. Molecular Microbiology, 13*, 719–732.

Stojiljkovic, I., Hwa, V., de Saint Martin, L., O'Gaora, P., Nassif, X., Heffron, F., et al. (1995). The *Neisseria meningitidis* haemoglobin receptor: its role in iron utilization and virulence. *Molecular Microbiology, 15*, 531–541.

Stojiljkovic, I., Kumar, V., & Srinivasan, N. (1999). Non-iron metalloporphyrins: potent antibacterial compounds that exploit haem/Hb uptake systems of pathogenic bacteria. *Molecular Microbiology, 31*, 429–442.

Stojiljkovic, I., Larson, J., Hwa, V., Anic, S., & So, M. (1996). HmbR outer membrane receptors of pathogenic *Neisseria* spp.: iron-regulated, hemoglobin-binding proteins with a high level of primary structure conservation. *Journal of Bacteriology, 178*, 4670–4678.

Suci, P., Kang, S., Gmur, R., Douglas, T., & Young, M. (2010). Targeted delivery of a photosensitizer to *Aggregatibacter actinomycetemcomitans* biofilm. *Antimicrobial Agents and Chemotherapy, 54*, 2489–2496.

Suits, M. D., Jaffer, N., & Jia, Z. (2006). Structure of the *Escherichia coli* O157:H7 heme oxygenase ChuS in complex with heme and enzymatic inactivation by mutation of the heme coordinating residue His-193. *The Journal of Biological Chemistry, 281*, 36776–36782.

Suits, M. D., Pal, G. P., Nakatsu, K., Matte, A., Cygler, M., & Jia, Z. (2005). Identification of an *Escherichia coli* O157:H7 heme oxygenase with tandem functional repeats. *Proceedings of the National Academy of Sciences of the United States of America, 102*, 16955–16960.

Szurmant, H., White, R. A., & Hoch, J. A. (2007). Sensor complexes regulating two-component signal transduction. *Current Opinion in Structural Biology, 17*, 706–715.

Tiedemann, M. T., Muryoi, N., Heinrichs, D. E., & Stillman, M. J. (2008). Iron acquisition by the haem-binding Isd proteins in *Staphylococcus aureus*: studies of the mechanism using magnetic circular dichroism. *Biochemical Society Transactions, 36*, 1138–1143.

Torres, V. J., Pishchany, G., Humayun, M., Schneewind, O., & Skaar, E. P. (2006). *Staphylococcus aureus* IsdB is a hemoglobin receptor required for heme iron utilization. *Journal of Bacteriology, 188*, 8421–8429.

Tullius, M. V., Harmston, C. A., Owens, C. P., Chim, N., Morse, R. P., McMath, L. M., et al. (2011). Discovery and characterization of a unique mycobacterial heme acquisition system. *Proceedings of the National Academy of Sciences of the United States of America, 108*, 5051–5056.

Verga Falzacappa, M. V., & Muckenthaler, M. U. (2005). Hepcidin: iron-hormone and anti-microbial peptide. *Gene, 364*, 37–44.

Vermeiren, C. L., Pluym, M., Mack, J., Heinrichs, D. E., & Stillman, M. J. (2006). Characterization of the heme binding properties of *Staphylococcus aureus* IsdA. *Biochemistry, 45*, 12867–12875.

Villareal, V. A., Pilpa, R. M., Robson, S. A., Fadeev, E. A., & Clubb, R. T. (2008). The IsdC protein from Staphylococcus aureus uses a flexible binding pocket to capture heme. *The Journal of Biological Chemistry, 283*, 31591–31600.

Villareal, V. A., Spirig, T., Robson, S. A., Liu, M., Lei, B., & Clubb, R. T. (2011). Transient weak protein-protein complexes transfer heme across the cell wall of *Staphylococcus aureus*. *Journal of the American Chemical Society, 133*, 14176–14179.

Walker, J. E., Saraste, M., & Gay, N. J. (1982). *E. coli* F1-ATPase interacts with a membrane protein component of a proton channel. *Nature, 298*, 867–869.

Wandersman, C., & Delepelaire, P. (2004). Bacterial iron sources: from siderophores to hemophores. *Annual Review of Microbiology, 58*, 611–647.

Wandersman, C., & Stojiljkovic, I. (2000). Bacterial heme sources: the role of heme, hemoprotein receptors and hemophores. *Current Opinion in Microbiology, 3*, 215–220.

Watanabe, M., Tanaka, Y., Suenaga, A., Kuroda, M., Yao, M., Watanabe, N., et al. (2008). Structural basis for multimeric heme complexation through a specific protein-heme interaction: the case of the third neat domain of IsdH from *Staphylococcus aureus*. *The Journal of Biological Chemistry, 283*, 28649–28659.

Webb, D. C., & Cripps, A. W. (1999). Immunization with recombinant transferrin binding protein B enhances clearance of nontypeable *Haemophilus influenzae* from the rat lung. *Infection and Immunity, 67*, 2138–2144.

Weng, J., Fan, K., & Wang, W. (2012). The conformational transition pathways of ATP-binding cassette transporter BtuCD revealed by targeted molecular dynamics simulation. *PLoS One, 7*, e30465.

Wilks, A., & Burkhard, K. A. (2007). Heme and virulence: how bacterial pathogens regulate, transport and utilize heme. *Natural Product Reports, 24*, 511–522.

Xu, Y., Sim, S. H., Nam, K. H., Jin, X. L., Kim, H. M., Hwang, K. Y., et al. (2009). Crystal structure of the periplasmic region of MacB, a noncanonic ABC transporter. *Biochemistry, 48*, 5218–5225.

Xu, Y., Sim, S. H., Song, S., Piao, S., Kim, H. M., Jin, X. L., et al. (2010). The tip region of the MacA alpha-hairpin is important for the binding to TolC to the Escherichia coli MacAB-TolC pump. *Biochemical and Biophysical Research Communications, 394*, 962–965.

Yukitake, H., Naito, M., Sato, K., Shoji, M., Ohara, N., Yoshimura, M., et al. (2011). Effects of non-iron metalloporphyrins on growth and gene expression of *Porphyromonas gingivalis*. *Microbiology and Immunology, 55*, 141–153.

Yukl, E. T., Jepkorir, G., Alontaga, A. Y., Pautsch, L., Rodriguez, J. C., Rivera, M., et al. (2010). Kinetic and spectroscopic studies of hemin acquisition in the hemophore HasAp from *Pseudomonas aeruginosa*. *Biochemistry, 49*, 6646–6654.

Zhu, H., Alexeev, D., Hunter, D. J., Campopiano, D. J., & Sadler, P. J. (2003). Oxo-iron clusters in a bacterial iron-trafficking protein: new roles for a conserved motif. *Biochemical Journal, 376*, 35–41.

Zhu, H., Liu, M., & Lei, B. (2008). The surface protein Shr of Streptococcus pyogenes binds heme and transfers it to the streptococcal heme-binding protein Shp. *BMC Microbiology, 8*, 15.

Zhu, H., Xie, G., Liu, M., Olson, J. S., Fabian, M., Dooley, D. M., et al. (2008). Pathway for heme uptake from human methemoglobin by the iron-regulated surface determinants system of *Staphylococcus aureus*. *The Journal of Biological Chemistry, 283*, 18450–18460.

CHAPTER FOURTEEN

The Mechanisms of Mg^{2+} and Co^{2+} Transport by the CorA Family of Divalent Cation Transporters

Albert Guskov and Said Eshaghi[1]
School of Biological Sciences, Nanyang Technological University, Singapore, Republic of Singapore
[1]Corresponding author: E-mail: said.eshaghi@ntu.edu.sg

Contents

Abstract

The metal ions Mg^{2+} and Co^{2+} are essential for life, although to different degree. They have similar chemical and physical properties, but their slight differences result in Mg^{2+} to be the most abundant metal ion in living cells and the trace element Co^{2+} being toxic at relatively low concentrations. Specialized transporters have evolved in living cells to supply and balance the Mg^{2+} and Co^{2+} need of the cells. The current knowledge of the molecular mechanisms of Mg^{2+} and Co^{2+} -specific transporters is very limited at this point. Recently, there has been remarkable advances to understand the CorA family, a family of transporters that are able to transport both ions. These new data have increased our insights in how Mg^{2+} and Co^{2+} are translocated across

Current Topics in Membranes, Volume 69
ISSN 1063-5823,
http://dx.doi.org/10.1016/B978-0-12-394390-3.00014-8

membranes. Presently, CorA is probably the best system to study the mechanisms of Mg^{2+} and Co^{2+} transport. This chapter discusses the mechanisms through which CorA selects, transports, and regulates the translocation of its substrate. In addition, we highlight the physical and chemical properties of the substrates, which are important parameters required for better understanding of the transporter action.

1. OVERVIEW

Mg^{2+} and Co^{2+} are absolutely essential for the existence of all known forms of life on Earth. Mg^{2+} is the most abundant divalent cation in all living organisms and is employed in different processes as ranging from stabilization of the cell membranes and its compartments to participating in the energy transfer in the chlorophylls, not to mention its omnipresence as cofactor in different proteins. Co^{2+} is a rare but essential element as it acts as cofactor for cobalamin, which in turn is a part of the cellular metabolic pathways especially in DNA and fatty acid synthesis (Barceloux, 1999; Koury & Ponka, 2004; O'Leary & Samman, 2010; Selhub, 2002). It has also been reported as the essential cofactor for many enzymes in some microorganisms (Epting, Vieille, Zeikus, & Kelly, 2005; Holden & Adams, 2003; Nakajima, Imamura, Shoun, & Wakagi, 2003; Song, Ahluwalia, Leduc, Delbaere, & Vieille, 2008; Sterner et al., 1996; Wu, Navarro, Pina, Quénard, & Mandrand, 1994). Nevertheless, Co^{2+} is a trace element thus known to be toxic already at fairly low concentrations. Obviously, the intracellular concentration of these essential divalent cations must be kept under tight control, as imbalance in their concentration will have major consequences that could lead to cell death. Specialized transporters have evolved to sustain the ion levels within the cells at optimal concentrations.

At the moment, we have limited knowledge about the function and architecture of the biological systems that transport and regulate the intracellular concentration of Mg^{2+} and Co^{2+}, which in turn limits our knowledge about their mechanisms of action. This is due to a number of reasons, all related to the nature of these transporters, which are integral membrane proteins. Hence, any step toward biochemical and structural studies of these transporters is a challenge and a time-consuming hurdle that must be overcome. The advances in technologies for isolating membrane proteins have considerably progressed over the years. For Mg^{2+} and Co^{2+} transporters, the most notable achievement has been determination of the crystal structures of two divalent cation transporters, CorA and MgtE. These structures have shed light on the Mg^{2+} and Co^{2+} transport mechanism. In the case of CorA, which is known to transport both Mg^{2+} and Co^{2+}, the crystal structures not only revealed some features about the architecture of the protein but also corrected several earlier erroneous assumptions about the structure and the function of this protein. Nevertheless, there are still some contradictions, where biochemical and structural data are in disagreement. Thus, there are a number of questions yet to be answered about the mechanism of Mg^{2+} and Co^{2+} transport across the membranes.

This chapter discusses the suggested mechanisms through which CorA selects, transports, and regulates the translocation of its substrate. In addition, we highlight the physical and chemical properties of the substrates, which are important parameters required for better understanding of the transporter action.

2. INTRODUCTION

Consistent with being the most abundant divalent cation in biological systems, the total cellular concentration of Mg^{2+} is estimated at 14–20 mM (Romani & Scarpa, 1990) and the concentration of free Mg^{2+} is, although much lower, still relatively high and typically around 1 mM (Hurwitz & Rosano, 1967; Romani & Scarpa, 1992). The large difference between the concentrations of bound and free Mg^{2+} demonstrates the widespread utilization of this ion for maintaining the stability and function of many cellular molecules and various complexes. Mg^{2+} is the cofactor of many enzymes, including some involved in DNA replication, transcription, and protein translation. Lack of Mg^{2+} in these enzymes has a direct negative effect on the cell cycle, metabolism, and cell survival. Mg^{2+} is also involved in stabilizing the membranes, such as grana stacks of the thylakoid membranes in chloroplasts. The stacking of the grana allows a larger number of photosystem II protein complexes to be placed efficiently within grana, thus markedly increasing the conversion of light energy into its chemical equivalent. Moreover, Mg^{2+} is the essential part of chlorophylls and the depletion of Mg^{2+} leads to chlorophyll bleaching, which eventually causes photosynthesis to halt.

There is a constant exchange between the free Mg^{2+} and its bound form, as well as between various cellular compartments, such as the cytosol, nucleus, mitochondria, etc. The rate of these exchanges is directly proportional to various cellular activities. For example, many of the Mg^{2+}-binding enzymes involved in the transcription and translation processes are activated during protein synthesis. They are also located in the nucleus, cytosol, mitochondria, and chloroplast. The activation of these enzymes demands the presence of Mg^{2+}. During the high demand period, Mg^{2+} needs to be imported into the corresponding cell compartment, and afterward the excess Mg^{2+} must be exported out. This traffic of Mg^{2+} is performed and regulated by the action of Mg^{2+} transporters, which not only import the extracellular Mg^{2+} into the cell but also ensure that the Mg^{2+} requirements of various organelles are met.

The most studied Mg^{2+} transporters belong to the superfamily of 2-TM-G*x*N, where the largest family is formed by CorA-like transporters (Knoop, Groth-Malonek, Gebert, Eifler, & Weyand, 2005; Papp-Wallace & Maguire, 2007). In addition to Mg^{2+}, the members of the 2-TM-G*x*N family can also transport Co^{2+}, Ni^{2+}, Zn^{2+}, Mn^{2+}, Cu^{2+}, Ca^{2+}, La^{3+}, Ga^{3+}, and Al^{3+} (Kehres, Lawyer, & Maguire, 1998; Knoop et al., 2005). Interestingly, the CorA family members have been shown to transport Ni^{2+}, Co^{2+}, and Mg^{2+}. Despite CorA's ability to transport these three metal ions, its physiological role as either Ni^{2+} or Co^{2+} transporter has been questioned. CorA from *Escherichia coli* and *Salmonella typhimurium* can transport Mg^{2+} at concentrations as low as 15–20 μM (Kehres & Maguire, 2002), which is at or below the Mg^{2+} concentration of the environment these bacteria grow in. However, when studying the Co^{2+} and Ni^{2+} transport activity of the CorA in the same organisms, Co^{2+} concentration of 20–40 μM and Ni^{2+} concentration of 200–400 μM were needed for the transport to occur (Kehres & Maguire, 2002). Not only these concentrations are toxic for these bacteria and other related organisms, the environment that they inhabit most likely lacks such high concentrations of these trace elements. These considerations have been the main arguments against the physiological relevance of Co^{2+} and Ni^{2+} as the true substrates of the CorA family. Indeed, in contrast to the Mg^{2+}, the Co^{2+} might be toxic to the most living organisms at relatively low concentrations. However, in some microorganisms that live in harsh and extreme conditions, such as at the extreme pH and temperatures, high intracellular concentrations of trace elements such as Co^{2+} and Cu^{2+} have been reported (Llanos, Capasso, Parisi, Prieur, & Jeanthon, 2000). In addition, a number of enzymes in such organisms require Co^{2+} instead of Mg^{2+} or Zn^{2+} as the cofactor, which puts a different demand for Co^{2+} in these organisms (Epting et al., 2005; Nakajima et al., 2003; Wang, Stieglitz, & Kantrowitz, 2005; Wojciechowski, Cardia, & Kantrowitz, 2002). Therefore, the physiological concentration of Co^{2+} is a relative term that strongly depends on the environment in which organism lives and functions. So far, there are no reports of such Ni^{2+}-rich cellular environments.

Recently, the CorA from the hyperthermophilic bacterium *Thermotoga maritima* was reported to select for Co^{2+} at concentrations two magnitudes lower than that of Mg^{2+}; it also lacks the ability to regulate the Mg^{2+} transport (Xia et al., 2011). The superfamily of hyperthermophiles *Thermotogales*, to which *T. maritima* belongs, are highly Co^{2+} resistant and can grow in the presence of >1 mM Co^{2+} (Llanos et al., 2000; Schumann, Wrba, Jaenicke, & Stetter, 1991). Altogether, these facts for the first time provided with clear evidence for the physiological relevance of the CorA's ability to transport Co^{2+}. It is clear that although within the CorA family some members, probably the majority, select for Mg^{2+} and regulate its transport, some members are mainly Co^{2+} transporters. Such functional heterogeneity within the CorA family renders it as a suitable system for understanding the mechanisms for Mg^{2+} and Co^{2+} transport, most importantly to find out how these ions are selected by different CorAs.

Recent advances in the structural studies of CorA revealed that many of the previous reported assumptions concerning the structure and, hence, the function of CorA were incorrect. For example, CorA was thought to be a tetrameric complex with three transmembrane helices and the large hydrophilic domain facing the extracellular compartment (Smith, Banks, Snavely, & Maguire, 1993; Warren et al., 2004). These incorrect earlier assumptions formed the framework for many functional studies and the interpretations that can no longer be used as the foundation for the mechanism of transport of Mg^{2+} and Co^{2+} by the members of the CorA family. Another experimental limitation of the earlier studies (which is still being reported) is the frequent utilization of the Ni^{2+} to dissect the mechanism of the Mg^{2+} transport through CorA. The CorA's ability to uptake Ni^{2+} has been reported to be 10–20 times worse than its ability to transport Mg^{2+}; thus, Ni^{2+} could be simply leaking through CorA. Therefore, in order to discuss the possible mechanisms of Mg^{2+} and Co^{2+}, we should mainly use the recent reports on biochemical and biophysical characterizations of CorA. Nevertheless, there is still a serious demand for accurate data for better understanding of this mechanism.

3. PROPERTIES OF MG^{2+} AND CO^{2+}

Although some CorAs have a preference for Mg^{2+} while other primarily select Co^{2+}, all known CorAs are capable of transporting both cations. This dual-substrate specificity of CorA raises the question of how the same protein can transport two different ions, in some cases with similar affinities, and what is the molecular basis for heterogeneity in a substrate selection within the family. To answer this question, it is useful to compare the physical and chemical properties of the CorA substrates.

3.1. The Physical and Chemical Properties of Mg^{2+}

As the element 12 in the periodic table, magnesium has three shells containing 2, 8, and 2 electrons, respectively, with the $[Ne]3s^2$ electron configuration. Thus, the only oxidation number of magnesium is two, leaving Mg^{2+} as the only ionic form. The atomic and the ionic radius of magnesium are reported to be 1.5 and 0.86 Å, respectively (Greenwood & Earnshaw, 1997). The crystal structure of magnesium is hexagonal close packed, which reflects its coordination number of six. The hydration shell of Mg^{2+} in various aqueous solutions has been intensively characterized by X-ray diffraction and computer simulations. The average interatomic distance between magnesium and oxygen (Mg–O) in an octahedral water coordination geometry has been measured at approximately 2.09 Å (Marcus, 1988; Ohtaki & Radnai, 1993), which is referred to as the Mg^{2+} hydrated radius of the first hydration shell (Fig. 1).

The formation of a hydration shell is caused by the orientation of the water molecules through the electric field of the ion; thus, it is the charge–dipole interaction. The strength of this interaction depends greatly on the size of the ion; therefore, small size of Mg^{2+} (0.86 Å radius) creates a higher charge density and allows it to bind water molecules in its first hydration shell very tightly. There are some indications that up to three layers of water are affected by the Mg^{2+} which gives very high value of $\Delta G_{hydration}$ ~ −455 kcal/mol (Permiakov, 2009).

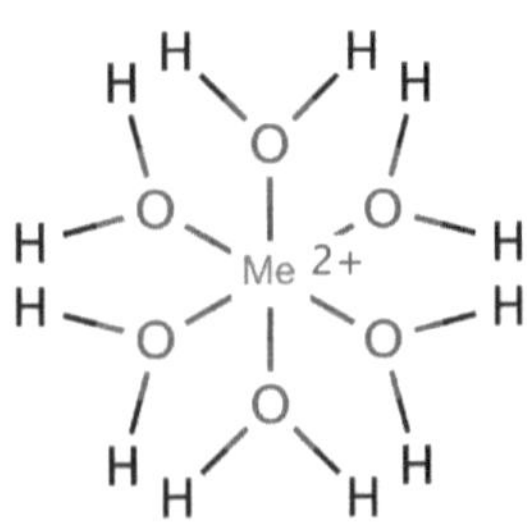

Figure 1 The structural arrangement of six water molecules in the octahedral coordination by the Mg^{2+} and Co^{2+} cations (Me^{2+}) with Me–O distances of 2.09 and 2.10 Å, respectively. For color version of this figure, the reader is referred to the online version of this book.

Because of this tight interaction, it is believed that Mg^{2+} as a cofactor binds to the enzymes through spatial coordination with water molecules (Cowan, 2002). Thus, it is possible that a transiently bound Mg^{2+}, e.g. an Mg^{2+} passing through a channel, is always present in a hydrated form.

Mg^{2+} eagerly forms the complexes with other ligands preferably the "hard" ones (i.e. hydrated Mg^{2+} behaves as a hard Lewis acid) with a low polarizability (Dudev & Lim, 2003; Pearson, 1988; Permiakov, 2009; Richens, 1997), mainly linked by oxygen atoms via electrostatic interactions. Therefore, the usual coordinating ligands for the Mg^{2+} are oxygen of the amino acids Asp and Glu as well as oxygen of the main chain carbonyls. Occasionally, oxygen of Ser, Thr, Asn, Gln, and even that of Tyr are reported to coordinate Mg^{2+}(Harding, Nowicki, & Walkinshaw, 2010; Permiakov, 2009). In rare cases, such as in chlorophylls, nitrogen from His can also bind Mg^{2+}. It is necessary to mention that ligation of Mg^{2+} always occur in the monodentate way.

3.2. Physical and Chemical Properties of Co^{2+}

Cobalt has four electron shells with the outer shell containing two electrons, electron configuration $[Ar]4s^2 3d^7$. As a transition metal, it has oxidation numbers ranging from −1 to 5+, but the most common ones are 2+ and 3+. Cobalt has an atomic radius of 1.35 Å (Greenwood & Earnshaw, 1997). The ionic radius of cobalt has been reported to be 0.84 and 0.72 Å for the 2+ and 3+ states, respectively (Greenwood & Earnshaw, 1997). Similarly to magnesium, cobalt has a hexagonal close packed structure; however, in contrast to the former, Co (II) may also have tetrahedral, square planar, trigonal or square bipyramidal, and octahedral (still the most common) ligand configuration (Hughes, 1981). The hydration shell of Co^{2+} in various aqueous solutions has been examined by X-ray diffraction and computer simulations, and the average interatomic distance between cobalt and oxygen (Co–O) in an octahedral water coordination has been measured as approximately 2.10 Å (Marcus, 1988; Ohtaki & Radnai, 1993). This value represents the hydrated radius of the Co^{2+} first hydration shell (Fig. 1). For the Co^{3+}, the Co–O distance is reported to be around 1.87 Å (Greenwood & Earnshaw, 1997). In contrast to Mg^{2+}, Co^{2+} belongs to the class of borderline cases (Pearson, 1963); hence, it might be coordinated by the "soft" polarizable ligands. Thus, in addition to oxygen-bearing side chains of Asp, Glu, Ser, and Thr, amino acids such as Cys and His are regularly found in Co^{2+}-binding sites.

3.3. Similarities and Differences between Mg^{2+} and Co^{2+}

As summarized in Table 1, cobalt is a much denser, but slightly smaller metal than magnesium. Both metals have the same hexagonal structure and coordinate with six water molecules in their first hydration shell, when their oxidation number is 2+. This quick glance at the properties of these two metals suggest that the main differences between these metals are only their weights and densities, features that are hardly important for their recognition by a transporter. However, as most other metal ions, both Mg^{2+} and Co^{2+} have a second hydration shell. The relatively short distance to the water molecules of the first hydration shell characterizes the second hydration shell of these metal ions. The O1–O2 distance (oxygen in the first and the second shell, respectively) is estimated by the X-ray diffraction to be 2.75–2.81 Å in both cases (Marcus, 1988; Ohtaki & Radnai, 1993). This is shorter than the average hydrogen bond between water molecules (3.69 Å) in the bulk water (Jorgensen & Madura, 1985). Thus, a fully hydrated Mg^{2+} or Co^{2+}, including both hydration shells, has a hydration radius of less than 5 Å. However, Mg^{2+} binds 12 water molecules in the second hydration shell, whereas from 6 up to 14 water molecules have been reported in different salt conditions to occupy the second hydration shell of Co^{2+} (Caminiti, Cucca, Monduzzi, Saba, & Grisponi, 1984; Marcus, 1988; Ohtaki & Radnai, 1993). The molecular dynamics simulations revealed about 12.5 water molecules in the second hydration shell for Co^{2+} (Spezia et al., 2006). Such variation could be mainly due to the difficulties in the experimental measurements, but it could also be an indication of the actual flexibility in the second hydration shell of Co^{2+}. Although the exact and detailed information about the second hydration shells of metal ions is not as reliable as the data obtained from the first hydration shell, the similar results from many measurements indicate a significantly tight and robust hydration shells surrounding Mg^{2+}. Thus, hydrated Mg^{2+} can be considered as a well-coordinated ion with a defined

Table 1 List of the properties of magnesium and cobalt

	Magnesium	Cobalt
Weight (Da)	24.305	58.9332
Density (g/cm^3)	1.74	8.9
Structure	Hexagonal	Hexagonal
Atomic radius (Å)	1.50	1.35
Ionic (2+) radius (Å)	0.86	0.84
Hydrated radius (first shell) (Å)	2.09	2.10

number of water molecules surrounding it tightly. Co^{2+}, on the other hand, has probably a less rigid second hydration shell considering such a variable number of reported coordinated water molecules. The flexible second hydration shell of Co^{2+} also suggests a less defined structure of hydrated Co^{2+} as compared to the structure of hydrated Mg^{2+}. Moreover, the flexibility of cobalt hydration shells is supported by the observation of $[Co(OH_2)_4]^{2+}$ formation with the tetrahedral coordination in Co^{2+} aqueous solutions (Swift, 1964). Furthermore, the water exchange from the first hydration shell for Co^{2+} (3.2×10^6/s) (Ducommun, Newman, & Merbach, 1980) is about 10 times faster than that for Mg^{2+} (5.3×10^5/s) (Neely & Connick, 1970). Altogether, we can summarize the main structural differences between these two ions, from a substrate point of view, as follows:

A fully hydrated Mg^{2+} has a defined and less flexible structure with 18 water molecules in its hydration shells that are bound tightly to Mg^{2+}, whereas the structure of fully hydrated Co^{2+} is much more flexible and can accommodate 12–20 water molecules in its hydration shells. Hence, Co^{2+} has the ability of mimicking Mg^{2+} to coordinate carboxyl groups, but Mg^{2+} cannot mimic Co^{2+}. Therefore, during transient interactions with a protein, the hydrated Co^{2+} can make similar interactions with the protein as a hydrated Mg^{2+} does, but the hydrated Mg^{2+} is unable to mimic a hydrated Co^{2+} to the same extent. Plausibly, these features of the two hydrated ions could be the key to understand why the CorA proteins can transport both Mg^{2+} and Co^{2+} (hereon defined as Me^{2+}), and yet there is a heterogeneity in the substrate selectivity within the family.

4. THE CORA FAMILY OF MG^{2+}/CO^{2+} TRANSPORTERS

The CorA family belongs to the 2-TM-G*x*N superfamily of metal ion transporters. The 2-TM-G*x*N family is characterized by the only two transmembrane helices, both located at the end of the C-terminus, and connected by a conserved loop containing the signature motif of the family, namely the amino acid sequence of G*x*N, where *x* could be M, V, or I (Knoop et al., 2005). The Mg^{2+} transporters, such as CorA, bear the GMN motif (Kehres et al., 1998; Knoop et al., 2005; Niegowski & Eshaghi, 2007). The CorA family members are found in both prokaryotes and eukaryotes (Knoop et al., 2005), but with the exception of the GMN motif, there is almost no sequence homology between the CorA members from prokaryotes and eukaryotes.

CorA is considered as the primary Mg^{2+} transporter in many bacteria and archaea. MgtE is the primary Mg^{2+} transporter in those organisms that lack CorA, e.g. *Thermus thermophilus*. However, in some organisms, such as *T. maritima*, both CorA and MgtE are found. The only known physiological role of MgtE is to import Mg^{2+} and no other substrates have been assigned to this transporter. Since CorA in *T. maritima* is highly competent in selecting and transporting Co^{2+}, it is believed that MgtE is the primary Mg^{2+} transporter in *T. maritima*, whereas CorA fulfills the Co^{2+} needs of the organism (Xia et al., 2011).

In prokaryotes, the GMN motif is extended to YGMNF (Niegowski & Eshaghi, 2007). The loop containing the YGMNF motif is facing the extracellular space, and the large N-terminal domain of CorA faces the periplasm. In most CorA proteins, another signature motif, MPEL, is also found at the C-terminus of the YGMNF motif (Niegowski & Eshaghi, 2007). In general, the sequence homology within the CorA family is the highest at the C-terminus and decreases toward the N-terminus. However, based on the amino acid sequence homology, and possibly also tertiary structure and functional homology, the CorA family can be divided into at least two subgroups A and B (Niegowski & Eshaghi, 2007). Within each subgroup, the members have higher level of homology at the N-terminus.

4.1. The Physiology of CorA

Biochemical characterizations of CorA have been performed on transporters from *E. coli*, *S. typhimurium*, *Haemophilus influenzae*, *Methanocaldococcus jannaschii*, and *T. maritima*. Most of them have shown CorA to be capable of transporting Mg^{2+}, Co^{2+}, and Ni^{2+}, with Mg^{2+} as the primary selection except for the *T. maritima* CorA that is a Co^{2+}-selective transporter (Kehres & Maguire, 2002; Xia et al., 2011). The capability of CorA in these organisms to unselect Ni^{2+} is very high under physiological concentrations of Ni^{2+}, and therefore, this ion, at this time, is ruled out as a substrate for any of these homologues.

As a housekeeping protein, CorA is continuously present in the membrane and maintains the optimal intracellular concentration of its substrate. Due to the relatively high concentration of free Mg^{2+} within the cells, the immediate needs for Mg^{2+} ions are met by those already available, so delay between the immediate need and import of Mg^{2+} can be tolerated by the cell. Also, Mg^{2+}, in contrary to, e.g. Ca^{2+}, is usually not directly involved in special cellular activities such as signaling or the regulation of voltage and current. Therefore, a sudden rise in the intracellular concentration to the same extent as Ca^{2+} and K^+ is not required. Likewise is the need for Co^{2+} in microorganisms that use Co^{2+} as a cofactor and/or for thermostability. Therefore, the current of Me^{2+} transport into these systems through CorA does not need to be high. In principal, a slow transport could be beneficial from the regulatory point of view. When the Me^{2+} concentration is suddenly elevated in the extracellular space, e.g. when an Mg^{2+}-rich diet reaches *E. coli* in the intestines or when there is a sudden rise in the Co^{2+} concentration around *T. maritima* after a volcano eruption, a slow transport of these substrates might be a protective mechanism of the regulatory process to prevent any toxic effects. Such a slow transport mechanism would probably be more favorable and also ease the regulation of the gating.

Whether CorA has the role of exporting its substrate is not known, but a distant homologue, ZntB, in which the GMN, the signature motif of the family (see below) is exchanged to GVN, is believed to export Zn^{2+} from the cytoplasm (Caldwell & Smith, 2003; Worlock & Smith, 2002).

4.2. The Structure of CorA

The most detailed crystal structure of CorA from *T. maritima* revealed CorA to be a pentamer, with the large intracellular domain making a funnel-like structure (Fig. 2) (Eshaghi et al., 2006). There are a total of eight helices shown by the crystal structure (Fig. 2). The seventh helix is a long α-helix that continues into the membrane and makes the first transmembrane helix, which is also the pore-forming helix (Fig. 2). The second transmembrane helices form a highly hydrophobic ring surrounding the first transmembrane helices and also stabilizing the position of the protein in the lipid bilayer. Two metal-binding sites, M1 and M2, were found near the top of the intracellular domain at the monomer interface (Figs 2 and 3). Interestingly, both Mg^{2+} and Co^{2+} could bind to the same sites with the same coordination (Eshaghi et al., 2006). M1 has a tight binding of about 2 Å with D89 and D253, followed by a loose coordination with E88 (~4.4 Å). It is therefore believed that the carboxyl groups of D89 and D253 have replaced two water molecules in the first hydration shell, and the carboxyl group of E88 has replaced one water molecule in the second hydration shell. The M2 coordination includes the carboxyl groups of E88, D175, and D253 and the carbonyl groups of L12 and P13. The distances between the Me^{2+} and the oxygen groups in the M2 are between 3 and 5 Å, suggesting that the ion holds its first hydration shell and the carboxyl and carbonyl groups replace

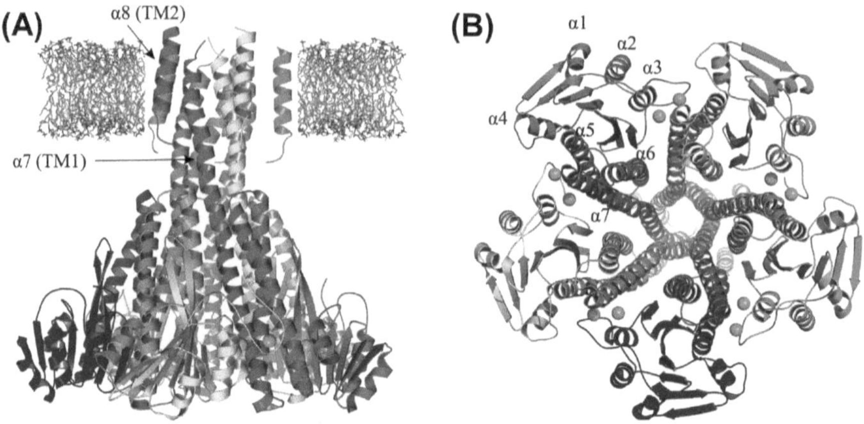

Figure 2 The general architecture of CorA from *Thermotoga maritima*: (A) side view, monomers are colored individually and shown as ribbons. Transmembrane helices forming the pore are on the top and embedded into the plasma membrane and large hydrophilic domain carrying regulatory binding sites at the bottom. (B) Top view from the cytoplasm, magnesium ions in the metal-binding sites shown as green spheres. The numbering of the helices is indicated. Note that α-7 is extended into the membrane and makes the first transmembrane helix, TM1. See the color plate.

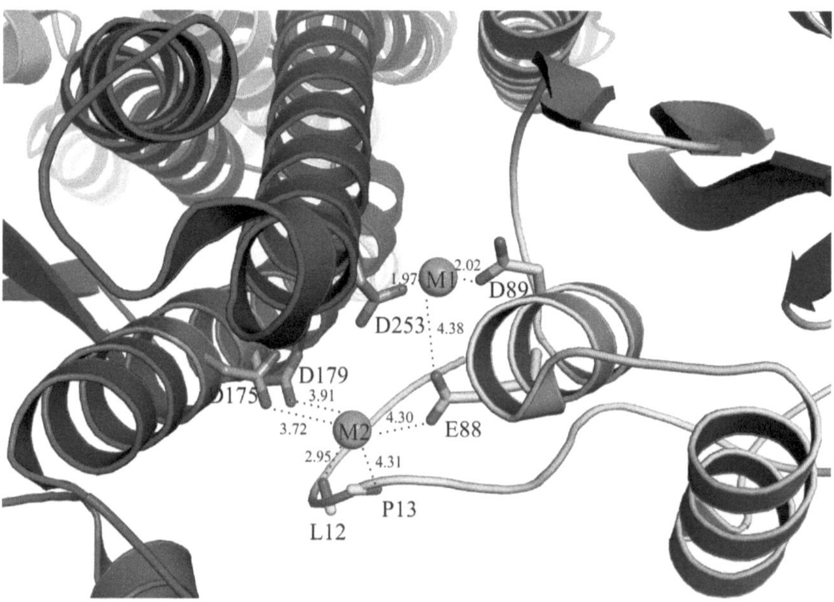

Figure 3 The detailed view on the binding sites M1 and M2. Coordinating amino acids are labeled and distances are in Å. See the color plate.

some of the water molecules of its second hydration shell. Altogether, M1 makes a much tighter and more stable binding than M2. The crystal structure also shows the N-terminal domain of CorA to cover a cavity where M2 is located. This cover seems to be stabilized by the presence of two adjacent prolines (P13 and P14) and through interactions between the L12/P13 main chain carboxyls with the Me^{2+} at M2 (Fig. 3).

Down the funnel and where the stalk begins, an aspartate ring is formed by D277 (Fig. 4). The O–O distances between the carboxyl groups that are opposing each other are around 8–9 Å. This space could accommodate a partially hydrated Me^{2+}, i.e. the ion with only the first hydration shell where the aspartate ring could substitute the second hydration shell. Two helical turns down this region, the serine ring (S284) is located with the similar O–O distances as within the aspartate ring, again a suitable environment for a partially hydrated Me^{2+}. However, between the two hydrophilic rings, there is a hydrophobic ring made by L280, where the opposing side chains are 6–7 Å apart. This leucine ring makes an environment that is too hydrophobic and narrow for a partially hydrated Me^{2+} and

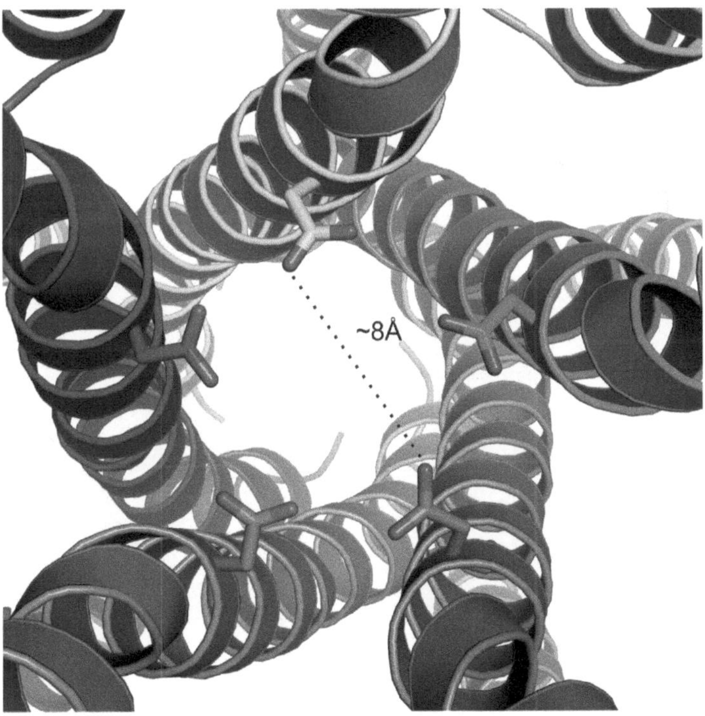

Figure 4 The aspartate ring inside CorA funnel. See the color plate.

thus resembles a closed or blocked passage. This blockage becomes clearer within the membrane-spanning region of the pore where M291, L294, and M302 create a hydrophobic girdle by pointing the side chains toward the interior of the pore (Fig. 5A). This girdle is tightest at M291 and L294 with distances of about 6.4–7 Å between the opposing side chains, leaving a very thin passage of about 2 Å in diameter (Chakrabarti, Neale, Payandeh, Pai, & Pomes, 2010). The M302 side chains are further apart from each other (~7.5–8 Å) yet creating an unsuitable environment for a partially hydrated Me^{2+} to pass through. The hydrophobic interaction between M302 residues seems to be further stabilized by additional hydrophobic interactions between the rings of F301 and P303 at the monomer interfaces (Fig. 5B). The latter has also created a bend in TM1. Thus, the narrow hydrophobic passage seen in the crystal structure of *T. maritima* CorA presents an apparently closed conformation, which is probably induced by the Me^{2+} bound at the N-terminal metal-binding sites.

Among the important features revealed by the crystal structure was the long and a fairly hydrophobic passage that extends 20 Å beyond the membrane. This feature is dissimilar to the common structural arrangement of the ion channels, which shortens the passage through the membrane by creating water-filled cavities (Doyle, 2004a, 2004b). Whether this long passage exists also in the open state of the protein is at the moment unclear.

Unfortunately, all reported crystal structures did not show the conserved loop containing the YGMNF and MPEL motifs; thus, no structural information is currently available for this highly conserved region.

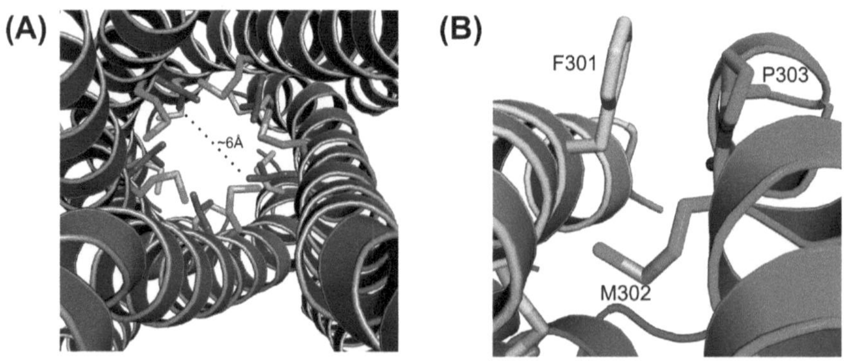

Figure 5 (A) The hydrophobic girdle formed by Met291 (in cyan), Leu294 (in blue), and Met302 (in green). (B) Additional hydrophobic stabilization of the girdle by Phe301 and Pro303 from adjacent monomers. See the color plate.

5. THE MECHANISMS OF MG^{2+} AND CO^{2+} TRANSPORT THROUGH CORA

5.1. Ion Transport

The amino acid sequences of the first transmembrane helices of CorAs show no carboxyl residues in the pore (Niegowski & Eshaghi, 2007). In fact, the first transmembrane helices of most CorAs are quite hydrophobic with only a few polar residues. The crystal structure of CorA did not show any carbonyl groups from the main chain that could potentially be involved in ion coordination. Although the crystal structure shows a closed state, it is highly unlikely that any carbonyl group could directly be involved in ion coordination considering the arrangement of the first transmembrane helices where only side chains are facing the interior of the pore.

Biochemical characterization of the pore region of *T. maritima* CorA has shown that if the side chain of L294 is either removed or changed to a charged amino acid (either positive or negative), the rate of the transport as well as the regulation of Mg^{2+} are severely disturbed (Svidova, Sponder, Schweyen, & Djinovic-Carugo, 2011). This observation is in agreement with the molecular dynamics simulation studies that also pinpointed this region, as the part of the hydrophobic girdle and along with L280, to be very important for allowing Mg^{2+} to pass through (Chakrabarti et al., 2010). In the latter study, the free passage of water molecules in a putative open form was also shown, which in turn would reflect the passage of at least a partially hydrated Mg^{2+}. Based on the simulation studies, the authors suggested that the hydrophobic girdle is widening in an iris-like motion, through the movement of the entire helix-7. In that work, an increase in the diameter of the pore by only 0.8 Å was described as sufficient to allow the passage of water. However, although such diameter appears wide enough to fit a water molecule, it is unlikely that water would go through. This is because a much wider pore of hydrophobic nature is needed to allow the passage of bulk water, as had been demonstrated earlier (Beckstein et al., 2003). Nevertheless, the computational studies suggest that in *T. maritima* CorA, the hydrophobic girdle leaves space that is just enough for a partially hydrated Mg^{2+} to pass through without any direct coordination with the protein. To stop the Mg^{2+} flow, the hydrophobic girdle is narrowed through an iris-like mechanism. Whether this is really the case remains to be confirmed.

These are the only studies reported so far, which aim to explain the mechanism of ion transport through CorA. In another study focusing on the

Mg^{2+} binding to the conserved extracellular loop, it was concluded that the fully hydrated ion interacts with and binds to this loop (Moomaw & Maguire, 2010). Thus, the ion passing through CorA is most likely either partially or fully hydrated rather than being completely dehydrated. However, most of these results are based on the *T. maritima* CorA Mg^{2+} transport ability, whereas *T. maritima* CorA is a selective Co^{2+} transporter and cannot regulate the Mg^{2+} transport (Xia et al., 2011). Therefore, additional studies using Co^{2+} as the substrate are needed before such mechanisms could be accepted.

As revealed by the structure of *T. maritima* CorA, the long hydrophobic passage extends the membrane region into the cytoplasm and thereby prolongs the distance the ion needs to travel in a fairly "dry" environment. Such an arrangement seems inconsistent with the idea that an ion transporter functions to facilitate the transport as fast as possible and thus releases the substrate into the aqueous phase immediately after the membrane region is crossed. Perhaps such delay in transport through CorA is part of the regulatory process, especially since the organism can tolerate a relatively slow import of Mg^{2+} and Co^{2+}.

5.2. Gating

The identification of the M1 and M2 sites in the intracellular domain of the *T. maritima* CorA crystal structure has led to an obvious notion that these sites might be the regulatory sites controlling the gating of the channel. The closed state of the structure in the presence of Me^{2+} in these sites suggested that when the intracellular concentration of the metal substrate is high enough the ions would bind to these sites and force the channel to close. In a follow-up report, biochemical studies were combined with the site-directed mutagenesis of the residues in the metal-binding sites in *T. maritima* CorA (Payandeh et al., 2008). Although the role of metal-binding sites was concluded to be the regulation of the pore gating, there were some inconclusive data presented. For example, in an attempt to mimic the bound state of M1, the D89K mutation was introduced to make a salt bridge with D253. The result was indeed a reduction in Mg^{2+} transport ability of CorA. The D253K showed also a reduction in the Mg^{2+} transport ability of CorA, although not to the same level as the D89K. However, when the D89F, which could not form any salt bridge with D253, was introduced, it also caused the same inhibitory effect as the D253K mutation. Although the authors concluded that both M1 and M2 are involved in the regulation of the protein gating, one could also postulate that perhaps M1 is (in addition) important to maintain a stable and intact protein, especially considering the high temperatures of *T. maritima*'s natural habitat. Furthermore, M2 could instead be the main regulatory site to gate the ion transport. This postulation is supported by the D253F mutation that resulted in dramatic improvement of Mg^{2+} transport ability of the protein, termed as "gain of function" (Payandeh et al., 2008). However, this study lacks some important information including the protein expression levels, the structural integrity, and the stability of the mutants, all which can have serious consequences on the correct interpretation of the results. For example, if the protein levels of various constructs (i.e. the mutations and the wild type) per cell were not the same, the differences in the results could be due to the differences in the amount of protein present. Despite some contradictions and uncertainties, these results do give the notion that the metal-binding sites may indeed have a role in the regulation of gating.

The hypothesis about the regulatory role of the M1 and M2 sites in the gating was further strengthened by molecular dynamics simulations, which suggested that the sideway movement of the helix-7 directly correlated with the occupancy of the metal-binding sites (Chakrabarti et al., 2010). According to this study, partially hydrated Mg^{2+} could pass through the hydrophobic gate in the pore region. In the presence of Mg^{2+} in the metal-binding sites, the helices 7 would move sideways and inward and thus tighten the hydrophobic gate, preventing the partially hydrated ion to pass through.

In such a scenario, where the Mg^{2+} binding to the M1 and M2 sites would result in closing the gate, the affinity of these sites for Mg^{2+} must be fairly low (in sub-millimolar range), reflecting the intracellular concentration of free Mg^{2+}. The same is probably also true for Co^{2+} in organisms, such as *T. maritima*, that require Co^{2+} at a higher degree, and the actual affinity will depend on the intracellular concentration of free Co^{2+}. Indeed, Mg^{2+} transport studies through *T. maritima* CorA liposomes did not cause the closure of the gate despite the presence of 10 mM Mg^{2+} (Payandeh et al., 2008). However, this is most likely due to the fact that *T. maritima* is not able to regulate Mg^{2+} transport, which means that the metal-binding sites in this protein have an extremely low affinity for Mg^{2+}. In fact, the Mg^{2+} binding to these sites has been observed only during crystallization with 100 mM Mg^{2+} present. No structural rearrangement of *T. maritima* CorA could be detected in the presence of up to 5 mM Mg^{2+} (Xia et al., 2011). Co^{2+}, on the other hand, could induce structural rearrangements at very low concentrations ($\sim$5 μM), but its *in vitro* thermostabilizing effect appears only at concentrations above 50 μM.

Thus, Mg^{2+} and Co^{2+} are believed to pass through a hydrophobic gate of CorA and the gating is most likely regulated through the weak binding of the Me^{2+} to the intracellular metal-binding sites.

5.3. Ion Selectivity

Currently, no conclusive reports are available which can explain the mechanism of ion selectivity of CorA. The main question remains on how certain CorAs, such as those from *E. coli* and *S. typhimurium*, are able to select for Mg^{2+}, while *T. maritima* CorA is highly selective for Co^{2+}. A hydrophobic gate that opens up upon Me^{2+} binding at the gating/regulatory sites may be a possible mechanism of transport for the partially hydrated ion, but it cannot explain how the selection process occurs. The conserved loop is believed to be involved in the ion binding and perhaps also the selection. This is most likely the case for unselecting metal ions other than Mg^{2+} and Co^{2+}, and an early location for ion selection is certainly logic. But since the loop is highly conserved and is almost identical between, e.g. *E. coli* CorA and *T. maritima* CorA, it is rather unlikely that this domain alone could distinguish between Co^{2+} and Mg^{2+}, unless the structural arrangement of the loop is different between the homologues. In fact, Nucleic Magnetic Resonance studies have indicated that both Mg^{2+} and Co^{2+} bind to the same extracellular site and that the Glu residue in the MPEL motif is important in the binding (Hu, Sharma, Qin, Gao, & Cross, 2009). The latter is also supported by another study, which proposes a general mechanism for the Mg^{2+} binding to this loop (Moomaw & Maguire, 2010; Payandeh et al., 2008). Nevertheless, CorA from *M. jannaschii*, which was shown by Maguire and colleagues to be an Mg^{2+} transporter, is one of the few exceptions in the CorA family that does not contain the MPEL motif. Therefore, the selectivity capacity of this domain remains unclear.

The substrate selectivity is totally dependent on chemical and/or physical differences between the actual substrate and possible and potential competitors. In the case of Mg^{2+} and Co^{2+}, as stated earlier, the two ions have many features in common, which are important for selection. Notably, the only such difference between these two ions is the fine structure of their second hydration shell. This would need to be complemented with the structural differences between CorAs. At the moment, no models for how Co^{2+} and Mg^{2+} are distinguished by different CorAs are available. Such model can only be given once more structural data are obtained. However, it should also be noted that all CorAs studied so far have shown the capability of transporting both substrates, which is most likely a consequence of high similarities between the two ions.

6. CONCLUDING REMARKS

CorA is a family of transporters for both Mg^{2+} and Co^{2+} and is thus a suitable system for studying the transport mechanism of these ions. In the recent years, there have been major breakthroughs in this field, especially concerning the structural information. The available three-dimensional structures of *T. maritima* CorA has not only corrected a number of wrong assumptions about the structure of this protein but also enabled a better and deeper understanding about how this protein may work. It has also paved the way for more directed biochemical and computational experiments, and thus, new ideas and models on the mechanism of Mg^{2+} and Co^{2+} transport have emerged. The current suggested mechanism is that Mg^{2+} (and perhaps also Co^{2+}) is transported as partially hydrated ion through a hydrophobic region that is enough open/wide to allow the transport. The transport is terminated by tightening up this hydrophobic region as a consequence of Me^{2+} binding in the intracellular metal-binding sites, which pulls the pore-forming helices sideways.

However, the picture is not yet complete and there exists a number of contradictory and inconclusive data. For example, if we consider the gating mechanism, CorA should close when intracellular concentration of Mg^{2+} is nearly 1 mM, which corresponds to the intracellular concentration of free Mg^{2+} (Hurwitz & Rosano, 1967; Romani & Scarpa, 1992). Then the question is raised on why the protein kept transporting Mg^{2+} in the presence of 10 mM in a proteoliposome experiment (Payandeh et al., 2008). The reason for this being most likely the mismatches within the system that is being studied: *T. maritima* CorA as a selective Co^{2+} transporter without the ability to regulate the Mg^{2+} transport has been vividly used to study how Mg^{2+} is transported and gated. Inevitably, this mismatch creates major problems observed as inconclusive and sometimes contradictory results. Therefore, we still need to obtain more functional data on *T. maritima* CorA using Co^{2+} as the substrate. Also, more structural information of other CorA homologues that are true Mg^{2+} transporters, such as *E. coli* CorA, are needed in order to find out about important structural differences concerning the transport and especially the ion selectivity.

ACKNOWLEDGMENTS

This work was funded by the National Research Foundation (grant NRF-CRP4-2008-02) and Biomedical Research Council of Singapore.

REFERENCES

Barceloux, D. G. (1999). Cobalt. *Clinical Toxicology, 37*(2), 201–216.

Beckstein, O., Biggin, P. C., Bond, P., Bright, J. N., Domene, C., Grottesi, A., et al. (2003). Ion channel gating: insights via molecular simulations. *FEBS Letters, 555*(1), 85–90.

Caldwell, A. M., & Smith, R. L. (2003). Membrane topology of the ZntB efflux system of Salmonella enterica serovar Typhimurium. *Journal of Bacteriology, 185*(1), 374–376.

Caminiti, R., Cucca, P., Monduzzi, M., Saba, G., & Grisponi, G. (1984). Divalent metal–acetate complexes in concentrated aqueous solutions. An X-ray diffraction and NMR spectroscopy study. *The Journal of Chemical Physics, 81*(1), 543.

Chakrabarti, N., Neale, C., Payandeh, J., Pai, E. F., & Pomes, R. (2010). An iris-like mechanism of pore dilation in the CorA magnesium transport system. *Biophysical Journal, 98*(5), 784–792.

Cowan, J. A. (2002). Structural and catalytic chemistry of magnesium-dependent enzymes. *Biometals, 15*(3), 225–235.

Doyle, D. A. (2004a). Structural themes in ion channels. *European Biophysics Journal, 33*(3), 175–179.

Doyle, D. A. (2004b). Molecular insights into ion channel function (Review). *Molecular Membrane Biology, 21*(4), 221–225.

Ducommun, Y., Newman, K. E., & Merbach, A. E. (1980). High-pressure oxygen-17 NMR evidence for a gradual mechanistic changeover from Ia to Id for water exchange on divalent octahedral metal ions going from manganese(II) to nickel(II). *Inorganic Chemistry, 19*(12), 3696–3703.

Dudev, T., & Lim, C. (2003). Principles governing Mg, Ca, and Zn binding and selectivity in proteins. *Chemical Reviews, 103*(3), 773–788.

Epting, K. L., Vieille, C., Zeikus, J. G., & Kelly, R. M. (2005). Influence of divalent cations on the structural thermostability and thermal inactivation kinetics of class II xylose isomerases. *FEBS Journal, 272*(6), 1454–1464.

Eshaghi, S., Niegowski, D., Kohl, A., Martinez Molina, D., Lesley, S. A., & Nordlund, P. (2006). Crystal structure of a divalent metal ion transporter CorA at 2.9 angstrom resolution. *Science, 313*(5785), 354–357.

Greenwood, N. N., & Earnshaw, A. (1997). *Chemistry of the elements* (2nd ed.). Butterworth-Heinemann.

Harding, M. M., Nowicki, M. W., & Walkinshaw, M. D. (2010). Metals in protein structures: a review of their principal features. *Crystallography Reviews, 16*(4), 247–302.

Holden, J. F., & Adams, M. W. (2003). Microbe-metal interactions in marine hydrothermal environments. *Current Opinion in Chemical Biology,* 7(2), 160–165.

Hu, J., Sharma, M., Qin, H., Gao, F. P., & Cross, T. A. (2009). Ligand binding in the conserved interhelical loop of CorA, a magnesium transporter from *Mycobacterium tuberculosis. The Journal of Biological Chemistry, 284*(23), 15619–15628.

Hughes, M. N. (1981). *The inorganic chemistry of biological processes edition.* John Wiley and Sons.

Hurwitz, C., & Rosano, C. L. (1967). The intracellular concentration of bound and unbound magnesium ions in Escherichia coli. *Journal of Biological Chemistry, 242*(16), 3719–3722.

Jorgensen, W. L., & Madura, J. D. (1985). Temperature and size dependence for Monte Carlo simulations of TIP4P water. *Molecular Physics, 56*(6), 1381.

Kehres, D. G., Lawyer, C. H., & Maguire, M. E. (1998). The CorA magnesium transporter gene family. *Microbial and Comparative Genomics, 3*(3), 151–169.
Kehres, D. G., & Maguire, M. E. (2002). Structure, properties and regulation of magnesium transport proteins. *Biometals, 15*(3), 261–270.
Knoop, V., Groth-Malonek, M., Gebert, M., Eifler, K., & Weyand, K. (2005). Transport of magnesium and other divalent cations: evolution of the 2-TM-GxN proteins in the MIT superfamily. *Molecular Genetics and Genomics, 274*(3), 205–216.
Koury, M. J., & Ponka, P. (2004). New insights into erythropoiesis: the roles of folate, vitamin B12, and iron. *Annual Review of Nutrition., 24*, 105–131.
Llanos, J., Capasso, C., Parisi, E., Prieur, D., & Jeanthon, C. (2000). Susceptibility to heavy metals and cadmium accumulation in aerobic and anaerobic thermophilic microorganisms isolated from deep-sea hydrothermal vents. *Current Microbiology, 41*(3), 201–225.
Marcus, Y. (1988). Ionic radii in aqueous solutions. *Chemical Reviews, 88*(8), 1475–1498.
Moomaw, A. S., & Maguire, M. E. (2010). Cation selectivity by the CorA Mg^{2+} channel requires a fully hydrated cation. *Biochemistry, 49*(29), 5998–6008.
Nakajima, M., Imamura, H., Shoun, H., & Wakagi, T. (2003). Unique metal dependency of cytosolic alpha-mannosidase from Thermotoga maritima, a hyperthermophilic bacterium. *Archives of Biochemistry and Biophysics, 415*(1), 87–93.
Neely, J., & Connick, R. (1970). Rate of water exchange from hydrated magnesium ion. *Journal of the American Chemical Society, 92*(11), 3476–3478.
Niegowski, D., & Eshaghi, S. (2007). The CorA family: structure and function revisited. *Cellular and Molecular Life Sciences, 64*(19–20), 2564–2574.
Ohtaki, H., & Radnai, T. (1993). Structure and dynamics of hydrated ions. *Chemical Reviews, 93*(3), 1157–1204.
O'Leary, F., & Samman, S. (2010). Vitamin B12 in health and disease. *Nutrients, 2*(3), 299–316.
Papp-Wallace, K. M., & Maguire, M. E. (2007). Bacterial homologs of eukaryotic membrane proteins: the 2-TM-GxN family of Mg(2+) transporters. *Molecular Membrane Biology, 24*(5–6), 351–356.
Payandeh, J., Li, C., Ramjeesingh, M., Poduch, E., Bear, C. E., & Pai, E. F. (2008). Probing structure-function relationships and gating mechanisms in the CorA Mg^{2+} transport system. *The Journal of Biological Chemistry, 283*(17), 11721–11733.
Pearson, R. G. (1963). Hard and soft acids and bases. *Journal of the American Chemical Society, 85*(22), 3533–3539.
Pearson, R. G. (1988). Absolute electronegativity and hardness: application to inorganic chemistry. *Inorganic Chemistry, 27*(4), 734–740.
Permiakov, E. A. (2009). *Metalloproteomics*. John Wiley & Sons. p. 786.
Richens, D. T. (1997). *The chemistry of aqua ions: synthesis, structure, and reactivity: a tour through the periodic table of the elements*. J. Wiley.
Romani, A. M. P., & Scarpa, A. (1990). Hormonal control of Mg^{2+} transport in the heart. *Nature, 346*(6287), 841–844.
Romani, A. M. P., & Scarpa, A. (1992). Regulation of cell magnesium. *Archives of Biochemistry and Biophysics, 298*(1), 1–12.
Schumann, J., Wrba, A., Jaenicke, R., & Stetter, K. O. (1991). Topographical and enzymatic characterization of amylases from the extremely thermophilic eubacterium Thermotoga maritima. *FEBS Letters, 282*(1), 122–126.
Selhub, J. (2002). Folate, vitamin B12 and vitamin B6 and one carbon metabolism. *The Journal of Nutrition Health and Aging, 6*(1), 39–42.
Smith, R. L., Banks, J. L., Snavely, M. D., & Maguire, M. E. (1993). Sequence and topology of the CorA magnesium transport systems of Salmonella typhimurium and Escherichia coli. Identification of a new class of transport protein. *Journal of Biological Chemistry, 268*(19), 14071–14080.

Song, S. H., Ahluwalia, N., Leduc, Y., Delbaere, L. T., & Vieille, C. (2008). Thermotoga maritima TM0298 is a highly thermostable mannitol dehydrogenase. *Applied Microbiology and Biotechnology, 81*(3), 485–495.

Spezia, R., Duvail, M., Vitorge, P., Cartailler, T., Tortajada, J., Chillemi, G., et al. (2006). A coupled Car-Parrinello molecular dynamics and EXAFS data analysis investigation of aqueous Co(2+). *The Journal of Physical Chemistry A, 110*(48), 13081–13088.

Sterner, R., Kleemann, G. R., Szadkowski, H., Lustig, A., Hennig, M., & Kirschner, K. (1996). Phosphoribosyl anthranilate isomerase from Thermotoga maritima is an extremely stable and active homodimer. *Protein Science, 5*(10), 2000–2008.

Svidova, S., Sponder, G., Schweyen, R. J., & Djinovic-Carugo, K. (2011). Functional analysis of the conserved hydrophobic gate region of the magnesium transporter CorA. *Biochimica et Biophysica Acta, 1808*(6), 1587–1591.

Swift, T. J. (1964). The kinetics of structural transformation of hydrated cobalt(II) and zinc(II) ions in aqueous solution. *Inorganic Chemistry, 3*(4), 526–529.

Wang, J., Stieglitz, K. A., & Kantrowitz, E. R. (2005). Metal specificity is correlated with two crucial active site residues in Escherichia coli alkaline phosphatase. *Biochemistry, 44*(23), 8378–8386.

Warren, M. A., Kucharski, L. M., Veenstra, A., Shi, L., Grulich, P. F., & Maguire, M. E. (2004). The CorA Mg^{2+} transporter is a homotetramer. *Journal of Bacteriology, 186*(14), 4605–4612.

Wojciechowski, C. L., Cardia, J. P., & Kantrowitz, E. R. (2002). Alkaline phosphatase from the hyperthermophilic bacterium T. maritima requires cobalt for activity. *Protein Science, 11*(4), 903–911.

Worlock, A. J., & Smith, R. L. (2002). ZntB is a novel Zn^{2+} transporter in Salmonella enterica serovar Typhimurium. *Journal of Bacteriology, 184*(16), 4369–4373.

Wu, L. F., Navarro, C., de Pina, K., Quénard, M., & Mandrand, M. A. (1994). Antagonistic effect of nickel on the fermentative growth of Escherichia coli K-12 and comparison of nickel and cobalt toxicity on the aerobic and anaerobic growth. *Environmental Health Perspectives, 102*, 297–300.

Xia, Y., Lundback, A. K., Sahaf, N., Nordlund, G., Brzezinski, P., & Eshaghi, S. (2011). Co^{2+} selectivity of Thermotoga maritima CorA and its inability to regulate Mg^{2+} homeostasis present a new class of CorA proteins. *The Journal of Biological Chemistry, 286*(18), 16525–16532.

INDEX

Note: Page numbers with "f" denote figures; "t" tables.

B

C

H

K

L

N

O

T

COLOR PLATES

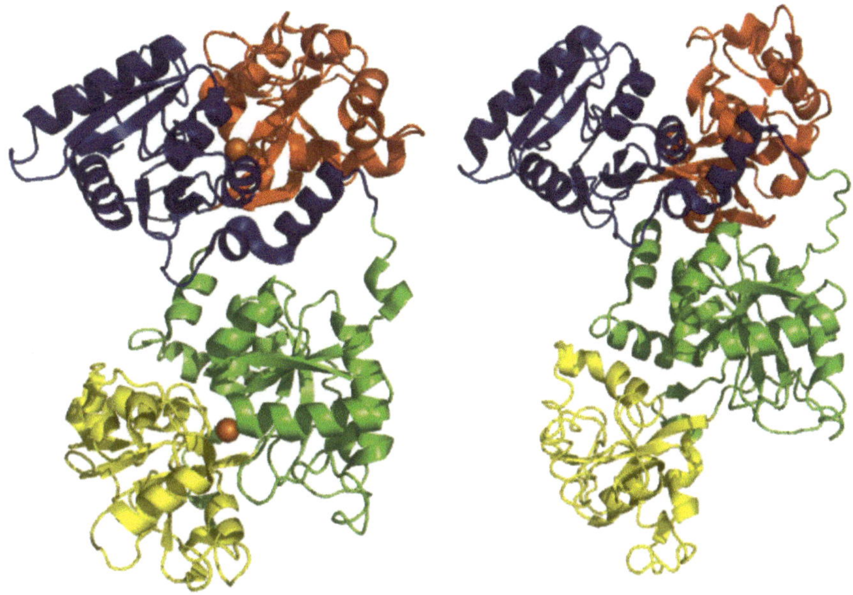

PLATE 1 (Figure 1 on page 10 of this volume)

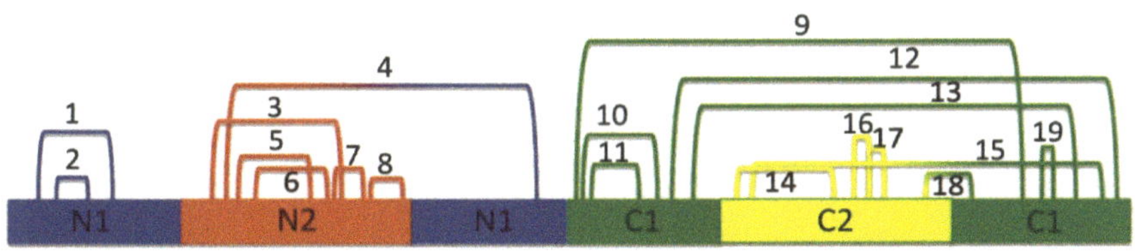

N-lobe			C-lobe		
Disulfide Bond #	**Cys Residue 1**	**Cys Residue 2**	**Disulfide Bond #**	**Cys Residue 1**	**Cys Residue 2**
			9	339	596
1	9	48	10	345	377
2	19	39	11	355	368
			12	402	674
			13	418	637
3	118	194	14	450	523
4	137	331	15	474	665
5	158	174	16	484	498
6	161	177			
7	171	179	17	495	506
8	227	241	18	563	577
			19	615	620

PLATE 2 (Figure 2 on page 11 of this volume)

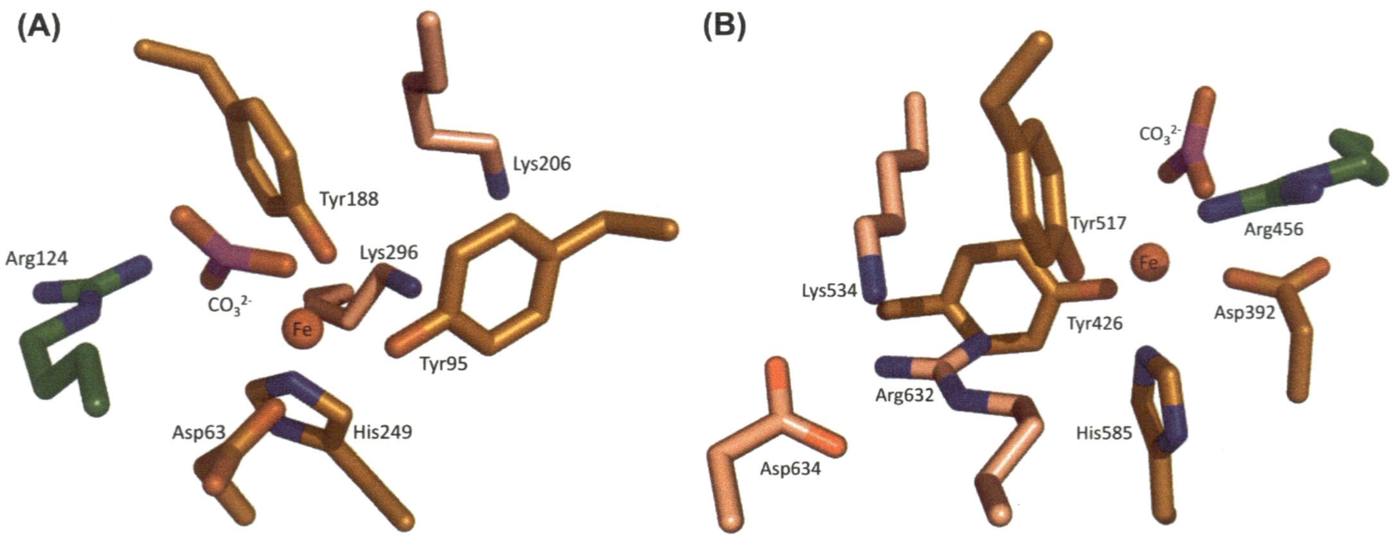

PLATE 3 (Figure 3 on page 12 of this volume)

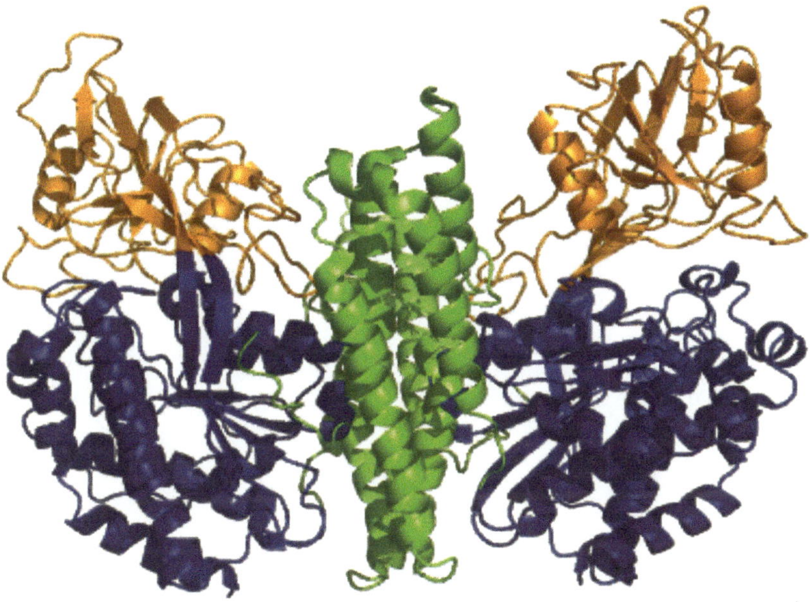

PLATE 4 (Figure 5 on page 18 of this volume)

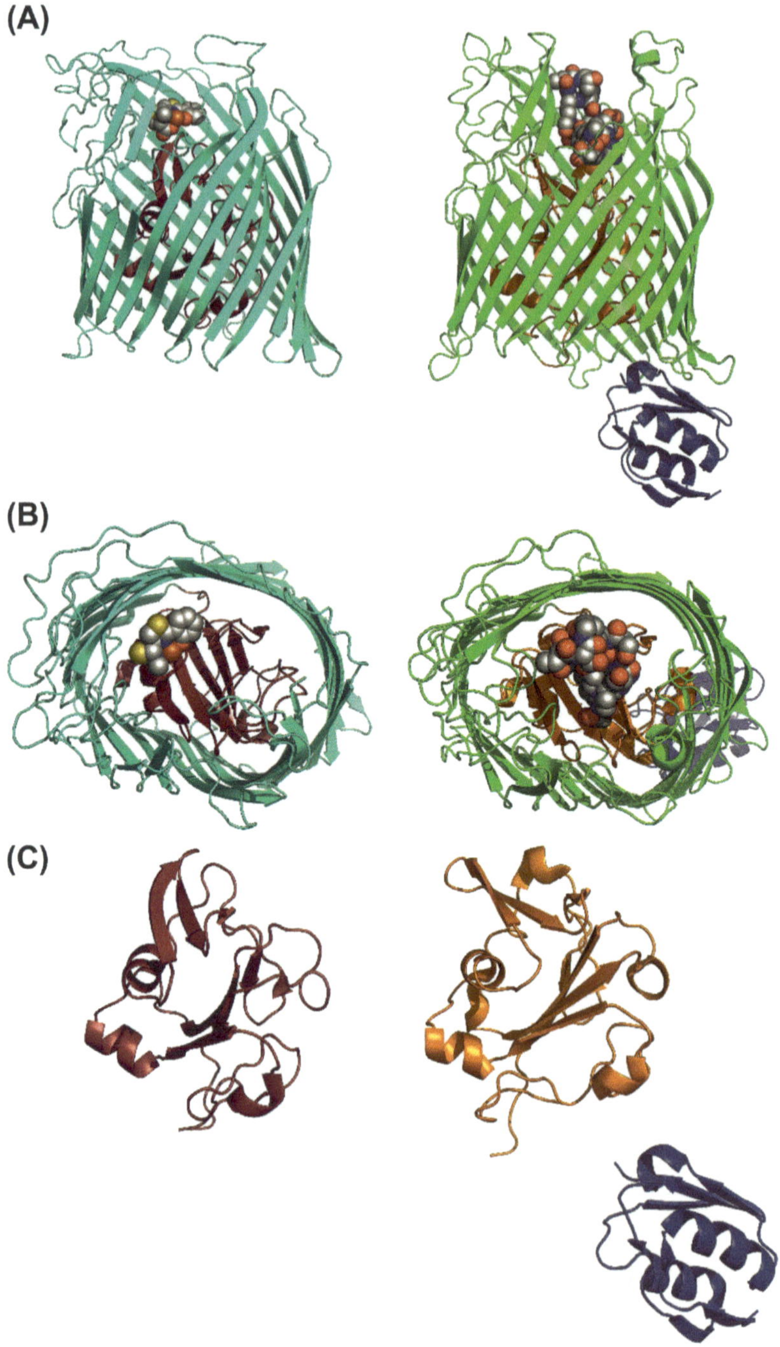

PLATE 5 (Figure 3 on page 44 of this volume)

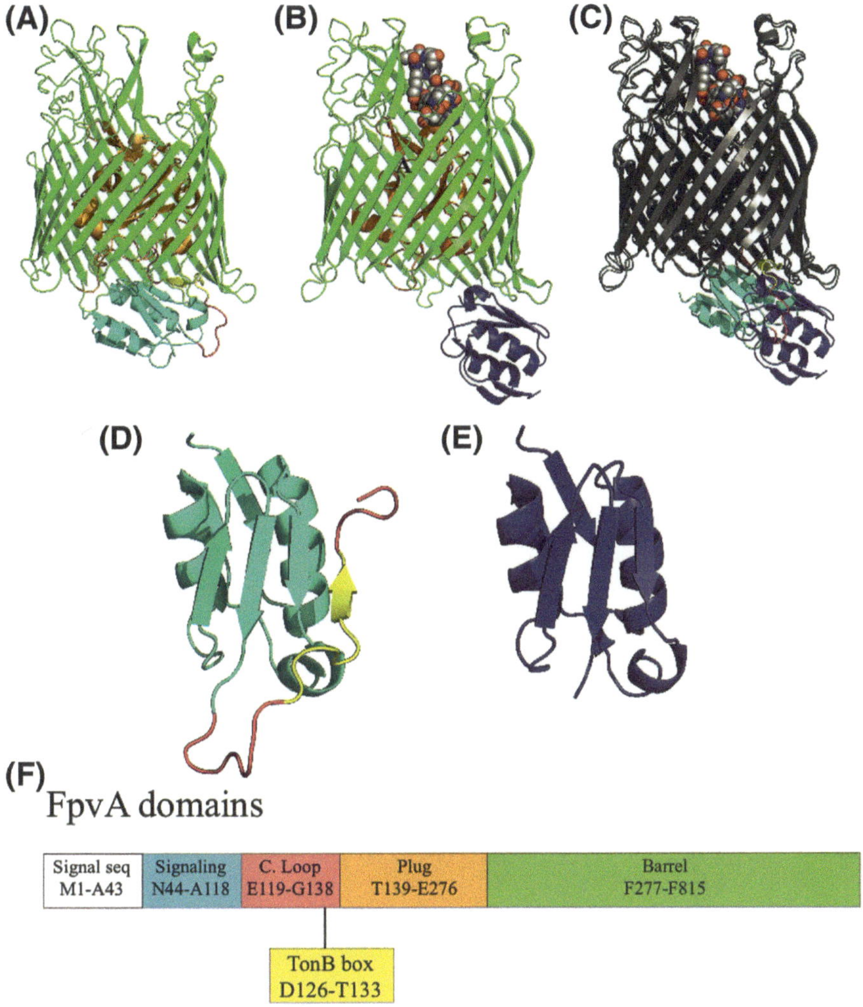

PLATE 6 (Figure 4 on page 49 of this volume)

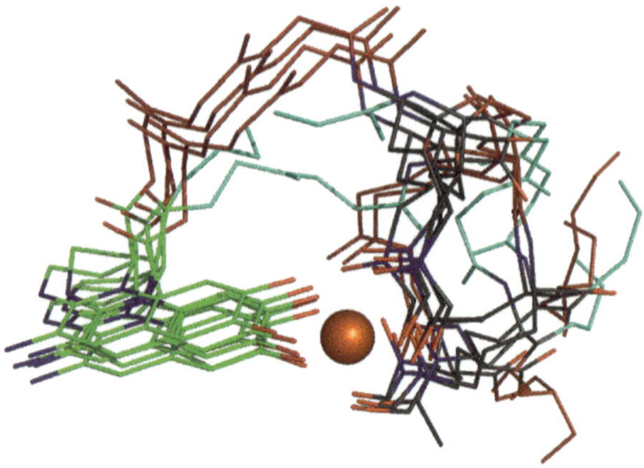

PLATE 7 (Figure 5 on page 51 of this volume)

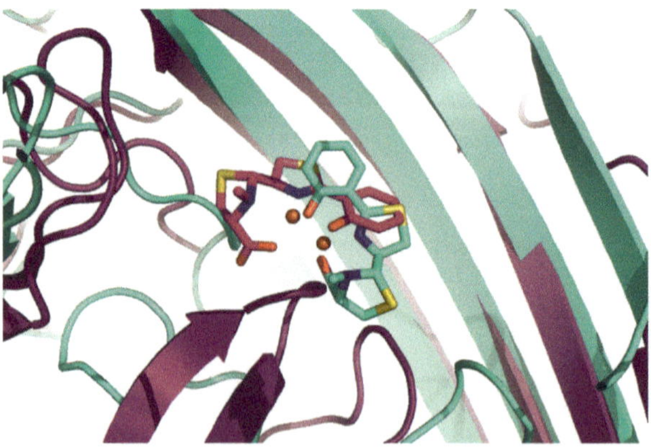

PLATE 8 (Figure 7 on page 54 of this volume)

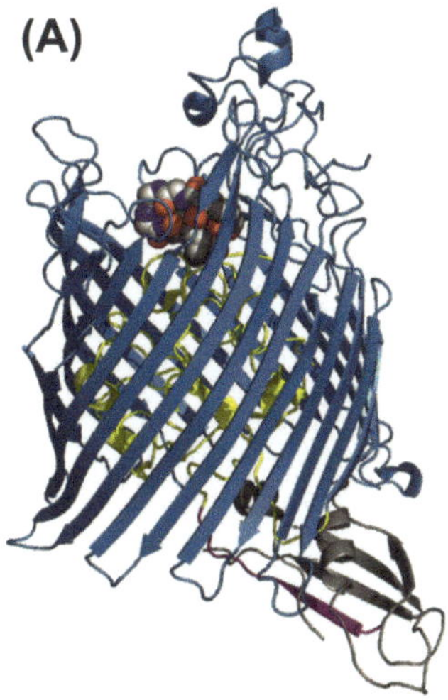

PLATE 9 (Figure 8 on page 58 of this volume)

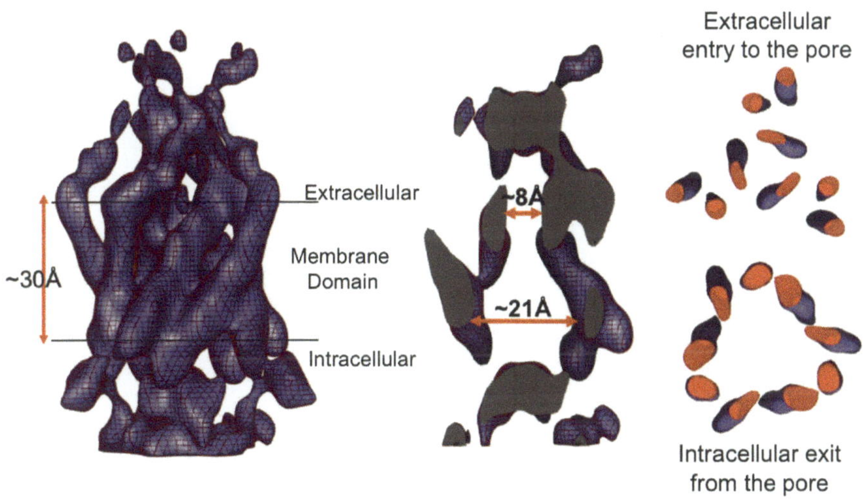

PLATE 10 (Figure 2 on page 102 of this volume)

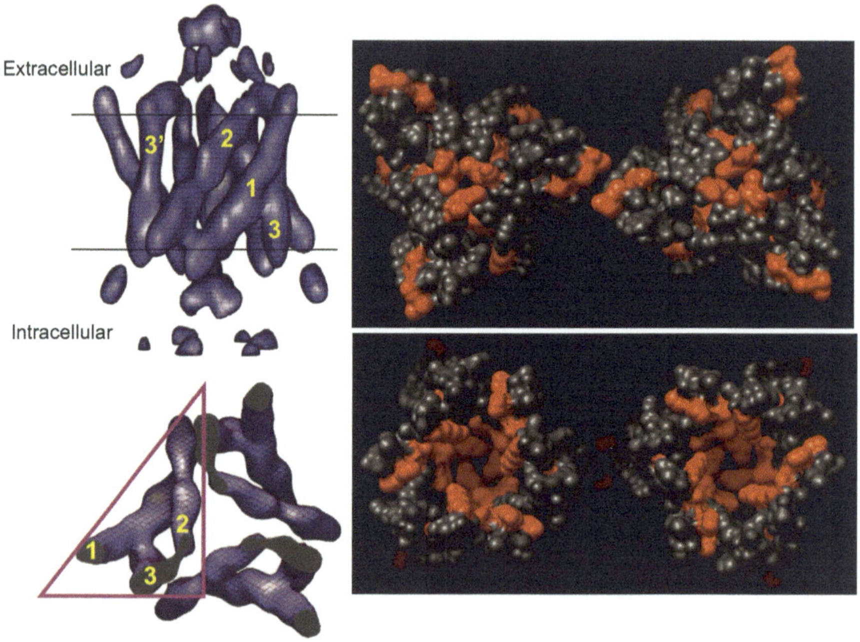

PLATE 11 (Figure 3 on page 104 of this volume)

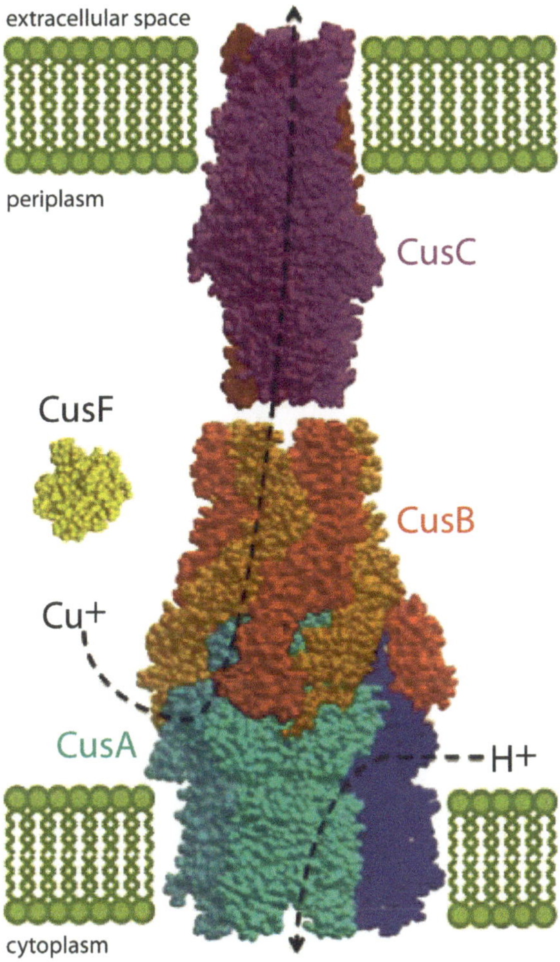

PLATE 12 (Figure 1 on page 166 of this volume)

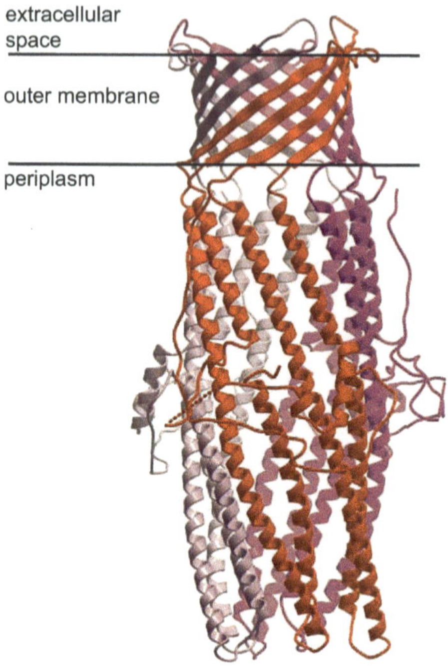

PLATE 13 (Figure 5 on page 179 of this volume)

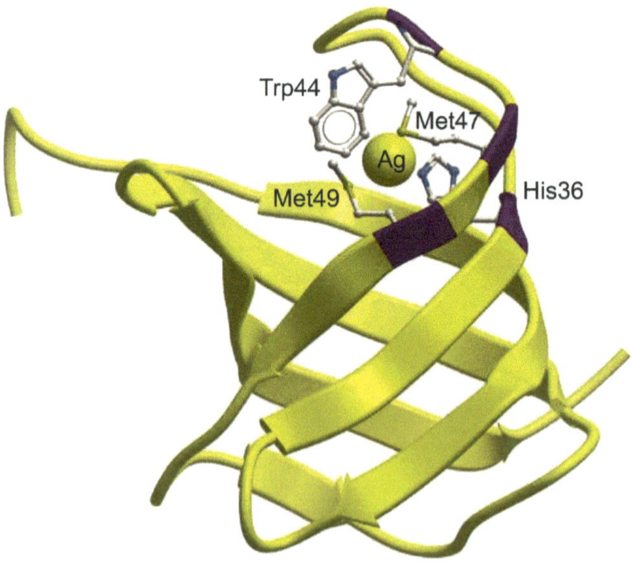

PLATE 14 (Figure 6 on page 182 of this volume)

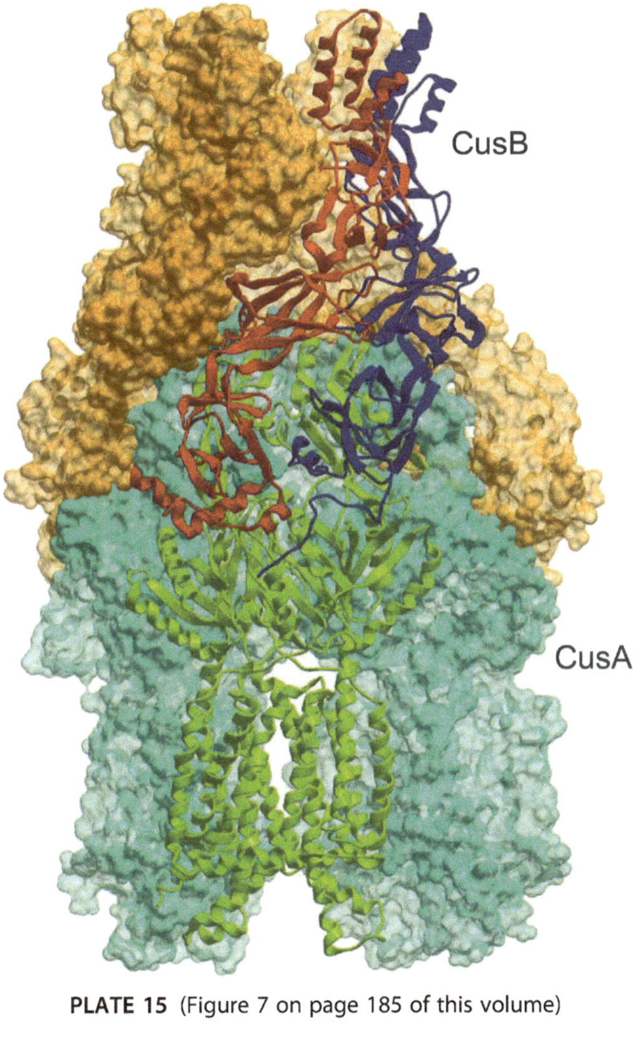

PLATE 15 (Figure 7 on page 185 of this volume)

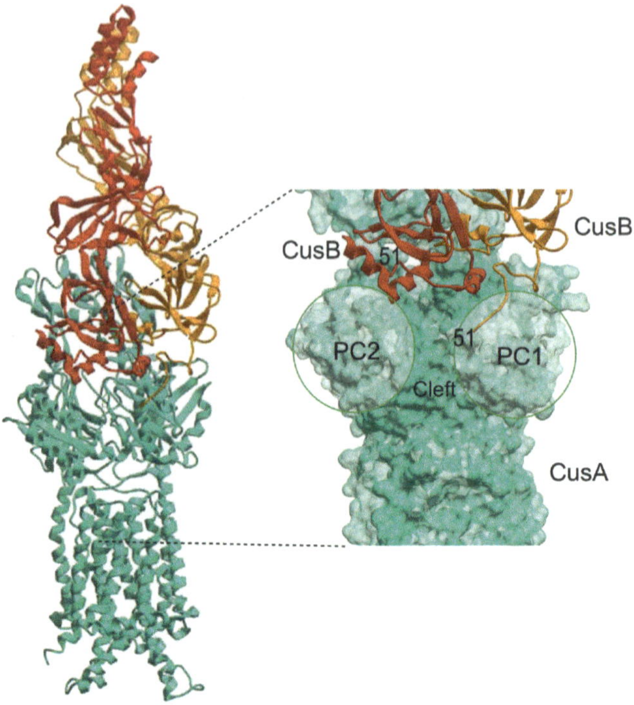

PLATE 16 (Figure 8 on page 189 of this volume)

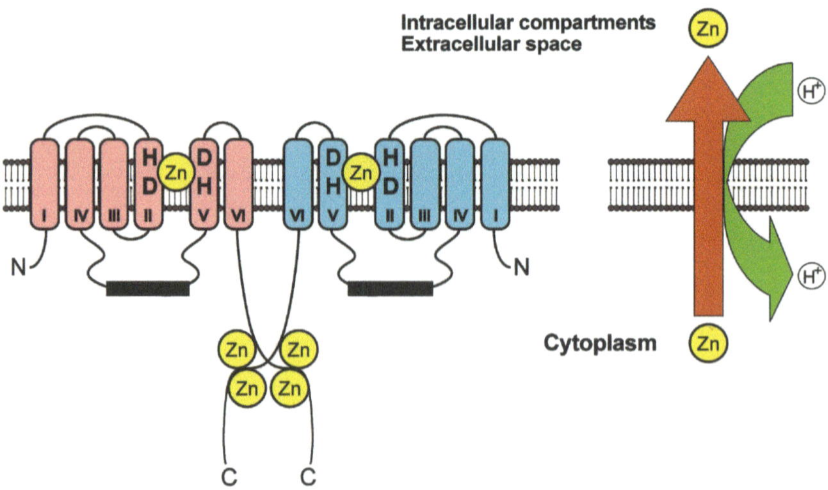

PLATE 17 (Figure 1 on page 201 of this volume)

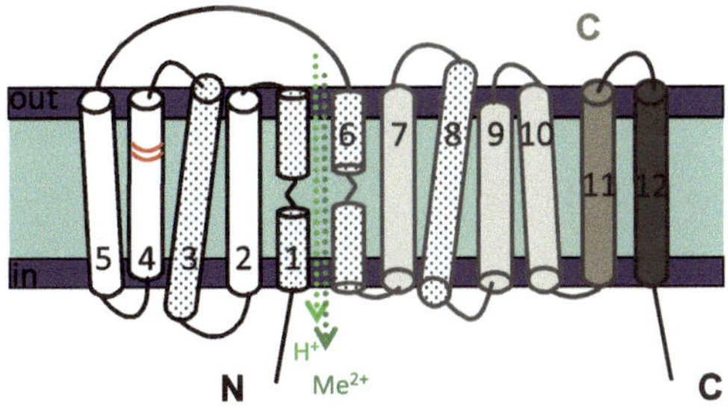

PLATE 18 (Figure 1 on page 252 of this volume)

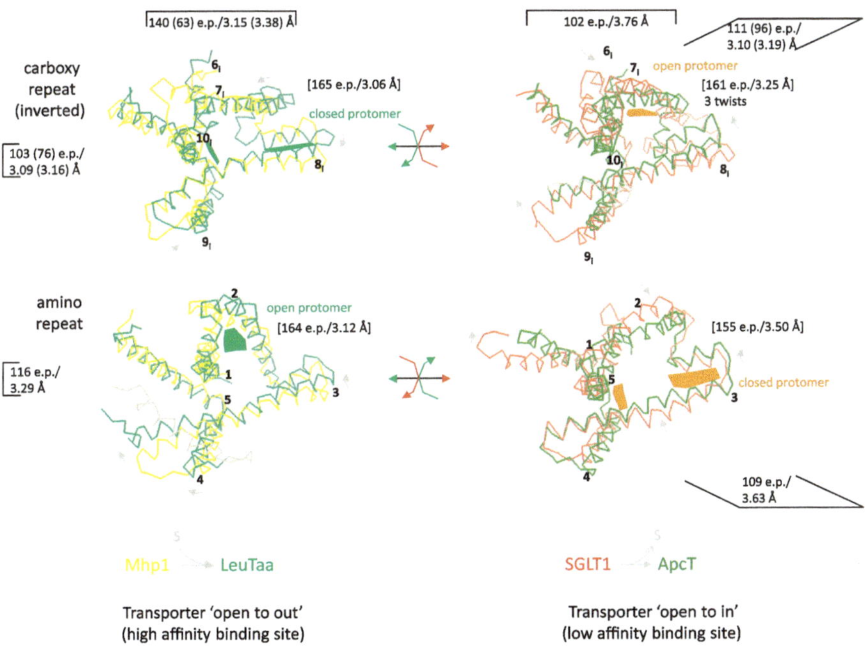

PLATE 19 (Figure 2 on page 258 of this volume)

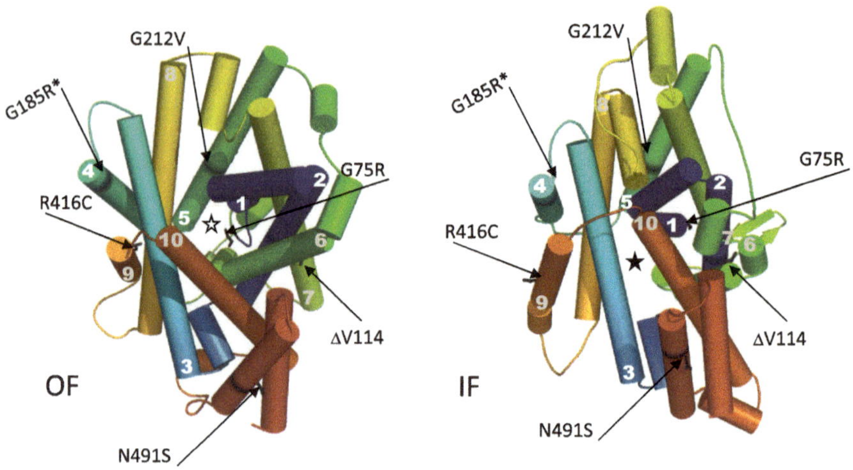

PLATE 20 (Figure 3 on page 259 of this volume)

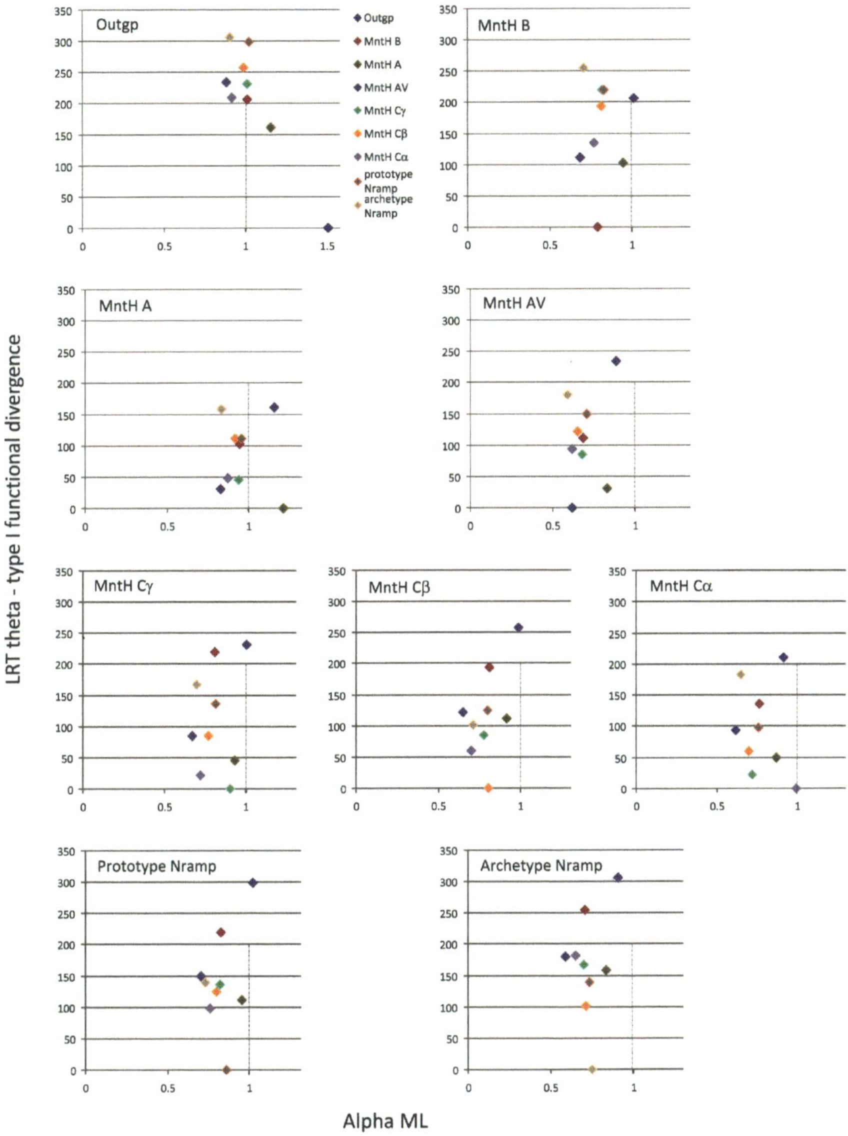

PLATE 21 (Figure 6 on page 263 of this volume)

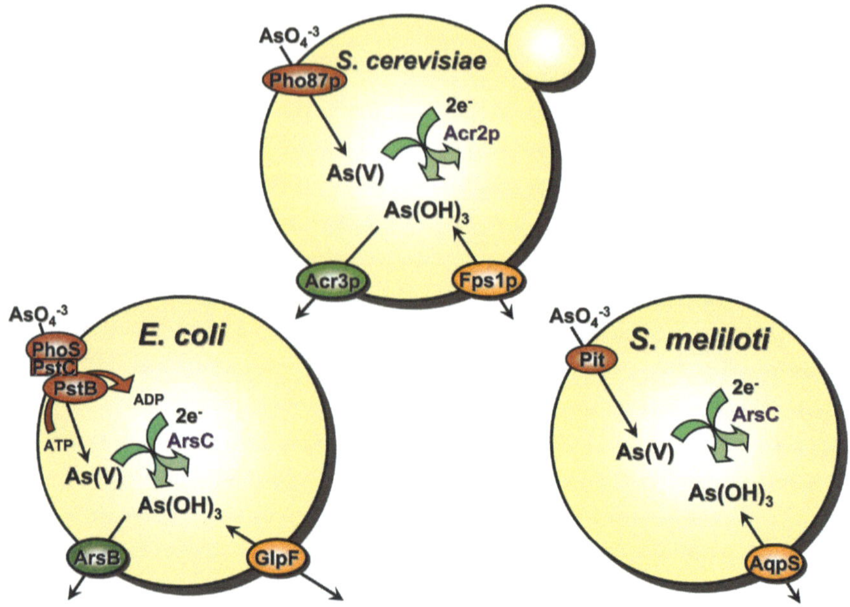

PLATE 22 (Figure 1 on page 330 of this volume)

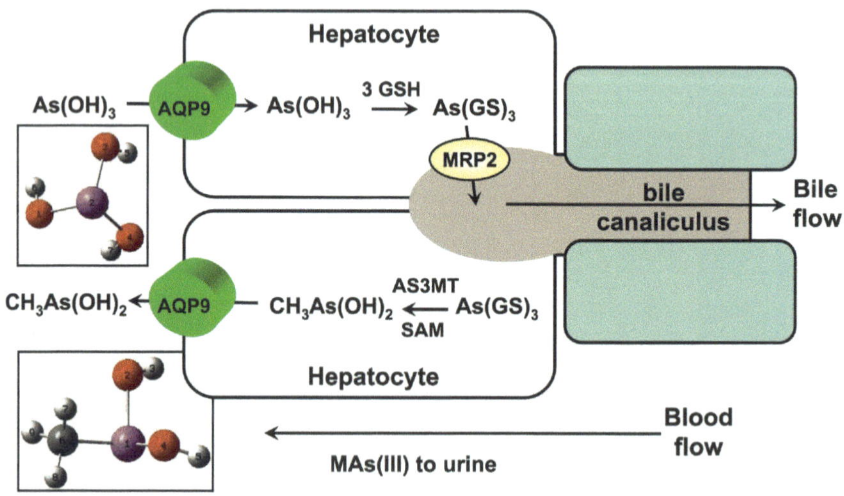

PLATE 23 (Figure 2 on page 333 of this volume)

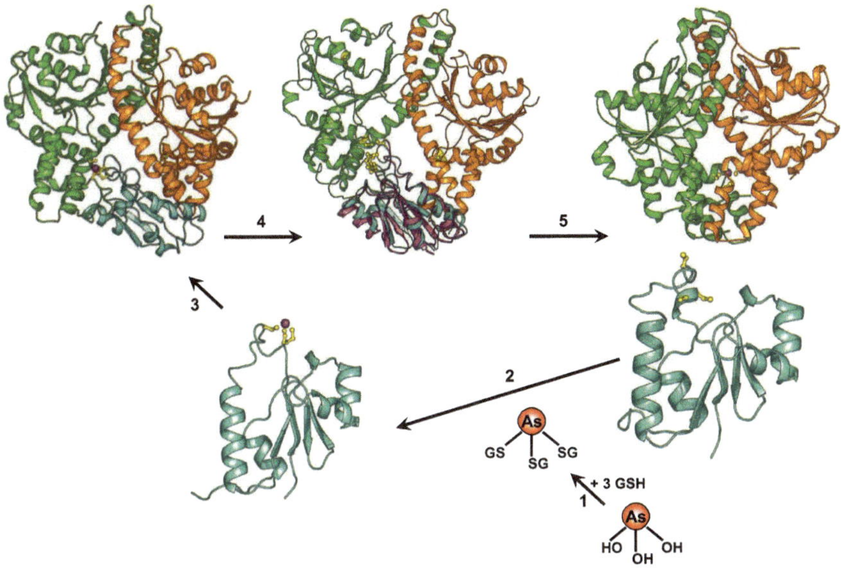

PLATE 24 (Figure 3 on page 347 of this volume)

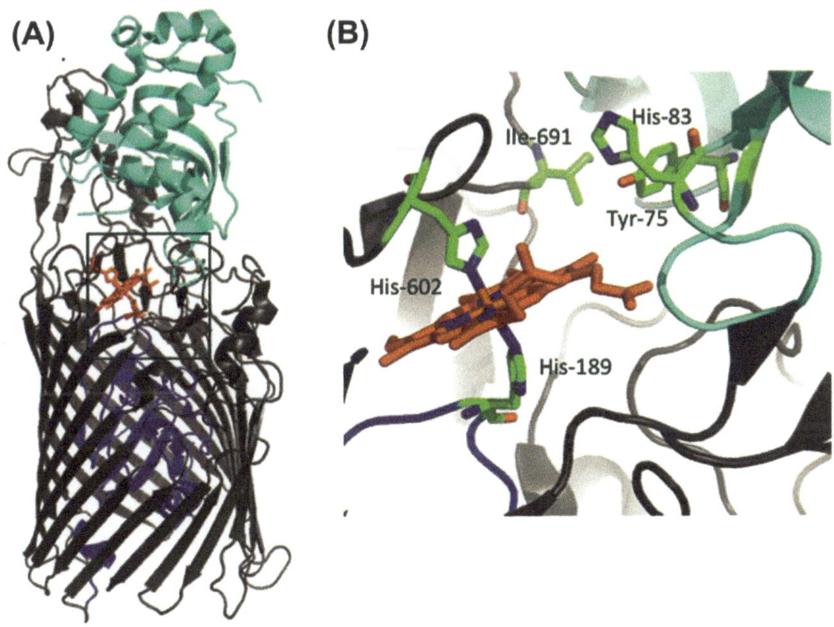

PLATE 25 (Figure 2 on page 363 of this volume)

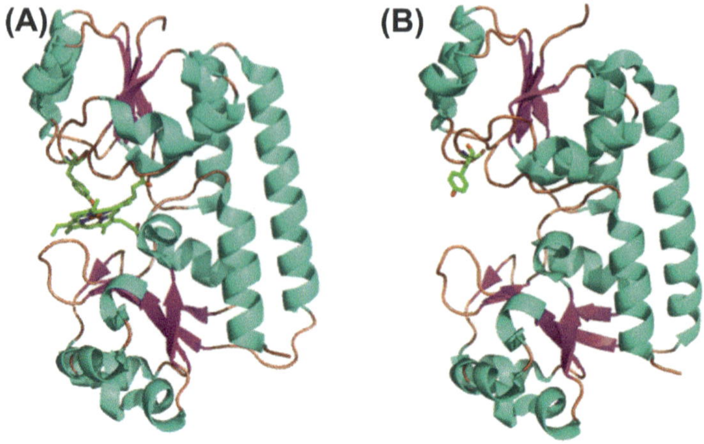

PLATE 26 (Figure 3 on page 368 of this volume)

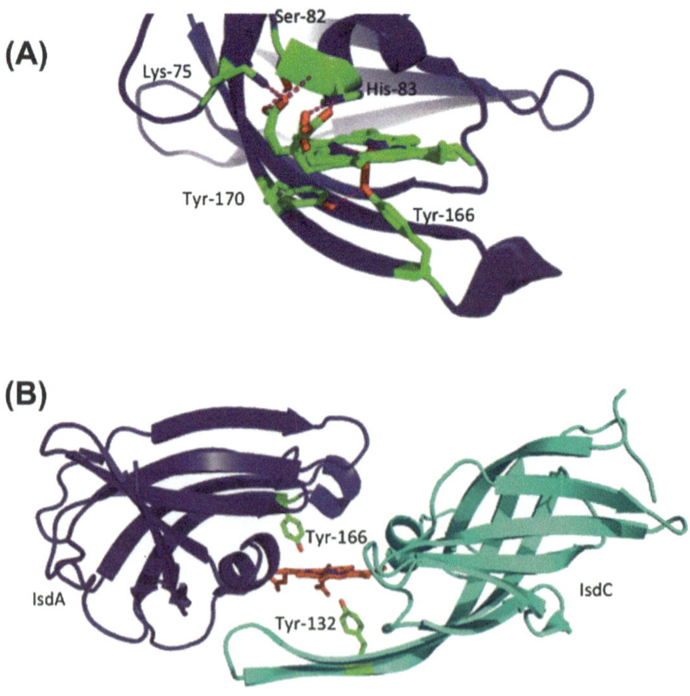

PLATE 27 (Figure 6 on page 374 of this volume)

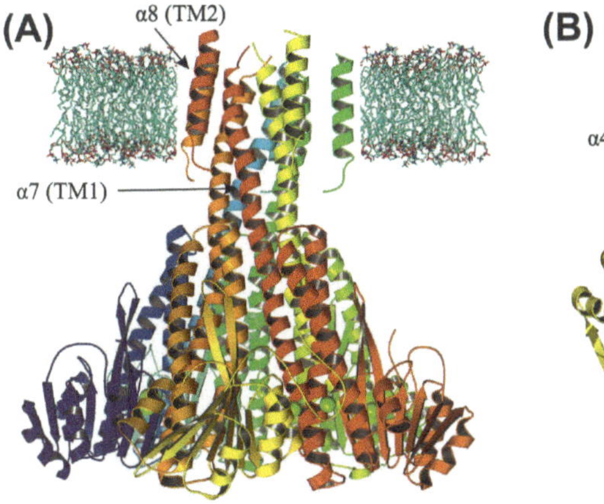

PLATE 28 (Figure 2 on page 404 of this volume)

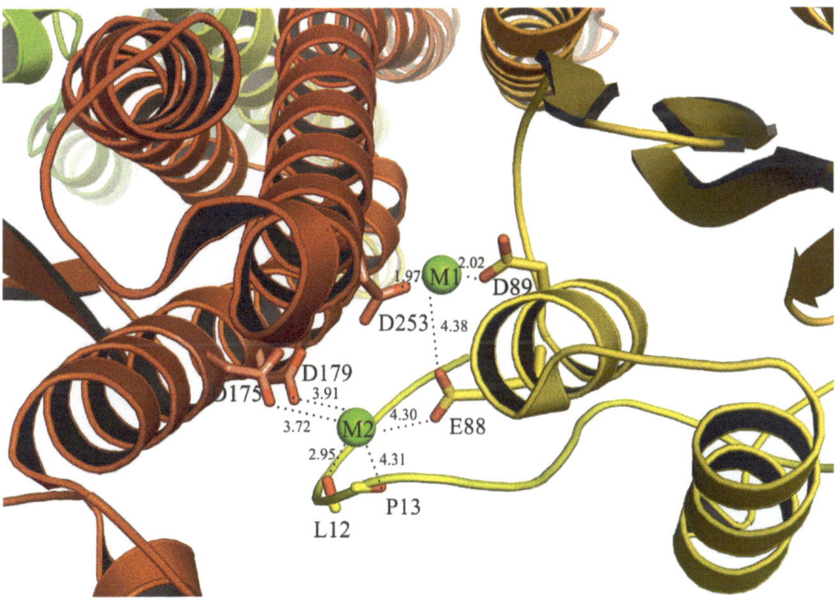

PLATE 29 (Figure 3 on page 404 of this volume)

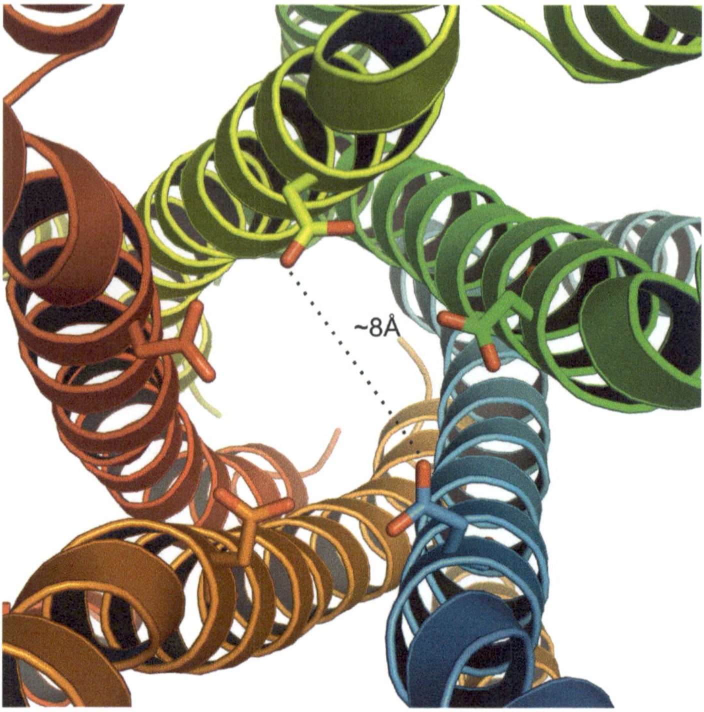

PLATE 30 (Figure 4 on page 405 of this volume)

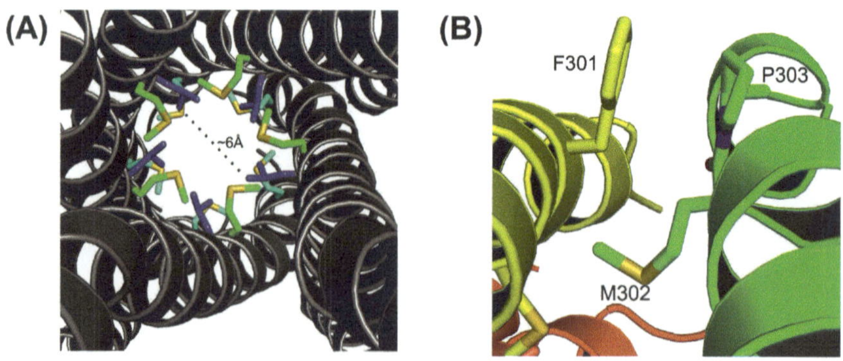

PLATE 31 (Figure 5 on page 406 of this volume)

FSC
www.fsc.org
MIX
Papier | Fördert gute Waldnutzung
FSC® C083411

Kolibri 360 GmbH
Ferdinand-Jühlke-Straße 7
99095 Erfurt, Deutschland
produktsicherheit@kolibri360.de